Management of Innovation and Product Development

Marco Cantamessa • Francesca Montagna

Management of Innovation and Product Development

Integrating Business and Technological Perspectives

Second Edition

Springer

Marco Cantamessa
Department of Management and Production Engineering (DIGEP)
Politecnico di Torino
Turin, Italy

Francesca Montagna
Department of Management and Production Engineering (DIGEP)
Politecnico di Torino
Turin, Italy

ISBN 978-1-4471-7530-8 ISBN 978-1-4471-7531-5 (eBook)
https://doi.org/10.1007/978-1-4471-7531-5

This Springer imprint is published by the registered company Springer-Verlag London Ltd., part of Springer Nature.
The registered company address is: The Campus, 4 Crinan Street, London, N1 9XW, United Kingdom

Preface

What is innovation? Everyone nowadays seems to be talking about this subject. Managers, politicians and journalists all look to innovation as either a cure-all way to solve problems in society and in the economy or—in other instances—as a threat for the future. On the one side, the popularity of the topic is comforting for those who are engaged in studying the topic, or in making it happen. On the other, it raises the challenge of making sure that innovation is rigorously discussed and professionally managed, instead of using common sense and simplistic reasoning, as it often happens, leading to decisions that are naïve from a business and/or a technical perspective.

The purpose of this book is to provide students, researchers, managers and entrepreneurs with a strong, well-founded and interdisciplinary understanding of this engaging and fascinating discipline. Specifically, it has the aim of enabling them to understand the dynamics and impact of innovation in an industry, to analyze and define an innovation strategy for a given company and, finally, to manage the development process for new products and services from a managerial and technical perspective.

Having chosen to cover the spectrum of activities that go from innovation strategy to design and development of products and services is a characterizing feature of this text. Most other textbooks tend to keep the two aspects well apart, covering the former from the perspective of Economics and Management, and the latter assuming an Engineering view. However, it is quite apparent that—in real life—innovation does not follow these academic distinctions at all, and a multidisciplinary perspective is therefore required. Now and again, when working with firms, interacting with academic colleagues and reading literature, we have noticed how this separation can indeed be artificial and misleading. Trying to cope with the management of innovation processes without knowledge of how products and services are designed from a technical perspective often leads to a superficial view of the phenomenon. Conversely, adopting a purely technical perspective on product and service design without having a strong understanding of the economic and business forces that shape this effort often leads to naïve assumptions and unsuccessful outcomes. Though this may be an imperfect analogy, a skipper that must safely lead a boat in a long passage must at the same time have strong navigation competencies and be a good helmsman.

The intended readership of the text is therefore—and quite ambitiously—very wide. As previously stated, it is aimed at people interested in Management and in Engineering, in the hope that it will contribute bridging these two fields. Moreover, it is directed at both academics and practitioners, since it is founded on rigorous research and evidence coming from the field and—at the same time—it provides a number of practical suggestions and insights. We therefore hope that academics may find useful ideas on how their findings translate to practice, while practitioners may understand the reasons that underlie many popular recipes suggested by professional literature and by consultants.

Concerning its use as a textbook, our experience suggests that the contents are particularly suitable for graduate students in the fields of Economics, Business or Engineering. Undergraduate students may also benefit from its multidisciplinary perspective, although the scientific depth with which topics are covered might make the book somewhat demanding to be read.

Readers of the book will notice that we have strongly associated innovation to technology. We are well aware that innovation is not only *technological innovation* and that it can therefore can assume many different forms. However, we are convinced that technology—intended in a broad sense—is nonetheless a central enabler to innovation of any kind, and must therefore be adequately covered. Conversely, the book also extensively covers innovation from the perspective of social sciences, in the acknowledgement that it is a social phenomenon indeed, both on the side of supply and on the side of demand. This nature of innovation as a jointly social and technological phenomenon suggests that—at least up to a given point—it is invariant with respect to the different disciplines of technology and industries.

Of course, having decided to adopt a multidisciplinary perspective on such a broad topic such as innovation has its own limitations. From the perspective of scholars in Economics and Management, this can be considered to be a textbook providing the foundations of the discipline. However, not being an advanced treatise, it reports outcomes and results of major contributions to literature without providing a full analytical and quantitative discussion. From the perspective of engineering scholars, the book reaches into the topic of engineering design processes, but stops at the threshold beyond which the discussion would have been discipline-specific. This can make it a useful complement to traditional and focused engineering textbooks, while it will provide students with the understanding that design processes are broadly common to all fields.

The book is thought as divided into four parts. The first chapters of this book consider innovation mainly as an economic phenomenon and look at it from the high-level perspective of an industrial sector and of society at large. It investigates the role of innovation in business and society, the determinants, the many types of innovation and its dynamics.

Then, the discussion enters a lower level of analysis and considers the definition of an innovation strategy for a given firm, as part of the overall corporate strategy. It relates innovation strategy to the most popular approaches to corporate strategy, and then focuses on the management of portfolios of corporate competencies and of projects, as the two main levers with which innovation strategy can be enacted.

Subsequently, the operational-level activities and the management of product development are debated. The book focuses on issues related to the organization of product development and of the resources involved, to the structuring and management of the product development process and, finally, to the use of project management techniques in the context of product development.

Finally, the book explores individual phases of the product development process and provides readers with the technical tools that can be used in this context, and make it happen. It starts from the initial phases of market research and product positioning and—after discussing the issues pertaining to the definition of product requirements and specifications—it leads to the conceptual design or redesign of products or services.

The book purposely does not include a glossary. In our teaching experience, we have noticed that students always and proficiently use Internet sources, such as Wikipedia, to find meanings of terms and concepts. We have chosen not to go against an innovative trend and, therefore, we have highlighted with an asterisk the terms and concepts that could be interesting to readers and for which—upon checking—Wikipedia or other Internet sources provide a sufficiently complete and reliable presentation, which makes it appropriate *in lieu* of a glossary entry.

Finally, and as it will be discussed in the first chapter, innovation studies heavily rely on the exploration and interpretation of past cases and events. Most concepts and topics are therefore illustrated by using short examples coming from real cases that have occurred in history. We have chosen not to provide fully fledged case studies, but simple pointers to these stories, each time extracting the interpretation that is relevant to the topic being discussed. Examples come from fields that readers should be easily familiar with, such as automobiles, consumer electronics, information technology and energy, so that each example may be readily understood thanks to personal experience, and without having to provide a thorough technical explanation. We have made this choice for three reasons. First of all, to reduce the size of the book and make it quicker to read. Secondly, to show readers how contemporary phenomena that have to do with ordinary life are impregnated with topics relevant to innovation management, and can therefore be interpreted through this perspective. Finally, in an educational context, lecturers may use each of these examples as a starting point for developing student assignments. We have observed that this approach is very helpful for students, who can personally experiment the work of collecting information and data on innovation-related phenomena and of deriving interpretations, ultimately acquiring stronger analytical skills and competencies.

We are conscious that the examples provided will naturally tend to age, making them less immediate and meaningful to students' memories and experiences. Due to this reason, we invite readers to visit and interact with the LinkedIn page 'IMPD

Group' associated with our research and teaching group (https://www.linkedin.com/company/71526281/), where we constantly post news that are relevant to the subject. These posts contain a short comment explaining the relevance of the news being shared, along with some critical questions aimed at stimulating a debate.

Turin, Italy

Marco Cantamessa
Francesca Montagna

Acknowledgements

The second edition of this book results from more than twenty years of experience in teaching, sharing and practicing the captivating subject of innovation management and product development in different contexts, having taught and supervised projects with university students from engineering, architecture and business; interacted with scholars in our research activities; consulted and run seminars with managers and entrepreneurs in both established and start-up firms; cooperated with policymakers at both regional and national levels. We therefore owe the insights behind the book not only to the authors of the huge underlying literature that we have attempted to systematize, but also to the thousands of people that over the years have given us feedback and suggestions, and have confirmed the value of the multidisciplinary approach we adopted.

Special thanks goes to the colleagues and friends who over the years have built and shared this path with us, and especially Luigi Buzzacchi, Mario Calderini, Gaetano Cascini, Paolo Neirotti, Emilio Paolucci, Giuseppe Scellato, Giulio Zotteri along with our Ph.D. students Stefania Altavilla and Alessandro Casagrande-Seretti, as well as with our numerous colleagues from The Design Society, the Alta Scuola Politecnica and the I3P Incubator.

Finally, we thank our families for their support and understanding throughout the long process of writing this book and keeping it up to date.

Contents

Part I
Understanding Innovation and its Dynamics

Innovation in Business and Society

1

Abstract

Innovation is an often used (and misused) word, which is tied closely to the substantial changes that characterize the evolution of technology and society. This chapter has the objective of providing a preliminary foundation on innovation as a social, economic and technological phenomenon, on its relationship with science, and on the main actors—public and private—that are involved. A short discussion is also provided on the relationship between innovation and geography and on the potentially negative impacts that innovation can have under societal, ethical and legal perspectives.

If we pause for a second to look at our daily lives and try to compare them with our lives of ten years ago, or even with our parents' and grandparents', the differences are striking. Just about any activity we tackle, be it in our family life, at work, or when relaxing, is so different from previous generations and benefits from completely new means.

Ever more effective drugs and medical devices, ever safer cars and planes, an increasing productivity in agriculture, and—most of all—an impressive progress in the field of information and communication technology, nowadays allow more than 7 billion people to live on our planet. Moreover, the wide majority of these human beings are living a longer and better life than the one that could be experienced even by the wealthiest persons of the nineteenth century.

We can spare the reader a long list of technological advances that have allowed such progress (everyone will have their own preferences!). Instead, we prefer raising the awareness that technology *per se* is not enough to create an impact, unless individuals and organizations do not cast their efforts to use this technology so to bring change of some sort to markets, industries and society and to make these changes spread. This is what we can call—quite loosely for the time being—'*innovation*'.

M. Cantamessa and F. Montagna, *Management of Innovation and Product Development*, https://doi.org/10.1007/978-1-4471-7531-5_1

Innovation is not exclusive to our times, of course. If we focus on the most recent of times, characterized by the advent of digital technology, we see computing systems started permeating society, while communication infrastructure allowed connections at all levels. First, the Internet allowed to connect people, initially through applications such as websites and email and later through social media, while the Internet of Things (IoT) technology led to a continual connection and flow of data between people and objects and between objects and other objects. Ever more powerful and cheaper IT equipment (processors, bandwidth and storage) enabled the virtualization of physical servers, leading to Cloud Computing and to storage of massive amounts of data, sometimes collected in real time. Then, tools from Data Science are enabling the identification of patterns and behaviors in phenomena and occurring within systems, as well as allowing new forms of automation and decision-making, based on the elaboration and assimilation of such amounts of data. Besides, we are now facing the possibility of further advances in the field of augmented and virtual reality, which will allow richer representations of objects and environments, merging the real with the virtual, and vice versa. The consequence is that smart products and digital processes in services have pervaded our lives, and if we take a glimpse into products and services that are currently being developed, the future pace of change shows no sign of slowing down. However, while the manifestation of innovation changes along history, the phenomenon of innovation has accompanied humankind since the days in which *Homo sapiens* started developing and introducing new tools to a then primitive society, from the Ancient times to the Middle Ages, and all the way to the Industrial Revolutions that have marked the last three centuries. We can therefore say that innovation is somehow inherent to humankind.[1]

Innovation, and the underlying progress of technology, is not only relevant to our lives as members of society, but also to our lives in business. If we compare the Fortune 100 or Fortune 500 lists of companies over the years, we find that many current members of these rankings were nothing but small firms, or fledgling

[1] Greek mythology brings to us the well-known story of Prometheus, the Titan who stole the fire from the gods and presented it to humankind. A particularly detailed narrative of innovation in ancient times also comes from the VIII book of the Aeneid. In this part of Virgil's poem, Aeneas, who has fled Troy for the shores of Italy, is in trouble fighting the natives of Latium. To his relief comes Venus, goddess of love and passion. She seduces Vulcan, the god of fire and metalworking, who then heads for the volcano where he has established his workshop and manufactures a new set of weapons for Aeneas. These weapons bear the images of future and glorious events in Roman history and allow Aeneas to win a crucial battle, which paves the way for the future foundation of Rome. This story, written more than 2000 years ago, is very interesting by our contemporary standards. In fact, it shows three key ingredients to innovation: Aeneas, as the symbol of an urgency to change, which generally is due to competition of some sort (quite often, of military nature); Venus, as the symbol of passion and commitment, making events unfold; finally, Vulcan, as the symbol of technical prowess and know how. Even today, we might argue that innovation arises if and only if all three of these ingredients are present. A further detail is interesting. Vulcan carves scenes of the future of Rome on the shield, which Aeneas does not understand and simply enjoys their beauty. In other words, innovation is a force so strong, that it shapes the distant future in a way that baffles even the ones who pursue it in the present.

startups just a few years ago. At the same time, large and successful firms have dropped off these lists. Competition between entrants and incumbents is such, that large firms are spending an ever more significant effort in evolving their technology base, their processes and their business models, accepting changes and risks of a magnitude that would probably not have been accepted in the past. Incumbents are also spending increasing amounts in acquiring entrant firms they perceive as a present or future threat to their competitive position. It is therefore unsurprising that the attention paid by managers to innovation is so high.

Innovation is also recognized as an enabling factor in contexts of severe crises (Archibugi et al. 2013). The COVID-19 pandemic has spurred an impressive innovative response both in the field of health care (e.g., the rapid development of mRNA vaccines) and in allowing parts of society to carry on life and business even during the severest lockdowns. Similarly, the urgency with which humankind prepares to cope with climate change leads to look for innovative solutions allowing a rapid change in the way energy is produced, stored and used. Of course, these possibilities depend on the ability of each actor and organization to adopt new technology and use it in a way that is appropriate to deal with these problems.

At a broader level, economists have shown that economic growth and innovation are strongly intertwined. Robert Solow* (1924–), the 1981 Nobel Prize in economics laureate, explained that GDP growth cannot simply be explained by looking at growth in production factors (i.e., capital and labor). What also matters is productivity growth (i.e., the economic value created by a better use of the same units of production factors) and that progress in knowledge and in technology is responsible for this. Policymakers are therefore quite keen on understanding innovation and in finding ways to foster it in the countries and regions they govern.

In this book, we tackle innovation from the perspective of a business manager operating in a specific company. She/he will have the goal of making choices that will ensure success to the company's products or services, and profitability to the firm as a whole. This perspective is therefore different—and entails more risk—than that of a government agency, whose objective in drafting innovation policy will be to ensure economic success to the firms of a region, but thinking in aggregate terms. However, a sound understanding of innovation policy is useful for business managers both when they consult with government officials at the time the policies are drafted, and then subsequently, to make profitable use of the same. This book can also be useful to those interested in managing innovation for social contexts. The book does not explicitly address the topic of *Social Innovation**, and this is done on purpose, since it would require a specific focus and a dedicated in-depth analysis. Nevertheless, it does provide knowledge on innovative processes and on the fundamentals that guide innovation dynamics, which remain valid also in this particular context.

The relatively narrow focus on the management of an individual firm might lead someone to question the actual possibility of managing innovation. In fact, innovation implies bringing to life something that did not exist before. So, this appears having much more to do with serendipity, luck and strokes of genius, rather than with the activities one would typically associate with management, such as planning and organizing. However, as Mark Twain famously said, "*The past does not repeat itself, but it rhymes*". The discipline of Innovation Management somewhat follows this witticism by the American novelist and is based on the idea that—by studying past cases occurring in industry, from ocean clippers to light bulbs and from automobiles to digital photography—the patterns that are uncovered might shed some light on current and future transitions, such as cloud computing, mobile payments, electric mobility or space travel. These insights on the future, fuzzy as they may be, might be used as a foundation for a rigorous and systematic management of the innovation processes that occur in a firm.

The study of Innovation Management is rooted in an adequate understanding of technology. In fact, we adopt a somewhat strong hypothesis that, while not all innovations are strictly technological innovations, technology will at least have an enabling role to innovations of any kind.[2] At the same time, no innovation can be fully understood from a purely technological perspective, and requires a deep understanding of economic, organizational and social elements. If one looks at the side of supply, technology is the outcome of collaboration and social interaction in companies and supply chains. Symmetrically, if one looks at the side of demand, innovation has to do with the arising of a need, and with the acceptance of a specific solution, within society. In the end, innovation is a societal phenomenon and a managerial activity whose object is strongly—though not exclusively—related to technology.

The study of innovation as a societal phenomenon dates back to the work of a number of pioneering academics, among which we especially remember Joseph Schumpeter* (1883–1950), who is commonly considered to be the founder of Economics of Innovation. Economists had always been aware that technology and innovation had an economic impact, but mostly considered them as *exogenous* to the economy, i.e., something similar to a disturbance, that could at most be studied from a technical perspective. Schumpeter was among the first to suggest that technology and innovation were instead *endogenous* to the economy, i.e., both determined by economic factors and impacting on them.[3] This observation is of fundamental importance, since it implies that a full understanding of the phenomenon is only possible by adopting multiple perspectives, deriving from both the social sciences and from technical subjects.

---

[2] For instance, think of the role of complex booking systems based on revenue management techniques* in the airline industry, which have enabled significant business model innovation. Or, the use of digital mapping systems that have enabled innovations ranging from food delivery in cities to medical logistics during epidemics in developing countries, such as Rwanda (UNCTAD, Technology Innovation Report 2018).

[3] The exogenous or endogenous nature of phenomena leads economists to interesting choices in their fields of study. For instance, it is obvious that climate and the weather impact the economy, but economists have only recently started studying these phenomena, i.e., since it has been recognized that climate change is likely to be determined by human activity or, in other words, that it is *endogenous* to society and the economy.

This societal nature of technology and innovation suggests that many of the related phenomena will be invariant with respect to the different branches of technology and the many industries in which economic activity is articulated.[4] Though this might at times seem strange to those who have been trained in a specific technical field, it implies that the main concepts of Innovation Management are applicable—at least up to a given level—to any field of technology and to any industry and sector. This statement might find an exception in the case of technological changes whose impact is outstandingly broad, such as the one of digital technologies. For this reason, the book will dedicate ample room to discussing digitalization, at times treating it simply as an innovation, like the others, but at other times discussing how digitalization is changing the phenomenon of innovation itself and therefore is requiring new and specific approaches for its management.

1.1 Defining Innovation

Defining innovation is not trivial. The Merriam-Webster dictionary provides two definitions. One claims "*a new idea, device, or method*", while the other states "*the act or process of introducing new ideas, devices, or methods*". The two definitions are somewhat different, with the former centered on the new artifact and the latter on its introduction. In the Innovation Management literature, it is this latter definition that matters, looking at how new ideas, devices or methods can be introduced in business and society and at their impact. This definition allows pointing out the differences between innovation and other related terms such as '*discovery*', '*invention*' or '*product development*'. In first approximation, it is possible to propose the following definitions:

- **Discovery**—Discovery is the act—or the achievement—of uncovering something previously unknown. Discovery is the usual outcome of the activity called science. *Science** has the aim of creating new knowledge on natural or social phenomena and—unsurprisingly—the main indicator with which individual scientists and research organizations are evaluated is the number and quality of publications reporting on this knowledge. Discovery therefore produces scientific knowledge, by building new results on top of prior parts of knowledge of the same kind.[5] This is evident by glancing at any academic paper and observing the central role that authors give to references to other previous works and to

[4] As an example, the strategies followed by personal computer makers in order to make their proprietary solutions standard might be of some teaching to electric car makers. Similarly, methods used to manage software development projects can be applied to the design of industrial products, or even to the redesign of services and the reengineering of operational processes of a non-technology-driven organization, as a social cooperative.

[5] Isaac Newton famously used the metaphor of scientists as "*dwarves standing on the shoulders of giants*" to express this idea.

pointing out the research gap they attempt to fill in with their original contribution. When working on their research projects, scientists are drawn to discovering and understanding phenomena at the highest possible level of abstraction, and—at least in principle—they are not immediately concerned with the applicability of their findings. Of course, a scientist who studies the impact of a given protein complex on the development of cancerous cells will probably wish that his/her results might one day lead to the development of a cancer treatment. However, the two activities of discovering the impact ('propositional knowledge', in Mokyr's terminology) and developing the treatment ('prescriptive knowledge') are not only quite far away in time, but conceptually different from one another, though they both aim toward a body of 'useful knowledge'.

- **Invention**—Invention is the act—or the achievement—of devising a new solution to a problem. Invention is the typical outcome of the activity called *technology** (from the ancient Greek *techné*), which has the aim of ideating and validating *artifacts*. The term 'artifact' is very broad and encompasses material entities (i.e., an airplane) as well as immaterial ones (i.e., a service or business process), simple entities (such as a vase) or complex ones (such as a spaceship). In principle, a technologist might wish the broadest possible applicability of his invention, but in practice, his concern will be to achieve a desired level of performance, given a specific functional goal and setting. For instance, an inventor dealing with corkscrews will typically try to develop an improved tool for pulling corks out of wine bottles and will not be likely to attempt the development a generic tool for separating any two objects that are wedged into each other. Technologists carry out their activity by using both scientific knowledge and technological knowledge. For instance, it is obvious that an engineer designing a new elevator will benefit by being familiar with Newton's laws of dynamics. However, she/he will also use a host of other empirical knowledge elements that a scientist would not normally care about, but are critical to developing artifacts that work effectively and safely when placed in the hands of users.
- **Innovation**—Finally, innovation can be defined as the '*economic exploitation of an invention*' (Roberts 1987). In simple terms, society moves from invention to innovation when an invention is marketed and bought. This means that a producer is able to develop an artifact that will give customers a *utility** that is greater than the cost of production and to offer it at a given price and in such a way that customers will effectively be able to recognize such value.[6] Moreover, the difference between invention and innovation is much deeper than usually thought. In fact, most people mistakenly credit the invention of important artifacts not to the (often many) inventors who have progressively and incrementally

[6] In principle, an innovation occurs when a technical solution becomes a product (or a service) in such a way that the firm who offers it is able to ensure that the utility to customers is greater than the market price (otherwise customers would not buy it) and that the market price is greater than the cost in which the firm incurred (otherwise the business would not be sustainable). A number of interesting technical solutions have not been able to become innovations, because of the inability to properly balance this triad.

worked on them, but to the innovator that first managed to create a successful business.[7] In some instances, innovation may occur decades after the original invention and in completely different environment and circumstances.[8] Innovation is then followed by *diffusion*, which is the process with which the market progressively adopts the new technology and makes it mainstream. Depending on the industry, diffusion can take from months (e.g., mobile messaging applications) to decades (e.g., industrial machinery). This wide variation in the time required by diffusion makes the financial attractiveness of innovations in different industries widely dissimilar. Economists and statisticians usually record the occurrence of an innovation by looking its launch on the market, irrespectively of the subsequent diffusion phenomenon. In other words, they do not consider key elements of market success such as sales volume and diffusion speed, which instead will be very important in the eyes of those who look at innovation from a managerial perspective.

- **Product development**—Product development is the name given to the business process that a company performs in order to deliver an innovation to the market. When compared to other business processes, such as payroll, purchasing or production, product development shows some striking differences. Not only its duration is measured in months and years instead of hours or days, but this process is also highly multi- and inter-functional, inter-disciplinary and knowledge-intensive. Virtually, all functions of a firm have a role in product development, but the marketing and the technical function process usually govern most parts of it. In fact, and as will be reflected in Chaps. 13–16 of this book, the main theoretical contributions to product development come from the academic communities that deal with Marketing and Design.

It is quite common to discuss innovation by looking at the objects it affects. This leads to distinguishing among *product, process* and more recently *service innovation,* but also between *organizational, business model* and *social* innovations. Literature often uses the term *organizational innovation* for process innovations that have a strong immaterial content and that—rather than changing the technology

[7] Many examples can be mentioned. One of the probably most striking ones is the incandescent light bulb, whose invention most people would credit to Thomas Alva Edison. However, what Edison really did, was to improve prior artifacts invented by a series of predecessors and make the light bulb commercially viable. To achieve this result, Edison blended technological and business acumen. First of all, he managed to identify beachhead markets to which he could direct his earlier and still underperforming products, such as ships, industrial facilities and wealthy individuals (Jonnes 2004). Then, he followed with street lighting and, finally, offices and households. Even more, he deeply understood the systemic nature of his invention and the need to develop complementary goods, such as generators, enabling the market to adopt his technology as a complete solution. Similarly, neither Henry Ford nor Bill Gates invented respectively the motor car or the personal computer. However, it is thanks to them if these artifacts have effectively become widely diffused innovations.

[8] For instance, Stirling engines* (a special type of steam engine) were invented in the eighteenth century and have hardly ever been used in practice. However, they are now being considered as interesting solutions for generating electricity from the heat obtained in solar concentration plants.

used by a company—modify the way it operates its business processes. Examples of such innovations can be pure methods or business practices (e.g., Lean Production and Six-sigma) or technological artifacts that have a strong impact on the organization (e.g., a CRM system). *Business model innovations*, instead, have to do with new ways to conduct a business, its relationship with customers and value chain, its revenue flows and cost structure. Business model innovations are often cited when talking about Internet companies,[9] but can apply to firms in other industries as well.[10]

In reality, this distinction is weak, since a same technical innovation may be classified quite differently, depending on the perspective one takes[11] and on the magnitude and breadth of its potential impact.

1.2 The Innovation Process and Its Actors

In very simple terms, it is possible to link discovery, invention and innovation/ product development as in Fig. 1.1.

This linear model of innovation is quite simplistic and often challenged by scholars. Authors such as Kline (1985) and Edgerton (2004) have developed more complex models, showing how the innovation process is not linear and exhibits many relationships of iterative nature, as well as not all innovation derives from the outcomes of scientific endeavor. Moreover, a single discovery might lead multiple innovations in different industries, while an innovation might require the development and integration of multiple enabling technologies. However, the linear model

[9] For instance, we can say that a company like Google owes its success to the business model innovation introduced in the Internet search industry, and not only to the process innovation they have introduced in the search process, thanks to sophisticated algorithms.

[10] For instance, Dell Computer has been an important innovator in the computer hardware business by selling directly to customers and assembling computers to order. We can safely say this has been a business model—and not simply a process—innovation, because of the change of upstream and downstream relationships. Similarly, low-cost airlines such as Southwest and Ryanair have introduced what we can define as business model innovations, because of the way they overturned traditional revenue flows and cost structures. For instance, meals and baggage handling have become a source of revenue instead of a cost. Sometimes, and with significant controversy, the same has happened for airport fees, thanks to the choice of flying to under-used airports owned by public authorities, who subsidize flights in order to stimulate local tourism.

[11] For instance, one might wonder how to classify a new videoconferencing software. If one takes the perspective of the software firm that has developed it, it is clearly a product innovation. From the perspective of a university lecturer that uses it to teach online, the same artifact becomes the tool to perform a process innovation. For the university that decides (as it happened during the COVID-19 pandemic) to set up online courses, this can become an organizational innovation. If the same university, once the health emergency has subsided, decides to leverage on the experience gained and to change its teaching model in order to cater to remote students, this can become a business model innovation. Finally, this same software tool can become an important social innovation both for the communities of origin of these students (who can now access quality education without having to relocate and sustain the associated costs) and for the communities in which the university is hosted, who will see a marked decrease in the inflow of young students.

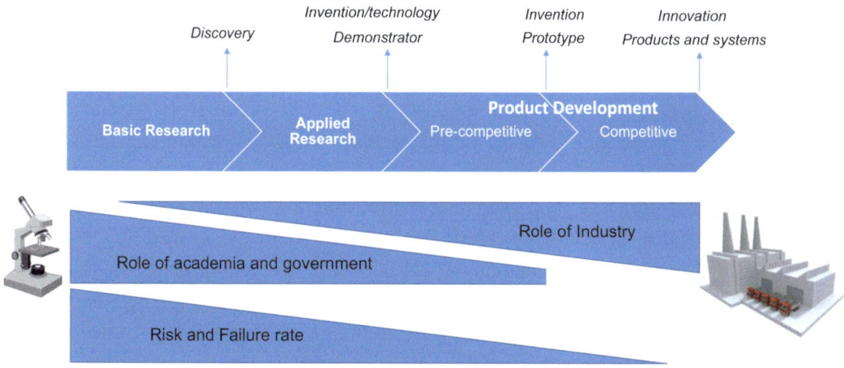

Fig. 1.1 Simplified 'linear' model of innovation

still is valuable, since it is adequate for providing a first explanation of the phenomenon, and because it serves as the foundation for many concepts that are widely accepted by managers and policymakers. To this purpose, we will follow the definitions that come from the Frascati Manual and the Oslo Manuals of OECD (2015), which are standard and globally accepted references for statistics and policymaking, respectively, in the fields of research and development and of innovation.

Starting from the upstream phases of the model, the activity responsible for discovery is called *basic research*, where basic means it is not directed to a specific application.[12] The activity that leads to the solving of a particular problem and the invention of a new technology is instead termed *applied research*.[13] The practical deliverable of this phase is generally a *demonstrator*, which is an artifact whose purpose is simply to show that the invention is technically viable. Demonstrators are usually very far from commercial products. At this phase, the main focus is technical and rests on showing the viability of the individual technology *per se*. Attention paid to potential application areas and target markets is still limited, and the same goes for choices related to complementary technology.

When moving into product development, the focus shifts quite dramatically, since the firm will start dealing in detail with commercial attractiveness (thus defining price) and industrial viability (thus defining cost and margin). This also implies dealing with seemingly menial and often neglected aspects such as usability, safety, manufacturability, logistics, product certifications, writing users' and field service manuals. The term 'product development' encompasses a broader set of activities than the 'development' that is meant when speaking of 'research and development'. In fact, 'product development' occurs in general whenever a

[12] The Frascati Manual states "*basic research is experimental or theoretical work undertaken primarily to acquire new knowledge of the underlying foundation of phenomena and observable facts, without any particular application or use in view*".
[13] Again, referring to the Frascati Manual, the definition is as follows "*Applied research is original investigation undertaken in order to acquire new knowledge. It is, however, directed primarily towards a specific, practical aim or objective*".

firm engages in a project leading to a new product or service, regardless of the innovative content and of the extent to which it aims to generate new knowledge. In some cases, product development goes beyond what is called a '*routine change to products and processes*' and aims at expanding the state of the art in the industry. In such cases, one can properly speak of 'experimental development'.[14]

It is customary, when dealing with experimental development, to split the product development process in a *precompetitive* and a *competitive* stage. The discriminating element is that the former leads to *prototypes* that do not have commercial value (i.e., they cannot be sold), while the latter leads to the actual product that will be launched on the market through a phase termed *industrialization*, which is aimed to understand how to make it producible. A prototype is clearly different from a demonstrator, because the focus is no longer on an underlying technology and its potential, but on the features and technical choices that the future product will incorporate.[15]

The bottom part of Fig. 1.1 shows that the progress along the linear model of innovation can be split in phases defined by the *Technology Readiness Levels* (TRLs)* achieved by the technology. TRLs, along with their definition, were proposed at NASA during the 1970s to estimate the maturity of technologies during the acquisition phase in space programs. The concept spread beyond this particular use and the European Commission since 2010 have adopted this scale for the definition of EU-funded research and innovation projects. Today, especially in the EU, the definitions contained in the Frascati Manual and in the TRL scale have assumed a quasi-legal status, since they are used to clearly define the limits of allowable state aid and to define the applicability of policy instruments. For instance, based on these definitions, an EU Region may launch a policy which

---

[14] Experimental development is defined as "*systematic work, drawing on knowledge gained from research and practical experience and producing additional knowledge, which is directed to producing new products or processes or to improving existing products or processes*". In order to aid in properly identifying R&D, the Frascati Manual proposes five criteria, which must jointly be present for a project to qualify as such:

1. Novel (i.e., whether the project entails an advancement of the stock of knowledge available to the reference industry, and not simply to the firm; while a significant advance in the products and services being developed is usually present in such cases, this is not a prerequisite, since the focus is on the generation of original knowledge);
2. Creative (i.e., the activity requires something more than the enactment of routinely professional work);
3. Uncertain (i.e., the project is characterized by risk, either with respect to the outcome, the technical solutions to be adopted, or the time and cost to be incurred);
4. Systematic (i.e., to be explicitly planned, budgeted and managed, thus excluding simple 'learning by doing');
5. Transferable or reproducible (i.e., it must lead to explicit and codified knowledge, and not simply to a growth of employees' personal knowledge).

[15] The Frascati Manual suggests a further difference between prototypes aimed at validating the new and uncertain technical solutions being developed through experimental development, and the 'final' prototypes aimed at achieving certifications or launching a sales campaign.

ensures grants or tax breaks to "firms who engage in Industrial Research projects associated to technology X, and which allow moving from TRL 2 to TRL 4".

The process depicted in Fig. 1.1 is quite lengthy. Depending on the field of research and on the industry, moving from basic research to the diffusion of products that will incorporate its results can take anything from a few years to decades. In some cases, the connection from research to innovation is quite straightforward, since a single result upstream ends up into one or more applications downstream. For instance, the development of laser beams led to a number of applications, from metal cutting to surgery and from bar code scanning to optical communications (and many other). In other cases, instead, an innovation is enabled by the confluence and integration of several technologies, whose development times may be completely different. An autonomous vehicle, for instance, involves technologies associated to the control of the vehicle, both traditional and based on machine learning, and on a number of sensor technologies, from cameras to Lidar. Most of these technologies have not originated specifically for the autonomous vehicle application and have evolved in other industries and applications, with completely different dynamics. Therefore, it is difficult to define a single linear model for the system that encompasses them all (i.e., the autonomous vehicle), and any such representation will explain the phenomenon only in very broad terms.

Risk and uncertainty are inherent to innovation process. This can be considered negative, but it also leads to a number of effects that have the desirable potential of stimulating the process itself. First, the risk of failure makes the process selective and competitive. Individuals and companies engaged in the innovation process will go at great lengths to propose new, alternative and improved solutions, thus creating a wide experimental arena in which the most promising solutions may eventually become successful. If this essentially 'Darwinian' process operates as it should, society will benefit since, due to the same uncertainty, it would not be possible to pick the best solution ex-ante.[16] Moreover, even if an enlightened decision-maker were able to do so, lack of competition would probably not create enough incentives to fully develop that same promising solution. Secondly, the selection process will cause firms espousing a losing solution to fail or to downsize. This improves the potential returns for the winners, who will gain market share. At the same time, this leads to knowledge spillovers, since at least part of the employees being laid off

[16] The reference to Darwin has not made by chance, since the innovation process is based on a very similar mechanism of experimentation, recombination and selection. In fact, economics of innovation is tightly connected, from an epistemological perspective, to the somewhat heterodox school of Evolutionary Economics*. As a side note, innovation economists are well aware that technological progress is somewhat brutal and irresistible, but this does not automatically create a connection with strict economic liberalism and *laissez-faire* policies. For example, an economist of innovation may recognize that e-books and music streaming services are supplanting the traditional media distribution industry. It is likely that he/she will dismiss suggestions to use public funds to support the industry in crisis, but may recommend policies aimed at reconverting its assets to other markets, or supporting displaced employees.

by a losing company are likely to be hired by a successful one, taking their know-how and experience with them, and thus enriching the latter even further.[17]

Risk in the innovation process might however be difficult to be appreciated as positive. In many cultures, both at national and at corporate level, failure is considered as a very negative outcome. A failed project or a failed company might translate in a broken career for a manager or an entrepreneur, and this might reduce the incentive to pursue innovative, but risky, initiatives. In some cases, public authorities may also step in to support entities that have made wrong choices with respect to technology or business models. Such a decision has obvious merits with respect to the immediate social benefit. However, propping up jobs instead of supporting the jobless may also lead to hindering the growth of stronger and more innovative firms and take away incentives to them.

Given the role of risk and failure, one can wonder which *economic actors* might bear the effort of investing in the activities represented here. Given the risk and time involved, it is obvious that private sector companies will not usually finance the entire process, and especially its upstream phases, which are riskier and further away in time from providing economic returns. This clearly represents a *market failure**, in which governments are called to step in and support the activities that private actors would not invest in.

The debate on public–private relationships in the innovation process is very strong, but it is mainly associated to the magnitude of the effort and the approach to be followed by public entities. In fact, even the most fervent proponents of economic liberalism would never suggest that the State should completely withdraw from the field. Besides the fact that the innovation process would simply stop working, it would be socially undesirable to have private investment cover the early phases of the innovation process. In such case, firms would attempt to keep the knowledge generated confidential, both concerning successes and failures, and this would lead to a huge duplication of effort. Conversely, public funding of research is usually considered to be efficient, because it brings the knowledge generated in the public domain. This allows researchers to drive science forward by incrementally building on top of each other's results and allows companies to use such results freely for the purpose of inventing and innovating. Therefore, even the most free-market-oriented countries throughout the world subsidize and support, often quite generously, the early stages of the innovation process.

The discussion here will examine the process from upstream to downstream, by first discussing public support to the innovation process and then moving to a discussion of privately led activity.

Governments usually have a direct role down to the so-called precompetitive phase of product development, whose end-products are prototypes that do not have commercial value. They do not usually act in the last phase of product development since—at this stage—the time required to recover investments is shorter, risks are lower, and markets are therefore usually able to work properly. Moreover, it would

[17] Well-known technology clusters such as the Silicon Valley are characterized by a very high 'churn rate' of companies and jobs, which leads to the collective growth of the entire system.

be questionable for a government to support product development activity for commercial products. This would distort competition both within the country (i.e., subsidized producer *X* would be favored over another producer *Y*) and in international trade (i.e., producers of a subsidizing country would be favored over the other ones). In fact, this practice is usually prohibited by international agreements, such as the ones promoted by the World Trade Organization (WTO). This general rule does not mean that governments cannot have a strong impact on this phase too, provided they refrain from direct financing.

The main means with which governments can support the innovation process are summarized in the following Table 1.1, which covers both financial and indirect support. Specifically, governments can stimulate the innovation process by supporting both the supply of innovative products and services, as well as the demand for the same. The table should be quite self-explanatory, and we will therefore not provide a detailed discussion on each policy measure. In general, the table clearly shows that each policy approach has specific pros and cons. This suggests that a sound support to innovation will probably require the enactment of a coordinated mixture of measures.

Table 1.1 Means of support from governments

Policy	Rationale	Typical tools	Pros	Cons
Supply side/direct funding	State provides financial resources that market does not supply	Grants and low interest loans (e.g., EU grants for R&D and for SME innovation), usually awarded by selecting projects submitted in response to a 'call for proposals'	Speed (provided procedures are run efficiently) Ex-ante definition of the budget	Potentially low allocational efficiency, since decisions are made on simple proposals and market involvement is weak (i.e., private actors who propose projects and especially public actors who select them have little 'skin in the game') 'Red tape' in submitting proposals and reporting on projects

(continued)

Table 1.1 (continued)

Policy	Rationale	Typical tools	Pros	Cons
Supply side/incentives	State amplifies the resources provided by market or reduces risks (ROI is improved by reducing immediate outflows)	Matching funds (e.g., public co-funding schemes for venture capital) Tax breaks (e.g., personal income tax credit on the amounts that business angels provide to startups) Guarantees (e.g., Italy's 80% state guarantee on bank loans to innovative firms)	Speed Market involvement in selecting beneficiaries should improve allocational efficiency	Needs a market capable of selecting beneficiaries wisely Risk of opportunistic behavior Cost of policy difficult to quantify ex-ante (makes it difficult to budget)
Supply side/boosting outcomes	State increases benefits to market, but when and if results are achieved (ROI is improved by boosting potential future inflows)	Rebates to income tax accruing from innovation (e.g., 10-year waivers on corporate income tax for innovative firms, rebates on capital gains achieved by business angels, 'patent boxes')	Very strong market involvement Lower risk to State (money only goes to those who actually gain from the investment and after societal benefits have accrued)	Needs a market willing and capable of selecting beneficiaries wisely Cost of policy difficult to quantify ex-ante
Demand side/procurement	State stimulates innovation by creating an early market for innovators	Public Technology Procurement (e.g., SBIR program, DoD procurement programs in the USA)	Award is given to effective results and not just to 'promises' Leads to the creation of a real market and supply chain State benefits from the adoption of improved goods and services	Requires a strong entrepreneurial culture and appropriate legal framework and procedures within civil service

(continued)

Table 1.1 (continued)

Policy	Rationale	Typical tools	Pros	Cons
Demand side/regulation, standardization	State creates and shapes the demand for innovation and reduces commercial and technical uncertainty	Product-specific regulation (e.g., biodegradable bags, CO_2 emissions for vehicles)	No immediate cost	State might impose technical choices/targets that are not neutral and/or are difficult to achieve Risk of technological lock-in Requires a credible State that will not renege and a market capable of delivering the innovation
Demand side/incentives	State reduces the cost/boosts the benefits of adopting the innovation	Direct incentives to adopters when purchasing (e.g., purchase rebates for electric vehicles, hyperamortization for investment in innovative capital goods) Incentives to the usage of innovative goods (e.g., feed-in tariffs for renewable energy producers)	Strong and immediate impulse given to adoption Award given to effective results and not to promises Creates a market and a supply chain Society benefits from rapid diffusion	Applicable to technology that is market-ready Risk of benefiting foreign producers Cost can be high and difficult to budget Risk of technological lock-in Can induce volatility in demand (diffusion can slow down, in anticipation of incentives, and decrease when these are lifted) Policy could be fiscally regressive, not generate additionality (i.e., money goes to customers who would buy anyways) and lead to unused goods

(continued)

Table 1.1 (continued)

Policy	Rationale	Typical tools	Pros	Cons
Demand side/complementary assets	State facilitates and shapes the demand for innovation by providing complementary infrastructure	Investment in complementary assets (e.g., direct provision or funding of recharging infrastructure for electric vehicles)	High impact on diffusion	No technological neutrality and risk of technological lock-in High investment cost
Non-specific	State improves general 'business friendliness'. This improves attitudes toward risky investment, amplifies upside and reduces downside	General reforms (taxation, education, infrastructure, access to justice, bankruptcy laws, red tape, etc.)	Stable and visible impact No need to prioritize an industry and/or field of technology	Time required

Moreover, the table shows that public action in the context of innovation is fraught with risks. The most obvious risk is associated to acting too little (and making the constituency being governed fall behind with respect to more competitive territories) or acting ineffectively (and wasting public money that could deliver better societal returns, if used elsewhere). The choice of the policy mix appropriate to a specific local context is risky, as well.

On the one side, the private actors operating in the constituency must be able and mature enough to respond to the policy, and the same holds for civil service. In fact, each approach requires peculiar competencies and processes, which might not be present within the local public administrations or might even be incompatible with the legal framework in which they operate. It follows that policymakers should be wary of simplistically imitating 'recipes' that have appeared to work well in other contexts.

On the other side, the type of choices made, and the timing or pace with which they are carried out, is highly critical. Policymakers generally wish to accelerate innovation processes in order to improve public welfare; nevertheless, this exposes them to the risk of hindering an orderly evolution of new technology, which does not only require money, but also time. More specifically, policymakers should take care of promoting measures that do not respect the principle of '*technology neutrality*'*. For instance, a hastened introduction of a specific and still immature technology risks leading into 'technological lock-in', which preempts the emergence of alternative technological solutions that might be better in perspective, even if are simply lagging behind in their evolution.[18]

If one now moves downstream in the linear model of innovation, with risks becoming lower and rewards getting closer in time, private actors progressively assume a more significant role.

The literature on innovation has widely discussed the actors who perform such activities; they can be summarized as in the following:

- **Innovators-entrepreneurs**. Alexander Graham-Bell, Thomas Edison, but also today's Bill Gates, Mark Zuckerberg, Jeff Bezos and Elon Musk are typical examples of innovators-entrepreneurs. This particular role has been studied since the early works of Joseph Schumpeter (colloquially termed 'Schumpeter Mark I'). These individuals are able to merge technical knowledge with business insight and to introduce dramatically new solutions to the market. Thanks to their efforts, they often lead what are small startup firms to business success and come to dominate the new industries they spawn. Dealing with these cases, Schumpeter introduced the term *widening innovations*, to emphasize the newness of related products, industries and markets and described the phenomenon as a '*gale of creative destruction*' that disrupts past economic entities and establishes new ones.
- **Large firms**. A few years later, Schumpeter noticed that large firms engage in innovative activity ('Schumpeter Mark II') as well, and commit considerable labor and capital to improving products, processes and services. Nowadays as well, large companies in industries such as automotive, aerospace and pharmaceutical invest very substantial resources in this type of activity. The nature of these improvements is significantly different from the previous one, since the aim is not to generate completely new solutions, but to improve existing ones. One could debate on whether this kind of 'minor' improvements should be considered as innovations or not. However, if we do follow the previous definitions, even a minor improvement of an existing product will be an innovation (albeit small) if it has an economic impact of some sort. Schumpeter aptly termed these *deepening innovations*, in order to emphasize their incremental nature.

[18] As an example of the issues associated to technological neutrality, one can think of the policies that are promoted worldwide to decarbonize the automotive industry. These include support to the rolling out of recharging infrastructure, a rapid reduction of allowable CO_2 emissions in new vehicles and citywide bans for the circulation of polluting cars. The former policy is explicitly not technology neutral, since recharging stations are specific to Battery Electric Vehicles (BEVs). The other two are technology neutral in principle, since they could also support Fuel-Cell Electric Vehicles (FCEVs). However, this neutrality is only apparent, since FCEV development is trailing BEVs' and the velocity of these new regulations is such that only BEVs can effectively qualify. Should markets adopt BEV technology and its complementary recharging infrastructure, this will reduce the market space available to FCEVs, together with the financial incentives to develop this technology, regardless of the possibility that it might be (in principle and in perspective) superior to BEVs'.

After Schumpeter, the debate on the types of actors involved in creating innovation has continued. Nowadays, at least three other categories have become important.

- **Networks of firms** (or *ecosystems*). In recent years, technology has become ever more complex, making it difficult for a single firm to master all the competencies required to develop modern products. At the same time, Information Technology (IT) has provided companies with unprecedented means to cooperate with one another and coordinate joint activities. Thanks to these two trends, more and more innovation is being created by groupings of complementary actors that not only include the innovative startups and the large firms identified by Schumpeter, but also customers and suppliers, up to universities and research centers.[19] This approach to innovation is often structured around the concept of *Open Innovation** (Chesbrough 2003), a term that highlights a strategy by which a company organizes to pursue innovation opportunities beyond its boundaries. These groupings of actors are generally associated to a focal firm, which acts as the coordinator of an *ecosystem*—a concept that extends the traditional concept of supply chain and includes customers, suppliers and complementors. In some cases, these innovation ecosystems operate thanks to a platform model, which leads to the definition a multisided market (e.g., in the case of app stores for smartphones).
- **Ecologies**—Dougherty and Dunne (2011) have recently suggested that ecosystems cannot adequately describe the context within which complex innovations emerge. *Complex innovations* are innovations that require an intricate and sometimes unforeseeable interplay between actors, which often do not have a previous record in cooperating among themselves. In such cases, one can speak about *ecologies*, which are a somewhat broader and looser concept than that of ecosystem and its relatively well-defined perimeter.[20]
- **Customer co-creation**—Firms have long been experimenting some degree of involvement of their customers in the innovation process. A close relationship,

[19] Many large firms, such as P&G and pharmaceuticals, have reduced their internal R&D expenditure and increased the cooperation with other entities that may already have developed solutions to relevant problems, or have the competencies to do so quickly.

Another striking example is Apple which is, as of today, one of the world's largest firms by market capitalization. However, it is a relatively small firm in terms of employment, and it certainly does not derive its market value from its fixed assets. Its success is mostly due to its capability of directing the innovative effort of a complex ecosystem of suppliers (e.g., component manufacturers who develop new products specifically for Apple) and complementors (e.g., application developers).

In a similar way, many other firms innovate by leveraging on a focal Internet company (e.g., games developers on Facebook) to deliver innovative products and services.

[20] The shift to electric mobility is a typical example of a complex innovation, since it heavily depends on the rollout of a recharging infrastructure. In turn, this requires disparate actors such as vehicle manufacturers, energy producers and distributors, municipalities, recharging service providers and manufacturers of recharging infrastructure, to cooperate around business models and interaction patterns that are still undefined.

up to co-design, between manufacturer and client, has in fact traditionally been carried out in B2B contexts or for make-to-order products. More recently, some firms have decided to limit themselves to providing semi-finished products or system platforms, so that customers may use them to tailor a product or service that best suits their own specific needs (Ramaswamy and Gouillart 2010). Customer co-creation can range from a simple customization of user experience to a technically deep customer involvement in innovation.[21] The topic of customer co-creation of innovation naturally points to the broader concept of the *prosumer,* i.e., the consumer who does not limit his action to using and consuming goods and services, but also contributes to their creation (Kotler 1986).

It would be quite lengthy to delve in a discussion on the classification of prosumer activities (see Xie and Bagozzi 2008 for a proposal) and their relationship to innovation. However, it is quite clear that prosumer activities can at times become innovations *per se* (e.g., think about the amateur bloggers who have led to the emerging industry of influencers) while, in other instances, prosumers' activities are carried out passively and/or become simple enablers of an innovation that will be carried out by producers. In the context of digitalization, for instance, consumers increasingly participate in the process of value creation by unconsciously transmitting data on their usage (Espejo and Dominici 2017) which may provide valuable information for producers.

When discussing the roles of private actors involved in innovation, one must also consider the private sources of financing they can access. Table 1.2 shows a summary of the main financing means available and of their pros and cons. This table too is quite detailed, and again we will not provide a further discussion of each approach listed. As a general consideration, one should notice how the choice among the means of financing is heavily dependent on the type of firm and type of innovation projects one wishes to engage in. Moreover, each approach requires the existence of specialized actors possessing the appropriate competencies not only in selecting the firms or the projects to be financed, but also to provide support. The existence of a such complex 'institutional and organizational infrastructure' in a given territory is not be taken for granted and might require significant effort and time before it is established.

Although not described in detail here, it is noteworthy to point out that the group of actors, the typology of ecosystem, the financing mechanisms, etc. change fundamentally when considering social innovation initiatives that involve the so-called non-profit sector. In this case, also due to the nature of the services provided and to the public interest attached, other mechanisms prevail, from impact finance to government grants. Typically, in fact, organizations working in the third sector are

[21] As a B2B example of simple customization, a paint maker can provide customers with primary pigments and a colorimeter and leave them the task of blending the colors in order to achieve the desired shades. In the case of B2C, one can think of the way each of us tailors in an original way our appearance on social media and the associated interactions. As an example of a deeper role of customers in innovation, one can think of including 'makers' communities' who produce their own gadgets thanks to domestic 3D printers and Raspberry PI microcomputers.

Table 1.2 Private sources of financing

Type of private financing	How does it work?	Pros	Cons
Bootstrapping	Use margins from current operations to finance innovation	No dilution in shareholding Total independence	Time required Dependence on entity and stability of margins
Debt	Get loans from bank or issue bonds	No dilution	Unacceptable for early-stage and high-risk projects, but might be used to support expansion (i.e., financing fixed assets and working capital)
Customer financing	Early customers finance a still immature innovation. In B2B, it may be a single customer that funds the development of a preliminary pilot. In B2C this is equivalent to 'reward crowdfunding'	Deep engagement with market needs No dilution in shareholding Risk sharing	Engagement with needs emerging in a narrow customer segment Reliance on the existence of such customers
Business angels	Wealthy people (usually with industry experience), acting alone or in 'club deals', provide 'smart money' in exchange for equity to newborn high-risk startups	High acceptance of risk 'Smart money' includes competencies and relations	Dilution of shareholding Potential intrusiveness by business angels
Venture capital	General Partners (GPs) raise money from investors (Limited Partners, or LPs), establish a VC fund and provide staged financing to high-risk startups in exchange for minority stakes GPs also support the growth of portfolio companies and look for an exit (IPO or trade sale to larger firms) in time for	High acceptance of risk Focus on high-return startups and exits creates a very competitive market focused on business impact and scalability	Dilution of shareholding (moderate at each funding round, but is cumulatively significant) The reliance on exits might hinder the financing of sound but slow-growth businesses. It also requires founders to act as 'serial entrepreneurs' and not as founders of family

(continued)

Table 1.2 (continued)

Type of private financing	How does it work?	Pros	Cons
	closing the fund (up to 10 years) A few significant successes (e.g., 10x returns) in the portfolio compensate for limited successes (e.g., 1-3x), failures and write-offs and provide attractive IRR to LPs		firms or 'lifestyle businesses'
Private equity	Like VC, but usually operates with firms having a more moderate risk and growth profile. PEs might focus on dividends rather than exits and usually entail greater involvement in management	Moderate acceptance of risk	Significant dilution of shareholding Likely loss of management control
IPO	Raise money in exchange for equity from public markets	Large amounts of funding can be raised Founders can keep control Liquidity of shares	Dilution of shareholding Costly and complex activities required at the time of IPO, and after, to ensure compliance with regulations for listed firms

characterized by being private organizational subjects, in some cases a public co-participation, aimed at the production of goods and services of public or collective value. These organizations can, in view of their nature, access to public as well as private funding also for market-oriented projects, and this practice is justified by the fact that what is supplied to the market (which is represented here by society at large) are just public utility services.

1.3 The Geography of Innovation

The previous discussion has highlighted the complexity of innovation as a societal phenomenon, which requires a sophisticated interplay between actors of many different types. It is therefore quite evident that this phenomenon will not arise identically in all contexts. Not only some regional contexts will be more favorable to innovation than others, but the places that are favorable will exhibit significant specificity with respect to the type of innovation activity they can lead to.

The geography of innovation is an important topic not only for academics, but for anyone who is involved in the process, be it policymakers, who always strive to make their territories more innovative; managers, who must decide where to locate their operations and how to structure them and individuals, who may wish to relocate to regions supporting their aims. The literature on this subject is very wide, and only a few main points can be discussed in the context of this book.

One first strand of the literature focuses on the 'innovation systems' that are located in a given geography, be it at national scale (Freeman 2004) or regional (Cooke et al. 1997). In first approximation, this literature can be associated to the more general literature on Marshallian districts and industrial clusters (Porter 1998) and to the key concept of agglomeration economies. *Agglomeration economies** posit that economic actors that operate around a common specialization and in a given geographic space can benefit from the interplay of a number of factors, such as:

- the common specialization around a field of activity and/or industry and its complementary activities;
- the possibility to pool labor and resources (e.g., raw materials, specialized service providers, facilities, etc.) among the various actors that operate in the region. In a district, these actors can be relatively small, thus gaining in flexibility, while the pooling effect compensates the lack of economies of scale at the individual unit of production. Moreover, this pooling effect increases the likelihood of knowledge spillovers among firms, which enhances their individual ability to explore new avenues and innovate;
- low transaction costs, due to the spatial, cultural and relational proximity between the actors.

In such a system, innovation occurs not only because of firms, but because there is a virtuous and coherent interplay between firms, academia (which produces new knowledge and a competent workforce) and public entities (i.e., government and policymakers) who can provide a generally favorable milieu, as well as a portfolio of specific actions, that facilitate innovative processes. This vision is commonly referred to as the '*Triple Helix model*' of innovation (Etzkowitz 2008), which also includes the presence of 'bridging institutions' (i.e., science parks, technology incubators, applied research centers) whose task is to facilitate the processes that occur across the boundaries of actors that are quite different from one another. The Triple Helix model has recently been extended to consider the role of the media and civil society (the so-called Quadruple Helix, Carayannis and Campbell 2009), as well as the role of a shared attention to environmental concerns (Quintuple Helix, Caryannis et al. 2012).

This literature points to actors and processes that are inherently complex, highly specific and path dependent and whose development requires significant time.[22] This perspective would therefore suggest that each local system of innovation is unique and that 'replicating Silicon Valley in our region' can be nothing more than a daydream for campaigning politicians. However, in recent years, thriving local innovation systems have emerged very quickly in many unexpected locations, such as small US cities (e.g., from Denver to Austin) and European capitals with only a moderate focus on technology (e.g., Berlin, London, Lisbon or Paris). A few authors (e.g., Autio et al. 2018) have started studying this phenomenon and have highlighted the role of entrepreneurial ecosystems focused on digital technology. In short, such local entrepreneurial innovation systems can develop rapidly if innovation is:

- driven by startups founded by entrepreneurs (which are a highly mobile community) and supported by complementary actors, such as venture capitalists;
- focused on pursuing highly scalable business models (thus ensuring rapid economic growth);
- based on a technology which has rapid development cycles and requires complementary assets that are generic and do not need to be rooted in a given location.

This perspective suggests that policymakers nowadays might have a chance of 'building the next Silicon Valley' in their cities and regions, if they are able to develop a milieu that is attractive to the entrepreneurs, thanks to institutional stability and friendliness, availability of talented professionals to recruit and a high quality of life (Florida 2002). However, at the same time, one should realize that lower barriers to entry increase the degree of competition among regions that attempt to emerge as innovation hotspots on the global map.

1.4 The Impact of Technological Innovation on Society

While technological innovation is generally considered as a desirable phenomenon, it does not only create positive effects. Anyone engaged in this field must be aware of this issue, especially because these negative effects can be quite significant. The debate on the negative—though generally unwanted—side effects of technological progress has started since the first Industrial Revolution, with the Luddites' protests against mechanization of the textile industry. With the advent of nuclear power, the debate has become stronger, with the realization that our society had, for the first

[22] For instance, narratives of the 'Silicon Valley' phenomenon (Castells 2011) point to a long sequence of events, from the foundation of Stanford University in the late nineteenth century, to the establishment of US Navy bases and technology centers in the early twentieth century and, finally, to the emergence of high-tech entrepreneurship funded by modern venture capital in the post-WWII economy.

time in history, created a tool with the potential of *irreversibly* changing the fate of humankind and of our planet. This topic is clearly very broad, and this book cannot but point out just a few main elements.

1.4.1 Innovation, Employment and Economic Disparities

A first element of this debate is concerned with *the impact of technological innovation on employment*. The most significant product and process innovations have the potential of disrupting established companies and to lead to the birth of new industrial sectors, which are likely to require a differently educated workforce (e.g., think of the impact of telephones on telegraph companies). The worry of technological displacement is probably as old as technology itself. However, economic history tells us that this impact is limited to the short or medium term, since humankind has always found new needs to be satisfied, and this leads to the growth of new industries and the creation of new jobs. Over the last two centuries or so, technological innovation has in fact determined a massive shift in employment at first from agriculture to industry, and then from industry to services.

In spite of this generally benign view, contemporary innovation does lead to some unprecedented worries (Brynjolfsson and McAfee 2011). First, the speed at which innovation is occurring might make it more difficult than in the past to substitute the ‘old’ jobs with the ‘new’. This substitution traditionally went hand in hand with demographics and generational change (e.g., a factory worker close to retirement would lose his job because of automation, but his daughter would become a programmer in the bank’s IT department). A higher velocity of innovation, combined with the lengthening of professional lives due to higher life expectancy might require significant effort in catering to displaced workers who are too young to retire.[23]

Secondly, technology traditionally displaced workers at the bottom of the pyramid, thus leaving room for new jobs with less manual and more intellectual content. This is not so true anymore, since automation and artificial intelligence are challenging all types of workers, from warehouse operators and factory workers to financial analysts and lecturers, making it quite difficult to foresee which jobs are more likely to be displaced by innovation and to define remedies (World Economic Forum, WEF 2020).

Third, digital job intermediation platforms have partially reshaped the worker-employer relationship, allowing the emergence of the so-called gig economy. In this model, freelance workers compete in quasi-perfect competition to provide occasional tasks to employers, from food delivery to translation services

[23] This can be done via upskilling (i.e., providing them with the new skills that are needed to go on with their existing job) or via reskilling (i.e., providing them with skills that might enable them to operate in a different professional role). At the same time, some authors argue in favor of measures such as Universal Basic Income, which implies a more or less permanent loss of a job.

(Woodcock and Graham 2020), which raises obvious issues associated to worker welfare, job security and equitable sharing of value.

Fourth, the innovation-driven dynamics that occur in the industries that make up the economy have consequences in the underlying labor market. On the one side, innovation can cause industry disruption, with the demise of incumbent firms in an industry and the emergence of new industry players. This is likely to lead to imbalances, with workers made redundant in the former and labor shortages in the latter. On the other side, innovation leading to increasing economic returns to scale can lead to the emergence of entrenched monopolistic players able to extract significant rents and share them to the workers involved in their value chain, increasing inequalities with the rest of the workforce. In fact, recent research has shown that a significant part of inequality in worker earnings is explained by between-firm effects, rather than by within-firm effects (Song et al. 2019).

Finally, differences in the adoption of innovations can lead to significant inequalities among countries, and among social strata in a population, due to the difficulty of poorer classes to access new technology. Distance teaching and learning during the COVID-19 pandemic has clearly shown the 'adoption divide' phenomenon, with children from rich households benefiting from the availability of high-bandwidth connections and individual personal computers, and children from poorer families having to attend class and study by using mobile devices over the cellular network. Public policy can quite easily provide means, such as economic subsidies, to reduce adoption divides. However, besides mere access to technology, adoption divides may also include the ability of adopters to make an effective and proper use of the technology, once this is available. For instance, one might wonder whether the children of a wealthy and highly educated family with parents working from home will make the same use of their broadband connection than those of a poorer and less educated family, whose parents are away from home all day to work.

1.4.2 Innovation, Ethics and the Law

A second element of the debate relates to the *power of technology* and its ethical and legal consequences. In the words of philosophers such as Martin Heidegger* (1889–1976) and Hans Jonas* (1903–1993), contemporary technology is no longer a tool that a fragile humankind uses to shield itself from a hostile nature, but has become a powerful force that can shape the world. With technology, man has a very powerful instrument that can be deployed to many means and in ways that can lead to long-term and far-reaching—if not even irreversible—consequences.[24] It should be imperative for managers involved in innovation to consider such consequences, which not necessarily arise out of ill intention, but can be due to carelessness, or

[24] Think of nuclear power or, more recently, the impact of plastics waste and microplastics on ecosystems. The latter might seem quite iconic, given that plastics were commonly considered, up to recent years, as the epitome of contemporary industry and human progress.

even to simple lack of knowledge, since side effects can be unintended, and manifest themselves only after a considerable amount of time.

Technology not only can have undesired consequences, but also raises the issue of humankind's capability of properly using it, whatever might be meant by the word 'proper'. Following two contemporary philosophers, Andrew Feenberg* (1943–) and Umberto Galimberti* (1942–), it is delusional to think of technology as a neutral tool in the conscious hands of rational humankind (a representation that—if true—would somehow allow technologists to ignore the debate). According to these authors, 'conscious use' of technology is a myth, because the appropriate use of a given technology develops with experience and over time. In the meanwhile, human beings will tend to follow the values that are intrinsic to technology itself, which are effectiveness and efficiency ("if it can be done, why not?").[25] This problem is aggravated by the fact that people let themselves be *identified* by the tools they use.[26] In other terms, technology shapes human behaviors as much as humans can pretend they can rationally and consciously determine the application of technology.

This discussion is of course relevant to philosophers and social scientists. Is it relevant to managers too? The answer is positive, since the business impact of these issues can be quite significant.

---

[25] The phenomenon can be observed quite starkly when modern technology is introduced in a primitive society and when technological progress is particularly fast. The number of contemporary examples is abundant. For instance, one could think about the addictive and compulsive effects of digital content, from social networks to games, especially on children and adolescents, with their potential impact on cognitive and social development, and whose long-term effects on the psychological, emotional and affective spheres will probably be discovered by the next generations. Similarly, one can consider genetic screening and editing of human embryos, with techniques such as CRISPR. Aside from the controversies associated to the nature of embryos ("is it a human being we are discarding and/or working on?"), the objectives can be debatable. These techniques were originally intended to avoid giving birth to babies with genetic diseases. However, when technology will allow it, and given the choice, what parents will not try to use it in order to have better looking and more intelligent babies too? How is society going to make sure that this will not develop into fully fledged eugenics, as in Huxley's "Brave New World"? Finally, similar controversial aspects can be tied to the use of artificial intelligence (AI) supporting decision-making in many different fields, from recruitment in firms to university admissions, and from granting bank loans to deciding bail for people charged in court, all the way to deciding whether a subject is friend or foe in military operations. AI algorithms are intrinsically 'opaque' to decision-makers and tend to reinforce the stereotypes that emerge from the datasets on which they were trained, thus opening the way for potentially unjust decisions. Moreover, even if procedures imposed by the organization or by regulators dictate that a human decision-maker is kept firmly in the loop, one might wonder to what extent she/he will be willing to assume the responsibility of going against the recommendations of an algorithm, lest this decision turn out to be a bad one.

[26] A caveman with a club is not simply a stronger man than his peers. According to his tendencies, he might become a man of order who protects the village from enemies, the village chief or a mugger. Similarly, a social media user of today might develop into a more informed and socially connected citizen or professional, but might also end up as a social outcast, or even be affected by mental illness. Companies know this phenomenon quite well and marketing effort often lever on the effect.

First of all, they are of immediate relevance to *business ethics**, *Corporate Social Responsibility** and the management of *Environmental Social and Governance* (ESG)* impact factors. Contemporary business is also witnessing a blurring of the boundaries between traditional firms and 'third sector' organizations, with the former caring more and more about the social impact they generate, and the latter seeking managerial and technological approaches making them more effective and efficient.

Secondly, the neglect of such issues can have major impact on profitability. If, in fear of its potential environmental or social impact, society does not trust a potentially far-reaching innovation, it might stop or slow down its adoption, ultimately reducing the return on investment for the firms that are proposing it.[27]

Finally, unforeseen consequences of a given technology can have dramatic consequences because of *product liability**. Most legal systems affirm the principle that producers are responsible for defects in the products they market and for the related consequences. In the case of minor innovations, this liability is commonly connected to mistakes made when designing or producing the product, and the management of liability does not generally raise significant problems.[28] However, one can wonder what happens in the case of major innovations, when the product is new to the world and not enough is known about potential side effects.[29] This issue is highly controversial. On the one side, customers have the right to be shielded from dangerous products. On the other side, should producers be fully responsible for any unforeseeable consequence of their technology, and this additional risk would likely lead them to the decision of not investing if not in minor innovations based on proven solutions.[30]

---

[27] In the case of GMO crops and food, European markets have long resisted adoption out of distrust toward producers. Even if, in the end, they will end up accepting this technology, this delayed adoption might have had significant financial impact.

[28] There are cases in which product liability can raise complex issues even for technology that is not particularly innovative. For instance, a pharmaceutical company might decide to withdraw a socially desirable product, such as a drug, if its profitability is not high enough to offset the potential liability that might emerge, because of rare but highly damaging consequences on particular customers.

[29] Dramatic and well-known examples of such phenomena are tobacco smoke and asbestos, which have led to countless casualties and decades of debate between the first detection of the problem, legislative action restricting their usage and corporate lobbying trying to resist.

[30] In order to cope with this tradeoff, the EU has introduced a rule termed 'Development Risk Clause'. The Clause states that "*The producer shall not be liable [...] if he proves that the state of scientific and technical knowledge at the time when he put the product into circulation was not such as to enable the existence of the defect to be discovered*" (Directive 85/374/EEC). However, this solution is not without problems. Besides leaving consumers with residual risk, firms are effectively taken away incentives to work on product safety, since ignorance on side effects shields them from liability. This leaves the advancement of the state of the art on potential dangers of new technology to public authorities alone, who might not be able to develop or access enough knowledge on the subject.

References

Archibugi D, Andrea FA, Frenz M (2013) Economic crisis and innovation: is destruction prevailing over accumulation? Res Policy 42(2):303–314
Autio E, Nambisan S, Thomas LDW, Wright M (2018) Digital affordances, spatial affordances, and the genesis of entrepreneurial ecosystems. Strateg Entrep J 12(1):72–95
Brynjolfsson E, McAfee A (2011) Race against the machine: how the digital revolution is accelerating innovation, driving productivity, and irreversibly transforming employment and the economy. Digital Frontier Press
Carayannis EG, Campbell DFJ (2009) "Mode 3" and "quadruple helix": toward a 21st century fractal innovation ecosystem. Int J Technol Manage 46(3–4):201–234
Carayannis EG, Barth TD, Campbell DFJ (2012) The quintuple helix innovation model: global warming as a challenge and driver for innovation. J Innovat Entrepreneurship 1(1):1–12
Castells M (2011) The rise of the network society. Wiley, New York
Chesbrough H (2003) Open innovation: the new imperative for creating and profiting from technology. Harvard Business School Press, Boston
Cooke P, Uranga MG, Etxebarria G (1997) Regional innovation systems: institutional and organizational dimensions. Res Policy 26(4–5):475–491
Directive 85/374/EEC. http://ec.europa.eu/enterprise/policies/single-market-goods/documents/internal-market-for-products/liability/index_en.htm. Retrieved 23 Oct 2022
Dougherty D, Dunne DD (2011) Organizing ecologies of complex innovation. Org Sci Informs 22 (5):1214–1223
Edgerton D (2004) "The linear model" did not exist: reflections on the history and historiography of science and research in industry in the twentieth century. In: Grandinand K, Wormbs N (eds) The science–industry nexus: history, policy, implications. Watson, New York
Espejo R, Dominici G (2017) Cybernetics of value cocreation for product development. Syst Res Behav Sci 34(1):24–40
Etzkowitz H (2008) The triple helix: university-industry-government innovation in action. Routledge, New York
Florida R (2002) The rise of the creative class: and how it's transforming work, leisure, community and everyday life. Perseus Book Group, New York
Freeman C (2004) Technological infrastructure and international competitiveness. Ind Corp Chang 13(3):541–569
Huxley A (1932) Brave new world. HarperCollins Publishers, New York
Jonnes J (2004) Empires of light: Edison, Tesla, Westinghouse, and the race to electrify the world. Random House Trade Paperbacks, New York
Kline SJ (1985) Research, invention, innovation and production: models and reality, report INN-1. Stanford University, Mechanical Engineering Department
Kotler P (1986) The prosumer movement A New Challenge for Marketers. Adv Consum Res 13:510–513
Mokyr J (2002) The gifts of Athena: historical origins of the knowledge economy Princeton. Princeton University Press, Princeton
Organization for Economic Cooperation and Development (OECD) (2015) Frascati manual 2015: guidelines for collecting and reporting data on research and experimental development
Porter ME (1998) Clusters and the new economics of competition. Harv Bus Rev 76(6):77–90
Ramaswamy V, Gouillart F (2010) The power of co-creation: build it with them to boost growth, productivity, and profits. Simon and Schuster Inc., New York
Roberts EB (1987) Generating technological innovation. Oxford, New York
Schumpeter JA (1911) The theory of economic development: an inquiry into profits, capital, credit, interest and the business cycle. Transaction Publishers, London
Song J, Price DJ, Guvenen F, Bloom N, Von Wachter T (2019) Firming up inequality. Q J Econ 134(1):1–50

United Nations Conference on Trade and Development (UNCTAD) (2018) Harnessing frontier technologies for sustainable development, technology innovation report
Woodcock J, Graham M (2020) The gig economy: a critical introduction. Polity Press
World Economic Forum (2020) The future of jobs, WEF 2020 Report
Xie C, Bagozzi RP (2008) Trying to prosume: toward a theory of consumers as co-creators of value. J Acad Mark Sci 36(1):109–122

Technological Knowledge and Organizational Learning

2

Abstract

Innovation occurs thanks to the ability of humankind to produce and exploit practical knowledge. This chapter provides a discussion on the different forms of knowledge, and—following Nelson and Winter's evolutionary theory of the firm—it shows the centrality of knowledge in studying the nature of organizations. Moreover, it provides a high-level perspective on organizational learning and on its role in determining competitive advantage. Finally, the chapter provides a discussion on creativity and on its relationship with knowledge and innovation.

Innovation and knowledge are intimately connected. Firms and individuals are able to innovate thanks to the knowledge they own, while the process of innovation they are engaged in generates new knowledge. As we will soon see, this emphasis on knowledge has led scholars of innovation to develop a very specific perspective on the nature of firms, based on knowledge assets and learning capabilities. So, the first point to be covered is the definition of knowledge and its role in science and technology.

2.1 Knowledge in Technology and in Science

The nature of *knowledge** is a central topic in many fields, such as Philosophy, Psychology, Sociology and Economics. Providing a single and exhaustive definition is probably next to impossible. Moreover, the perspectives from which knowledge has been studied vary significantly, with a debate in the Western world that is split between empiricist and rationalist approaches, as well as an Eastern perspective, in which rationality and emotion are blended under the concept of 'whole personality' (Nonaka and Takeuchi 1995).

M. Cantamessa and F. Montagna, *Management of Innovation and Product Development*, https://doi.org/10.1007/978-1-4471-7531-5_2

For the time being, we can follow the traditional 'tripartite' definition, of knowledge as a '*justified true belief*' or as a '*belief that is in accordance to the facts*'.[1] Given the aims of this book, we can avoid attempting a philosophical discussion on the nature of knowledge. Instead, it is certainly useful to define some characterizing features of the different forms of knowledge, since this is highly relevant from an economic and managerial perspective.

- **Incorporation in people versus incorporation in capital**. Knowledge typically has to do with people, and we can consider it quite natural to claim that knowledge is 'incorporated' in the human beings who possess it. However, it is also possible to conceive knowledge being incorporated in a material asset, such as a book, a computer algorithm or a mechanism.
- **Codifiability versus non-codifiability**. Codifiability means the possibility of representing knowledge through a language (or *code*). For instance, the knowledge required to bake a cake can be expressed in natural language, as you will find in a recipe book, while the knowledge required to run a statistical test can be expressed in a mathematical formula. Codifiability cannot generally be complete. A recipe usually will retain some degree of ambiguity that can be solved only by an experienced cook (e.g., 'how much is a pinch of salt'?) or by using costly equipment (e.g., a precision scale for weighting the same salt).
- **Explicit versus tacit**. Knowledge can at times be easy to articulate and express and will therefore be said to be explicit. In other cases, articulation of knowledge may be very difficult. Quite often, someone may not even be aware of the knowledge she/he possesses, let alone find a way to express it to others. This tacit knowledge is inherently non-codifiable and will tend to be incorporated in a person. The only way to transfer it to others will be through a lengthy process of observation and apprenticeship. Therefore, a person may be a very skilled and experienced cook, without being able to articulate the actions and choices he makes when working in the kitchen. Conversely, another person may read a recipe but—lacking the tacit knowledge that is required to correctly interpret it—might not be able to cook a decent meal.
- **Private versus public**. This definition generally refers to whether knowledge is kept secluded, or whether it is given to the public domain. However, it should be noted that this terminology leaves some ambiguity regarding the strict economic definition of knowledge as a *private* or *public good**.[2] In general terms,

[1] However, as Bertrand Russell* (1972–1970) pointed out, the problem is that "*no one knows what a belief is, no one knows what a fact is and no one knows what sort of agreement between them would make a belief true*".

[2] In Economics, a private good is characterized by the fact that it is *rivalrous** (i.e., consumption by an individual prevents the simultaneous possibility of others to consume it) and *excludable** (i.e., it is possible to prevent someone from consuming it). An example is a video taken with a smartphone and that is residing in its memory. Conversely, a public good is non-rivalrous (consumption by an individual does not prevent simultaneous consumption by others) and non-excludable (i.e., it is not possible to prevent someone from consuming it) as, for instance, would it be for the same video, were it broadcasted by the local TV station. The two categories do not cover all combinations. On the one side, you can have club goods, which are non-rivalrous and

knowledge can be considered to be a public good but, in reality, this will depend on its technical embodiment, which determines its accessibility, both physical (i.e., whether it is rival vs. non-rival) and legal (i.e., whether it is excludable vs. non-excludable).[3] Moreover, the legal arrangements that regulate the use of the knowledge might be relevant. For instance, the knowledge contained in a patent is public knowledge and is a public good, because anyone can freely access it at any time and use it properly as knowledge. The technology described by the patent's claims itself, however, is not a public good, because the same patent excludes anyone but the assignee of the patent (or its licensees) from the right of making commercial use of it.

With these definitions in mind, it is now possible to describe four main forms of knowledge.

- **Factual knowledge (or 'know that')**. An example of this kind of knowledge is a record of the results of a scientific experiment, or of the raw data of a phenomenon. They could be for instance the failures observed in a fleet of vehicles during the last year, as well as the daily reports of COVID-19 data during the pandemic period. This is a very basic form of knowledge, evidently easy to represent and structure, as for instance in a report and/or a database. It is usually explicit, codifiable and simple to incorporate in capital. This type of knowledge is very close to the basic concept of *information*, but can be considered as knowledge, because raw information must have been structured and described to some extent, thus making it fall in the definition of 'justified true belief'.
- **Causal knowledge (or 'know-why')**. This is a higher level of knowledge and is concerned with understanding, or trying to infer, the reasons behind facts. In its quest to explain natural and social phenomena, science is particularly involved in the creation of causal knowledge, generally via inductive reasoning. A paper in a scientific journal will typically describe a phenomenon, propose a theory that explains it and then justify it on rational and logical grounds, analytically, or through empirical or experimental evidence. Epidemiologic models able to explain how and why a virus diffuses and to infer contagion spread were an example of this knowledge owned by medical researchers in 2020. A technologist too will attempt to generate causal knowledge, though in a way that will tend to be situation-specific. For instance, a designer may analyze the database of

excludable (e.g., the same video, but shown on pay-TV), and common goods, which are rival and non-excludable (e.g., the same video, shown on my smartphone to the passersby on the street, which is limited to the ones that are able to huddle around it).

[3] For instance, the tacit knowledge possessed by an individual or the explicit knowledge contained in a personal recipe book would clearly be a private good. The digital version of the same recipe book placed on Facebook would certainly be non-rival from a technical perspective, but its excludability would depend on the choices of who posted it. Should she/he share it on a public profile, this would make it a public good, while posted on a private profile, it would be a club good reserved for friends (provided they don't send copies around, of course). Finally, should she/he place the recipe book hanging from a chain in front of the doorstep, this would make it a common good (until someone steals it, or takes pictures and makes a pdf version to share).

product failures in vehicles, extract patterns, try to infer the reasons behind them and maybe trace them back to a specific component of the car. Causal knowledge is relatively easy to codify and then to be incorporated in capital (e.g., an academic paper, or a technical report written in natural language, and making use of mathematical formulae, tables and graphs). When dealing with causal knowledge, a distinction may be made between *empirical* and *theoretical* knowledge. The former correlates observations on the basis of experience, but without yet making claims on the underlying reasons. The latter attempts to explain the reasons behind these phenomena. A cook may empirically observe that boiling water is not hot enough for cooking rice when working in the mountains, but not have a clue about the reason. She/he may attempt a theoretical explanation, such as "probably in the mountains it's much colder", which may or not be the real one (as a matter of fact, it isn't), but nonetheless can support practical behavior (e.g., allowing more time for the water to boil). Empirical and theoretical knowledge are nonetheless tightly related. Following the work of Poincaré* (1902), it is impossible to draw knowledge from experience '*without prior ideas*', since it is background knowledge that allows a person to structure, name and classify observations into knowledge elements.

- **Procedural knowledge (or 'know-how')**. This is a still higher level of knowledge and is what everyone employs in order to achieve a given result in private and professional life. This knowledge derives from a mixture of deductive reasoning, schooling, experience and, possibly, innate talent. To a scientist, developing procedural knowledge is not a core objective, but is a useful tool that she/he will use in professional life, i.e., when setting up experiments, analyzing results or presenting findings at a conference. To a technologist, whose professional objective is to solve specific problems, the development of procedural knowledge is instead of central importance. In our examples, once the designer has developed the causal knowledge that product failure is due to a particular component, procedural knowledge will enable him to redesign the component in order to improve reliability. Similarly, once epidemic diffusion models are validated, procedural knowledge can lead to the definition of public health measures for contrasting the contagion. Know-how is typically tacit, difficult to codify, incorporated in people and, therefore, naturally tends to be private.
- **Positional knowledge (or 'know-who')**. Finally, one may not directly own knowledge, but might know where it is stored, or who possesses it. Even though this is an indirect type of knowledge, know-who is indeed very powerful. This is especially true for organizations, where it is not expected that everyone knows everything but, on the opposite, economic value is created by linking together the knowledge elements that are possessed by its many members. To the extent that underlying knowledge can be categorized and classified, positional knowledge is relatively easy to codify and incorporate in capital.

The previous discussion has purposely highlighted a few key differences between technological and scientific knowledge. However, we know from the previous chapter that science and technology are related to one another in the innovation process. One can therefore wonder how this relationship works, and how scientific knowledge can contribute to technological progress.

The issue is not trivial since, following Fleming and Sorenson (2004), the strength of the connection between science and technology varies quite significantly across disciplines and industries.[4] Moreover, the contribution provided by scientific knowledge to technology is not direct, since the objectives for which it has been developed (i.e., providing broad explanations of phenomena) are somewhat different from those of technology (i.e., finding solutions to specific problems). So, these two forms of knowledge can be viewed as complementary to one another.

Technologists, having to stay close to the requirements cast by a specific situation, tend to operate on a 'local search' basis, applying small variations to known solutions. By operating through trial and error, they progressively build a body of technological knowledge based on experience and empiricism. Technology is sometimes represented as a rugged terrain where solutions correspond to the position on the xy plane and the associated performance to the elevation of the terrain on the z-axis (Gavetti and Levinthal 2000). Technological progress can therefore be illustrated as an exploration of the terrain in the dark, in which one progressively looks for solutions having an improved performance. Following this analogy, scientific knowledge can be equated to a relatively rough map of the terrain, possibly not detailed enough as to allow a precise definition of solutions, but able to provide technologists with the 'wide picture' they would miss by myopically exploring the terrain step by step. Thanks to this 'map', technologists can therefore operate with greater efficiency. When operating incrementally, the map shows the most promising directions along which to search, thus reducing the amount of experiments required. The map can also allow technologists to break loose from 'incrementalism' and look for better and globally optimal solutions, even if they are significantly different from traditional ones. Finally, the map can provide organizations with an objective rationale for justifying the initiation and continuation, or the rejection of projects, from a financial and organizational point of view.

Given that scientific knowledge has to a technologist the same role that a map has to an explorer, we still need to achieve a better understanding of technological knowledge. Up to now, we only have shown that this latter kind of knowledge is tacit, difficult to codify and transfer, since it is rooted in experience.

[4] In the case of Life Sciences, the relationship is very tight and it is at times difficult to discriminate between activities, actors and languages. An academic researcher will be likely to have a good understanding of industrial needs, while a researcher in the pharmaceutical industry will routinely read academic papers. Conversely, if one looks at Computer Science, a programmer engaged in writing code will use a substantial amount of practical know-how, but with a loose connection with the knowledge developed in the academic domain.

This topic has been amply studied by Vincenti (1990) in his seminal work, based on a deep analysis of the evolution of knowledge in the field of aeronautical engineering. Vincenti's main finding is that technological knowledge is broader than simply 'applied science' and can be classified in the following categories:

- **basic concepts**, such as working principles and common product configurations that can be immediately learned when entering the field;
- **design criteria and specifications**, such as requirements, technical standards and regulations;
- **theoretical tools**, such as mathematical formulae and mathematical models;
- **quantitative data**, coming from prior experience, purposely generated through experimentation or collected from the many processes that occur within the value chain (which is nowadays particularly relevant);
- **practical knowledge**, deriving from experience, 'rules of thumb' and lessons learned from past failures;
- **design instrumentalities**, such as procedures and judgmental rules shared by the firm or by the wider professional community.

The four former categories can be considered to be explicit and fairly easy to codify, while the latter two are typically of tacit nature. These same categories can be found in about any technological field or industry. However, the knowledge residing within each category will emerge out of the experience gained by repeatedly solving the problems encountered by each industry and firm and will therefore be quite specific to them. This specificity raises the issue—that will be covered in the next section—of understanding how differences in knowledge can determine differences in the behavior and the performance recorded by firms.

When dealing with technological knowledge in contemporary industry, the issue of multidisciplinarity must also be considered. Current products and services integrate multiple technologies that require the involvement of the corresponding knowledge sources (e.g., a collaborative robot requires the cooperation among mechanical engineers, control engineers, AI experts and cognitive scientists). Even more, the trend toward the development of products and services from an integrated lifecycle perspective needs a strong interplay between domain experts in designing the product, the process, aspects pertaining to environmental and social impact, etc. Coping with these disparate knowledge domains raises multiple issues. On the one side, it leads to the dilemma between operating via the cooperation of individual domain experts (e.g., the mechanical engineer and the control engineer working in a multidisciplinary team) and using a single domain expert (e.g., a mechatronic engineer) characterized by an interdisciplinary profile. Furthermore, and especially when multidisciplinary teams are involved, it leads to the need that individual domain experts be able to cooperate with one another both by having a minimal understanding of their colleagues' fields and by having soft skills that enable a fruitful cooperation. In this case, literature often uses the term '*T-shaped*' to describe the worker's knowledge profile. In recent years, the growing and pervasive importance of using data in developing new products and services has led a few authors to suggest that data science competencies should not be simply provided by adding experts in this domain to multidisciplinary teams, but that all professionals should be proficient enough to use data science methods in their vertical domains. In this case, the knowledge profile is labeled '*Π-shaped*' (Ceri 2018).

2.2 An Evolutionary Theory of the Firm

Since Schumpeter, economists of innovation have struggled in trying to define the nature of firms as actors of technological change. To them, it was not possible to settle with the neoclassical vision of the firm as an economic agent that, with perfect rationality and full information, observes market demand, chooses production factors (capital and labor) and defines a production function in order to maximize profits. In fact, economic environments affected by innovation are inherently uncertain and turbulent, and this clashes with the hypothesis of perfect rationality and full information. Moreover, turbulence will likely make competing companies behave in significantly different ways. This too diverges from the neoclassical idea that differences between firms may exist only in the short term, since less successful firms will quickly imitate successful ones. So, a theory of the firm that satisfies students of innovation should be able to answer the question "why are firms different?".

The question has gone unanswered for a long time, until Nelson and Winter (1982) proposed that a theory of the firm could be developed by observing organizations inside-out, instead of outside-in.[5] Following this theory, a firm can be defined by a *static* and by a *dynamic* view.

The static view posits that a firm can be defined as an *organized association of complementary resources*. In this definition, resources are intended as human resources or physical assets that can be acquired at a cost, and who create an economic value that is hopefully greater than their cost. Complementarity means that the economic added value that is jointly produced by resources is greater than the sum of the economic added values that would be separately created by each resource. Complementarity justifies the very existence of firms, i.e., the coexistence of a set of resources in the same organization, instead of the interaction between autonomous entities through market mechanisms. Thanks to complementarity, resources in a firm will generate a combined super-additive economic value that is sufficient to cover the fixed costs of the organization (e.g., management, support

[5] This perspective is coherent with other scholars' views of firms. For instance, Hayek (1945) identified the role of tacit knowledge as a source of competitive advantage, and Penrose (1959) emphasized the role of learning in organizations.

functions, etc.) and to provide economic returns to its shareholders.[6] Finally, 'organized association' means that the firm is able to put these resources to work and therefore ensure the production of economic value.

The dynamic view builds on the static view and deepens this concept of 'organized association'. Here, the organization is not viewed as the simple allocation of duties and responsibilities that one could represent with an organizational chart, but as *a bundle of organizational routines*. As anyone who has worked in a company will have noticed, a wide majority of the work being carried out by employees at all levels is routine, and a large portion of this routine work is performed without having been explicitly defined or formalized. This observation is obvious when workers deal with similar problems all the time (e.g., assembly work, processing accounts payable, etc.), use a common language and share procedural knowledge. However, this observation is also true for employees who daily encounter different problems (e.g., a designer, a medical doctor, etc.) but end up tackling them in the same way, as suggested by experience or dictated by the formalized protocols.

These organizational routines define the steps that the firm will follow in order to reach a given objective and the resources to be involved. Routines can therefore be viewed as the actions that firms enact in order to 'activate' resources and 'extract' a contribution from them, based on the knowledge they possess, thus progressively building up economic value.

Looking at the example in Fig. 2.1, we can suppose that a firm is made up of four people who individually own the knowledge elements represented in the rectangles. We can therefore say X is part of corporate knowledge, not because the company directly knows X, but because a routine is in place, which allows the firm to 'use X' being provided by its employees. Conversely, we cannot say M or Z to be part of the firm's know-how, since no routine has been set up for using these knowledge elements. Therefore, routines represent a firm's knowledge base, which can be defined as the organized collection of the knowledge elements used by resources involved in routines. A major part of this organizational know-how will be tacit, being based on individuals' tacit knowledge, and because of the informal nature of the routines themselves.

Should an employee with knowledge of X leave, the firm would lose X, but not its capability of using it. Therefore, managers will respond by recruiting a new employee possessing X and integrating her/him in the appropriate routines (Argote et al. 1997). From the perspective of the employee who leaves the company, she/he would take out her/his personal knowledge, as well as the experience of using it within that specific firm's routines. However, it would be difficult for her/him to join a new company and replicate the routine, since an individual is not likely to

[6] Based on this perspective, management's duty is to respectively recruit and purchase people and equipment, so that the difference between the economic value created and the cost is maximized. In the case of human resources, this does not simply mean recruiting highly effective people, but especially teams that will work well together and made up of individuals whose salaries are not too high with respect to their ability to produce economic value.

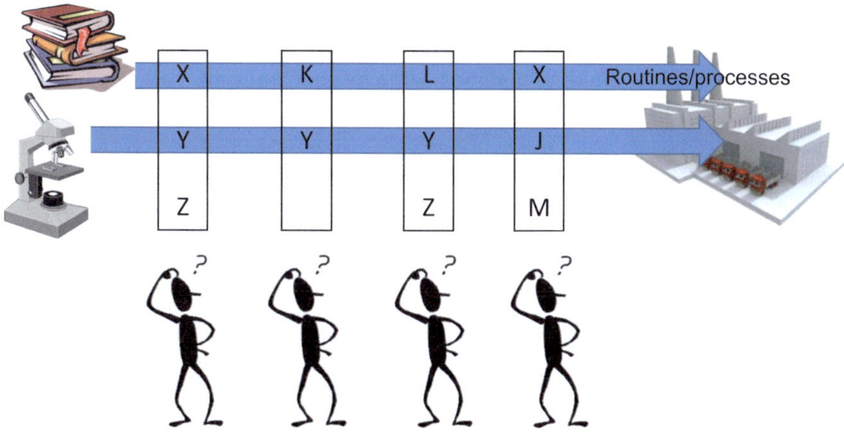

Fig. 2.1 Firm as a 'bundle of routines'

have a general view of the routine, and also because the effectiveness of the routine depends on the context. Even if this employee did have a broad vision of the routine, the *ex-novo* replication in the new firm would be difficult, given the different set of resources and routines already in place there. This explains why individuals' performance within organizations is said to be *institutionally specific* or *idiosyncratic* (Groysberg et al. 2006). In plain terms, an employee's performance will depend on the complementary resources that are present in the company and on the routines within which she/he works. An employee who performs well in one firm may therefore prove a disappointment if hired by another firm, and vice versa.

Therefore, we can conclude that there is a very tight connection between what a firm is (i.e., the 'organized association of complementary resources tied together by a bundle of organizational routines'), what it does (i.e., the routines) and what it knows (i.e., the routines' use of knowledge brought by the resources).

The concept of routines is broader than that of *business process**, which can be defined as a routine that has been recognized and formalized. Routines mostly emerge by trial and error, and because of the effort the organization spends in finding effective and efficient ways to reach its objectives and to deal with the problems it encounters.[7] This search process will not lead to an optimal solution but, rather, to a satisficing[8] one. As a matter of fact, there can be nothing like an optimal routine, first of all because the iterative process with which a firm looks for

[7] By claiming that routines emerge mostly out of a continuous process of trial and error, this implies that they are not designed 'top down' by management. As shown by Argote (1999, 2003), organizations develop knowledge thanks to a process of knowledge creation, retention and transfer.

[8] The difference between optimizing and satisficing economic behavior is a cornerstone of the theory of *bounded rationality* * (1991) of Herbert Simon* (1919–2001).

new ways of doing things will slow down when the performance reached is 'good enough'. Second, the very concept of optimality is impossible to define, since the appropriateness of a routine is highly specific to the firm itself, i.e., to the market, technology, workforce and environment the firm operates in.

Routines are beneficial to firms, because they represent an effective, efficient and predictable way of working, in a given environment. However, if the situation changes, the firm will have to adapt its routines and possibly its pool of resources in order to regain a satisfactory level of performance. If the change in the environment is limited, the new desired state will likely be close to the previous one, and it will not be difficult to achieve it. However, if the change is significant (e.g., if a completely new technology emerges, or if the government introduces a sudden change in regulations), the firm will find it quite difficult to adapt and regain a satisficing state. The term *organizational inertia** is often used to this purpose, but should not be conflated with the concept of 'being lazy', since it has to do with the difficulty encountered in modifying resources and routines.

Since organizational knowledge is embedded in the firm's resources and routines, it will also be very hard to imitate. An imitating firm will likely be unable to observe and understand resources, their complementary relationships, the routines they are involved in and then replicate them. Moreover, since resources and routines emerge out of adaptation to a given context, the imitating firm should not simply replicate what it eventually observes, but also adapt it to its own specific situation. Authors often refer to *stickiness* of organizational knowledge when expressing the difficulty of translating, transferring and imitating it (Von Hippel 1994).

2.3 Competitive Advantage Explained

The theory of the firm described in the previous section is not only quite fascinating, but also provides considerable explanatory power for interpreting a number of phenomena that characterize corporate life, as we usually see it.

First of all, according to this theory, firms are not designed 'from the top down' but—rather—tend to *self-evolve*, mostly in reply to stimuli coming from the environment (in fact, this is often termed the '*evolutionary* theory of the firm'). This is not to understate the role of management, but to frame it in realistic terms. Managers are not lofty strategists and decision-makers who live somewhat separately from their firms. Instead, they are actors who are heavily involved in the organization, trying to understand the way it works and the landscape it lives in. Their key role is to progressively modify the pool of resources and the set of routines that make up their firm, in order to make it more competitive in a changing environment. Since they too are part of the organization, their work is carried out by following managerial routines that influence underlying operational routines, and they too can suffer from organizational inertia, whenever the competitive environment changes in a substantial way.

Secondly, since firms evolve, they will be path-dependent (Kogut and Zander 1992). *Path dependency** means that what a firm is today (in terms of resources and routines) is a function of its past history, and that the state in which it will be tomorrow will be a function of what it is today and of the specific challenges it is encountering. As a consequence of path dependency and organizational inertia, firms are usually unable to make big jumps, but will tend to move in small steps, eventually running into trouble if the conditions would instead call for radical change.

A third and very important consequence is that, since firms progressively evolve in a path-dependent way, they will all tend to be different from one another. This seems to be a fairly obvious observation but, as mentioned earlier, leads us significantly far away from the view of neoclassical economics. One particularly interesting difference between firms is related to profitability. Some firms, thanks to their differences in resources and routines, will be able to enjoy higher profitability than their peers in the same industry, thus achieving what is called *competitive advantage**. The evolutionary theory of the firm therefore suggests that the competitive advantage of a firm is not only due to managers having superior decision-making ability, but will mostly come from its different history.[9]

Competitive advantage can also depend on luck within an uncertain environment (Barney 1991). This apparently dull statement can have a rigorous explanation. In order to enjoy superior profits, a firm must be able to extract economic value from resources, above the cost of the resources themselves (economists would call this an *economic rent*). If it weren't for the uncertainty that surrounds the actual economic potential of resources, this rent would be extracted by the owners of resources, who would simply increase prices and draw the rent for themselves. From another perspective, and building on the fact that routines emerge out of a cumulative phenomenon, Denrell (2004) has shown that cumulative phenomena can lead to significant differences among actors, even when subject to a sequence of completely random choices.

Finally, the evolutionary theory of the firm explains the *sustainability* of competitive advantage over time. This occurs because would-be imitators will not be able to observe the tacit and sticky knowledge, the set of resources together with

[9] As an example, one can think of the competitive advantage gained by Dell Computers in the '90 s, by running a business model based on direct sales and Assemble-To-Order operations, which was significantly different from the prevailing model, based on Make-To-Stock operations and a complex distribution chain. Dell Computers developed this successful business model because its founder, Michael Dell (1965–), started this way in 1984, by selling computers to his friends from his college dorm room and understood that this model could be scaled. There were countless garage-based computer assemblers like Michael Dell in the world, but he was probably unique in understanding the potential value of this model and having the ambition and the determination to realize his vision.

Another example is the competitive advantage enjoyed by Toyota Motors Company, based on the so-called Toyota Production System (TPS) set of techniques. TPS has been progressively developed within Toyota, first by Kiichiro Toyota (1894–1952) and then by Taiichi Ohno (1912–1990), as a way to improve manufacturing operations under the peculiar circumstances of Japanese culture and economic environment.

their complementarities and the routines that determine competitive advantage. Moreover, because of significant causal ambiguity, they might not be able to fully understand how all this leads to competitive advantage. Finally, and even if they had successfully overcome the previous obstacles, it would be unlikely for them to successfully replicate them. As discussed above, this is due to contextual dependency, which would require adaptation of resources and routines and not simple replication, not counting the fact that similar resources might not be readily available, and that routines might require time to be developed.[10]

These problems found in imitation will also be encountered by a firm that discovers some form of best practice in one of its units and attempts to replicate it throughout its organization. Managers will therefore have to decide on the tradeoff between the cost of understanding and codifying knowledge, in order to allow its transfer to other business units, and the opportunity cost of not doing so. At the same time, they will have to deal with the so-called *replication paradox* (Kogut and Zander 1992). Codification will make knowledge easier to transfer within the organization, thus potentially increasing competitive advantage, but will at the same time make it more accessible to imitators, thus endangering its sustainability.

2.4 Organizational Learning

If one follows the evolutionary theory of the firm, organizational learning consists in changing the firm's knowledge base and, in turn, this implies modifying its pool of resources and the routines that link them. Firms will not exhibit the same ability to modify their resources and competencies. Therefore, it is possible to suggest (Knott 2008) that there can be something similar to an organizational Intelligence Quotient (IQ) for firms, intended as a measure of the organization's ability to learn, as opposed to its inertness.

Organizations are also characterized by *absorptive capacity* (Cohen and Levinthal 1990), which is the ability of the firm to put external knowledge to effective use. In turn, absorptive capacity depends on the existence of *gatekeepers*, who are employees who are particularly apt in accessing knowledge that exists outside the firm and in interpreting it in a way that is of practical use. This allows the transfer of this knowledge to relevant employees and to its integration within routines. Moreover, absorptive capacity depends on the stock of related knowledge. Just as it happens for individuals, organizations do not learn in jumps, since the acquisition of new knowledge does not consist in the simple addition of new information but—rather—in the creation of new connections between previously acquired elements. Therefore, the ability of an organization of acquiring new

[10] It is noteworthy that established competitors to Dell, such as Hewlett-Packard, did not try to replicate its business model, but competed against Dell by improving their own traditional model. Similarly, Toyota never kept its business practices secret, but publicized them, possibly betting that imitators in the West would not have been able to reach the same level of performance, but would have facilitated diffusion of these practices, making it easier to find adequate suppliers.

knowledge will be proportional to the extent with which this new knowledge is related to the knowledge it already possesses.

In general, it is possible to define four generic modes of organizational learning (Huber 1991). In Chap. 8, the discussion will be expanded in order to identify specific strategies and activities, given the managerial relevance of decisions pertaining to what and how to learn.

- **Innate learning** is what gives rise to the first knowledge base in a company, at the time of its foundation. Since there are no routines yet, this knowledge base will come from the contribution of its founders, tied together by a common vision of the world and a common objective, such as the one that can be found in a business plan.
- **Experiential learning** is the typical process by which firms modify their routines thanks to 'learning by doing' and 'learning by failing', by following a trial and error mechanism. In the case of developing technological knowledge, this typically occurs by performing in-house research and development (R&D) activities. An important distinction can be made between *exploitation learning* and *exploration learning* (March 1990). The former occurs when a company operates normally and runs its routines as usual (i.e., it exploits its existing knowledge) and, in doing so, discovers some form of improvement, thus generating new knowledge. Instead, exploration learning happens when a company attempts something that is purposely new (e.g., it ventures in a new market or explores new technology) and, in doing so, it builds new experiences and routines. Path dependency will of course constrain the ability of the firm to achieve experiential learning. Both exploitation and exploration learning can occur unconsciously or can be deliberately pursued by the firm.
- **Vicarious learning** happens when the firm attempts to tap into an external source of knowledge, such as advice provided by a consultant, or a book. The first issue at stake in this process is the identification of the knowledge required, which can happen through *scanning* (i.e., searching across a wide spectrum of possible sources, which is more expensive, but potentially more rewarding) or *focused search* (i.e., searching in a specific direction). Once this knowledge has been identified and acquired by the firm (e.g., a consultant has been hired, or a book has been bought), it has to be diffused in the organization and put to use in its routines. It is quite apparent that vicarious learning will be heavily dependent on the organization's absorptive capacity.

  There is a subtle relationship between experiential and vicarious learning. Since absorptive capacity is related to previously held knowledge, companies who wish to be effective in vicarious learning should also practice experiential learning. In practice, this means that firms who want to tap into external knowledge sources must nonetheless perform some amount of internal R&D, in order to achieve and maintain a sufficient alignment with the state of the art.
- **Learning by grafting** occurs when the company gains new knowledge by hiring an individual, or by acquiring another organization, and integrating them within the organization. This approach to learning is apparently faster than the previous

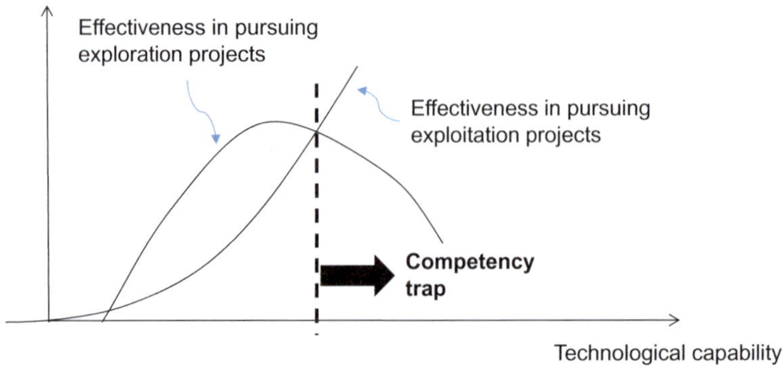

Fig. 2.2 Competency trap

ones, since knowledge does not need to be developed. However, integration of the acquired resources is by no means costless and trivial, since it requires to alter existing routines and/or to create new ones.

Though valuable, organizational knowledge can be a double-edged sword for firms (Zhou and Wu 2010). As depicted in Fig. 2.2, the stronger an organization's knowledge, the larger the benefits it will achieve by pursuing 'exploitation' activities that build on that same knowledge. If one considers activities that require 'exploration' of adjacent knowledge, the benefits will at first grow with the strength of the firm's knowledge, due to increasing absorptive capacity. Beyond a given point, benefits accruing from exploration projects will however start decreasing, because of organizational inertia created by the strong knowledge base. Beyond this point, the firm will find it easier and more efficient to stay on the same path and follow exploitation-based activities. This phenomenon is called a *competency trap*, since it leads a company to become locked-in, or imprisoned, to its own strengths.

The consequences of a competency trap can be quite dramatic, if the environment in which the firm operates changes significantly and the firm is unable to adapt to it.

As discussed by Lee and Van den Steen (2010), the risk of falling into a competency trap becomes even greater for firms that actively involved in knowledge management practices aimed at identifying best practices and codifying them. In fact, the firm might end up codifying and formalizing practices that are only moderately successful, or might even be damaging, should the environment change.

2.5 Technological Knowledge and Creativity

The previous discussion has focused on knowledge in organizations and the different features that characterize knowledge in the two domains—scientific and technological—that are fundamental to innovation. At this point, it is convenient to add another element to the discussion, which is the role of *creativity**. In fact, many times we have heard that there is no innovation without creativity, and many academic domains, including Architecture and Industrial Design, have founded their cultural traditions on the link between innovation, imagination and creativity.

Therefore, what is the link between creativity and individual knowledge? What is the role that creativity plays, in addition to technological knowledge, to generate innovation? In order to answer to these questions, it may be worth exploring the links between creativity, human thought and knowledge.

Cognitive scientists studying creativity have shown that creativity consists in recognizing the relationship between things or concepts by producing a new concept or idea, sometimes giving solution to a problem. Creativity can be *combinatorial* if it uses attributes of known ideas/concepts, *exploratory/investigative* if it starts by carrying out further investigation of ideas/concepts, as well as *transformational*, when some dimensions of the conceptual structure are modified to create new conceptual structures.

The individual engaged in creative activity can use *perception* of the different dimensions characterizing the problem under hand, *memory retrieval* to recall what is known and a variety of forms of thinking (such as *deductive*, *inductive* or *abductive** reasoning) to combine, investigate and (re)structure these elements of prior knowledge.

Creativity is also founded on the prior knowledge of an individual, which is the starting point of the creative process; however, the possession of such knowledge is not a sufficient condition to carry out creative tasks, since not all individuals are equally able to manipulate it toward creative ends. In fact, the same knowledge that might enable creativity may also hinder it, when it binds the individual into routine, limits its interpretation of the problem (*framing**) or anchors the individuals to previous concepts (*anchoring**). This aspect is essential in defining an individual's attitude toward creativity.

In the end, as shown in Fig. 2.3, the outcome of a creative process will be the combination of the individual's greater or lesser attitude toward creativity, and of her/his greater or lesser ability to use creative mental processes for the purpose of generating novel solutions or concepts. Everyone is potentially creative, especially during childhood and early adulthood; however, this aptitude can be developed and the ability exercised. As it will be shown in Chap. 15, firms use a variety of techniques to stimulate creative abilities and outcomes, so to help people to break out of their usual mindset and widen the range of solutions being conceived.

At the same time, in the context of innovation, creativity is not an end to itself, simply aiming at a limitless generation of new concepts that diverge from the present ones. Creativity is a mean to an end that consists in generating and developing new technical concepts that must have an economic and societal impact. Therefore, in order to lead to innovation, creative processes must be enacted in a context aimed at having such an impact, such as to find a feasible solution to a

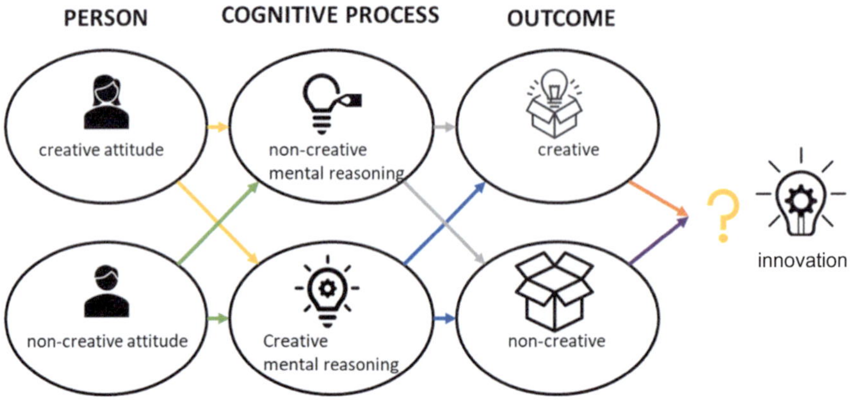

Fig. 2.3 Circumstances that make creativity a determinant of innovation

problem. To this aim, creative processes take place within a broader process, which is made up of a complex combination of *divergent and convergent* phases, called *design* (see Chap. 15).[11]

2.6 Data, Technological Knowledge and Innovation

Since a few decades, we currently have been living in what has been called the 'information society', a society that is ever more reliant on the ability to generate, store, communicate and analyze an impressive amount of data. One might wonder what impact this has on innovation.

On the one hand, one might argue that information technology and data science (or IT in short) are technologies as all the others, and that their contribution to innovation is, albeit possibly more significant, of the same nature as what is brought by other developments. Such a position may appear overly dismissive of the impact brought by IT, but should not be neglected, since it is true that—under many perspectives—the contributions brought by IT can be framed by existing conceptual structures. Therefore, one should be wary of overly broad—and often heard—claims such as 'with digital nothing will be like in the past', without actually explaining what and why this may occur. For instance, we have just discussed different form of knowledge, and 'big data' easily fits in the 'factual knowledge' category, while current machine learning algorithms can be seen as a novel way to generate 'causal' and 'procedural' knowledge.

[11] Creative processes are typically considered only those divergent in the creativity literature; while in design, both divergent and convergent thinking can lead to creative outcomes (Basadur et al. 2000).

On the other hand, it is also true that the features and the potential of contemporary IT does have the potential to change substantial parts of the processes by which the innovation occurs, getting close to changing the nature of the process itself. Specifically, this impact may be seen in three different dimensions:

- in the technical architectures of contemporary and digitally intensive, products and services;
- in the processes that link the generation and usage of data, information and knowledge[12];
- in the organizational processes and the industrial value chains by which products and services are developed.

The following part of this section will deal with an introductory discussion on the process-oriented dimensions, while the latter two will be discussed quite amply in the following chapters, considering the changes brought by digitalization, but without any freewheeling discussions that neglect the fundamentals of the discipline.

Companies have always collected data. However, the main contemporary change, with the advent of digitization, is that data are collected at very low cost, in real time, in very large quantities and from all the many processes within value chains. Data, which can be classified into demand side (if coming from the market and from users) and supply side (if coming from the firm and its supply chain), can become a key element in supporting innovation processes (Cantamessa et al. 2020).

Demand-side data are gathered from customers. They are mainly collected by recording customer purchasing behavior, through the observation of individual choices (Lesser et al. 2000), as well as from stakeholders (Cantamessa et al. 2016). Nowadays, and increasingly in the future, they are collected from the actual use of smart products or digital services (Kagermann 2015; Bharadwaj and Noble 2017; Li et al. 2019) or even from the sheer presence of smart products in our environment (e.g., a smart speaker picking up conversations or a manufacturing system in a B2B industry). Walmart, for instance, gathers more than 2.5 petabytes of data every hour from its customer transactions (McAfee et al. 2012), and Facebook or LinkedIn continuously collect data to suggest new personal contacts (Davenport et al. 2012). General Electric has become the manufacturing industry leader in managing customer product data, to the point where it is becoming a leading developer of service offerings based on big data (Davenport and Dyché 2013). Tesla, similarly, collects more data than most car companies, such as vehicle location, user settings and video clips from the car's external cameras (Muller 2019).

The value of demand-side data is that it allows companies to understand customer/user needs fully. Not only companies will be able to have a real, and not simply estimated, understanding of mission profiles, but they will also be able to

[12] It comes quite natural, when dealing with this topic, to think about 'The Rock', the visionary poem that T.S. Eliot wrote in 1934, and remember that we always neglected the fourth element cited by Eliot, i.e., wisdom.

adapt continuously, based on market stimuli. This adaptation might consist of newly added functions identified through the analysis of data generated by product that are on the market (Lewis and Van Horn 2013), up to a real-time adjustment of the offering to address the evolving customer needs (Brown et al. 2011) or up to self-customized products (Porter and Heppelmann 2014).

In addition to demand-side data, companies collect reams of useful data from their production systems and operating environments. Data can be generated from a multitude of sources, such as production machinery, supply chain management systems and monitoring systems (Noor 2013; Ma et al. 2017; Ghobakhloo 2020), also using systemic approaches (Alfieri et al. 2012).

Data from these sources are often collected and analyzed with a focus on opportunities for reducing cost and increasing efficiency. Smart manufacturing, for instance, is often conceived as an environment in which electronic sensors, microprocessors and other components allow a high degree of automation (Noor 2013) and provide real-time warnings about potential problems or opportunities for cost reduction (Jenab et al. 2019). The effects of digitization of production processes can lead even to broader benefits, especially if this information is fed early in the development process (Schuh et al. 2016). Tesla Model 3, for instance, turned out to be particularly critical during the assembly phase, due to its many weld points and rivets that were not suitable for heavy production automation (Welch 2018). Data from manufacturing, in that case, provided information useful for industrialization, suggesting product architecture alternatives in view of the consequences on assembling.

The use of this data to support innovation is not simply an addition to processes that are carried out in a traditional way. On the opposite, the availability of data has an impact on researchers', product developers' and designers' work. Specifically, there are operational consequences, which have to do with the activities they have to carry out and the tools they can use, as well as organizational consequences, if one looks at the way these actors must operate within organizations (Cantamessa et al. 2020). Later in this book, we will analyze each of them step by step.

References

Alfieri A, Cantamessa M, Montagna F (2012) The SoS approach for lean manufacturing systems. Int J Technol Manage 57(1–3):149–165

Argote L (1999) Organizational learning: creating, retaining, and transferring knowledge. Kluwer, Academic Publishers, Boston

Argote L, Epple D, Rao RD, Murphy K (1997) The acquisition and depreciation of knowledge in a manufacturing organization: turnover and plant productivity. Graduate School of Industrial Administration, Carnegie Mellon University, Working paper

Argote L, McEvily B, Reagans R (2003) Managing knowledge in organizations: an integrative framework and review of emerging themes. Manage Sci 49(4):571–582

Basadur M, RuncoVega MA (2000) Understanding how creative thinking skills, attitudes and behaviors work together: a causal process model. J Create Behav 34(2):77–100

Barney J (1991) Firm resources and sustained competitive advantage. J Manag 17(1):99–120

Bharadwaj N, Noble C (2017) Finding innovation in data rich environments. J Prod Innov Manag 34(5):560–564
Brown B, Chui M, Manyika J (2011) Are you ready for the era of 'big data.' McKinsey Q 4(1): 24–35
Cantamessa M, Montagna F, Altavilla S, Casagrande-Seretti A (2020) Data-driven design: the new challenges of digitalization on product design and development. Des Sci 6:e27
Cantamessa M, Montagna F, Cascini G (2016) Design for innovation—A methodology to engineer the innovation diffusion into the development process. Comput Ind 75(1):46–57
Ceri S (2018) On the role of statistics in the era of big data: a computer science perspective. Statist Probab Lett 136:68–72
Cohen WM, Levinthal D (1990) Absorptive capacity: a new perspective on learning and innovation. Adm Sci Q 35:128–152
Davenport TH, Barth P, Bean R (2012) How 'big data' is different. MIT Sloan Manag Rev 54 (1):43–46
Davenport TH, Dyché J (2013) Big data in big companies. Davenport and SAS Institute Inc., International Institute for Analytics, T.H
Denrell J (2004) Random walks and sustained competitive advantage. Manage Sci 50(7):922–934
Eliot TS (1934) The Rock. Faber & Faber, London
Fleming L, Sorenson O (2004) Science as a map in technological search. Strateg Manag J 25(8–9):909–928
Gavetti G, Levinthal D (2000) Looking forward and looking backward: cognitive and experiential search. Adm Sci Q 45(1):113–137
Ghobakhloo M (2020) Determinants of information and digital technology implementation for smart manufacturing. Int J Prod Res 58(8):2384–2405
Groysberg B, McLean A, Nohria N (2006) Are leaders portable? Harv Bus Rev 84(5):92–100
Hayek FA (1945) The use of knowledge in society. Am Econom Rev 35(4):519–530
Huber GP (1991) Organizational learning: the contributing processes and the literatures. Organ Sci 2(1):88–115
Jenab K, Staub S, Moslehpour S, Wu C (2019) Company performance improvement by quality based intelligent-ERP. Decis Sci Lett 8(2):151–162
Kagermann H (2015) Change through digitization—value creation in the age of industry 4.0. In: Albach H, Meffert H, Pinkwart A, Reichwald R (eds) Management of permanent change. Springer, pp 23–45
Knott AM (2008) R&D Returns causality: absorptive capacity or organizational IQ. Manage Sci 54(12):2054–2067
Kogut B, Zander U (1992) Knowledge of the firm, combinative capabilities and the replication of technology. Organ Sci 3(3):383–397
Lee D, Van den Steen EJ (2010) Managing know-how. Manage Sci 56(2):270–285
Lesser E, Mundel D, Wiecha C (2000) Managing customer knowledge. J Bus Strateg 21(6):34–37
Lewis K, Van Horn D (2013) Design analytics in consumer product design: a simulated study. In: ASME 2013 international design engineering technical conferences and computers and information in engineering conference (IDETC-CIE2013), Portland, Oregon, USA, August 4–7
Li Y, Roy U, Saltz JS (2019) Towards an integrated process model for new product development with data-driven features (NPD 3). Res Eng Des 30(2):271–289
Ma H, Chu X, Lyu G, Xue D (2017) An integrated approach for design improvement based on analysis of time-dependent product usage data. J Mech Des 139(11):111401
March JG (1990) Exploration and exploitation in organizational learning. Organ Sci 2:71–87
McAfee A, Brynjolfsson E (2012) Big data: the management revolution. Harv Bus Rev 90(10): 60–68
Muller M (2019) What Tesla knows about you, Axios, https://www.axios.com/what-tesla-knows-about-you-1f21d287-a204-4a6e8b4a-0786b0afac45.html. Accessed Oct 2022
Nelson RR, Winter S (1982) An evolutionary theory of economic change. Harvard University Press, Cambridge

Nonaka I, Takeuchi H (1995) The knowledge-creating company. Oxford University Press, New York

Noor A (2013) Putting big data to work. Mech Eng 135(10):32–37

Penrose ET (1959) The theory of the growth of the firm. Wiley, New York

Poincaré H (1902) La science et l'hypothèse. Flammarion, Paris

Porter ME, Heppelmann JE (2014) How smart, connected products are transforming competition. Harv Bus Rev 92(11):64–88

Schuh G, Rudolf S, Riesener M (2016) Design for Industrie 4.0. In: 14th International design conference (DESIGN 2016), Dubrovnik, Croatia, May 16–19

Simon H (1991) Bounded rationality and organizational learning. Organ Sci 2(1):125–134

Vincenti W (1990) What engineers know and how they know it. John Hopkins University Press, Baltimore and London

Von Hippel E (1994) "Sticky information" and the locus of problem solving: implications for innovation. Manage Sci 40(4):429–439

Welch D (2018) Tearing apart Teslas to find Elon Musk's best and worst decisions, Bloomberg. https://www.bloomberg.com/news/features/2018-10-17/tearing-apart-teslas-to-find-elon-musk-s-best-and-worst-decisions. Retrieved 23 Oct 2022

Zhou KZ, Wu F (2010) Technological capability, strategic flexibility, and product innovation. Strateg Manag J 31:547–561

The Many Types of Innovation 3

Abstract

Innovation is not a linear phenomenon, and this chapter has the objective of providing readers with a complete understanding of the different types of innovations that occur along the technological lifecycle. A detailed discussion is at first provided on the way with which evolutionary and revolutionary innovations alternate along s-curves, and on the way with which such innovations shape technological paradigms, viewed as the outcome of a complex interplay between supply and demand. Then, the discussion focuses on the mechanisms that can lead to—or can prevent—disruptive innovations, along with the ensuing managerial implications, both for incumbent firms and for aspiring entrants.

After having covered the main theoretical foundations, we can now fully delve into the topic of innovation. The first main issue, which will be covered in this chapter, is to understand the different types of technological innovation, their determinants and their implications for individual firms and industries.

3.1 The Determinants of Innovation

When dealing with technological innovation, the first significant question that has to be answered is related to its causes, or *determinants*. Over the years, two possible and contrasting determinants have been identified. According to advocates of *technology push*, innovation occurs when a technological development is generated independently from a specific market need and is eventually deployed in a given industry, thus matching a latent demand. Such a technological development can be generated internally to the innovating firms or can occur elsewhere, in which case the innovating firms will realize the potential behind the technology and incorporate

M. Cantamessa and F. Montagna, *Management of Innovation and Product Development*, https://doi.org/10.1007/978-1-4471-7531-5_3

it in products and services.[1] Conversely, *demand pull* posits that firms observe the demand for improved products that come from the market and from society at large, and explicitly direct the development of technology in order to respond to these needs.[2]

The debate between these two determinants has been quite active, until scholars have realized that technological innovation can be due to either of the two, depending on the stage and type of innovation.

This conclusion comes from the observation that technology does not follow a linear process, but it is subject to distinct and alternating phases of *evolutionary* and *revolutionary* progress (Tushman and O'Reilly 1997; Iansiti 2000). If we pick an industry and identify a relevant performance indicator for its products, the evolution of this indicator will not proceed in a straight line, but it will follow a sequence of *s-curves*.

As an example, we can study the evolution of passenger airplanes and collect data on cruising speed, range, capacity and fuel efficiency for a selection of models that have been particularly important in civil aviation history. Table 3.1 shows that the industry has worked to improve the performance of aircraft, and that this improvement has been initially played around the speed and the size of the same while, in more recent times, fuel efficiency has become more important. In order to achieve this progress, the industry has introduced a few discontinuous changes in the underlying technology used for airframes (from wood to metal and then to composites) and for engines (from internal combustion engines and propellers, to jet engines and then to turbofans, while recent experiments look toward other propulsions, such as hydrogen and electric).

We can also make a qualitative representation of this evolution with *s-curves*, with performance indicators on the y-axis and time on the x-axis, as in Fig. 3.1.

S-curves show that, when a technology emerges, performance is usually quite low, until a sufficient degree of maturity is reached. At this point, performance starts growing at a significant speed, until it eventually reaches a *technological limit*, i.e., a performance level that cannot be overcome, due to the limitations that are intrinsic to the technology itself.[3]

[1] For instance, think of the development of laser technology. This technology had originally been developed without a clear commercial application in mind. Over the years, it found multiple applications that have revolutionized entire industries, from CD players to metal cutting and from telecommunications to barcode scanners. Similarly, one could think of 3D printing and the progressive application in the aerospace sector, and then in industries such as jewelry, food or constructions.

[2] Today's internal combustion engine cars have a significantly lower environmental impact than their predecessors, and most of the innovations introduced have been directly developed by the automotive industry in response to market and societal needs.

[3] For instance, an aircraft powered by an internal combustion engine and a propeller cannot practically go faster than a given speed, because of cavitation and because of the speed reached by propeller blades. Moreover, this technical solution cannot work if air pressure is too low, and will therefore operate only at altitudes below a given threshold.

Table 3.1 Passenger airplanes and the underlying technologies

Airliner	Year	Airframe	Engines	Cruise speed (km/h)	Range (km)	Max passengers	Fuel efficiency (km seat/l)
Flyer	1903	Wood + fabric	1 piston + propeller	48	–	1	–
Farman Goliath	1919	Wood + fabric	2 piston + propeller	120	400	14	13.4
FIAT C.R. 32	1933	Metal	1 piston + propeller	315	750	1	2.2
Douglas DC-3	1935	Metal	2 piston + propeller	333	2400	32	8.4
Boeing 314- flying boat	1938	Metal	4 piston + propeller	340	5900	74	23.5
Vickers Viscount 700	1948	Metal	4 turboprop	496	2220	48	15.3
De Havilland Comet	1949	Metal	4 turbojet	740	2400	44	3.56
Boeing 747–100, 200	1970	Metal	4 turbojet	893	12,690	550	26.9
Concorde	1976	Metal	4 turbojet	2158	7222	120	6.4
Helios	2001	Composites	Solar powered	300	7200	–	–
Boeing 787	2011	Composites	2 turbofan	913	15,400	440	36.6

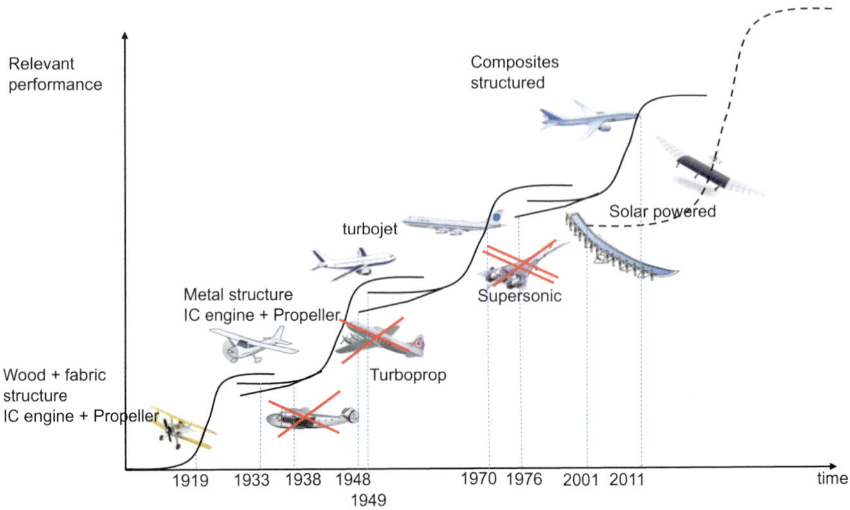

Fig. 3.1 A qualitative depiction of the progress of civil aviation history using s-curves

Once the limit has been reached, firms that want to improve their products will have to embrace new technical solutions (e.g., moving from piston and propeller power plants to jet engines). In doing so, firms must choose among a number of new candidate technologies available.[4]

The former aspect is critical, since only one technology will generally emerge (ref. the concept of *paradigm* in the next section and of *dominant design* in Chap. 4). Figure 3.1 for instance lists a few choices that have not been successful in the civil aviation history, because the technology did not prove to be good enough, because it did not match customer needs, or because it was not possible or profitable to develop specific complementary infrastructure.[5]

The timing aspect is also important, since a premature investment may lead to unattractive returns from a financial point of view or—even worse—to erroneous choices in technology. Conversely, a late investment may forfeit gaining experience with the new technology and an early access to the market. In some cases, early adoption of the new technology might prove to be advantageous and lead to what is commonly termed *first mover advantage** (more on this concept will be covered in the next chapter too).

[4] This is what is currently happening in the case of powertrains for passenger cars. Electrification of these powertrains is underway, but it is not yet clear whether—in the long run—it will be based on batteries alone, on hybrid solutions, or on hydrogen fuel cells.

[5] For instance, Boeing 314 flying boats could only connect cities that were located close to a suitable stretch of water. Supersonic flight with the Concorde was so expensive, that it did not appeal to a sufficiently high number of passengers. The Airbus 380 required specific terminal buildings enabling a fast turnaround of a very high number of passengers and was designed for hub-to-hub transport in hub-and-spoke networks, exactly at the time in which airlines were moving towards point-to-point networks.

In general, *evolutionary progress* occurs when moving along a given s-curve, while an industry will experience *revolutionary progress* when transitioning from an old s-curve to a new one. Once again, it is appropriate to stress that this revolutionary phase is characterized not only by conflict among old and new technology, but also among alternative new technologies. Demand pull will be the prevailing determinant of innovation that accompanies evolutionary progress, during which a given technology is improved and fine-tuned according to customer needs. Conversely, technology push will mainly occur in the revolutionary phases, when firms are forced to 'look around' for new solutions that may overcome the technological limit characterizing the *incumbent**[6] technology.

S-curves are a fundamental instrument in the toolbox of those who deal with technological innovation, as scholars or as practitioners. However, they should be treated with some caution. First, one should choose a performance indicator that is *valued by customers*, though their preferences might change over time.[7] Secondly, using time as the independent variable can be misleading, since technology does not improve because time goes by, but because of the investment and of the effort that companies spend on it. Therefore, s-curves should not be plotted as a function of time, but of *cumulated industry-wide R&D expenditure*, under the hypothesis that technological improvements occur because firms imitate each other, and therefore build on each other's investments.

Data on R&D expenditure are not always easy to find and, therefore, cumulated sales at industry level are sometimes used as a proxy. The rationale of this choice is that R&D expenditure usually is a fairly constant and industry-specific percentage of sales, the two proportional. However, using either time or cumulated revenues as an independent variable is risky, since s-curves might flatten not because a genuine technological limit has been reached, but simply because companies in the industry have reduced their R&D spending. In this case, one might be drawn into thinking that the old technology has reached its limit and make a premature investment in the further technology, only to discover that, by increasing the investment in R&D, the old technology gains a strong lead again.

In spite of these limits, s-curves remain a fundamental tool for orienting oneself in a sector and understanding its dynamics. For instance, because of environmental beliefs it is commonplace to hear that the era of fossil-fuel plastic materials is over and that the future lies in bioplastics. Such a statement should be pondered with care, by looking at the specific technical performance in view of the various industrial applications. For instance, if one considers yield stress/$ as one of the main customer-valued performances of PLA biopolymers, despite recent improvements in

[6] The term 'incumbent' refers to an entity that is currently operating. It is commonly applied to a firm that is operating in an industry or applied to a technology that is being used.

[7] The case of civil aviation provides a clear example. Having taken for granted that speed was the main performance to be pursued, led to the commercially disastrous development of supersonic passenger planes such as the Concorde*. Instead, fuel economy and passenger comfort have progressively become the main performance being sought by airlines. This explains why composite materials, allowing the production of lighter and comfortable airplanes, are currently emerging as the characterizing technology in passenger aircraft.

Fig. 3.2 (top), this performance has not reached the fossil-based polymers' one yet (bottom panel in Fig. 3.2). Performances therefore should be considered in association with, and as a function of, the specific application: for shopping bags the current performance is acceptable, whereas for water bottles it is not, suggesting the use of alternative solutions, such as for instance Tetra Pak packaging.

This does not mean that bioplastics will not supplant the incumbent technology one day, but this possibility is tied to the different applications and to a required change in the way society values the performance, the uses and the impact of these materials.

3.2 Technological Paradigms

Each of the s-curves described in the previous section does not simply represent a technology, but rather a key component of a broader concept that can be termed *technological paradigm*.[8] Figure 3.2 attempts to represent the key components of a technological paradigm, viewed as a framework that brings together supply-side and demand-side elements merging in a coherent whole and which give birth to a technological *trajectory* (i.e., the s-curve), which is at the same time viable for companies and appreciated by the market.

From the side of supply, a paradigm is based on a systemic view of products and services. There will be a focal product, composed of subsystems, some of which will be generic, and other ones specific and critical to the paradigm. Moreover, the focal product will likely depend on the existence of complementary systems which, again, might be generic or specific.[9] In order for these systems to emerge, it is necessary to have a set of theories, knowledge, tools, and methods that allow transforming a given technology into actual products and services. This is often called a *technological regime* (Nelson and Winter 1982). The exploitation of a technological regime in order to make a paradigm emerge also requires the coherent coming together of a number of actors, who must be able to master the regime and find a way to establish profitable business models. Among such actors, one can view producers of the focal product, their key suppliers and *complementors*. The distinction is of course not absolute, since it depends on producers' choices with respect to vertical integration and a single or integrated company may cover all such roles. Finally, the technological regime requires research centers and universities,

[8] Giovanni Dosi (1953–) introduced this term in 1982 by borrowing it from Thomas Kuhn's concept of scientific paradigm*. The term 'paradigm', intended as a coherent framework of broadly accepted elements, can of course be applied to a wide range of cases (e.g., you can have 'social', 'religious', 'political paradigms' and many more).

[9] For instance, if an analyst attempts to study the likelihood of a paradigm emerging around battery electric vehicle (BEV) technology, he/she will be drawn to focus on batteries as the key subsystems. Concerning complementary systems, he/she will look at the many possible modes with which the batteries can be charged, from domestic wallboxes to superchargers, and from battery swapping to roads equipped with inductive charging systems.

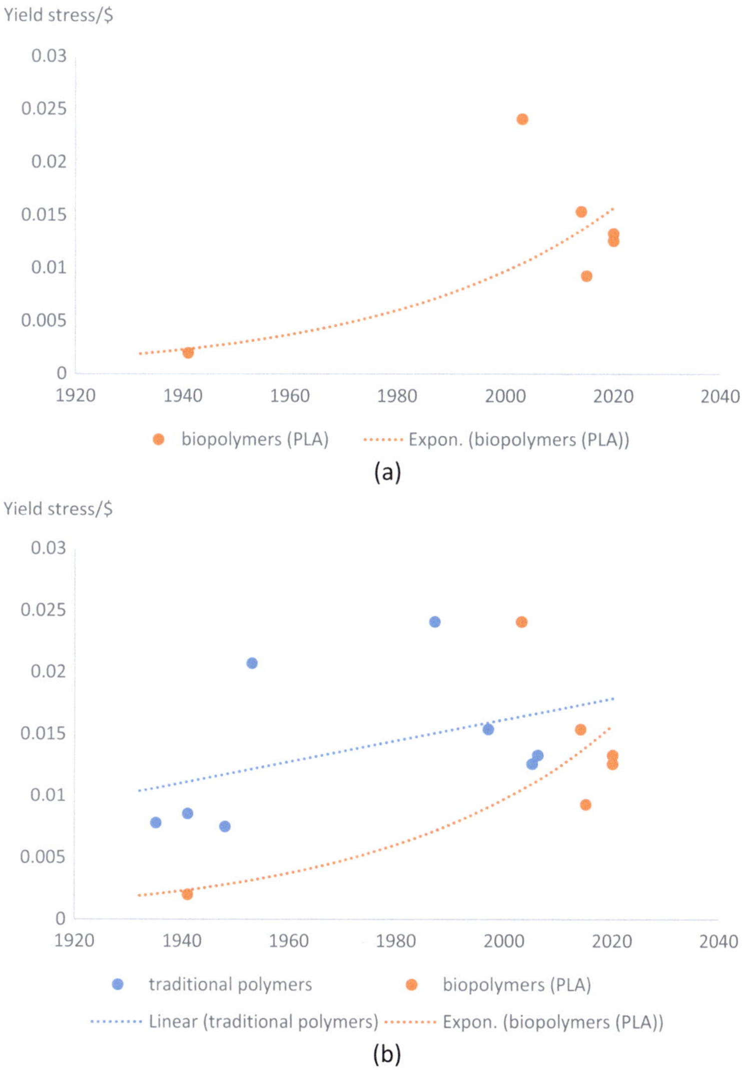

Fig. 3.2 Absolute performance of PLA biopolymers (**a**) and fossil-fuel polymers versus the PLA biopolymers performance (**b**)

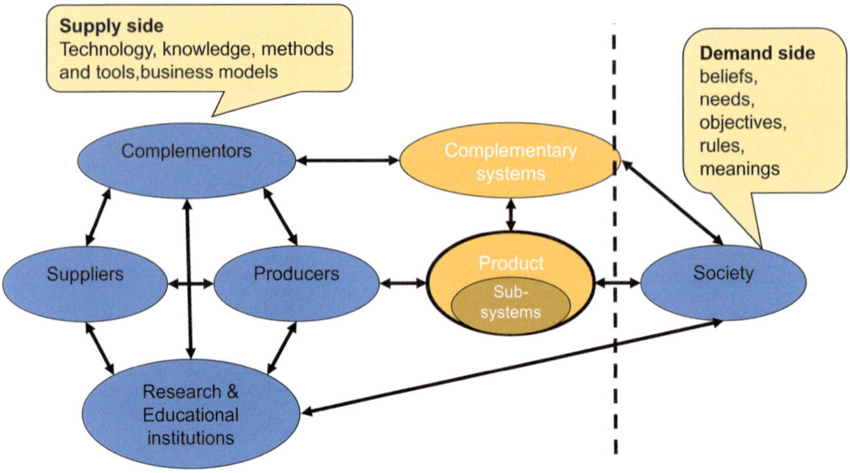

Fig. 3.3 Constituent elements of a technological paradigm

whose role is to provide the underlying practical knowledge and, especially, an appropriately trained workforce.[10]

From the side of demand, a paradigm is defined by beliefs, needs, objectives, rules and *meanings*[11] that customers attach to the products and services. If these are met by the technology, customers will choose them and the paradigm will emerge.

[10] In the case, of BEVs, an analyst should consider the role and the sustainability of the business models for players such as BEV makers, battery manufacturers, recharging services providers, and DSOs (i.e., electric power distribution system operators). Moreover, he/she should question whether research and educational institutions are providing enough technical experts able to operate in the new technology, which often recombines traditional technical disciplines (e.g., managing the rollout of charging systems in a city requires a blend of electrical, civil and traffic engineering).

[11] The concept of 'meaning' is central to many products. Especially when dealing with consumer markets and products associated to industries such as food, fashion and Industrial Design, evaluation is not carried out on the basis of technical performance alone, but by looking at subtle and deeply personal aspects (Verganti 2009). This so-called *co-constructivist* perspective on innovation states that products stay alive as long as there is a social network of actors that see value and meaning in producing and purchasing them. For instance, one can think of the technological innovation behind high-end mechanical watches, which is generally targeted towards achieving greater 'complications'. The relevant customer segment assigns very high economic value to these complications because of the very peculiar cultural meaning attributed to reaching complex results with traditional mechanical technology. It is clear that this meaning goes well beyond the main function of a watch (i.e., telling time), which could very well be covered by an inexpensive quartz movement.

This element is also important for reasoning about the possibilities of a technology shift. Think about consumption of meat and about intensive farming and CO_2 consumption. There are alternative technologies that could meet the demand in terms of nutritional performance (i.e., in vitro cell reproduction) and shape and appearance (i.e., food 3D printing), but their acceptance would require a cultural transition to drive perception and taste toward this type of food consumption.

Instead, if the features of the product or service conflict with what customers believe or need, or with social or legal norms, the paradigm will not emerge or, at the least, will be delayed.

The concept of paradigm, and the interplay between the bundle of supply-side and demand-side elements that define it, can help explain why a specific technology may emerge or fail during a technological revolution.[12] Of course, it is always easier to use theory for retrospectively interpreting history, than using it to forecast what will happen in the future.

In some cases, a new technology can overcome an existing one, without the latter actually having to reach its limit, but simply because the former provides higher performance and/or a combination of features that the market considers to be superior. In this case, one can speak about *discontinuous technology* (Schilling 2008). Quite often, discontinuous technology does not simply lead to advancing a specific technical performance, but to a radical change in the way with which the product is conceived for the customer, enabling new uses and behaviors.[13] In such sense, discontinuous technology can lead to a new technological paradigm.

From a managerial perspective, the concepts outlined above suggest that firms must constantly be aware of which phase of technological progress they are operating in and take appropriate actions, be it the development of known technology, the scouting for new technologies, or trying to make a newly chosen technology prevail over competing ones being backed by competitors.

More specifically, these concepts suggest that the Marketing and R&D functions have highly complementary roles in the innovation process (Cooper and Kleinschmidt 1990; Calantone et al. 1993). Marketing and R&D must also behave differently when operating in evolutionary or revolutionary phases. During evolutionary

[12] For instance, the cloud computing paradigm has overcome the client–server paradigm that dominated computer systems since the '80s. The reason behind this success is tied to the convergence of supply-side and demand-side factors. Among the former, the economies of scale that can be achieved by centralizing data processing and storage, and the widespread availability of cheap and reliable bandwidth for connecting servers with clients. Concerning the latter, users have valued the possibility of accessing online services through multiple devices, both fixed and mobile. Moreover, users have been confident that an Internet connection will always be with them, while they have been quite unconcerned of privacy issues associated to the personal data they entrust to service providers.

Similarly, the technology that will power the car of the future will depend on technological progress, but will also depend on users' attitudes. Specifically, one can wonder whether users will really be 'range anxious', as many current surveys suggest, or whether they will realize that driving electric cars that have a limited range perfectly suits most driving patterns. Moreover, many analysts suggest that, since electric and self-driving vehicles will be more expensive to build, this may be offset by not selling cars to individuals, but by providing mobility as a service, thus allowing a more efficient use of these assets. This is possible, only if cars cease to be valued as status symbols, and if societal patterns of usage will allow sharing (e.g., if people will not need cars all at the same time during the day to commute, or during the week or the year to go on holiday).

[13] For example, the shift from film-based photography to digital has not simply been a matter of image quality and cost, but has most of all led to a deep change in the way with which photography is used and enjoyed by consumers.

times, Marketing has the focal role of achieving a deep understanding of customer needs and directing R&D efforts accordingly. In revolutionary phases, R&D must identify promising technology, while Marketing can provide selection criteria and manage communication actions to explain the technology to customers and drive cultural changes. Then, Marketing will also have the responsibility of convincing customers that this technology meets their needs, both explicit and latent.[14]

3.3 A Taxonomy for Technological Innovation

The previous section has allowed us to identify evolutionary and revolutionary innovation shifts, by looking at their relationship with the concept of technological paradigms. Innovations can also be classified according to a number of other perspectives, which have interesting managerial implications. As the title suggests, we now are focusing on technological innovation only, while other types of innovation will be introduced in the following chapters. For the moment, innovations can be defined as:

- **incremental versus radical** (Dutton and Thomas 1984), by adopting a technical perspective. One can look at the technical features of the product (i.e., the underlying technology, its functions and performance) and—more specifically—at whether the innovation significantly changes the technical tradeoffs that define the product. In general, both producers and customers look at products as 'bundles' of functions and performance indicators, in which one can readily identify the design tradeoffs. An incremental innovation will consist in an improved product that does not significantly alter these design tradeoffs. Conversely, radical innovations alter such design tradeoffs, leading to products that introduce completely new functions or performance values, clearly setting them apart from their predecessors.[15]
- **competence-enhancing versus competence-destroying** (Tushman and Anderson 1990), by adopting the perspective of the knowledge required to pursue it. When dealing with a competence-enhancing innovation, a firm leverages its existing stock of knowledge and—in pursuing the innovation—adds to it, thanks to experiential learning (as discussed in Chap. 2). Conversely, a

[14] One may think of autonomous vehicles or in vitro meat. Both of these technologies raise fears or distrust; in both cases, Marketing must explain the technology in order to show that it fits with explicit or latent needs, thus enabling a cultural change leading to acceptance of the technology. When dealing with revolutionary change, using input from market research can be highly misleading. Henry Ford* (1863–1947) is credited to having said that, if he had asked potential customers what they wanted, the answer would have been a faster horse, and not an affordable motor car.

[15] For instance, a smartphone is characterized by the tradeoff between the computing power, the battery life and the size of the battery. The shift from NiMH to Li-ion batteries can be considered to have led to a radical innovation, since it has significantly reduced the difficulty of managing this tradeoff.

competence-destroying innovation will require the firm to set its existing knowledge aside, and gain new and possibly unrelated competencies.
- **core versus peripheral**, depending on whether they affect a core functionality of the product, or an ancillary one.
- **sustaining versus disruptive** (Christensen 1997), looking from the perspective of the business impact on the entire industry, innovations can generate. A sustaining innovation will not lead to significant change in competitors' positions and market shares, while a disruptive one will lead to major changes. With disruption, incumbents may fall behind, or even exit the industry, while market leadership may be achieved by firms that previously had a minor role, or by new entrants. Among the latter, one may have existing firms that operated in other industries (*de alio* entrants), or by startups (*de novo* entrants). This topic is of extreme managerial interest and will be the subject of the following section.[16]

When studying a given innovation, its classification will not only depend on the criterion being used, but also on the position in the value chain of the firms being considered (Afuah and Bahram 1995).[17] Therefore, it is possible to state that the type of innovation is to some extent 'subjective' with respect to the position in the value chain and the strategy undertaken by the firm. This is an important point to be considered, since the way with which innovation unfolds is not deterministic, but depends on the situation and on the choices made by firms, at times leading to counterintuitive effects.

Finally, it is possible to classify innovation in a slightly more complex way, by observing it along the two axes of underlying technology on the one side, and of *product architecture** on the other (Henderson and Clark 1990).

In order to explain this concept, it is at first necessary to define product architecture. For the time being (a fuller definition will be provided in Chap. 16), it is possible to state that product architecture is defined by the main components that make up the product and their mutual relationships (see the *block diagram** in Fig. 3.4, top left). These relationships may be due to functional interactions,

[16] To be precise, the definition of disruptive innovation given here is more extensive than the one proposed by Christensen (2015) and focuses on the type of impact, rather than on the reasons why it occurs. Christensen often held the position that the term 'disruptive innovation' should be used only when the phenomena unfolded according to his model, as described in Sect. 3.4. We chose to use the term in this broader sense, given that Christensen's theory is not the only possible explanation for innovations having a significant industrywide impact. After all, 'disruptive' is an adjective describing an effect, and not the process because of which it occurs.

[17] As mentioned, the shift from NiMH to Li-ion batteries allowed an energy density that was significantly greater than before. The innovation can be considered to be radical for both battery makers and device makers. For the battery maker, it significantly changes the underlying technology, the performance achievable, and the tradeoff between energy storage and battery size. For the device maker, it alters the tradeoff between features, size, computing power and battery time of the device. The innovation will probably be competence destroying and core for battery makers, but not for device makers. Similarly, it will probably be disruptive for battery makers but sustaining for device makers, unless it is supplied under some kind of exclusivity clause to a single device maker.

physical proximity, or even unintended influences. The design of each component will generally be assigned to a specific organizational entity (be it an individual or a team) and these entities will have to communicate with each other in a way that replicates the pattern of inter-component relationships (Fig. 3.4, top right). Therefore, product architecture and the organization will share a similar structure; as some authors state (Henderson and Clark 1990; Sosa et al. 2002, 2003), "*organization architecture mirrors product architecture*", some specific cases apart (Sosa et al. 2004). As discussed in Chap. 2, most of these communications flows will not be explicit and formalized, but will occur in the form of organizational routines that emerge out of experience and trial and error.

Now, should the firm change product architecture (Fig. 3.4, bottom left), it will not be straightforward for the organization to adapt to this change by identifying new routines and communication flows, and discarding obsolete ones (Fig. 3.4, bottom right). Therefore, innovations leading to change in product architecture will be difficult to manage.

Based on this, it is possible to identify four types of innovations (ref. Figure 3.5):

- **incremental innovation**, neither the underlying technology nor product architecture will change (e.g., a product development team that is working on the geometry of an automotive braking system in order to improve heat dissipation). This kind of innovation can be costly to achieve—in terms of effort—but is not difficult to manage, since it completely replicates the experience gained in the past.
- **modular innovation**, underlying technology does change in one or more functional elements, but product architecture does not (e.g., an automotive company that switches from a traditional instrument panel to an LCD screen with virtual indicators). In this case, the innovation in the affected modules and may be significant and difficult to pursue, since it requires changing the competencies used. However, these problems will be limited to the affected modules, and the development of the complete product will therefore be relatively easy to manage.
- **architectural innovation**, underlying technology does not change, but inter-component relationships do (e.g., a carmaker specializing in luxury saloons that starts developing a small city car). Architectural innovation is generally hard to manage, since the organizational routines that allow the development of a consistent product will not be immediately available, but will take some time (and probably some costly mistakes too) to develop. In some cases, architectural innovation may take companies 'by surprise' because seemingly minor innovations lead to unexpected change in inter-component relationships.
- **radical innovation**, both underlying technology and product architecture change (e.g., a carmaker switching from a traditional vehicle to an electric car). This kind of innovation is of course the most difficult to pursue, even though it is not very common, and firms will also be well aware of the problems they will face when tackling it. When dealing with radical innovation, firms are often so wary of architectural change that they will try treating it as modular, by changing

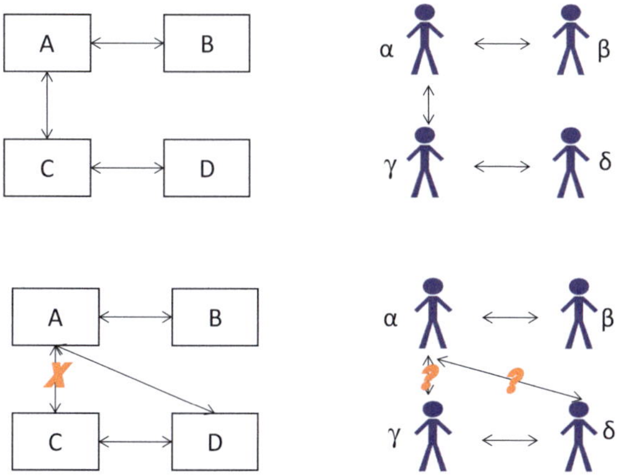

Fig. 3.4 Organization mirrors product architecture

Fig. 3.5 Taxonomy proposed by Henderson and Clark (1990)

Relationships between components / Reference technologies	Do not change	Change
Change	Modular innovation	Radical innovation
Do not change	Incremental innovation	Architectural innovation

product subsystems one by one, but trying to steer away from altering their mutual relationships.[18] This hesitation in embracing radical innovation opens up interesting opportunities for firms that—instead of making piecemeal changes to the existing architecture—accept the challenge of jointly changing technology and product architecture.

[18] For instance, early cars were built in the guise of horseless carriages (and were even named as such) and it took some time before the architecture of cars evolved into something that could make the best use of the new technology. Similarly, if one looks at the first electric cars being marketed in the 2010s, their architecture mimics very closely the one of traditional vehicles, with a single engine in the front, a complete transmission with gearbox and differential, and batteries located in the rear of the car. If one looks at electric cars being developed in the 2020s, one can see firms' attempt to develop a new architecture that fully exploits the potential of the new technology, with batteries spread out in the underbody of the vehicle and possibly electric wheels (i.e., motors directly coupled to the wheels).

In a different field, early digital cameras were developed by taking a traditional camera body and adding a 'digital back' with a sensor located on the film plane, and it took some time before this architecture changed in order to fully exploit the potential of digital sensors.

3.4 The Puzzling Nature of Disruptive Innovation

When studying innovation, one can wonder why radical innovation can ever be disruptive and why this occurs also in technology-intensive sectors, in which incumbents should be expected to be quite well-prepared against disruption.[19] In fact, it is quite puzzling to observe incumbent companies, which are generally large, well-established and well-run, act somewhat helplessly when facing the challenge arising from de alio or de novo entrants. This is quite striking given that the latter, which are often smaller in size and lack experience in the industry,[20] must overcome *barriers to entry**.

Understanding the mechanisms behind disruptive innovation is of course important both for incumbents, who want to stave off competition, and for entrants, who instead want to gain a foothold in the market. Literature suggests three main reasons for disruptive innovation: *inability to join an emerging paradigm, neglect of an emerging market, inability to adopt ambidextrous strategic behaviors.*

3.4.1 The Inability to Join an Emerging Paradigm

The first reason behind disruptive innovation is the most dramatic, though probably not the most common, and occurs when two conditions are met. First, there is an old technology that does not keep pace with growing or new user needs, while the new technology does, and this creates the opportunity for paradigm change. Then, the new technology requires competencies and assets that are far from the ones relevant to the old paradigm. Incumbents are unlikely to be able to deal with a competency-destroying innovation, and may be unsuccessful in pursuing paradigm change.[21]

The inability to switch paradigm depends on both objective and subjective factors. The main objective factor is the distance in technology and associated competencies between the two paradigms. Subjective factors are instead due to incumbents' responses to radical innovation, that can suffer from inertia both in

[19] The Bain Technology Report 2020, including company data from 1996 to 2018, shows that technology-intensive firms are 12% more likely to be disrupted than retail companies and 25% more likely than those in financial services.

[20] For instance, one can think of the emergence of Microsoft when personal computers and client–server architectures entered the scene and—when IT shifted to Internet-based services and the cloud computing paradigm—the fast growth of a company like Google.

[21] A well-known example is the case of refrigerators, who progressively led to the demise of a very successful ice trade industry* in the nineteenth century. Physical assets and competencies that made an ice production and trading company successful were clearly useless for producing refrigerators. The main incumbents tried staving off competition by lowering the price of harvested ice (a simple strategy that often allows to 'buy time'), but this simply accelerated their fall. A few other incumbents left the industry and redeployed part of their assets and competencies to industries where they could be of value (i.e., fresh produce logistics).

Similarly, we can think of Blockbuster, whose assets and competencies made them structurally unable to counter the mail-based DVD rental service offered by Netflix.

understanding the new situation (*cognitive inertia**) and in reacting effectively to it (*action inertia**). While action inertia is essentially due to path dependency in the routines that define a firm, cognitive inertia derives form managers' tendency to follow a mindset that was applicable in the past paradigm, but is no longer valid in the new one. This phenomenon can be attributed to a number of *cognitive traps*:

- the first trap incumbents may fall in, is to look at the sources of competitive advantage and to the strategies that have granted them success in the past. Companies that have successfully survived until the end of a paradigm usually have a strong understanding and confidence in the weapons that made them win many battles, and will tend to heavily rely on them. However, the value of these assets and strategies within the new paradigm may be very low indeed.
- a second trap is related to incumbents' unwillingness to abandon an old technology because of previous investments that represent *sunk costs**.[22] The marginal cost of developing new products based on the old technology can effectively be much lower than switching to the new one, which would instead require new investments. However, assets connected to the old technology usually represent a sunk cost in the short term only since, in the medium-to-long term, they will have to be replaced and/or maintained. Therefore, the cost of remaining with the old paradigm tends to be underestimated.
- a third trap is related to observing the *status quo* and not thinking ahead, which can lead incumbents to reply to paradigmatic change when it's too late. If one focuses on the status quo in the early days of paradigm shift, the new technology will generally exhibit low performance and will not appear particularly attractive. However, while the old technology has probably reached its limit, the new technology still has plenty of upside, and its sudden growth is likely to catch incumbents by surprise.

3.4.2 Incumbents Tend to Neglect Emerging Markets

A second and very common reason for disruptive innovation is due to incumbents' focus on their reference markets, which is sometimes called 'Christensen effect' (Christensen 1997). Incumbents are strongly influenced by what they perceive as the needs of their current customers and take care that the performance they offer exceeds what is required by the market (and this, of course, is good business practice in normal times).

As shown in Fig. 3.6, when a new technology emerges, it is likely that its initial performance will be inferior to the one achieved by the established technology and to what current customers expect. Incumbents are well aware that performance of

[22] A sunk cost is an investment that has already been made and that is not recoverable, because the material or immaterial asset that has been bought is highly specific to a given technology or use, and can therefore neither be resold nor put to an alternative use.

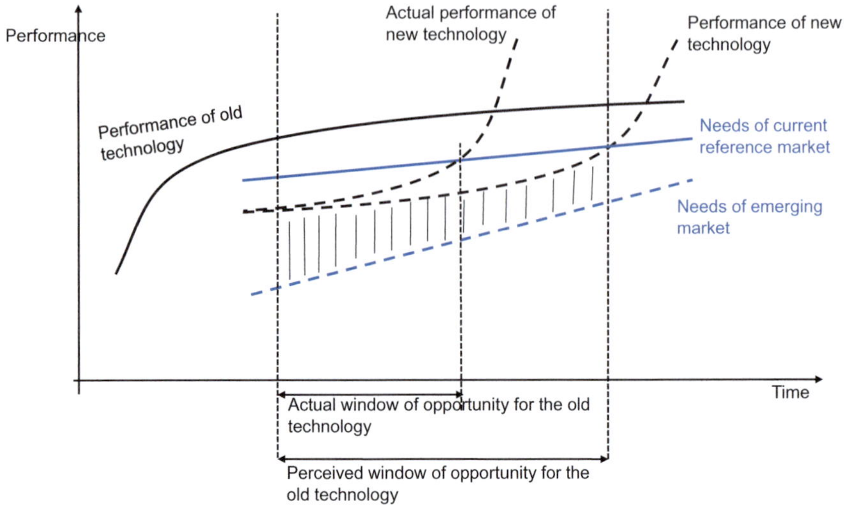

Fig. 3.6 Christensen's effect

the new technology will grow at a significant rate and eventually make it threatening to the old one, but not immediately. Therefore, they tend to keep an eye on it, and they sometimes even make R&D investments to pioneer it. However, they also tend to delay its introduction in new products.

As it often plays out, the inferior performance of the new technology may be perfectly satisfactory for a completely new market that the incumbents overlooked, which may have different needs than the existing one, and which favors the radical innovation. Sometimes this new market may be looking for a low-performance and low-cost solution. At other times, the new market may appreciate a different performance mix. This new and 'untapped' market may sometimes be a small niche that nonetheless allows entrants to establish a beach-head, reinvest margins from sales in R&D, accelerate technological development, and bring the day at which the new technology will become competitive forward. In other cases, the new market may be even larger than the existing one, and of course this leads to an even quicker disruption of the industry.[23]

[23] There are many examples to this explanation of disruptive innovation. When personal computers appeared in the '80 s, their performance was significantly inferior to the mainframe computers in use by firms, but was more than acceptable for home or small-office use, while their price was considered to be affordable.

In the case of digital photography, Kodak was a pioneer in digital imaging technology. However, due to the excellent resolution and low cost of film, it mostly considered digital technology as a complementary tool aimed at professionals. It did not foresee that somewhat pricey and low resolution (e.g., 3 MPixel) digital cameras would diffuse rapidly in the consumer market, since the use of digital photography was completely different from traditional photography, and provided consumers with a different bundle of benefits, in which convenience of use outclassed image quality.

This explanation of disruptive innovation leaves some open questions regarding what seems to be a very myopic behavior by incumbents. Again, subjective factors have a strong role in explaining the rationality of this behavior:

- given the known principle "customers pay our bills", managers will naturally tend to give great importance to what they think customers need and overweight market risks with respect to technological risks. Therefore, incumbents may even pioneer radically new technology, but then hesitate in bringing it to the market, "because that's not yet what our customers want". In some cases, they may do this in a half-hearted way, developing 'hybrid', even if suboptimal, products that exploit the new technology, but at the same time attempt to leverage on existing assets (Wu et al. 2014). As a managerial implication, incumbent firms should consider both the current and the potential customers that might be attracted by the radical innovation.

  Moreover, they should not be afraid of cannibalizing their existing business, in the knowledge that if they hesitate, or they remain 'stuck in the middle', other firms will. Of course, such a move may imply substantial costs, writing off existing assets, or having to manage heterogeneous and conflicting business models, with the ensuing diseconomies of scale (Greenstein 2017).[24] This requires adopting a wider stance when dealing with market research, using it to envision new markets and not only to exploit existing ones. Moreover, it implies the need for managers to go beyond the usual cognitive frames they employ in decision making (Raffaelli et al. 2019).
- the governance processes in incumbent companies also can have a strong influence. Incumbent companies usually enjoy high profitability and this creates expectations in the minds of shareholders. Even though this profitability is bound to be abruptly ended by disruptive innovation, the status quo makes it difficult for top management to justify what will be perceived as the cannibalization of a successful business model and the pursuit of a risky endeavor in an uncharted territory. Given the alternative between having to cannibalize their own business model, or be cannibalized in the future by competitors, managers easily choose the third way of 'waiting for the next quarter', before making an uncomfortable choice, which might put their position at risk. This is especially true in the case of public companies, with shareholders that might not be fully aware of the situation (Oemichen et al. 2021).
- in other cases, the internal organization might play a significant role. Managers and shareholders might be inclined toward carrying out radical choices, but proposals for doing it are simply not placed on their agenda. This generally happens when proposals for radical innovation emerge bottom-up instead of being championed top-down. If a proposal for a radical innovation comes from

[24] Examples of firms able to cannibalize an existing and successful business model in order to accelerate the transition to a new technology are somewhat uncommon. One is Netflix, when it phased out its profitable web-based rental of DVDs in order to shift its customers towards streaming services. The move was quite costly and ill-received by shareholders, but was ultimately successful.

the lower ranks of management (e.g., a project leader in the R&D unit), it is likely that it will be culled by the managers to which it is immediately presented (e.g., a middle manager). These may in fact appreciate the idea, but realize that they are not in a position to make strategic choices of this kind, which would require deep changes to parts of the business model they do not control. Moreover, they may hesitate in bringing such a proposal to top management, which is the only entity that could make decisions of this kind. Therefore, top management should always have a strong awareness of potentially radical innovations, in order to be able to challenge both internal and external stakeholders who may exert pressure in keeping the status quo. Moreover, top management should create communication channels exposing them to innovative ideas, thus avoiding that these same ideas may be filtered away by intermediate layers of management.

3.4.3 Incumbents' Need for Ambidexterity

The previous discussion should have made clear that it is not easy for an incumbent to successfully deal with radical and competence destroying innovation. There will be an obvious aversion to embrace it, since it will be perceived to be risky for the existing business. Furthermore, the radical innovation will succeed only if financed by the proceeds coming from that same technology that will be supplanted by it. It is therefore no wonder that paradigmatic change leads to involved and time-consuming decision-making processes within incumbent firms.

In order to tackle successfully this challenge, scholars have introduced the concept of organizational *ambidexterity* (Tushman and O'Reilly 1996, 2004; Gibson and Birkinshaw 2004).

An ambidextrous organization is able to maintain efficiency in the current business models and technology, at the same time reading itself to adapt for the future. Ambidexterity is not only a cultural attitude, but requires specific strategies and organizational actions, to make sure that both incumbent and radically innovative projects and units are able to progress under close coordination by top management. The former will have to be managed in a more formalized and standardized way, while the latter will have to be coordinated with greater freedom. More on this will be discussed in Chap. 4.

3.5 When Radical Innovation Does not Disrupt

The previous discussion should not lead to the popular conclusion that radical innovation always and invariably leads to disruption. On the contrary, radical innovation often does not lead to disruption at all, though this outcome is seldom discussed in both professional and academic literature. This asymmetry is probably due to the fact that it is quite difficult and certainly less attractive to study a non-event, than a visible outcome. However, both incumbents and entrants need to be well aware of the reasons because of which disruptive innovation might not materialize.

3.5.1 The Existence of Markets for Technology

One possible reason is due to the existence of *markets for technology*. Given the fact that incumbents have clear competitive advantages, a threatened incumbent and an entrant may find a deal, with the former acquiring a license for using relevant technology from the latter, or directly acquiring the entrant company. This latter case is especially common in the case of venture-backed startups. As previously discussed, venture capital firms have the objective of achieving returns on the capital invested in startups within a given time frame through an exit. From their perspective, it is often worthwhile to have a trade sale early on, rather than starting a lengthy and uncertain path, leading to competition with the incumbents and, only if successful, achieving an exit through an *initial public offering* (IPO)*. At the same time, the acquiring company takes a potentially disruptive competitor out of the market and absorbs its technology, thus allowing a radical innovation to let the reference industry unaltered, without any disruption.[25]

3.5.2 S-curves Can Be Misleading

Truly disruptive innovation usually requires that the old paradigm has really reached an endpoint, and evidence from 'flattening' s-curves is commonly used for this purpose. However, the use of s-curves if quite critical and can be misleading.

- To begin with (left-hand diagram (a) in Fig. 3.7), s-curves are not usually smooth, but are made up of nested s-curves that represent what are familiarly called *product generations*. Some degree of saturation will be observable from time to time in the s-curve, and it will not be easy to tell whether this saturation is soon to be overcome by a new product generation, or whether it represents a real technological limit that is inherent to the technological trajectory. In the former case, there might be some industry disruption when transitioning between

[25] Antitrust authorities have recently started monitoring and acting against this phenomenon (e.g., the FTC action against Meta's acquisition of Within in 2022). The underlying hypothesis is that technology acquisitions may prevent future competition and therefore constitute an illegitimate way for powerful incumbents to make their dominant positions unassailable. At the same time, one might wonder what might be of the venture capital industry, should this type of 'exit route' be barred, or simply made more uncertain.

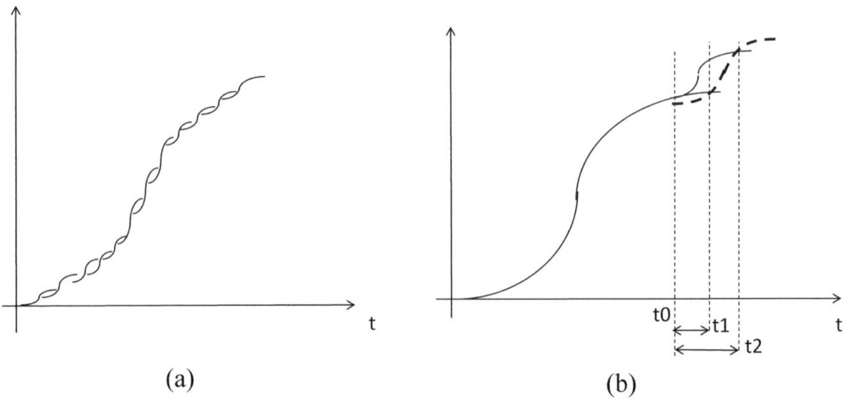

Fig. 3.7 **a** Nested paradigms; **b** the 'sailing ship effect'

product generations, but this will be on a minor scale to what would be expected from a full transition between paradigms.[26]

- Moreover, s-curves are usually drawn by using time as an independent variable. What seems as saturation may therefore not be due to a technological limit, but simply to the unobserved slowing of R&D effort. Faced with the threat of an emerging technology, incumbents are likely to react by restarting a new investment cycle on the old technology and quickly regain a performance lead. In some cases, elements of the new technology may *spill over* and make it easier for incumbents to further improve the existing technology. As shown in the right-hand side (b) of Fig. 3.7, this reaction allows incumbents to extend their lead with respect to the emerging technology from time t_1 to t_2. In some cases, this reaction will simply 'buy some time' and defer the date at which the new technology will displace the old one. In other cases, the advantage gained may be so significant as to effectively discourage entrants and stop the development of the new technology in its tracks. In some cases, incumbents' reaction may give them a lead across all market segments. In other cases, incumbents may retreat to segments where the old technology can maintain this lead for a longer time.[27] This phenomenon is often termed the *sailing ship effect** (Geels 2005),

[26] In fact, one of the case studies introduced by Christensen in his theory of disruptive innovation was related to successive generations of hard disk drives, which did not represent real paradigm shifts. Similarly, we can refer to the subsequent generations of microprocessors, each characterized by smaller sizes of the component transistors.

[27] A more recent example of sailing ship effect is represented by the television industry. The technology based on the cathode ray tube (CRT) was developed in 1897 by the German physicist K. F. Braun and represented the paradigm for television screens up to the twenty-first century. With this technology, electron beams are conveyed from a cathode in the rear, deflected by electromagnetic coils, and projected on a photosensitive surface on the front. This technology requires CRTs to be deep and the screen to be rounded. A liquid–crystal display (LCD) instead uses the light-modulating properties of liquid crystals laid out on the screen, which allows LCDs to be flat and thin. As LCDs began to diffuse, traditional TV manufacturers reacted by investing in

following the significant performance increase in long-range sailing ship technology that came along in the nineteenth century, when competition from steamships started to arise.

- Finally, it has to be verified whether the radical innovation affects the entire product (in Henderson and Clark's terms, if it really is a radical innovation, involving architectural change as well) or whether it is mostly localized at the level of individual components (i.e., if it is closer to modular innovation). In the former case, the innovation is more likely to be disruptive, favoring new entrants. In the latter case, the innovation will probably not be disruptive at all at product level, since incumbents will find it relatively easy to embrace it. If any disruption might occur, it will be at the level of affected components and subsystems, and therefore upstream in the value chain. Therefore, it is very important to be aware about the relationship between product architecture, industry structure, and the type of innovation.

3.5.3 Localized Technological Change

A further reason why radical innovation might fail to lead to disruptive effects lies in the adoption process carried out by customers. When a new technology emerges, each potential customer continuously evaluates whether to switch, or whether to keep on using the old technology. This evaluation will be carried out by considering a variety of elements. Rogers (2003) suggested five main factors:

- **relative advantage**, in terms of costs and utility deriving from both alternatives;
- **compatibility** between the new and the old technology, in terms of the possibility of reusing old complementary assets and experience when adopting the new one;
- **complexity of adopting** the new technology and learning how to use it;
- **trialability** of the new technology, considering the switching costs and the risk that one might not be satisfied, eventually having to incur in additional costs in order to revert to the old;
- **observability** of the actual advantages of the new technology.[28]

the CRT technology, leveraging on its temporary advantages (cost and image quality) by introducing flatter screens. These products dominated the market for about a decade, until the early 2000s.

[28] For instance, in the case of battery electric versus internal combustion cars, one might view relative advantage in terms of a comparison between the costs per km of both solutions, along with the discomfort and time needed to recharge the car, compared to simple refueling. Compatibility probably is a minor issue for individuals (while it is significant for society as a whole). Complexity has to do with changing habits in recharging versus refueling, especially for long road trips, plus the possible need of installing a domestic charger at home. Trialability is associated to the high cost of purchasing an electric car, if trial or rentals solutions are not available. Observability may have to do with the difficulty of evaluating relative costs in the presence of high price volatility for gasoline and electricity and with the difficulty of making sure such a choice is indeed beneficial for the environment, taking the whole product lifecycle into account.

Within this process, costs, benefits and *switching costs** in fact are not only based on the objective performance of the two technologies, but also on factors that are highly subjective to each user. The theory of *localized technological change* (Antonelli 1995) suggests that this can have a dramatic impact on the adoption of the new technology and, therefore, on the possibility that it disrupts the old.

Specifically, utility gained by the users of the old technology will depend on the investment that has been progressively accumulated in complementary assets and in the experience gained by using it. If these assets and such experience are highly specific to the old technology, and not usable with the new one, the subjective attractiveness of the new technology will consequently be reduced. Therefore, the performance gap that would justify switching increases, and adoption is delayed. In these cases, such customers are said to be locked-in the old technology.

Figure 3.8 compares the 'objective' utility provided by the new technology with the 'subjective' utility provided by the old one, which of course depends on the individual customer's previous behavior and decisions. Customer 1, who has made little prior specific investment in the old technology, will be the first to switch to the new one. He/she will be followed by customer 2, who has made a higher investment in the old technology but is not investing further. The last to adopt the new technology will be customer 3, who has not only made prior investments, but is continuously increasing them.

Should the market be completely made up of customers similar to customer 3, this could be enough to stop the emergence of the new paradigm. This is a somewhat extreme case. However, the presence of a significant share of customers who delay adoption because of localized technological change might determine a 'false start' and delay diffusion by a few years, until technology reaches a level of performance that will justify switching costs for the majority of customers, thus reducing the likelihood of industry disruption.

3.5.4 Appropriability Regimes and Complementary Assets

Finally, a reason explaining why radical innovation might not end up to be disruptive is connected to the roles of *appropriability regimes* and of *complementary assets* (Teece 1986).

Appropriability measures the degree to which the proponents of the radical innovation can keep the economic value created by this innovation to themselves, instead of having to share it with customers and suppliers and/or being imitated by incumbents. If imitation is difficult, either because of legal protection (see patenting in Chap. 6) and/or because of the tacit nature of technological knowledge, the appropriability regime will be high, and disruption by entrants will be likely. Conversely, a weak appropriability regime might not lead to disruption at all, since entrants may quickly and successfully be imitated by incumbents.

Concerning complementary assets, successful innovation requires not only to design new products and services, but also to create the infrastructure that is required to produce them, bring them to the market, provide customer service, and

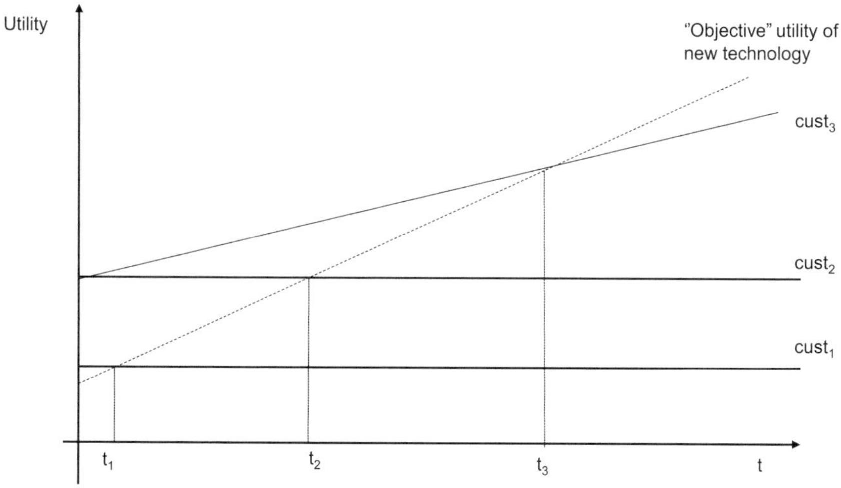

Fig. 3.8 Impact of localized technological change

so on. Significant differences may arise from the nature of the complementary assets that make up this infrastructure and from the possibility that this leads to a typical ‘chicken and egg’ problem, where the overall system does not develop because the actors developing one part wait for the actors who might develop the other, and vice versa. These complementary assets might be generic, specific to the new technology (i.e., given the technology, one must define an asset that will not be of use for other technologies) or co-specific with the technology (i.e., the technology and the asset have to be defined at the same time and are of value only when used jointly).

In the case of generic complementary assets, it will be easy for the entrant to contract these assets from suppliers, therefore enter without significant investment, thus opening the door to disruption. In the latter two cases, the specific nature of the complementary assets will either force the entrant to develop the assets himself (i.e., to integrate) or to finance suppliers in order to have them take care of it. In both cases, this raises a significant financial and managerial hurdle to entrants and might lead to sharing the proceeds with asset holders, therefore reducing the attractiveness of the decision.[29]

[29] In the case of electric cars, the obvious implication of this discussion is associated to the need of developing battery manufacturing facilities and recharging infrastructure.

Table 3.2 Strategies for incumbents and strategies for entrants in the context of radical innovation

Strategies for incumbents	Strategies for entrants
Pre-emptively lock customers in your technology Try 'killing' the new technology or buy time • improve the old technology • start a price war Segment the market by 'needs' and closely monitor (or serve) them all ('incumbent's advantage') Be wary of 'sunk cost' and 'status quo' traps Jump on the new technology • Listen to 'people on the edges' • Bring new champions on your top management team on time and allow some tension between old and new • Fund R&D top-down and/or go for acquisitions (possibly of small firms that are easier to integrate) • Always look for the market and business model that suits current and emerging technology Retreat to a niche or relocate to a different market where you can still use your assets	Find the right beach-head market that is • right for your immature technology • likely to be overlooked by incumbents Make sure that you have enough cash to withstand incumbents' reactions aimed at 'gaining some time' Stay away from markets that are overserved and locked-in the old technology Make the right vertical integration choices Carefully look after intellectual property Rights in order to maximize appropriability Be careful if you don't own the required complementary assets Be wary of possible improvements in the existing technology Consider selling your business to an incumbent

3.6 Strategies for Incumbents, Strategies for Entrants

The previous discussion has broadly explained the reasons because of which radical innovation might, or might not, lead to disruptive innovation. The topic has clear managerial implications that of course depend on whether the firm being considered is an incumbent or an entrant.

Table 3.2 draws from the previous discussion and provides a non-exhaustive summary of the strategies that either kind of company might follow when dealing with radical innovation in order to respectively avoid, or favor, disruption.

References

Afuah AN, Bahram N (1995) The hypercube of innovation. Res Policy 24:51–76
Anderson P, Tushman ML (1990) Technological discontinuities and dominant designs: a cyclical model of technological change. Adm Sci Q 35:604–633

Antonelli C (1995) The economics of localized technological change and industrial dynamics. Kluwer Academic Publisher, Dordrecht

Bain and Company (2022) Tech companies eat disruption for breakfast. Technology Report 2022

Calantone RJ, Di Benedetto CA, Divine R (1993) Organizational, technical and marketing antecedents for successful new product development. R&D Manage 23:337–351

Christensen CM (1997) The innovator's dilemma: when new technologies cause great firms to fail. Harvard Business School Press, Boston

Christensen CM, Raynor ME, McDonald R (2015) What is disruptive innovation? Harv Bus Rev 93:44–53

Cooper RG, Kleinschmidt EJ (1990) New products—the key factors in success. American Marketing Association, Chicago

Dosi G (1982) Technological paradigms and technological trajectories. A suggested interpretation of the determinants and directions of technical change. Res Pol 11(3):147–162

Dutton J, Thomas A (1984) Treating progress functions as a managerial opportunity. Acad Manag Rev 9:235–247

Geels FW (2005) Technological transitions and system Innovations: a co-evolutionary and socio-technical analysis. Edward Elgar Publishing, Northampton

Gibson CB, Birkinshaw J (2004) The antecedents, consequences and mediating role of organizational ambidexterity. Acad Manag J 47:209–226

Greenstein S (2017) The reference wars: Encyclopædia Britannica's decline and Encarta's emergence. Strateg Manag J 38(5):995–1017

Henderson R, Clark KB (1990) Architectural innovation: the reconfiguration of existing product technologies and the failure of established firms. Adm Sci Q 35:9–30

Iansiti M (2000) How the incumbent can win: managing technological transitions in the semiconductor industry. Manage Sci 46(2):169–185

Nelson RR, Winter S (1982) An evolutionary theory of economic change. Harvard University Press, Cambridge

Oehmichen J, Firk S, Wolff M, Maybuechen F (2021) Standing out from the crowd: dedicated institutional investors and strategy uniqueness. Strateg Manag J 42(6):1083–1108

Raffaelli R, Glynn MA, Tushman ML (2019) Frame flexibility: the role of cognitive and emotional framing in innovation adoption by incumbent firms. Strateg Manag J 40(3):1013–1039

Rogers EM (2003) Diffusion of innovations. Free Press of Glencoe, 5th edn, New York

Schilling MA (2008) Strategic management of technological innovation, 2nd edn. McGraw-Hill Education

Sosa ME, Eppinger SD, Pich M, McKendrick D, Stout S (2002) Factors that influence technical communication in distributed product development: an empirical study in the telecommunications industry. IEEE Trans Eng Manage 49(1):45–58

Sosa ME, Eppinger SD, Rowles CM (2003) Identifying modular and integrative systems and their impact on design team interactions. J Mech Des 125(2):240–252

Sosa ME, Eppinger SD, Rowles CM (2004) Misalignment of product architecture and organizational structure. Manage Sci 50(12):1674–1689

Teece DJ (1986) Profiting from technological innovation: implications for integration, collaboration, licensing and public policy. Res Policy 15:285–306

Tushman ML, O'Reilly III CA (1996) Ambidextrous organizations. Calif Manage Rev 38(4):8–30

Tushman ML, O'Reilly CA III (1997) Winning through innovation. Harvard Business School Press, Boston

Tushman ML, O'Reilly CA III (2004) The ambidextrous organization. Harv Bus Rev 82:74–82

Verganti R (2009) Design driven innovation: changing the rules of competition by radically innovating what things mean. Harward Business Press, Boston

Wu B, Wan Z, Levinthal DA (2014) Complementary assets as pipes and prisms: innovation incentives and trajectory choices. Strateg Manag J 35(9):1257–1278

The Dynamics of Innovation

4

Abstract

This chapter focuses on the phenomena that occur along technology s-curves, looking at both demand side and supply side aspects. The dominant design model by Abernathy and Utterback is presented in detail, followed by a discussion on its variants and on a number of aspects that are emerging in recent research, from the role of vertical integration to the phenomena that typically occur in the fluid phase. Finally, the chapter introduces the topic of technological standards and provides a discussion on the associated decisions and strategies.

The previous chapter has focused on the different types of innovation, mostly contrasting the extremes of evolutionary and revolutionary change. Not much has been said about evolutionary change and on the transformations that take place within a given technological paradigm. This chapter has the aim of covering this topic from a dynamic perspective, considering the interplay between supply and demand, and introducing the topic of standards.

4.1 Technological Lifecycles and Diffusion

S-curves have up to now been introduced to represent the evolution of a performance indicator against an independent variable such as time or, better still, R&D effort or investment (Fig. 4.1a). However, a second s-curve that roughly follows the performance s-curve can also be drawn, representing the *diffusion* of the innovation into the market (Fig. 4.1b).

Diffusion (or *penetration*) can be defined as the fraction of potential users that, at a given time, have decided to adopt the technology. In absolute terms, the diffusion curve is represented by *cumulated adoption sales*, these being the sales to users who have adopted the technology for the first time. Adoption sales in fact are only part of overall sales, which also include subsequent purchases made by adopters. In turn,

M. Cantamessa and F. Montagna, *Management of Innovation and Product Development*, https://doi.org/10.1007/978-1-4471-7531-5_4

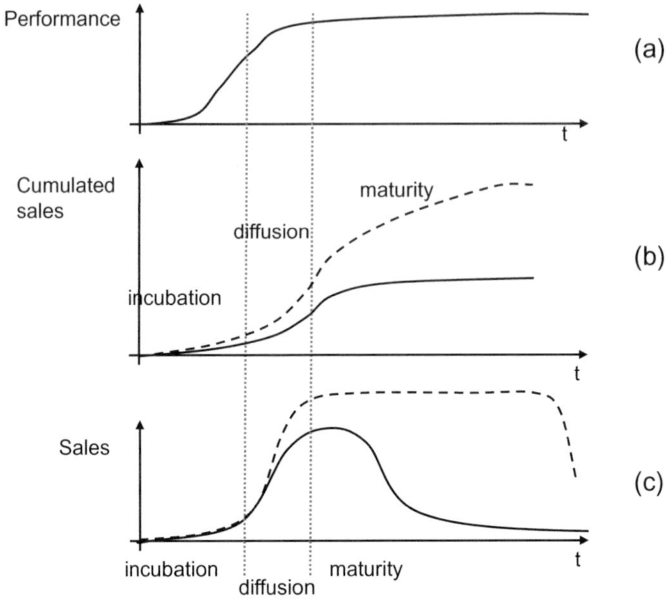

Fig. 4.1 Performance and diffusion s-curves

these are given by *additional sales* (i.e., sales to adopters who wish to own more than one such product) and *replacement sales* (i.e., sales to adopters who decide to replace their product with a new one because of wear or obsolescence). For the time being, we will draw adoption sales as a bell-shaped curve (Fig. 4.1c); in mathematical terms, this curve represents the derivative of the diffusion s-curve.

When progressing along the s-curve, it is common to define three main phases, which are respectively termed *incubation* (during which both performance and diffusion still have to 'take off'), *diffusion* (when performance and diffusion grow significantly) and *maturity* (when they approach saturation).

Being aware of the current phase in the product lifecycle is very important for firms, since it determines the type of technological effort that has to be pursued and the way with which it can be conveyed to the market. During the incubation phase, it will mostly be important to master the new technology that is giving birth to the new trajectory and integrate it in new products and services. In the diffusion phase, adoption sales are dominant and the firm will mainly have to convince the market of the utility associated with the new product. During maturity, adoption sales fade away and the firm will have to rely on additional and replacement sales. Therefore, it will have to focus its effort on convincing the market to increase the product replacement rate. At the end of the maturity phase, the firm will have to expect a period of revolutionary change, which will eventually lead to a new paradigm.

4.2 Diffusion s-curves and Customer Segments

Diffusion s-curves tell us that customers adopt at quite different moments along the product lifecycle. One can therefore wonder whether customers are all similar to each other and differences in time to adoption are purely due to chance, or whether customers are individually different in their propensity to adopt. In the previous chapter, we have already discussed how Rogers' five adoption criteria and the Theory of Localized Technological Change may explain why a specific customer will be an early or a late adopter. For a firm, the realization that customers are different leads to the need of understanding the different customer segments that can be found along the technology lifecycle.

The most popular segmentation has been proposed by Rogers (1962) and it is shown in Fig. 4.2. To this purpose, the sales curve can be approximated to a normal distribution curve and then split at the midpoint and at approximately −2, −1 and +1 standard deviations. The resulting segmentation is common to most cases of technology diffusion; the characterizing features of individual segments are discussed in Table 4.1.

By studying this segmentation, Moore (1991) proposed the existence of a very significant gap (or *chasm*) between the early adopters and the early majority segments. Customers in the former segment adopt because they 'look into the future' and are therefore ready to accept a relatively immature technology. Conversely, the latter will not do so at all and will only adopt a mature product that fully satisfies them. Therefore, a product that is highly successful with early adopters is likely to be unsuccessful when the early majority segment kicks in. Firms that are market leaders in the early phases of the product lifecycle may fail to understand the new user requirements that characterize the early majority segment if they do not change perspective on their customers, thus falling behind when attempting to 'cross the chasm'.

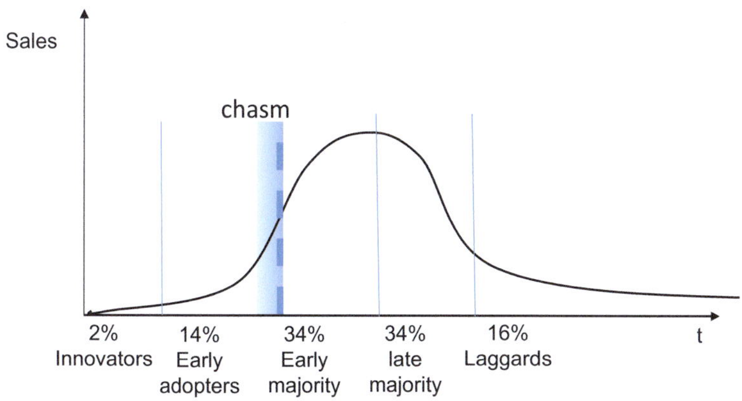

Fig. 4.2 Market segments along the technological lifecycle

Table 4.1 Characterizing features of Moore's market segments

Segment	Typical size of segment (%)	Segment description	Reasons to serve/not to serve the segment
Innovators (enthusiasts)	2	In the case of consumers, innovators are technology enthusiasts, who love technology for its own sake and might even find interest in its shortcomings due to immaturity, since it gives them the opportunity to 'play' with it In the case of businesses, innovators are companies with very specific needs, who might even pursue the innovation on their own, because of their peculiar requirements	The segment is very small and therefore not so attractive for large firms from an economic viewpoint Startups might find it interesting as a small beach-head market Innovators also often act as *lead users* (von Hippel 1986) and provide valuable technical feedback
Early adopters (visionaries)	14	Customers who think the technology will be important in the future and want to gain some early experience with it, even though a direct cost–benefit analysis might suggest waiting some more time Early adopters are not excessively concerned of adopting a technology that is still somewhat immature and/or difficult to use	The segment is *somewhat* attractive because of its size and because it occurs early, thus making it financially interesting Entrants may find it attractive for establishing their brand, since early adopters are also trendsetters and spread 'word of mouth'
Early majority (pragmatists)	34	Customers who decide on adoption, based on costs and benefits. If the technology is immature, too costly or difficult to use, they simply postpone adoption and wait	The segment is interesting to all firms because of its size, and because being active in this segment warrants being active during the rest of the product lifecycle as well
Late majority (conservatives)	34	Customers who decide on adoption based on costs and benefits, with an added hesitation due to an increased perception of risks and/or to localized technological change	The segment is interesting because of its size. Being successful at this stage usually requires having survived the earlier phases
Laggards (skepticals)	16	Customers characterized by very peculiar needs or hindrances to adoption and are therefore very late	The segment is of marginal interest, because of its limited size and because of the delay with which it occurs

Table 4.2 Suggestions for making the product attractive to the different categories

Making the product attractive to innovators	Making the product attractive to early adopters	Making the product attractive to the early majority
Understand who beach-head innovators are and their specific needs and find ways to contact them Make sure that initial technical shortcomings can be overcome by users, eventually with some help from the firm Engage in technical dialogue with innovators	Start creating a solid 'adoption network' (distributors, complementors, etc.) Start working on ancillary features such as ease of use and design Define a satisfactory (if not perfect) price–benefit ratio Ensure that the product is reliable and 'cool' enough, so that early adopters can start some positive word-of-mouth effect Lower purchasing risk ('get your money back' offers, razor–blade business models, etc.)	Work hard on ease of use, design, etc Consolidate the 'adoption network' (ensure that complementary products and services are made available) Sell as a bundle with other goods with which the user is familiar Make technical features 'just good enough', but do not launch overengineered products Look carefully at pricing (not too high to discourage, not too low to kill margins) Lower risk in purchasing

From a managerial perspective, what constitutes a successful product and what strategy should be followed depends on the segment that is being reached by the diffusion curve. Table 4.2 proposes a few possible recommendations. It must be recognized that Rogers' approach (usually labeled as the 'Diffusion of Innovations' or DoI approach) does not delve deeply into the adoption process carried out by individuals or organizations.

Further discussions on this topic may refer to the rich—but nowadays relatively neglected—literature on Technology Acceptance Models (Davis 1989a, b; Venkatesh 2000a, b). This literature originated in the field of Information Systems, but the concepts are applicable to other industries too, and try to explain the adoption process as the outcome of an interaction between the objective features of the innovation, the subjective features of the adopter and the influences cast by the organizational and/or societal milieu.

4.3 The Dominant Design Model and Its Implications

When observing s-curves, a question may arise on the shape of s-curves. Why is progress not linear and what causes the initial incubation phase, typified by a hesitant start, followed by rapid performance growth and diffusion? While a general answer to this question can be found by studying diffusion phenomena and their typical nonlinearities, as it will be done in Chap. 13, a more specific explanation has been proposed by Abernathy and Utterback's theory (1975), which has served as the foundation of a number of other contributions (e.g., Tushman and Rosenkopf

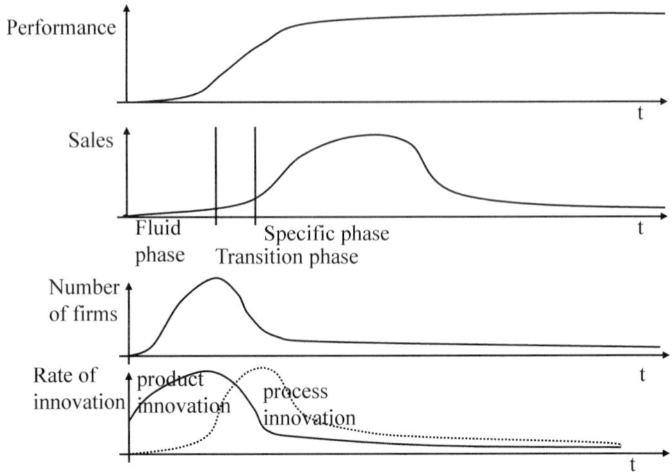

Fig. 4.3 Abernathy and Utterback's model

1992). We will initially introduce the Abernathy–Utterback theory in the case of discrete products (i.e., products that are made up of component assemblies, and for which it is therefore possible to define an architecture). Later on, the theory will be adapted to the case of continuous products and services.

4.3.1 Dominant Designs and the Technology Lifecycle

The main elements of Abernathy's and Utterback's theory in the case of assembled products are represented in Fig. 4.3.

According to this theory the initial, or *fluid*, phase is characterized by a still immature technology with low performance, for which demand will consequently be low as well. Even though it is difficult to make a profit in this stage, the number of firms entering the industry is high and keeps growing, since the industry is viewed as very promising in perspective. Each firm that enters the market carries a stock of prior competencies and assets that it believes will be useful, and this leads to an extremely diversified range of technical solutions. The rate of product innovation[1] is also very high, but—quite surprisingly—does not lead to significant progress. This occurs because firms are not concentrating their efforts in the same technical direction and cannot therefore generate cumulative progress based on the imitation and improvement of competitors' inventions.

At a given point, the industry enters what is termed the *transition phase*, during which one product architecture emerges as the frontrunner and becomes recognized as *dominant design*. The dominant design defines the technical solutions,

[1] The rate of product innovation can be measured through a number of indicators. For instance, one can count the yearly number of granted patents associated to the technology.

components and features (i.e., the architecture) that become widely accepted by the industry and the market, overcoming other competing technological solutions and designs. Around this dominant design, it will be possible to establish the technical and business connections between product architecture, subsystems and complementary systems, which will eventually lead to integration choices and the establishment of a paradigm, as discussed in the previous chapters.

The emergence of the dominant design triggers a sort of avalanche effect, which determines the growth segment of the s-curves. Product performance takes off, since the common design now allows competitors to imitate each other and to generate cumulative performance improvements. In turn, improved performance and a clearly recognizable system induce more customers into adoption, which lifts sales and margins (which may be reinvested in R&D) and also starts determining specific demands around which firms will focus further performance improvements.

At the same time, the number of active firms starts to decline. At first, this occurs because the companies who had proposed a different architecture, and whose assets and competencies do not match the dominant design, are forced to either leave the industry, or retreat to specific niches where they can still be competitive. Increasing sales and this initial industry shakeout imply that growth rates for a surviving firm will be higher than for the market in aggregate. While this might seem to be good news to them, it also represents a significant challenge, and not all firms will be able to sustain such high growth rates. Among the problems to be faced are raising capital to finance production facilities and working capital, managing a growing workforce, creating a sustainable and formalized organization, finding managers to run it, and so on. Many companies will therefore fail in this effort, thus continuing the shakeout and further stimulating the growth of survivors.

During this process, the lifecycle enters the so-called *specific phase*. During the specific phase, firms strive to compete over cost and quality in a scenario characterized by increasing demand. They therefore must shift their focus from product to process innovation. Before the emergence of the dominant design, uncertainty in the product and limited sales discouraged companies from caring too much about production. However, with a now-stable dominant design in the product and a growing demand, it becomes technically possible and strategically important to innovate the production process. Firms therefore develop optimized, product-specific and capital-intensive machinery, allowing higher quality and significant economies of scale. In turn, this leads to lower product cost, lower prices and increased demand. Moreover, it furthers the shakeout, because the new processes determine a *minimum efficient scale** and this generally leads the industry into an *oligopoly** which, not by chance, is a most common feature in mature industries.

4.3.2 The Emergence and Lock-in of Dominant Designs

In the study of dominant designs, one may wonder what are the reasons that lead the market and society into selecting a given technology and architecture, instead of other competing solutions.

A dominant design clearly is one of the key elements that make up a technological paradigm, and a partial answer to this question has therefore already been given in previous Chap. 3. More specifically, dominant designs tend to emerge because of technical superiority leading to better product performance. However, this superiority must not be intended in an 'absolute' sense. What matters is the superiority perceived by that first cohort of customers who will lead the adoption process into the transition phase and who base their choices on their specific needs. Referring to Rogers' segmentation, these early customers roughly coincide with innovators (whose needs are by definition *very* specific) and an initial portion of early adopters.

As we will discuss later in this section, a dominant design leads to a lock-in effect. Therefore, it is possible that the choices carried out by these early customers may not be beneficial to society, if one assumes a long-term perspective.[2] It is therefore possible that governments may decide to step in the adoption process and steer the selection toward a technology and design that they consider to be socially preferable. As mentioned in Chap. 1, the means for achieving this intervention include regulations, taxation, incentives and the development of specific and complementary infrastructure.

The dominant design selection process is in fact also driven by the role of specific complementary goods or assets that may facilitate the diffusion of a particular technology and architecture over competing alternatives.[3]

A further element that may influence the selection of the dominant design is due to the high degree of uncertainty that is inherent to the firms' adoption process (Pistorius and Utterback 1997). Given this uncertainty, decision-makers may try to 'play it safe' and quite blindly follow the choices made by those firms which are widely perceived to be market leaders (or consultants who are recognized to be key opinion leaders) and enjoy a very strong reputation.[4] This effect can be

[2] For example, early adopters of motor cars at the beginning of the twentieth century could not (and did not) foresee the long-term environmental and geopolitical consequences of adopting vehicles powered by fossil fuels.

[3] One of the reasons explaining the emergence of gasoline-powered vehicles was the availability of gasoline, which was then a cheap by-product of oil cracking, together with the willingness of the oil industry to invest in the development of a network of refueling stations. Today, the lack of recharging infrastructure and the difficulty of upgrading power distribution grids to support them is instead a significant hindrance to the establishment of a dominant design centered around Battery Electric Vehicle (BEV) technology.

[4] Imagine a manager from an automaker having to decide on whether to pursue BEV technology, in the wake of a highly reputed firm such as Tesla, or take an opposite position and go in the direction of Plug-in Hybrid Electric Vehicle (PHEV) technology. This 'bandwagon effect' can be explained quite rationally, especially in the case of a professional manager. Choosing differently from a recognized thought leader and its followers is very risky and impossible to justify in the likely event that he/she was wrong. Conversely, in the less likely event that he/she was right, the upside for the firm may be very large, but shareholders may easily ascribe the event to simple luck. In fact, it is more common for entrepreneurs than managers to take these bold and contrarian decisions.

enhanced if the industry leader allows some technological spillover to its competitors, since this locates the latter in a position where it would be even less reasonable to adopt an opposite choice. Such spillovers can be due to chance, but can sometimes be the consequence of a clever strategic decision. In fact, for the leading company it's better to erode part of its competitive advantage, but ensure that competitors stay trailing behind in its technological wake, than let them wander elsewhere in the technological landscape, and risk that they find better technological alternatives that might become dominant designs.[5]

To summarize, the emergence of a dominant design appears to be a somewhat chaotic process, characterized by a high degree of ambiguity and uncertainty. For companies operating in the early phases of the technological lifecycle, betting on the right technological and architectural choices is at the same time not only very important for their future, but also very difficult (ref. Technological Forecasting in Chap. 4).

Once a preferred architecture has emerged and the industry has entered the specific phase, the dominant design will tend to remain stable (or *lock-in*) for a very long time. This is mainly due to:

- **specialization and learning effects***, which allow firms to operate with increasingly higher quality and lower cost. Should alternative and objectively promising solutions arise later on, producers will not find them attractive, because of the larger benefits they gain by sticking with the current dominant design;
- **economies of scale***. During the specific phase, firms surviving the shakeout will have adopted equipment that is capital-intensive and specific to the dominant design. This constitutes a sunk cost that would make the switch to technical solutions that depart from the dominant design unprofitable.
- **complementary assets** and **network externalities***. Once an industry has entered the specific phase, customers, and society at large, will also invest in specific complementary assets that increase the utility deriving from the dominant design. A technical solution that does not use the same complementary assets (e.g., CNG-powered cars if we refer to example in the footnote 4) will therefore be at a disadvantage and find it hard to unseat an established architecture. In some cases, complementary assets trigger network externalities, and this leads to switching costs and lock-in on the customer side of the paradigm.

Due to lock-in, stability of a dominant design can last decades and will give way only when reaching a state in which a radical and potentially disruptive innovation becomes able to launch a successful challenge to it. In some cases, it will be possible to observe a few modular or architectural changes also during the transition

[5] This might be the reason why Tesla in 2014 announced a sort of free licensing agreement for its entire patent portfolio (even though, technically, they framed it as a decision to refrain from litigating against infringers).

and the specific phases, which may be interpreted as final 'settling movements' of the dominant design.[6]

In some cases, lock-in due to complementary assets is so strong that it might influence the evolution of the succeeding dominant design. Therefore, when coming up with a radical and potentially disruptive innovation, firms will not be able to work as if from a clean slate, but may have to come up with a technology that—at the least at the beginning—fits well with the previously installed base of complementary assets.[7]

Table 4.3 provides a summary of what happens along the lifecycle phases, looking at both the supply and the demand side of the market (Montagna and Cantamessa 2019).

4.3.3 Dominant Design and Vertical Integration

During the fluid phase of innovation, firms face the critical problem of deciding on the degree of vertical integration to adopt. Upstream vertical integration decisions have to do with the development and/or production of subsystems and components, while downstream vertical integration decisions are concerned with the role to be covered in the development of the 'supra-system' in which the product will operate, including its complements.[8]

The discussion can start from upstream vertical integration. In the case of products with a relatively low degree of modularity (i.e., components are significantly coupled to one another, so that the design of each strongly depends on the design of the other ones), it is likely to observe a high degree of vertical integration until the dominant design has emerged (Christensen et al. 2002). In fact, it is unlikely that other firms may enter the business of becoming a specialist supplier of an architecture-specific component, unless there is sufficient certainty that this architecture is indeed going to become the dominant design. Therefore, firms working on the final product will have to take up responsibility on component

[6] For instance, in the 1930s, the dominant design of passenger cars underwent a further evolution, with the adoption of unibody architectures (instead of the previous chassis + body) and of front-wheel drive technology.

[7] As an example of this case of strong path-dependency, one can think of LED lights, which require to be compatible to the sockets developed by Thomas Edison* (1847–1931) for light bulbs and fit in existing lamps (which were conceived for bulbs that had to be replaced and—incidentally—were shaped in a way that comes back from the era of candlesticks). By forcing LED lights to fit in a concentrated bulb-like shape, one forgoes much of their potential (e.g., think of the possibility of having flat panels with distributed lighting) and forces to introduce complex solutions for cooling the light-emitting elements. Only recently, manufacturers have started redesigning lighting systems from scratch, with non-replaceable LEDs distributed on a broad surface.

[8] For instance, a maker of an electric vehicle has to decide whether or not to integrate upstream and deal with the design and manufacturing of batteries, or simply source them from specialist suppliers. Similarly, it may decide to deal with recharging stations (downstream vertical integration with respect to a complementary good) or operate a service business for electric car sharing in urban areas (downstream vertical integration of the product supra-system).

Table 4.3 Implications of the emergence of a dominant design along the lifecycle phases

Lifecycle phases		Supply-side challenges			Demand-side challenges
		Supra-system	System	Subsystem	Customers
Fluid phase	Still immature technology, with low performance Many firms enter the fledgling industry, because of its future promise Each entering firm brings prior competencies and assets, which leads to a diversified range of technical solutions High rate of product innovation, but progress is limited, since efforts are not concentrated in the same direction	In many cases, the system cannot operate unless inserted in a coherent supra-system. It is therefore likely that the supra-system too must undergo radical change	The focal system undergoes *radical innovation*, which implies change in the underlying technical principles and in the architectures	In some cases, components and subsystems might be carried over from existing stock, but it is likely that they too must undergo radical change	Innovators and technology enthusiasts with very specific needs, and tolerant to shortcomings of the technology They can act as lead users and/or suggest technical improvements The technology often becomes overhyped
Transition phase	One product architecture emerges as the dominant design This event triggers the takeoff of product performance, market diffusion and shakeout of firms	Unless it has already been established, the freezing of a dominant design in the system enables the design and diffusion of a complementary and specific supra-system	Once the dominant design is established, innovation in the system will typically shift to *incremental* In some cases, there might be architectural innovations that 'optimize' the dominant design	Unless they have already been established, the freezing of a dominant design in the system enables the design and diffusion of complementary and specific subsystems	Early adopters who want to gain early experience with emerging (and possibly still immature) technology Word of mouth supports diffusion The hype cycle starts to deflate

(continued)

Table 4.3 (continued)

Lifecycle phases		Supply-side challenges			Demand-side challenges
		Supra-system	System	Subsystem	Customers
Specific phase	Firms compete on cost and quality while demand increases Economies of scale kick in, leading to process innovation (specific to the dominant design) and to oligopolistic competition	Due to path dependency, *incremental innovations* prevail	Due to path dependency, incremental innovations prevail At times, *modular innovations* that do not alter system architecture might be introduced	Subsystems might be subject to *radical innovation* when new technology emerges, as long as they can be easily integrated within the dominant design	Early majority and later customers require mature technology and adopt based on costs and benefits (eventually postponing adoption until they feel safe) Additional and replacement sales become prevalent with respect to adoption sales

development as well. The opposite may happen for highly modular products (i.e., products whose components are uncoupled from one another, so that the design of each does not significantly influence the design of the others). In this case, components may be developed and produced by independent companies, and simply assembled by the firms who are working on the final product.

A higher degree of vertical integration gives tighter control on components from both a technical and business perspective. However, it also requires a much greater amount of investment, which can make this choice impossible, unless the company is sufficiently well-funded. This strategic choice is always quite difficult and requires a deep understanding of the product being developed. In some cases, firms might try to develop a 'dominant design' out of off-the-shelf components, only to discover that the performance they achieve falls short of what the market requires. On the other side, other firms might undertake the difficult and costly road of vertical integration, simply to discover that other firms were able to operate successfully, and cost-efficiently, by using readily available components.[9]

When dealing with these issues, one can go beyond looking at specialization in its stricter sense (i.e., on who makes what), but also at the way knowledge is partitioned between producers and component makers, which can moderate vertical integration choices (Lee and Veloso 2008). In fact, it is possible to achieve results similar to vertical integration if the former develops a sufficient amount of component-level knowledge and the latter some product-level knowledge.

If one now looks at downstream vertical integration, the problem lies in the value that complementary goods and/or the supra-system provide to customers, therefore driving diffusion. If this value is significant, the issue cannot be neglected. However, a firm may decide not to have an active role, assuming that a market for these complementary goods will emerge independently. Alternatively, it may engage in partnerships with 'complementors' (i.e., firms producing complementary goods), or decide for full vertical integration and provide these goods. A higher degree of vertical integration requires the financial capability to sustain the related investment. At the same time, a lower degree of vertical integration will be possible only if the relationship between product and complementary assets is non-specific and if there is a clear business model allowing complementors to invest and then gain from their effort.[10]

[9] In Chap. 14 we will provide a deeper discussion on how all this depends on the type of performance being sought by the market. If the performance the market looks for can be associated to individual components, this implies a higher degree of modularity and a lower degree of vertical integration will be required. Conversely, if the market looks for performance that is related to system integration, this will require a lower degree of modularity and higher vertical integration. For instance, personal computers' and mountain bikes' performance are mostly associated respectively to microprocessors and gear shift systems. It is not surprising that these industries have been vertically disintegrated since the beginning with respect to these modules, and that leading component manufacturers (e.g., Intel and Shimano) have become more important in the value chain than the producers of the end item, to the point that their presence is certified and clearly communicated.

[10] More on this will be discussed in Chap. 7, when discussing multi-actor business models.

Therefore, firms must be well aware of the consequences of architectural design choices. Choosing a modular architecture implies defining a lower level of vertical upstream integration, looser links within the technological paradigm and non-specific relationships with complementary assets. In some cases, this leads to lower costs thanks to the use of standard components. Choosing an integrated architecture implies greater investments for higher levels of vertical upstream integration as well as for closer partnerships with complementors. Tackling a high degree of vertical integration both upstream and downstream may in principle provide the firm with a significant control over the evolution of the technology and of the paradigm. It is however apparent that this choice can not only be financially unfeasible, but also difficult to manage, because of the organizational complexity that it entails.

4.3.4 Dominant Designs in Process Industries and in Services

As previously mentioned, the model discussed above holds for assembled/discrete products. However, the model can also easily be adapted to the case of the continuous process industry and to services.[11] In these latter cases, the roles covered by product and process innovation are simply reversed with respect to what happens for discrete products. The fluid and the transition phase will be characterized by intense process innovation, during which firms experiment alternative technical solutions, and until a *dominant design for the process* emerges, thus leading into the specific phase. During the specific phase, firms will slow down the pace of process innovation and start working on product innovation (in the case of the service industry, this implies developing new services),[12] in the attempt of exploiting the previously developed processes and infrastructure.

In this case too, because of the dominant design-specific investments that have been sunk in the process and in related infrastructure, the dominant design will tend to lock-in and remain stable for years. This raises an important tradeoff related to the openness, flexibility and cost of processes. Knowing that infrastructure and processes will remain unchanged for years, if not for decades, decision-makers can on the one side decide to invest a higher amount of money in an open, powerful and flexible process, in view of the many potential products that might be developed and produced on it. However, this is a risky choice, since this variety of products might not occur, thus delivering disappointing financial returns. Conversely, firms may make a conservative investment, but risk that a limited infrastructure may not be able to accommodate future products with high demand. This would lead to

[11] No wonder the services have a similar behavior to continuous process industry. Services are in fact mainly composed of processes and are enabled, as continuous process industry, by technologies that act prevalently on them.

[12] Concerning the service industry, one can think of the many services, such as bills payments and cell phone recharging, that nowadays run on Automated Teller Machine (ATM) systems. These services have been added to the dominant design of ATM systems, which dates back from the'70s, when these new services were not even conceivable.

significant opportunity costs and to a clear disadvantage against competitors that have made more far-sighted choices.

It must be mentioned that—since observing a process is harder than disassembling a discrete product—studying dominant designs in the process and service industry is somewhat more difficult than for discrete products. In services, additionally, innovation is based on organizational processes, the inertia of which makes dominant designs in the service industry potentially more stable than in the product industry.[13]

4.4 The Limitations of Abernathy and Utterback's Model

The model proposed by Abernathy and Utterback is a key reference in the field of Innovation Management. However, many years have gone by since it was first proposed and, over this time, research has highlighted a number of issues that the model might not be able to fully explain.

One first weakness of the model lies in the assumption that, once a dominant design has emerged for the discrete product, efforts cast by the industry on process innovation will lead to production processes that enable greater efficiency and economies of scale. Since product technologies may have different degrees of manufacturability, it might occur that what seems to be the most promising dominant design might fail prematurely because, eventually, no manufacturing technology turns out to be satisfactory. In such cases, the transition phase of Abernathy–Utterback model may include one or more 'false' dominant designs and therefore take a longer time than expected.[14] The previously discussed bandwagon effect can also exacerbate this phenomenon, because participants to the bandwagon will prematurely abandon the exploration of technological alternatives (Greve and Seidel 2015).

A second issue with the model is associated to the fact that it assumes that product architectures are characterized by a significant degree of rigidity. The current shift to digitization is leading industry to work on highly modular and flexible architectures, made up of components that are functionally independent, programmable and that can interoperate through well-specified and non-specific interfaces (Hermann et al. 2016). As a consequence, digital technologies allow firms to widely differentiate their product offering through modularity and by combining a selection of functional elements with limited integration effort.

[13] Think of the postal service. Born in its architectural conception in the nineteenth century, and focused more on correspondence than packages, it remained stable until the advent of the electronic mail and e-commerce. In these years the focus has shifted to package delivery, though traditional postal services find it difficult to modify their architecture in the direction of decentralized points for the collection and distribution of packages.

[14] For instance, in recent years, many promising battery designs, both for mobile and utility scale applications, have failed because of the difficulty of 'scaling up' manufacturing operations from pilot-level facilities to mass production.

Therefore, the economies of scale and the organizational learning that used to lead to a dominant design might not be found at the level of the product any more but, rather, at the level of their components. In other terms, if the world is making up of Lego bricks, the dominant design lies in the brick and not any more in the projects you can build out of them.[15]

A third issue is related to the emergence of product–services. Given that Abernathy–Utterback model views assembled products and services as two somewhat opposing worlds, it does not provide a clear interpretative tool for the case of product–services. Research on this subject provides just a few contributions (Cusumano et al. 2006), which suggest that service innovation constitutes a third 'wave', following product and then process innovation. One possible hypothesis is that, in a product and service bundle, one of the two will prevail and determine the dominant design and the related lock-in effects, while the other will act as a complementary commodity. Of course, it is not trivial to understand which of the two elements will have a central and which will have an ancillary role, but interesting clues may be provided by examining where aspects such as standardization (see the following section in this chapter), modularity, economies of scale and learning might have a greater role.[16]

A final weakness of Abernathy–Utterback model lies in the fact that it focuses mainly on the transition phase of the technology lifecycle, while the fluid phase is left somewhat unexplored. A significant amount of recent research has been recently dedicated to this phase and will be discussed in the following section.

4.5 The Incubation Period

Academic research has recently started studying the incubation period of a technology. Moeen and Agarwal (2017), Agarwal et al. (2017), Moeen (2017) have proposed a model in which incubation follows a 'trigger' and eventually ends up in commercialization. The authors specifically identify three types of triggers (i.e.,

[15] To be more precise, modularity and programmability may push in two different directions. While modularity abates the role of the dominant design, programmability may lead to a dominant design that best suits the future possibility of altering the functions and/or the performances through this avenue. This has become quite apparent during the transition from traditional feature phones to smartphones. Nokia adopted a strategy based on modularity, offering a variety of differentiated mobile phones, which combined different modules as a function of the target customer (e.g., business smartphones had a physical keyboard and a smaller screen, while gaming smartphones had a larger screen and accelerometers). Conversely, Apple adopted a strategy based on programmability, placing all the possible functional components in its products, so that the user could select which functions were needed later on, via the apps that he/she would decide to install).

[16] For instance, in the industry dealing with shared mobility via electric scooters, one may wonder whether one should view the scooter as a mere component of a service and focus on the dominant design that lies at the level of the service architecture. Conversely, one may view the scooter as the element around which the dominant design emerges and the rest of the service architecture as a further wave of innovation that integrates it and allows to bring it to the market.

Table 4.4 Triggers and their influences in the incubation phase

Trigger	Incubation (pure investment)		Commercialization
Type of trigger	Key actors	Technological actions	Market actions
Scientific discovery (e.g., GMOs)	Academia and Research organizations Firms	Developing core and complementary technologies in absolute terms and for the 'right' applications	Finding the 'right' applications Building identity and acceptance
Unmet user needs (e.g., dishwashers)	User inventors, Entrepreneurs, Communities	Moving from prototype to product	Fine-tuning the user requirements
Mission-oriented 'grand challenges' (e.g., antibiotics)	Government agencies, Nonprofit organizations, Academia and Research organizations, Firms	Moving from prototype to product Developing core and complementary technologies	Moving from 'generic' to 'commercial' need Finding new applications

scientific discovery, unmet user needs, mission-oriented 'grand challenges') that strongly affect the dynamics of the incubation phase. Table 4.4 provides a summary of such influences.

In all three cases, the incubation phase will see a flurry of entry and exit activities in the nascent industry, along with mergers and acquisitions among firms and a constant reshuffling of the knowledge that each player contributes. What still remains elusive, however, is the mechanism that leads the industry to adopt a given variant of the technology, which will become the basis for the future dominant design.

A few researchers (e.g., Roy et al. 2019) hypothesize that this mainly occurs prior to commercialization, when the many firms operating in the industry experiment on variants of the technology learn from each other, until one 'best' solution is 'frozen', so that diffusion will start only at that point in time. However, historical evidence also points to another possible phenomenon, in which public awareness of the technology is quite strong, commercialization has already started (though with very small volumes) and the selection of the technology occurs thanks to the interplay between industry players and initial customers.

In this process, the new technology easily suffers from *hyperinflated expectations*, or simply *hype*, which can drive waves of investment. A well-known representation of this phenomenon is proposed by the Gartner consultancy, for technology related to ICT, in the form of *hype cycles* (Fig. 4.4).

Hype cycles show the current positioning of technologies in the incubation phase, according to five quite imaginatively labeled subphases, along with a forecast of their 'time to maturity'. According to this representation, soon after a technology appears (*technology trigger*), markets will tend to 'fall in love' with it, and expectations will grow until they become unrealistic, from the two perspectives of absolute performance and of breadth of potential applications (*peak of inflated expectations*). When these expectations eventually fail to materialize, the market

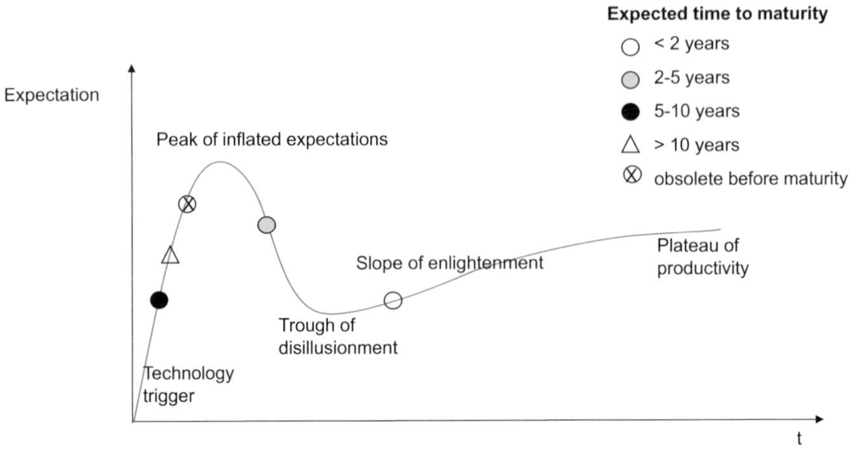

Fig. 4.4 A simplified representation of a hype cycle

will quickly lose interest in the technology, which will tend to be dismissed as a failure (*trough of disillusionment*). At this point, companies who persist in working on the technology will invest to identify the specific applications in which the technology might really create value, and by directing technical developments in order to address the corresponding customer needs (*slope of enlightment*). At this point, the technology can find a solid foothold in the market,[17] diffusion can start and, eventually, this will trigger economies of scale that may lead to a broader diffusion also in other applications areas (*plateau of productivity*).

The incubation phase (and, more specifically, the slope of enlightenment) is therefore very important for technological progress. It allows companies to improve technology in absolute terms, to find applications that are reasonable from a technical perspective and that might lead to a profitable market exploitation, to steer technology toward such realistic applications and to raise awareness and trust in customers.

The main strategic objective in this phase is to avoid the delusion that the new technology "is great, because you can apply it to about any industry" while, in reality, it may be more like "a solution in search of a problem to solve". While the technology might indeed have a broad potential in the long term, it will not be of any economic value, unless firms are able to define a set of specific applications and successfully tackle them one by one.[18] This process includes not only

[17] In this phase, it is crucial for innovators to target a significant market share in a small and well-defined market niche than to look for a small market share of a broad and weakly defined market.

[18] For instance, Radio-Frequency Identification (RFID) tags are widely used in supply chains and lead to significant advantages in the traceability of goods of appreciable value (or multi-item packages of goods of lesser value). However, their potential had been overstated around the year 2000, due to unreasonable expectations on the possibility of manufacturing them at minimal cost, which could have allowed placing a tag on the packaging of goods of any value, albeit small.

selecting the applications where the technology can provide significant economic value, but also the subvariants of the technology that are most suitable for the same.[19]

4.6 Reverse Innovation

Traditional models of the dynamics of innovation usually neglect the role of geography, under the tacit hypothesis that innovation occurs in the country or region that is most advanced and—eventually—the technology will be transferred to less advanced regions. This perspective has been challenged in recent years, given the way with which globalization has evolved and the new role of less advanced countries in developing and exporting—not only passively receiving—innovation. This phenomenon is usually called '*reverse innovation*' (Govindarajan and Ramamurti 2011). In essence, the idea is that original innovations can occur in developing countries because of local technological talent and of the specific needs found in the local market, and that these innovations can be of interest to developed markets as well.

Specifically, there are two main avenues by which this occurs. On the one side, markets in emerging countries may allow the development of products characterized by specific features (e.g., low cost, ruggedness, lack of worthless features) that are appealing also to specific segments in developed countries.[20] On the other side, and probably the more interesting phenomenon, reverse innovation can occur when the developing country actually pioneers the development and diffusion of a radical innovation because the predecessor technology simply has not had enough diffusion and cannot therefore generate any lock-in, neither on the side of supply nor on the side of demand. This phenomenon is often called *technology leapfrogging** and might actually lead to a significant technological lead for the emerging country. Technology leapfrogging can also benefit if the regulators in emerging countries—either because of lack of resources, or even intentionally—play a lighter hand on the emerging technology, than their colleagues in developed countries.[21]

[19] For example, the emergence of the internal combustion engine in motor vehicles did not only have to do with the overcoming of electric and steam engines but also implied a long selection process among the many variants of the technology, with respect to number of strokes, types of pistons, fuels, etc. Similarly, when thinking of BEVs today, many alternative solutions exist with respect to battery chemistry and to the way batteries can be charged (e.g., through recharging stations that operate at many possible speeds, induction charging, battery swap, etc.).

[20] For instance, poorer segments of the European population may be quite happy to buy a low-cost and rugged car that was originally developed for middle-class customers in emerging countries.

In another well-known example, doctors working as general practitioners in the West were reluctant to buy complex and costly ultrasound scanners, but were happy to buy the cheaper versions that had been conceived for basic screening in emerging countries.

[21] There are many interesting examples of this phenomenon. For instance, digital payments have diffused very rapidly in African countries, where the majority of people do not have access to banks and credit cards, but do own a mobile phone.

4.7 The Many Types of Innovation, Beyond Performance

To this point, we have considered product performance, as represented by s-curves, as a proxy to innovation. If one looks at innovation with a broader perspective, it is evident that innovation can touch many aspects other than performance as well. A number of classifications and taxonomies have been proposed over the years in order to provide this wider view of innovation. What will be proposed in the following, inspired by Keeley et al. (1999) and by Keeley et al. (2013), ought to be taken as one possible proposal, but by no means as a universally accepted classification. Among the additional types of innovation that can therefore be considered, one can mention:

- **Product systems**. The value of a product can be enhanced not only by increasing its performance but also by adding functions. In a number of instances, these functions do not have to be incorporated into the core product, but can be assigned to complementary goods that lead to the offering of a more complex 'product system'. When dealing with a product system, the focal company (i.e., the firm developing the core product) will have a tradeoff to solve when deciding on the openness of this offering (Boudreau 2010). By opting for a closed system, the focal firm will integrate horizontally in order to provide these goods and will therefore appropriate all returns. However, it will also have to make significant investments, including ones in fields that might not fit with its core competencies. Conversely, by opting for an open system, allowing the entry of complementors and the development of what is often called a *corporate ecosystem*, the focal firm will appropriate only part of returns. This might become an attractive option since the overall returns might be significantly higher (i.e., a slice of a very large cake can be bigger than an entire cake of smaller size) and the required investment may decrease.[22] Product systems may be considered also from the perspective of potential complementors. In this case, the strategic choice is related to deciding toward which of the competing ecosystems it should direct its offering. The problem can be quite critical if the investment has to be specific, because of either technical or contractual reasons. In this case, it will have to opt for the ecosystem that is expected to provide the highest overall returns combining expected margins granted to the complementor and volume.

Similarly, automated drone delivery has been pioneered in countries such as Rwanda, thanks to the lack of road infrastructure, the somewhat reasonable risks that are due to low population density, and to the local government's idea of using this as a form of industrial policy.

[22] For instance, one can think of the many accessories that go with smartphones and tablets, from docking stations to in-car holders to travel pouches. A consumer electronics producer can benefit by allowing other specialist firms to produce these accessories for its devices, because they add value and credibility to the core product, thus increasing demand, in a way that the core producer could not match with its own competencies and resources. Consumers would not buy a 1000 € smartphone cover sold by the OEM, but might gladly spend this same money if it comes from a brand that is well known as a luxury leather goods maker.

- **Services**. Services currently make up the bulk of economic value added in developed countries. A number of manufacturing industries have progressively shifted their attention from the simple sale of goods to the provision of related services. In many of these contractual agreements, the producer must manage the product during its lifecycle and is paid on the basis of its actual usage. As examples, photocopiers, passenger cars and jet engines are ever more leased or rented both to businesses and to consumers. Similarly, telecommunications companies nowadays provide bundles that include connectivity, devices and content. From the side of customers, the proposition is attractive because fixed investment costs can be transformed into variable operational costs. From the supply side, producers providing a product–service bundle enter a downstream phase of the value chain in which it might be easier to achieve higher and more stable margins, together with customer lock-in. This trend toward *servitization** does not only imply an innovation in business practices, but also a substantial degree of technological innovation. In fact, when shifting from the sale of a good to the provision of a service, the entire product lifecycle (including maintenance, repair and disposal) has to be internalized by the producer. The producer will consequently see lifecycle events, such as product failure and the subsequent maintenance activities as a cost, and not any longer as a source of revenue. Therefore, products will have to be redesigned together with a number of complementary systems (e.g., onboard diagnostics, the equipment to be used by field engineers, etc.) so that this lifecycle cost is minimized.

  Servitization can be designed in a number of forms, which differ in the purpose of the service components and on the extent to which the manufacturing company intends to change its business model. Furthermore, as proposed by Cusumano and Suarez (2015), these strategic intents can cover different roles according to the position in the technological lifecycle. Table 4.5 provides a summary representation of the different approaches possible.

- **Channels**. Goods can be sold through different channels, and alternative choices can lead to significant innovations. For instance, one can contrast the sale of computers through a traditional Make-To-Stock (MTS) distribution system, with direct Assemble-To-Order (ATO) sales to customers through an online system. Innovations of this kind should not be superficially considered as being limited to the business idea. Shifting from MTS to ATO can entail significant innovation at product level, for instance requiring a completely different product architecture that might enable the producer to add or withdraw functions and components without too much effort and with low lead time. In other instances, the shift might require significant innovation at process level (e.g., developing a complex online configurator that supports the sales process, or redesigning the production equipment and the factory layout in order to make it more responsive to individual orders).
- **Brands**. Marketing experts define a *brand** as a "*[...] feature that identifies one seller's good or service as distinct from those of other sellers*" (American Marketing Association dictionary). This distinctiveness includes technical

Table 4.5 Different approaches to servitization

Type	The service complements product purchase		The service substitutes product purchase
	Smoothing	Adapting	Substituting
Definition	The service does not significantly alter product functionality (e.g., insurance, technical support, EV charging stations)	The service significantly expands or alters product functionality (e.g., customization of solutions, consulting)	X as a service (e.g., SaaS, 'pay by the hour', etc.)
Role in fluid phase	Provide services needed to solve 'chicken and egg problems'	Educate the market to introduce the innovation	Reduce risks of ownership
Role in transitional phase	Capture value from the 'profit pool'	–	–
Role in specific phase	Capture value	Capture value, educate the market to new sophisticated functions	Capture value, stimulate replacements

features and their evolution, which determines an obvious—but sometimes neglected—relationship with innovation. The role of brands in innovation is very important, because customers are usually unable to fully understand the complex technical features and tradeoffs that define a product and its performance, and to make rational choices among them. A brand therefore becomes a summarized and tacit description of these features and tradeoffs, that Marketing can readily convey to customers. The managerial implication of this concept is that the values inherent to the brand and the objectives toward which the producer directs its innovation strategy must be coherent. Not doing so will result in improvements that the target customer will not be able to appreciate or, worse still, will end up with a confused brand identity.[23] The coherence between brand and technical content is not only true for incremental improvements, but also for radical innovation which customers will find even more difficult to conceptualize, and which will require a solid anchoring of the message conveyed by a brand. Producers will therefore have to decide whether a radical innovation requires the launch of a new brand (which clearly requires a significant effort), or

[23] From the side of demand, a customer that is choosing which car to buy will not normally study raw technical data on rigidity of suspensions, interior noise at different engine speeds and power and torque curves. It is more likely he/she will look for a brand that encapsulates the right values in the tradeoff between an 'elegant and comfortable' and a 'sporty' car. From the side of supply, a producer of a 'sporty' car brand will direct its innovation efforts in order to enhance the features that are tied to its brand (e.g., acceleration, speed, braking performance, etc.) and will accept some tradeoffs in other ones (e.g., comfort, size of trunk, etc.).

whether it can be marketed as an extension to an existing one (provided there if a good fit between customers' perception of the innovation and of the brand).

- **Customer engagement and experience.** The trend toward intangible goods that was discussed when talking about service innovation can be brought to an extreme, when the value that customers perceive in the product or service is highly intangible and emotional. Amusement parks, vacation resorts, video-games, but also automobiles and motorbikes and high-end food become special kinds of products and services in which functionality and technical performance do not matter *per se*, but in connection to—and as enablers of—customer experience. In this context, a significant role is covered by aspects pertaining to taste, culture, and meaning (in this case, one can also speak of *design-driven innovations*, Verganti 2009). Innovation is therefore involved with defining new or improved forms of customer experience and in defining the technical elements that allow it. In some cases, the value of these products may be associated to an apparently obsolete technology (e.g., mechanical watches, instant photography), and producers' efforts must be aimed at improving the technology and/or blending it with components coming from contemporary technology.
- **Process**. Process innovation has been defined in previous chapters of this book. Generally, process innovation occurs in a way that is not apparent to customers, who can only observe the outcome, in terms of lower costs and prices and/or improved products or service levels. Process innovation can entail the adoption of new technology and/or changes in the organization and in business processes. Moreover, process innovation can be related to both core and enabling processes of the firm.
- **Network and inter-firm relationships**. Innovation in this case has to do with a change in the relationships that the firm establishes with its customers, suppliers or complementors. In economic terms, this aspect of innovation has to do with vertical and horizontal integration choices. In some cases, these may be aimed at increasing efficiency in the production of ordinary goods and services. In other cases, changing inter-firm networks might make it possible to develop and produce innovative products and services that a single firm would not be able to provide by itself.
- **Business models**. Business model innovation can be viewed as a superset of the previously discussed forms of innovation. A company that pursues business model innovation is also likely to change the cost structure and the revenue streams that normally characterize the firms operating in that industry. A typical example is the innovation introduced by low-cost airlines. Practical methods for describing and designing business models and related innovations will be introduced and discussed in depth in the following Chap. 7.

If one considers the many forms of innovation discussed above and their deep impact, it is clear that the problem of managing innovation is not limited to following a sequence of s-curves, with an eye on a single performance indicator. Due to the many dimensions of innovation and strategic options involved, firms can maneuver in the innovation landscape in many different ways. For instance, disruptive innovation—as defined in the previous chapter—might occur because of a process innovation, but without actually causing paradigm shift and moving to a new technological trajectory.[24] Given the variety of aspects with which innovation may be introduced and their potential competitive impact, firms ought to constantly be on the lookout for opportunities as well as for the threat coming from innovating competitors. As stated by Nunes and Breen (2011), this requires a continuous adaptation of operational and managerial competencies, a process that requires quite long time constants. Therefore, this adaptation must be managed proactively, knowing that it is generally too late to react to radical innovation when its impact is becoming apparent on the performance of the firm.

4.8 Standards and the Dynamics of Innovation

The previous discussion has purposely tackled the topic of dynamics of innovation in general terms, without yet dealing with standards. The topic is very important when studying innovation, since standards have the potential of determining the way with which a product diffuses in the market, as well as industry structure.

Standards are generally important when communication or interaction between systems (components, processes, actors, etc.) occur, since they state the set of specifications, in terms of formats, metrics, rules, procedures, etc. that must be uniform and coherent to allow such communication and interaction. Therefore, it is not surprising that standards are particularly important in those industries where communication and interaction are relevant, such as for instance in IT. Many leading companies in these industries may simply be recognized as the ones who understood the significance of standards better than their competitors and crafted a cleverer strategy around them. In other industries, standards appear to have a minor role. However, this may simply be due to the fact that their relevance has not been fully recognized, or that IT components have not fully emerged. In that case, future leaders in these industries will simply be the first to have successfully played the 'standards game'.

[24] For instance, in the DVD rental industry, Netflix disrupted the pioneer and market leader Blockbuster thanks to a mixture of radical channel, process and business model innovations (i.e., Netflix used the postal system to deliver DVDs that were chosen by customers on the Internet on the basis of a monthly subscription, while Blockbuster followed the traditional model of having customers come to a store and pay on a 'per night' rental basis). Though radical (from a technical perspective) and disruptive (given the impact on the industry), these innovations did not lead to a paradigm shift. The paradigm shift started later, when online streaming technology and related subscription services started challenging the very idea of using physical DVDs as a medium for enjoying audiovisual content.

4.8.1 Defining a Standard

A *standard** can be defined as "*a set of specifications that provide value to the product because of its conformity to the standard*". The definition might sound a mixture between a tongue twister and a self-fulfilling prophecy and therefore needs some explanation.

The main idea behind the standard is that a product based on these specifications achieves greater value because of conformity, and not because these specifications define a better product from a technical perspective. In order to understand if a specification may be conducive to a standard, one can apply counterfactual thinking and ask "if product X did not follow standard Y, would it be of lesser economic value? If so, would it be because of non-conformity or because of lesser performance?" A standard is (or could be) present if the answer to this question would be "yes, it would be inferior, exactly because of non-conformity". A standard may actually define a product that is technically inferior to a non-standard one, yet of greater economic value because of conformity.[25]

This apparently odd phenomenon of value creation through conformity can be caused by one or more of the following mechanisms:

- **positive network externalities**. Compliance to specifications enables the creation of a network linking customers to one another. Every time someone adopts a product that is compliant to a standard, this creates a positive externality on the entire stock of existing users, since each of them can now link to the new user.[26] So, the greater the number of users adopting a compliant good, the greater the number of potential links, and the greater becomes the economic value of the good to each user. Conversely, a non-compliant good would not fit in the network and would therefore have very limited economic value.
- **complementarity with other goods**. Specifications can create economic value by allowing interoperability between complementary goods, such as software applications that run on operating systems, DVDs on disk players and gasoline formulations that can be used to refuel motor cars. A non-compliant good would not be able to interoperate with the complementary good and economic value would therefore be destroyed.
- **modularity**. As previously mentioned, modular product architecture is characterized by subsystems that can be developed independently from one another, thanks to standardized specifications that define their interfaces and allow interoperability between components. As it will be amply discussed in Chap. 16,

[25] A well-known historical example is the VHS* format that—though technically inferior to the Betamax* format—managed to become the standard for domestic videotapes.

[26] These network externalities arise in cases such as computer network protocols, data exchange formats used in electronic documents and e-mails. The value V of a network with n nodes is given by the number of potential links that are enabled, or $V(n) = n(n-1)/2$. When the $(n+1)$-th user joins the network, the value grows to $V(n+1) = (n+1)n/2$, with a relative increase in value given by $V(n+1)/V(n) = (n+1)/(n-1)$.

modularity, such as in the case of personal computer hardware components, creates economic value by enabling economies of scale and learning, and by allowing users to customize the products they buy.

- **specific learning**. This aspect is related to user interfaces, such as the icon-based interface that characterizes commonly used personal computer operating systems, or the layout of computer keyboards. A standardized user interface creates economic value, since users will not have to spend time in learning how to use the product and be immediately proficient with it.
- **economies of scale**. For very simple products, such as screws and pipe joints, standardization enables the use of highly efficient production technology that could not be used if each individual product had to be designed according to a specific application (e.g., standardized screws can be forged, instead of turned on ca lathe).

A standard has been defined 'as set of specifications' and does not necessarily represent the product involved in its entirety. A standard is therefore a different concept to a dominant design, which represents the architecture of the product. However, the two concepts are linked and Table 4.6 summarizes the possible relationships between dominant designs and standards.

In some cases, a standard can define product architecture and will therefore overlap with a dominant design. One can think of Blu-ray DVD players, where the standard dictates both the specifications and the technical architecture of the disk and player.

In other cases, the standard may be distinct from (but possibly based on) a dominant design. For instance, in current passenger cars, the pedals controlling clutch, brakes and throttle are laid out left to right in this same order. This is clearly a standard, because a car with a mixed-up pedal layout would be quite dangerous to drive and therefore of low value. However, the layout of pedals is a specification and not an architectural description and cannot therefore be defined as a dominant design. What is part of the current dominant design of passenger cars is the existence of these three controls and the choice of controlling them with pedals, but this is independent of their layout.

Finally, dominant designs may exist independently of standards. Most passenger cars are currently based on the unibody* architecture, with the steel body providing multiple functions such as aesthetics, rigidity, aerodynamic performance and protection to occupants. This is a dominant design, but not a standard, since an

Table 4.6 Relating dominant designs to standards

	Is a dominant design	Is not necessarily a dominant design
Is a standard	Blu-ray DVDs	Specifications of the Edison screw The specific layout of the keys on the keyboard
Is not necessarily a standard	Monocoque automobiles Writing on a PC via a keyboard	–

alternative architecture (e.g., based on a chassis such as in commercial vehicles, or on a space frame like in sports cars) can be better or worse, but only because of their intrinsic performance. It would not be possible to state that an alternative design is worse "because it is different from the usual unibody".

4.8.2 How Do Standards Arise?

The previous discussion has presented five economic mechanisms through which a standard can generate economic value. A standard with a high market share will give rise to a large network, attract a higher number of suppliers of complementary products and components and lead to greater economies of scale and learning. Regardless of which of these mechanisms are in place, the amount of economic value essentially depends on the size of the customer base. Therefore, a standard with significant market share will generate greater economic value than a technically superior standard with low market share, and the maximum possible value will come from a standard that completely dominates its market. As a consequence, if multiple alternative standards emerge, a selection process will start, in which weaker standards will progressively fail until very few (at the limit, a single one) emerge as winners (i.e., a *winner-take-all* competition).

When the winning standard will have emerged, the market will obtain a significant amount of economic value, but will also be *locked-in*, since switching to a new standard will become very difficult and costly. The economic value created by the dominating standard will be so high, that only a very significant performance increase would convince customers to switch to a new standard.[27]

In this winner-take-all environment, producers must decide whether to *agree* on a common standard, or compete against each other in a *standards war* and try to impose their own solutions as a *proprietary* and *de-facto* standard, with the potential of becoming monopolists with an attractive long-term perspective, due to lock-in. The decision is a highly strategic one for the companies involved (Ullmann and Margalit 1977; Shapiro and Varian 1999; McNichol 2006).

By starting a *standards war*, competitors engage in a very risky path. They try to become monopolists, which would of course provide quite attractive returns. At the same time, they bear the risk of being excluded from the market, should their standard lose. Moreover, a standards war is usually quite costly from a financial perspective. As we will discuss in the next subsection, winning a standards war requires specific strategies that significantly reduce profitability, at least until the war is ended. In addition, when a standards war starts, customers usually hesitate in buying the new product, because they want to avoid choosing the 'wrong' one. By

[27] An often-cited example is related to the QWERTY* standard for typewriter (and now computer) keyboard layouts. The standard had been designed around 1870 in order to minimize the likelihood of typebars jamming. When improved designs for typewriters made obsolete this requirement, other improved and faster layouts were not able to unseat the QWERTY standard, because typists' ability with the standard keyboard created lock-in.

deferring adoption until the situation is clearer, diffusion slows down, and this too contributes to making the financial return less attractive for producers.

Conversely, in case of *agreement* (for instance, by carrying out active cooperation in the relevant Technical Committee of a standardization body such as ISO*), competitors follow a low-risk path. By avoiding the standards war, they do not risk being excluded from the market, but at the same time they give up the opportunity of becoming monopolists. The agreed standard will in principle become a common playing ground for all players in the market, and the underlying technology will not provide competitive advantage.[28] Finally, the agreed standard will have a faster diffusion in the market, and the financial return will therefore be higher for the industry as a whole.

The choice between agreement and competition between standards depends on the risk aversion of the players, industry structure and on the existence of mechanisms and organizations that may support the process of finding an agreement. An industry with a high number of recognized players of comparable strength will be quite likely to look for agreements, since the risk and cost of starting a standards war will be perceived to be much greater than the expected monopolistic returns. Conversely, an industry with very few competitors, or undergoing continuous entry of new players with different strengths, might find it difficult to find an agreement. The technological nature of the standard may also matter. If the alternative standards are based on the same technology (i.e., think of the shape and pin layout in a connector for electronic circuits), it will be easy to find some kind of agreement. Conversely, if the competing standards are based on different technologies and each player has made a specific investment on one such technology, agreeing becomes close to impossible. Any agreement would in fact imply choosing one technology over the other ones, thus favoring one company and placing the others at a disadvantage.[29]

Policymakers usually follow standardization processes with great attention, especially when industry players lean toward starting a standards war. From a societal perspective, standards wars are not generally desirable. They are costly, since losing companies waste a significant amount of financial resources. They slow down the diffusion process and therefore reduce the utility that society can draw from a quick adoption of the innovation. They have the potential of leading to entrenched monopolies. Moreover, a government may feel the risk that the firm winning the standards war might come from abroad, which implies the demise of local industry players and a monopolistic position that national antitrust or tax authorities will manage with some difficulty. Finally, the risks and returns inherent

[28] In practice, market share will depend on the capability of each firm of developing attractive products based on the agreed standard. So, market share might be higher for the companies that managed to steer the standard closer to their technological base and to their competencies.

[29] This was the case of the standards war between Sony's Blu-ray and Toshiba's HD-DVD technology. The technological choices behind the two were completely different and it would have been irrational for the companies to agree on any one. For Sony, agreeing over HD-DVD would have implied writing off the very significant investment (a sunk cost!) made to develop Blu-ray technology. For Toshiba, agreeing over Blu-ray would have required costly licensing of Sony's intellectual property.

to a standards war create a huge incentive for companies to launch products based on proprietary standards as soon as possible, in the quest to be the first player to achieve critical mass and therefore make it easier to attract subsequent customers. Firms will therefore tend to launch products even if their technology is not yet completely mature. If this strategy works, the market will be locked-in an immature technology and unable to switch to an improved one.

Therefore, when spontaneous agreement by industry players is not possible and if policymakers are wary of the risks of de-facto standardization following a standards war, standards may be imposed *de iure* (i.e., by law). In this case, the public authority sets the standard—usually by asking for some form of technical cooperation with the industry—and finds a way to inhibit the diffusion of the product until the standard is defined. This process usually requires some kind of authorization procedure with which the government may effectively force producers to wait before launching (e.g., telecommunication standards require a license for using a public good such as the electromagnetic spectrum, vehicles must be licensed for roadworthiness, etc.).

4.8.3 Strategies for Imposing Proprietary Standards

When a company is involved in a standards war, making its proprietary standard emerge as a winner requires a very specific approach to product strategy. Being a winner-take-all process, strategic mistakes can cost very dearly and the firm must therefore not only make the right strategic choices, but also execute them correctly.

A list of possible strategies that have proved themselves successful is provided in the following. Among them, the first strategy is to be the first firm to achieve a critical mass of users, so that future customers will find greater utility in adopting the leading proprietary standard over competing ones, thus initiating a positive feedback mechanism. Ways to achieve this include:

- **Accelerating time to market** and being the first firm to launch a product, even if some further developments might lead to better performance.
- Using a **penetration pricing***[30] strategy, which may require selling the product below marginal cost or even giving it away for free, with the perspective of

[30] Penetration pricing means selling a product at a low price in order to stimulate demand, and increasing it when it becomes possible to do so. A penetration pricing strategy can be made explicit, i.e., through publicly revealed prices, or implicit, i.e., by informally granting deep discounts, or even allowing free giveaways (for instance, a software company may informally allow some degree of piracy in order to increase diffusion of its product). Penetration pricing runs counter to a *price skimming** strategy, with which a company would start with a high price and progressively lower it. Price skimming is a very popular strategy, since it allows achieving high margins thanks to inter-temporal price discrimination (i.e., the product is sold at a high price to customers with high willingness-to-pay and progressively to customers with lower reservation prices). Moreover, it allows companies to keep a constant margin, since prices will follow declining production costs. However, a price skimming strategy could be suicidal for a company trying to impose its proprietary standard.

recouping losses thanks to future economies of learning, and after a monopolistic condition has been reached.

- Using a market segmentation and **price-discrimination strategy***,[31] so that all market segments are induced into adopting the firm's products from the very beginning of the diffusion curve.
- **Licensing** the product to competitors. This way, the firm can increase supply of its proprietary standard without having to invest in the necessary assets (e.g., production equipment, salesforce, etc.) and can steer licensees away from pursuing their own—or other competitors'—standards. By following this strategy, the licensor forfeits being a monopolist for as long as the licensing agreements are valid. However, it might choose to stop renewing them when the standards war has been eventually won and licensees' help is no longer needed, thus becoming a monopolist.[32]
- Exploiting the so-called **'word-of-mouth' effect** between adopters and potential adopters through pressed communication actions, thus stimulating *viral* or *epidemic* diffusion in the market.[33]
- If the standard is based on complementarity, it lends itself to generating a *two-sided market**.[34] In this case, the firm can make sure that diffusion progresses evenly on both sides, so that both types of customers are attracted to the standard thanks to the number of customers of the other type. Similarly, a firm may support the provision of complementary goods either by itself or by relying on complementors.

A second commonly used strategy is to try creating a high market expectation that the proprietary standard will win, so that customers will be drawn in adopting it *in lieu* of competing ones. Companies having an established reputation as market leaders and tough competitors have the upper hand in this kind of strategy. However,

[31] A price-discrimination strategy implies providing different versions of the same product, targeted at corresponding customer segments, starting from a low-cost entry-level version with reduced performance, up to highly priced high-end versions with enhanced performance. In the software, Internet and media industries, this is often referred to as a *freemium** strategy. In such cases, a basic product is given away for free in order to promote diffusion, but customers who need additional features and have a higher willingness-to-pay must pay a fee in order to use the 'premium' versions.

[32] Antitrust authorities are nowadays quite wary of this kind of behavior. In general, they do require that the technology behind proprietary standards is made available to competitors on reasonable terms, that licenses are not opportunistically withdrawn, and that renewal is not refused at expiry. However, a licensor may deceive this requirement by making small upgrades to its technology, in a way that new agreements are required. Negotiations could then be artfully slowed down, and it would not be easy to prove that the licensor is 'dragging its feet' and is not sincerely trying to settle a complex deal.

[33] The word-of-mouth phenomenon will be covered analytically in Chap. 13.

[34] For instance, a smartphone operating system creates a two-sided market between customers and application developers. By increasing the number of the former, it becomes more attractive for the latter to create applications for that specific operating system. Conversely, the availability of many applications makes the operating system more attractive for customers.

other companies can attempt it too, by crafting a careful communication strategy and working on key influencers on customer choice (e.g., the media, consultancy companies, salespeople, etc.). In fact, standards wars are often won by the firms who manage to bring the players who operate closer to customers to their side.

A third strategy is based on creating an irreversible commitment to a specific technology and standard, based on a significant specific investment that therefore becomes a sunk cost. This action implicitly signals to competitors the readiness of starting a price war, should a competing technology arise.[35] The threat may be effective even if the company simply announces the investment. Of course, the announcement must be followed by facts since—failing to do so—the company will lose its credibility and will also not be able to repeat the strategy again.

A fourth strategy has to do with working on the lock-in effect. A firm trying to entrench its standard will try to increase customer lock-in by supporting the purchasing of specific complementary goods and helping them become proficient users of the standard. Conversely, a firm trying to fight an established standard will have to decrease customers' switching costs. This can be done with actions such as ensuring compatibility with the old standard, offering discounts aimed at *competitive migrations**.

4.9 Timing of Entry and Firm-Mover Advantage

The previous discussion allows us to shed some initial light in what is popularly known as *first-mover advantage**. This is the competitive advantage that a firm may enjoy by making an early entry in a market. Among the benefits of early entry one can consider:

- **Technological leadership** gained by developing a product and its underlying technology, since the very beginning. This may be particularly interesting during diffusion to innovators, due to the technical dialogue that may occur with them. Technological leadership may also enable firms to protect their inventions with *patents*, which act as barriers to entry against follower firms.[36]
- The possibility of securing **scarce resources**. Examples of such resources may be raw materials, suppliers with specific competencies, key personnel, etc. This scarcity may be due to an actual dearth of such resources and may be enhanced

[35] Suppose company X makes a non-recoverable investment I in developing a specific technology, with the expectation of producing a volume V at a variable cost VC. In normal conditions, X would ask a price $P > VC + I/V$ in order to recoup the investment. However, since the investment is a sunk cost, should a competitor Y try to enter the market, X would find it rational to start a price war and lower the price to $P > VC$. In turn, by knowing this would be a rational behavior for X, it would not be rational for Y to make a similar subsequent investment, knowing that it would never be recovered. This behavior is similar to Hernando Cortez's military strategy of 'burning boats on the beach'.

[36] This topic will be covered in depth in Chap. 6.

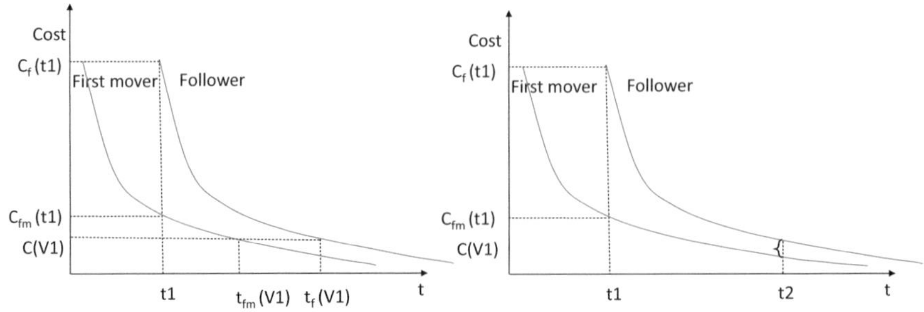

Fig. 4.5 First-mover advantage due to learning effects

by the use of exclusive rights and especially favorable contracts that only an early entrant may be able to obtain.

- **Learning effects** due to the experience gained in production. As it will be discussed in Chap. 14, it is commonly observed that production cost decreases as a function of cumulated volume, and that this effect is quite significant during the earlier production runs (see Fig. 4.5, where a constant production rate is assumed, and learning curves are therefore traced with respect to time). At time t1, when the follower enters, the first-mover will enjoy significantly lower costs. Taken from another perspective, the follower will narrow the gap with the first-mover only after a very significant time.
- **Cash flow** due to margins earned during early sales. These may not only help a company survive, but may be reinvested in the improvement of goods and services, thus accelerating sales and profitability.
- Early entrants, by definition, secure a high **market share** because competitors have not entered yet. Aside from higher margins due to lower competition, this may yield additional positive returns due to customer lock-in, the establishment of brand visibility and loyalty, and the possibility of imposing proprietary standards.[37]
- Early entrants are likely to be the first to reach significant production volumes. This enables them to invest earlier and profitably in facilities that will grant them **economies of scale.** These will allow them cost advantages and act as further barriers to entry.

At the same time, the choice of being a first-mover may have some drawbacks, which are associated to the technological and competitive risks involved. More specifically:

- **Research and development** may require higher investment. An early entrant will have to explore the technological landscape and probably end up pursuing a number of unsuccessful technical alternatives. At the same time, it will be unable

[37] The relationship of proprietary standards and early entry will be discussed in depth at the end of this chapter.

to learn from competitors' mistakes. In the early phases of the diffusion curve, it will also be difficult to finance R&D with margins from sales, since sales volumes will still be low.

- In order to be successful, an early entrant will need **complementary technologies, products and resources** (e.g., supply and distribution channels) that may not yet be available, or may be inadequate. In many instances, the firm will be unable or unwilling to provide these complementary assets. This being the case, it will have to wait for other firms to do so. However, this will occur only when partner companies (i.e., suppliers, customers, complementors) will find it profitable and this will usually happen when a sufficient level of diffusion will be reached. Alternatively, complementary assets may be provided jointly with competitors on a cost-sharing basis but—of course—this can happen only when these competitors will actually enter the market.
- **Customer needs** may still be uncertain and unknown. Moreover, needs change when moving from innovators to early adopters, and then jumping the chasm that leads to the early majority. The early entrant will therefore have to work through the process of understanding the needs of different customer segments, sometimes in a relatively short time frame. Recent research (Cantamessa et al. 2021) has shown that a faster diffusion trajectory reduces the time available to redesign products for the upcoming segment and that this may favor incumbent firms over new entrants, given the former's broader knowledge of the market.

Following these concepts, it appears that first-mover advantage is not an absolute concept at all, and might occur or not, depending on the technology and on the firm being considered. For a startup or a relatively weak entrant, it would be unreasonable to delay entry, since getting a foothold in the early diffusion segments is important for financial reasons, to obtain technical feedback and to establish brand equity. However, these are necessary, but not sufficient, conditions for ensuring the firm's enduring success. This will depend on its ability to secure initial funding and to adequately change its product offering when it will have to serve a fast-growing mainstream market made up of highly demanding 'early majority' customers. Conversely, a company with a strong technology, an established brand and significant financial resources, might decide to wait for other firms to 'open the market', skip the first two segments and reach directly for the early majority, with products targeted specifically to that segment.[38]

Therefore, entry decisions will have to be taken by carefully evaluating advantages and disadvantages, and understanding their relevance to the technology, to the firm taking the decision, as well as looking ahead for the actions that may be taken by competing firms.

[38] A typical example is the highly successful iPod from Apple, which was a relatively late entrant in the field of MP3 players, and whose features (ease of use, stylish design, complementary music downloading service) were aimed at early majority consumers. Conversely, market leaders in the early phases of the product lifecycle were displaced by the iPod's entry because of their focus on purely technical features that appealed to early adopters, such as disk drive capacity.

Finally, firms must be continuously able to simultaneously deal with *technological uncertainty* and with *behavioral ambiguity* in the upcoming market segments (Adner and Kapoor 2010). This ambiguity has to do with customers' actual needs, which at the beginning of the lifecycle are often latent, and with the purchasing process they follow before they decide on adoption, which is likely to be quite convoluted and slow. Moreover, especially when the innovation requires a restructuring of the dominant paradigm, firms must balance their attention between improving the performance of the product (which implies a focus on its critical subsystems and/or on integration issues) and improving the performance and/or availability of complementary goods.

It is probably impossible to define a general criterion specifying how a firm ought to split its attention between these different aspects while progressing along the s-curve. Firms should therefore continuously keep a close eye on what is likely to be the current or the upcoming bottleneck that might slow down diffusion, and make realistic evaluations on the likelihood with which they will be able to overcome it. Behavioral ambiguity and availability of complementary assets have the potential of significantly slowing down diffusion of new products. This can have a strong and negative impact on first-mover advantage, since a hesitating market 'buys time' for imitators and prevents pioneering firms from gaining a significant advantage based on economies of learning.

References

Abernathy W, Utterback JM (1975) A dynamic model of process and product innovation. Omega 33:639–656

Adner R, Kapoor R (2010) Value creation in innovation ecosystems: how the structure of technological interdependence affects firm performance in new technology generations. Strateg Manag J 31:306–333

Agarwal R, Moeen M, Shah SK (2017) Athena's birth: triggers actors, and actions preceding industry inception. Strateg Entrepreneurship J 11:287–305

American Marketing Association dictionary. https://www.ama.org/resources/Pages/Dictionary.aspx?dLetter=B. Retrieved 23 Oct 2022.

Boudreau KJ (2010) Open platform strategies and innovation: granting access vs. devolving control. Manage Sci 56(10):1849–1872

Christensen CM, Verlinden M, Westerman G (2002) Disruption, disintegration, and the dissipation of differentiability. Ind Corp Chang 11(5):955–993

Cantamessa M, Montagna F, Casagrande-Seretti A (2021) Speed of diffusion, rethinking time and firms' strategy: analysis of the interactions to leap across the chasm. Int J Innov Manag 25 (7):2150083

Cusumano MA, Kahl SJ, Suarez F (2006) Product, process, and service: a new industry lifecycle model, MIT Sloan School of Management, Working Paper, p 228

Cusumano MA, Kahl SJ, Suarez F (2015) Services, industry evolution, and the competitive strategies of product firms. Strateg Manag J 36(4):559–575

Davis FD (1989) Perceived usefulness, perceived ease of use, and user ac-acceptance of information technology. MIS Q 13(3):319–340

Davis FD, Bagozzi RP, Warshaw PR (1989) User acceptance of computer technology: a comparison of two theoretical models. Manage Sci 35(8):982–1003

Govindarajan V, Ramamurti R (2011) Reverse innovation, emerging markets and global strategy. Glob Strateg J 1:191–205
Greve HR, Seidel MD (2015) The thin red line between success and failure: path dependence in the diffusion of innovative production technologies. Strateg Manag J 36(4):475–496
Hermann M, Pentek T, Otto B (2016) Design principles for industrie 4.0 scenarios. In: 49th Hawaii international conference on system sciences (HICSS), Koloa, HI, USA 5–8 Jan
Keeley L, Pikkel R, Quinn B, Walters H (1999) The ten types of innovation, Doblin Inc
Keeley L, Walters H, Pikkel R, Quinn B (2013) Ten types of innovation: the discipline of building breakthroughs. Wiley and Sons, Hobcken
Lee J, Veloso FF (2008) Interfirm innovation under uncertainty: empirical evidence for strategic knowledge partitioning. Product Innov Manage 25:418–435
McNichol T (2006) AC/DC: the savage tale of the first standards war. Jossey-Bass-a Wiley Imprint, San Francisco
Montagna F, Cantamessa M (2019) Unpacking the innovation toolbox for design research and practice. Des Sci 5:e8
Moeen M (2017) Entry into nascent industries: disentangling a firm's capability portfolio at the time of investment versus market entry. Strateg Manag J 38(10):1986–2004
Moeen M, Agarwal R (2017) Incubation of an industry: heterogeneous knowledge bases and modes of value capture. Strateg Manag J 38(3):566–587
Moore GA (1991) Crossing the chasm. Harper Business Essentials, New York
Nunes P, Breen T (2011) Reinvent your business before it's too late. Harv Bus Rev 89(1–2):80–87
Pistorius CWI, Utterback JM (1997) Multi-mode interaction among technologies. Res Policy 26:67–84
Rogers EM (1962) Diffusion of innovations. Simon and Schuster Inc, New York
Roy R, Curba ML, Sarkar MB (2019) The pre-commercialization emergence of the combination of product features in the charge-coupled device image sensor. Strateg Entrepreneurial J 13 (4):448–477
Shapiro C, Varian H (1999) Information rules: a strategic guide to the network economy. Harvard Business Press, Boston
Tushman ML, Rosenkopf L (1992) Organizational determinants of technological change: towards a sociology of technological evolution. Res Organ Behav 14:311–347
Ullmann-Margalit E (1977) The emergence of norms. Oxford University Press, Oxford
Venkatesh V, Davis FD (2000) A theoretical extension of the technology acceptance model: four longitudinal field studies. Manage Sci 46(2):186–204
Venkatesh V (2000) Determinants of perceived ease of use: integrating control, intrinsic motivation, and emotion into the technology acceptance model. Inf Syst Res 11:342–365
Verganti R (2009) Design driven innovation: changing the rules of competition by radically innovating what things mean. Harward Business Press, Boston
Von Hippel E (1986) Lead users: a source of novel productconcepts. Manage Sci 32(7):791–805

Fundamentals of Technology Forecasting 5

Abstract

This chapter provides an introduction to the field of technology forecasting, which is instrumental to most of the studies that have been presented in the first four chapters of the book. Specifically, it discusses the main approaches, quantitative and qualitative, that can be used when dealing with both revolutionary and evolutionary change.

The previous chapters have looked at innovative phenomena from the privileged perspective of the management scholar, commenting on history after the facts have unfolded, trying to come up with rational explanations and to find recurrent patterns in s-curves and similar representations. However, practicing managers find themselves in the unenviable position of having to make critical decisions while events are happening and while there is substantial uncertainty on the direction and the values that will be taken by variables such as prices, demand and performance indicators.

*Technology forecasting** is a discipline that tries to provide some support by using quantitative and qualitative forecasting principles to the dynamic behavior of specific technologies and innovations. It can be viewed as a complement to *technology foresight**, which instead aims to provide a broad understanding of future technological scenarios, based on the possible exogenous evolutions and on alternative decisions that might be taken.

Technology forecasting is a specialist discipline, and it is usually practiced by scholars and by analysts working in specialist consultancy firms. However, managers who will never develop technology forecasts themselves will very often be readers and users of analyst reports. This makes it worthwhile to dedicate a short chapter on the main approaches being used, so that readers may become acquainted on the techniques with which they are developed and also on their limitations. Readers interested in the subject may refer to specialist sources (e.g., Martino 1983; Firat et al. 2008).

M. Cantamessa and F. Montagna, *Management of Innovation and Product Development*, https://doi.org/10.1007/978-1-4471-7531-5_5

5.1 Forecasting Revolutionary Change

As we have discussed previously, technological progress follows a sequence of evolutionary and revolutionary phases. While the former exhibit a substantial degree of continuity, the latter correspond to significant discontinuities. Making a forecast of discontinuous events is intrinsically difficult, especially when it has to do with phenomena characterized by strong social content such as innovation. In such cases, past experience cannot provide a solid basis on which future phenomena may be anticipated. More specifically, it is not feasible to use statistical methods in order to extrapolate future trends, since there is no prior data on which to run these methods.

When future events are characterized by discontinuity, analysts must use qualitative information and expert opinion on the future and then try to make a synthesis providing a vision of the future. The kind of insight being asked for can be highly qualitative and open-ended (i.e., "please tell me under what circumstances it is likely for technology *X* to emerge in the future") or defined in more precise and quantitative terms (i.e., "please provide a probability estimate that, by the year 2040, technology *X* will have reached a performance of *Y*").

When performing this kind of exercise, it is customary to interview *panels* made of multiple experts coming from different areas and disciplines and try to see whether a consensual view emerges. Analysts may interact with experts by organizing rounds of individual interviews or organize workshops requiring the physical presence of the entire panel, as well as online debates and fora in which experts interact at a distance. However, it is very unlikely that a consensus will be spontaneously reached. The obvious reason is that, when trying to imagine a medium- to long-term vision of highly uncertain events subject to discontinuity, experts will rely on their individual experience and background and neglect potentially relevant factors that fall outside the scope of their attention. In addition, experts will often tend to express somewhat extreme views, partially to emphasize their point and partially to express their ego and 'stand out of the crowd'.

In order to steer discussion toward a consensual view, analysts often use the so-called *Delphi method**, or some of its variants. In the basic Delphi method (Dalkey and Helmer 1963), multiple rounds of discussion are carried out. Quite often, experts are asked to provide a quantitative estimate for a future state. At each round, analysts focus on the experts who provided the most extreme values and ask them to write short statements explaining their reasons to the other participants. Experience shows that either these explanations are convincing and allow moving the estimates of the majority of the panel toward the extreme, or these 'extreme' experts will soon fall back to the intermediate values shared by other members. In any case, variance in the estimates will tend to decrease round after round. Analysts will therefore run the exercise until they feel that variance is acceptable, and that an additional round would be annoying to ask for, without substantially improving the result further (Fig. 5.1).

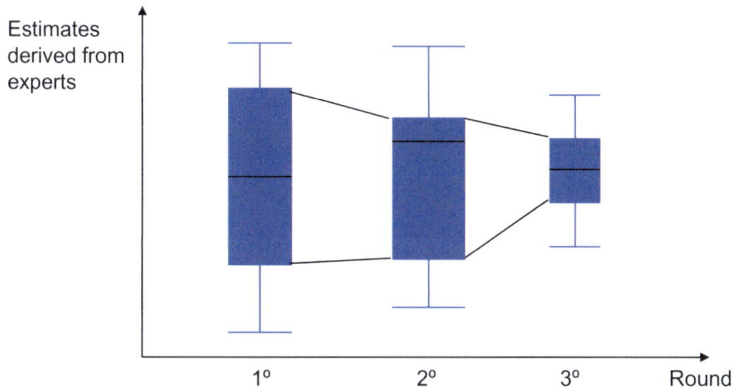

Fig. 5.1 Expert estimates in the Delphi method

When examining a report based on panel interviews, the reader should therefore understand what methodology was followed (e.g., in the case of individual interviews, were extreme views are considered and how?) and the composition of the panel, checking whether relevant fields of expertise were all included or whether some were omitted.

Recent literature also proposes a few quantitative approaches for forecasting in the context of radical change. The main idea underlying these methods is that features and functionalities associated to radical innovations are usually associated to specific technical processes and decisions that—in turn—are based on *repetitive patterns of system evolution* (refer to the Theory of Inventive Problem Solving in Chap. 15). Specifically, these methods consider that if there is a common and repetitive historical pattern in the evolution of technology leading to successful radical innovation, there will be an underlying pattern of similar design decisions. For instance, Cascini et al. (2015) propose, as shown in Fig. 5.2, to consider a vector of technical features $v = [v_1 \ldots v_n]$ that can be associated to the current prevailing technology and a set of similar vectors $V_j = [V_{j1} \ldots V_{jn}]$ that can be associated to the performance of the j-th candidate technology, which may lead to radical innovation. In view of that, the analyst can list a limited set of design decisions that could lead from v to V_j.

These design decisions are not infinite, but are associated to an abstract list of common design actions (e.g., the addition of a new feature, the shift from one technology to another in a module, the adoption of new material, etc.). A combination of design actions will be termed '*transformation function*' from v to V_j. Based on the existing technical constraints and on available resources, only a subset of these design decisions will be feasible and—therefore—only a subset of transformation functions will be admissible. Once this screening has been carried out, it is possible to look at past history in the relevant industry and identify the transformation functions (i.e., the combinations of design decisions) that have effectively taken place and estimate the probability with which each has led to a successful radical innovation. Statistical analysis then allows to retrace this probability to the

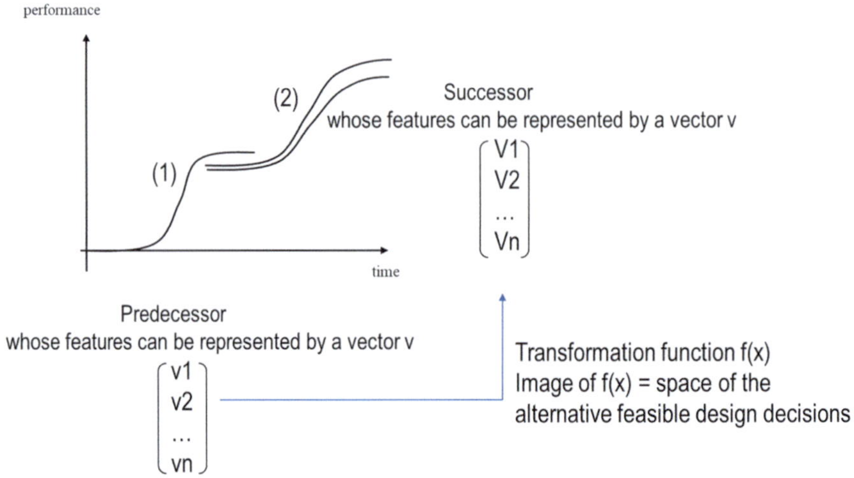

Fig. 5.2 Quantitative approaches for forecasting in the context of radical change

contribution of design actions (individually and through their joint effects) and—finally—to estimate the success probability of a candidate transformation function.

Of course, similar methods have to be calibrated to a specific industry, so to identify a catalogue of admissible design decisions and to carry out a sound statistical analysis of past dynamics (Casagrande-Seretti et al. 2019).

5.2 Forecasting Evolutionary Change

When dealing with evolutionary change, it is instead possible to use quantitative methods from statistics and time-series analysis change. As mentioned multiple times, the prevailing evolutionary phenomenon in innovation is well represented by an s-curve, or logistic curve. In general, a complete s-curve representing the evolution of a parameter v (such as a performance indicator) over an independent variable t (such as time) corresponds to a differential equation in which the instantaneous speed of variation of v is proportional to the same level v, with a saturation effect given by an inverse quadratic relationship to v. In algebraic terms:

$$\frac{\mathrm{d}v}{\mathrm{d}t} = kv - bv^2 \tag{5.1}$$

By solving the differential equation, the evolution $v(t)$ is as follows:

$$v(t) = \frac{k/b}{1 + e^{-k(t-t_0)}} \tag{5.2}$$

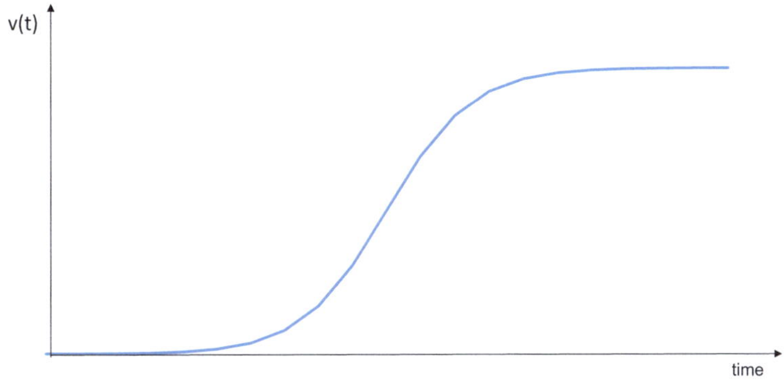

Fig. 5.3 Logistic curve

Equation (5.2) represents what is also commonly known as a logistic curve, where k/b is the asymptote (or the limit L) of the s-curve (Fig. 5.3).

When trying to represent an entire s-curve by using the logistic model, a number of problems arise in identifying parameters. In order to work on a time series, the model has to be transformed in a second-order finite difference Eq. (5.3):

$$v_{t+1} = (k+1)v_t - bv_t^2 \tag{5.3}$$

One way with which parameters k and b may be identified is to view the finite difference Eq. (5.3) as an autoregressive formulation in which v_t and v_t^2 are the two independent variables. With this assumption, multiple linear regression analysis can be carried out in order to identify values for k and b. This approach is very weak from a statistical point of view. Equation (5.3) is not linear and, by treating it as such, one neglects the obvious collinearity between v_t and v_t^2. In practice, it is possible to use this shortcut if one has enough data points and if the s-curve has evolved for a sufficient time. Conversely, it would be risky to use the approach if the s-curve has just started. In this latter case, minor measurement errors would lead to large swings in the values of the parameters identified through linear regression, and the resulting forecasts would be quite unreliable.

An alternative approach is to transform Eq. (5.2) in (5.4)

$$\frac{v_t}{L - v_t} = e^{k(t-t_0)} \tag{5.4}$$

where L = k/b and then into (5.5):

$$\ln\left(\frac{v_t}{L - v_t}\right) = -kt_0 + kt \tag{5.5}$$

This approach requires an *ex-ante* estimate of the limit of the s-curve L and works fairly well for short data series. This ex-ante estimate can be obtained through expert interviews, as in discussed in the previous subsection.

When one is operating in the early stages of a phenomenon, it might be better to make reliable forecasts on a relatively short horizon than highly uncertain ones on the entire evolution of the s-curve. In such cases, it is possible to use exponential and even simple linear growth models.

The differential equation that represents an exponential growth model is (5.6):

$$\frac{dv}{dt} = kv \tag{5.6}$$

which can be integrated to yield the Eq. (5.7):

$$v(t) = v_0 e^{k(t-t_0)} \tag{5.7}$$

Parameters for (5.7) can be identified by applying linear regression to the finite difference Eq. (5.8):

$$\ln(v_t) = \ln(v_0) - kt_0 + kt \tag{5.8}$$

As shown in Fig. 5.4, the exponential model is a good representation of an s-curve, until its inflection point is reached. After that point, it obviously provides a misleading forecast. Exponential growth models are often expressed in verbal terms in statements such as "technology X will see its performance increase by $y\%$ per year for the foreseeable future". When reading such statements, managers should be well aware of the 'foreseeable future' part of the sentence. This limit is represented by the inflection point, and—when that point will have been reached—growth in the s-curve will slow down in a way that is roughly symmetrical to what has

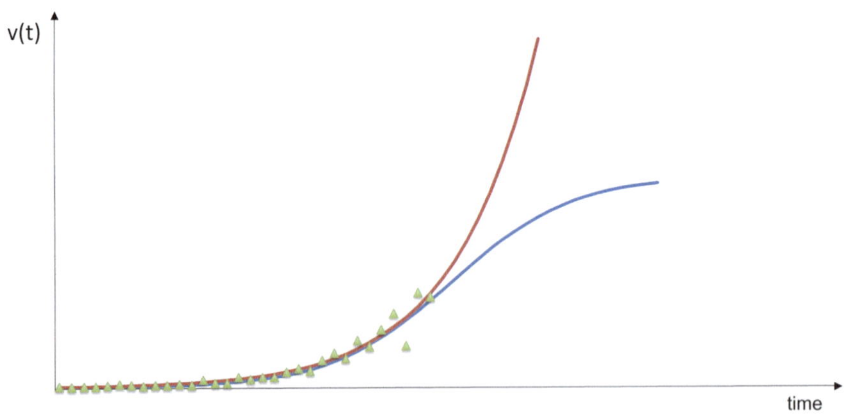

Fig. 5.4 Exponential approximation

happened up to that point. A rough rule of thumb for figuring out the position of the inflection point is to view it as the moment at which the s-curve will have reached 50% of its limit *L*. If one can rely on expert opinion about this asymptotic value, it is possible to make an educated guess on the time at which the exponential growth model will cease to be valid.

If the number of data points is very small, it is possible to use a linear model. In this case, the differential equation is (5.9)

$$\frac{dv}{dt} = k \tag{5.9}$$

Which can be integrated to yield the Eq. (5.10):

$$v(t) = v_0 + kt \tag{5.10}$$

Parameters for (5.10) can be identified by using the finite difference Eq. (5.11):

$$v_t = v_0 + kt \tag{5.11}$$

The linear growth model is the simplest one and can obviously be used for a short-term forecast, but it has the advantage of being usable even with a time series with very few data points. When referring to s-curves, it is therefore possible to think of three linear models being used, one for the incubation period, another one for the growth period and the last one for the maturity period (Fig. 5.5).

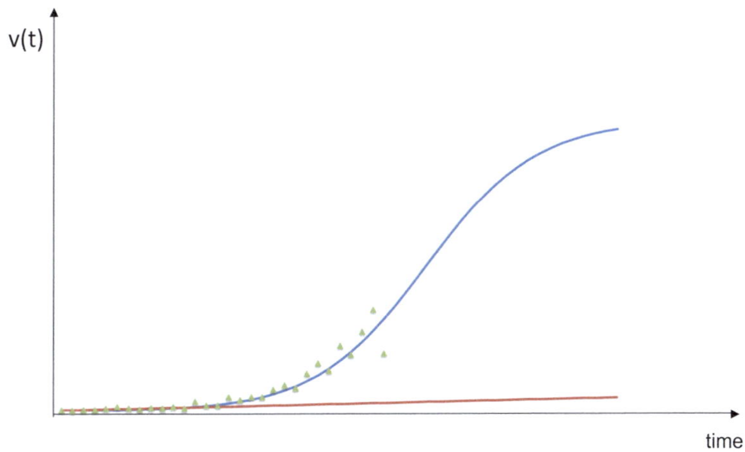

Fig. 5.5 Linear approximation

References

Casagrande Seretti Montagna Cascini Gaetano (2019) A decision support model to assess technological paradigms. Int J Technol Manage 80(1–2):61

Cascini G, Becattini N, Kaikov I, Koziolek S, Kucharavy D, Nikulin C, Petrali P, Slupinsky M, Rabie M, Balachandar R, Ruggeri L, Vanherck K (2015) FORMAT–building an original methodology for technology forecasting through researchers exchanges between industry and academia. Proc Eng 131:1084–1093

Dalkey N, Helmer O (1963) An experimental application of the delphi method to the use of experts. Manage Sci 9(3):458–467

Firat AK, Woon WL, Madnick S (2008) Technological forecasting—a review. Composite Information Systems Laboratory, Sloan School of Management, MIT, Cambridge

Martino J (1983) Technological forecasting for decision making. Elsevier, New York

Part II
Formulating an Innovation Strategy

The Many Approaches to Innovation Strategy

6

Abstract

Innovation strategy is a core part of corporate strategy. This chapter has the objective of defining a rigorous link between the main approaches and schools of thought in corporate strategy, and innovation. This connection allows the formulation of a sound innovation strategy, including considerations pertaining to intellectual property. The chapter focuses especially on the resource-based view approach, which leads to the suggestion that innovation strategy be jointly expressed as a dynamic capabilities strategy for developing competencies and as a complementary project portfolio strategy, aimed at enacting and exploiting the former. These two strands will then be described in detail in Chaps. 8 and 9.

Part I of this book has considered innovation mainly as an economic phenomenon and has looked at it from the high-level perspective of an industrial sector and of society at large. With Part II we enter another level of analysis, in which we 'take the driving seat' of a company to define its innovation strategy.

Now, what is an innovation strategy and how can it be defined? Drafting an innovation strategy obviously means creating a plan describing what kind of innovations the firm will pursue in the planning horizon, together with the means for doing so. Given its significant and wide-ranging implications, innovation strategy will be an integral part of strategy at corporate level. We can therefore define innovation strategy as the '*part of corporate strategy that has to do with the way with which the firm will pursue innovation in its many possible facets*'.

The formulation of corporate strategy can follow a number of approaches and schools of thought. Choosing among these approaches has to do with the specific type of company, its attitude, the environment it operates in and the availability of information that decision-makers can rely on. Therefore, given the close link between corporate strategy and innovation strategy, there will not be a single way for defining the latter.

M. Cantamessa and F. Montagna, *Management of Innovation and Product Development*, https://doi.org/10.1007/978-1-4471-7531-5_6

This not being a textbook on corporate strategy, we will limit ourselves to a quick description of the most popular approaches to corporate strategy and outline the way with which they relate to innovation strategy. Specifically, we will discuss the product portfolio, the competitive advantage and the resource-based approaches, as well as we will provide some comments on so-called shaping strategies. When discussing competitive advantage, a short section will also be dedicated to the management of *intellectual property rights** (IPR).

6.1 Choosing an Approach

Firms differ substantially when we look at the way they approach strategy, and even more so when we observe their attitude toward innovation.

In Miles and Snow's classical distinction (1978, 1984), companies can act as *prospectors*, *analyzers*, *defenders* and *reactors*. Prospectors are the ones who proactively and aggressively pursue opportunities for change and innovation, both incremental and radical. Defenders are firms that settle in a somewhat stable industry and avoid engaging in diversification or innovation, if not of incremental type. Analyzers are somewhat in the middle between the previous two and generally engage in diversification or radical innovation only after extensive reflections, and usually not as first movers. Finally, reactors are the most conservative firms and accept change only when external pressure to change becomes impossible to resist.

The membership of a firm to one of these classes is embedded in the culture shared by its shareholders and managers and throughout the organization. Therefore, pursuing an innovation strategy that is alien to this prevailing culture is bound to be very difficult and will require an extensive effort in order to change the culture itself.

A second perspective that can help understand which approach to strategy formulation is coherent with a given firm, has to do with the stance that the firm has toward *uncertainty*, which of course is a key aspect to be considered. In Fig. 6.1, companies may have different attitudes with respect to their capability to foresee where their environment is heading and to their capability to actively contribute to such change.

In the case of firms who perceive a high capability to predict future trends, but a low capability to influence them, strategy takes the form of planning. After analyzing these trends, the company will define a sequence of actions that leverage on its existing assets and involve future investment, with the objective of maximizing financial returns. The portfolio management approach and—to some extent—the competitive advantage approach lend themselves quite well to this type of firm. When a company perceives a low capability to both predict and influence the future, strategy will take the form of adaptation. The typical approach to corporate strategy will be resource-based, and the firm will progressively move in the competitive landscape by leveraging and pivoting on its assets and competencies. Firms who perceive a low capability to predict future trends, but feel confident of being able to

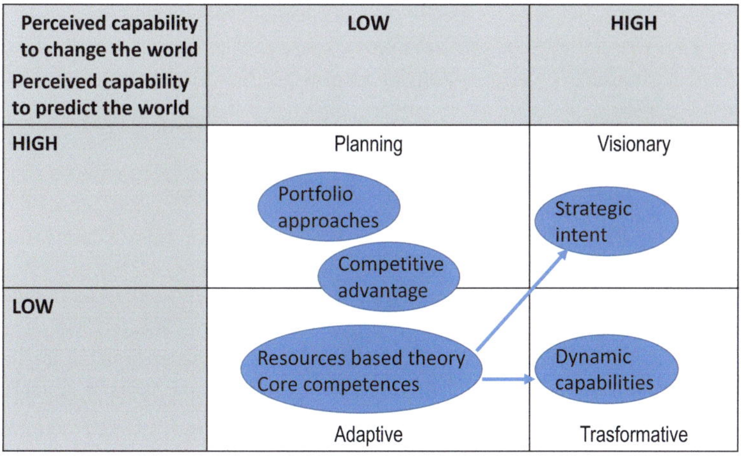

Fig. 6.1 Relationship between approaches to strategic management and uncertainty

exploit opportunities and play a major role in determining them, will view strategy as transformation. They may use the dynamic capabilities approach, which consists in working on current competencies and building new ones that fit the future scenario, thus allowing proactive moves in the competitive environment.

Finally, a few firms may exhibit a high confidence in foreseeing future scenarios, simply because they believe that the future is going to be determined by their own actions. In such cases, strategy is based on a visionary approach and will follow so-called shaping strategies (Hagel et al. 2008).

The strategic stance taken by a specific company will depend not only on its internal culture but also on its nature and its position with respect to the industry. A well-funded and fast-growing startup operating in an emerging industry might coherently pursue a visionary strategy, while a large and diversified multinational might find it rational to follow a much more cautious planning or adaptive approach.

The degree to which a firm will take a bold strategic stance will also depend on its ability to play a key role in emerging technological domains or application areas. Industry analysts quite often try to carry out an evaluation of this sort by locating competing firms on the two axes of 'strategic vision' and 'ability to execute'. Figure 6.2 contains an example of this type of analysis.

6.2 Innovation and the Product Portfolio Management Strategy

The product portfolio management approach views the firm as a collection of *business units** (BUs) and/or product families, directed to a variety of markets. Strategic decisions are associated to entry (i.e., opening new business units,

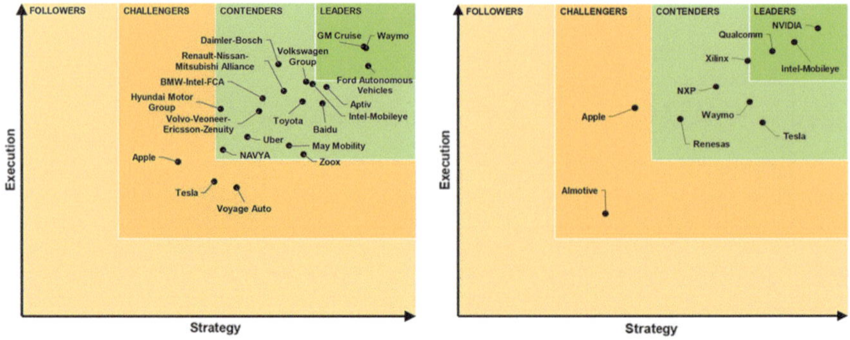

Fig. 6.2 An example of strategic evaluation of players in the emerging sector of autonomous driving (from Guidehouse Insights, 2021)

launching new products and entering new markets) or exit (i.e., closing business units down, taking products out, or exiting markets). A metaphor for the portfolio approach is the typical card game, in which players routinely discard an unwanted card and pick a new one from the deck, trying to maximize the value of the set of cards they hold in hand.

From a theoretical perspective, the product portfolio approach considers the firm as an *internal capital market*, in which cash flow from mature and successful products finance the development of up-and-coming ones.[1] The best-known method related to this approach is the BCG matrix (from the Boston Consulting Group, the consultancy firm that introduced it in 1970), which is depicted in Fig. 6.3. The BCG matrix is generally represented as a bubble chart, in which each BU is a circle whose diameter is proportional to the economic importance of the unit (e.g., sales). The two axes of the chart are associated to cash flow generation and cash flow absorption of each BU.[2]

[1] This approach is typical of highly diversified firms, or *conglomerates** and contributes to explaining their existence. In fact, why should an investor prefer buying a share of a conglomerate, and take ownership of the underlying activity portfolio defined by corporate managers, instead of asking an investment fund manager to pick shares and create a financial portfolio of smaller and more focused firms? The standard answer to this question (Kopp 1968; Faulkner and Campbell 2006) is that financial markets are imperfect. Lack of information and a high degree of uncertainty place a corporate manager in a better position to make investment decisions than a fund manager, especially when dealing with emerging business opportunities. A second answer is that there can be synergies between the business units of a conglomerate, together with common assets such as R&D leading to technological knowledge shared by BUs, a well-known brand, reputation and management skills.

[2] In practice, one must use proxies for cash flow generation and absorption. For the former, it is common to use relative market share (i.e., the market share of the BU divided by the market share of the largest competitor), since this provides a rough indication of profitability. For the latter, sales growth rate is generally used, since a growing business must generally be supported by adequate investment.

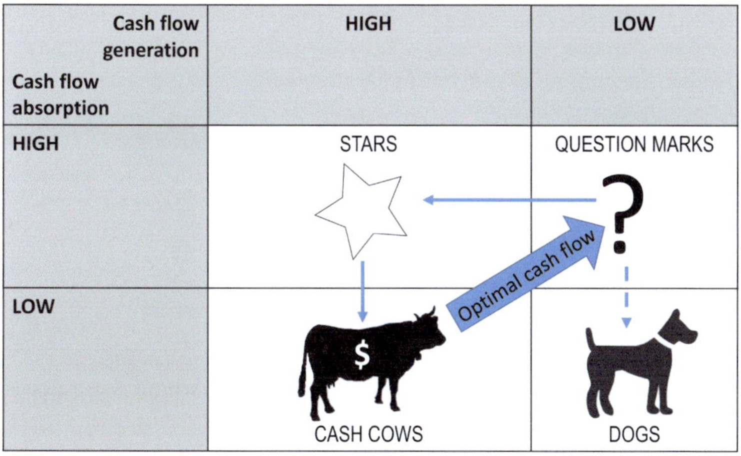

Fig. 6.3 BCG matrix

A fledgling and early-stage business that requires significant investment, but is still unable to generate any cash flow will be considered to be a *question mark*. A BU will be considered to be a *star* if it is at an intermediate stage, where it does generate cash, but still requires further investment. A *cash cow* will be a successful, cash flow positive and mature business. Finally, a *dog* is an unsuccessful business that does not require investment, but does not return significant cash flow. Strategic decisions consist in selecting a 'balanced' portfolio of BUs with no dogs, a set of question marks and stars offering opportunities for future growth, together with some cash cows providing dividends to investors and adequate financing to question marks and stars.[3] A portfolio leaning too much on the former or on the latter might, respectively, have problems in securing financing for growth, or in ensuring future competitive advantage.

The relationship between the BCG matrix and innovation strategy should be quite evident to readers. In fact, the cells in the matrix are closely related to the s-curve representation of innovation, with question marks being products belonging to the incubation phase of the curve, stars to the growth phase, and cash cows to the maturity phase. When following the BCG approach, innovation strategy therefore appears to be directly coupled to corporate strategy. However, one should notice that this innovation strategy is spelt out at a very high level of abstraction, since it does not go deeper than the decision to enter, invest and exit businesses. The way

[3] What constitutes a 'balanced' portfolio depends on shareholders' preferences. A startup company is likely to be made up of a single question mark and venture capital shareholders will explicitly require this, given the mandate they have received from their own investors (i.e., to look at rapid growth in valuations and not at dividends, and to diversify risk by investing in a number of individually well-focused startups). Conversely, a large conglomerate that has traditionally treated its investors to a steadily rising flow of dividends will need to have many cash cows and just enough question marks and stars to ensure sustainability of the business for the future.

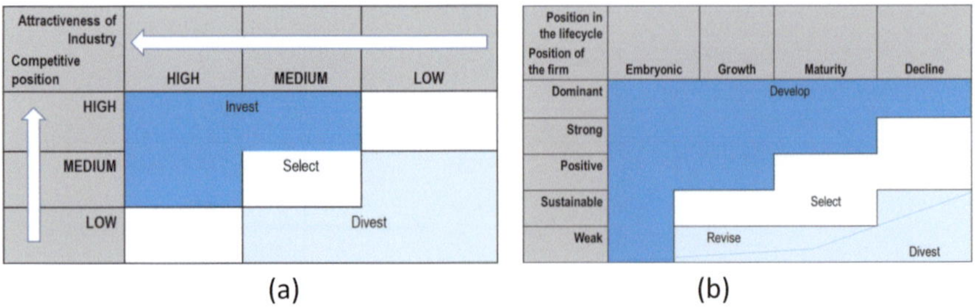

Fig. 6.4 GE/McKinsey (**a**) and A.D. Little (**b**) product portfolio matrices

with which the firm or its business units should operate within these decisions is instead neglected.

For the record, similar approaches to the BCG matrix have been proposed over the years, such as the McKinsey/GE matrix and the A.D. Little matrix (Fig. 6.4a, b, respectively). In these cases, there is a connection between the mapping of products and business units with their positioning on the innovation s-curve, as well. In fact, the two axes of the A.D. Little matrix are the 'Position on the lifecycle', which explicitly relates to technological s-curves and the 'Competitive positioning of the firm'. The 'competitive positioning' axis can also be found in the McKinsey/GE matrix, which uses 'attractiveness of industry' as the other axis. 'Attractiveness of industry' is related to s-curves too, since emerging technologies and markets are usually considered to be more attractive than mature or declining ones.

6.3 Innovation and the Theory of Competitive Advantage

The Theory of Competitive Advantage is a well-known approach to strategy formulation, proposed by Michael Porter* (1947–) in 1979. We will at first provide a short explanation of this theory and then relate it to innovation strategy.

Porter's approach to strategy has enjoyed a very wide diffusion, but the association to Innovation Management is not always straightforward. First, it relies on a somewhat deterministic perspective that associates competitive advantage to the objective decisions taken by a firm, rather than to the subjective situation and history of the same. Secondly, it strongly relies on the idea that firms operate within linear value chains, which is only partially true in a number of contemporary industries. Examples that deviate from this traditional model include platform-based businesses, where what matters the most is the dynamics of multisided markets (e.g., food delivery platforms), industries characterized by the role of complementors (e.g., electric vehicles and recharging infrastructure), and firms who work in the so-called circular economy*.

6.3.1 Competitive Advantage, the Five Forces and Generic Strategies

According to Porter (1985), firms must not be considered as isolated entities, as in the previous product portfolio approach. The firm, instead, must be viewed as part of a value chain that links it to upstream suppliers and downstream customers, and as part of a competitive context made of current and prospective rival firms. Following Porter, the competitive environment can be studied from the perspective of industrial economics and specifically following the so-called *Structure-Conduct-Performance (SCP) paradigm** (Bain, 1959). In very simple terms, this paradigm states that, given an industry structure, rationality in economic agents' behavior will determine the way with which firms operate and—in turn—that this behavior will determine their performance.

As a direct consequence of this perspective, a firm's main strategic competence will at first consist in spotting a potentially favorable industry and entering. Due to the determinism that is inherent to the SCP paradigm, it would be quite unlikely to earn money by operating in an industry in which profitability is structurally low. Secondly, the firm must structure its operations (i.e., it must invest in tangible or intangible assets and organize itself) in order to create as much economic value as possible.[4] Finally, it must maneuver in the environment in order to retain the value created, without having to share it with other actors. When a firm is able to do so better than its competitors, it will record higher contribution margins and profits and we will say that it enjoys *competitive advantage* with respect to rival firms.[5] If the firm is able to avoid imitation by competitors[6] and keep this lead over time, competitive advantage will be said to be *sustainable*. In order to systematize this game of value creation and retention, Porter identified the following *five forces* that act on the firm and attempt to reduce its competitive advantage (Fig. 6.5):

- **competitors** exert a first powerful force because they constantly try to gain market share by reducing prices, improving their product offering, and imitating the firm. This is associated to the degree of existing *rivalry* between firms, which has to do with the number of competitors, their similarity, and the existence of exit barriers forcing rival firms to fight for their position and survival.

[4] By economic value we intend the gap between the price that customers are willing to pay for the firms' products and the costs required to produce them.

[5] Following Grahovac and Miller (2009), competitive advantage is in its stricter sense connected to the above-normal *contribution margins* (i.e., the difference between unit price and variable cost) the firm earns in its product market. This is a necessary, but not a sufficient condition for profitability, since significant profits will arise only if the firm does not spend too much money in *factor markets,* when acquiring or developing its assets. These assets may in fact be expensive either because they might be inherently costly, or because whoever sold them to the firm (e.g., an equipment maker, or an employee negotiating her salary) might be able to foresee their economic potential and raise prices accordingly.

[6] Always following Grahovac and Miller, imitation will be more likely if profitability is very high (i.e., if competitors see very attractive contribution margins and/or believe the cost of observing and replicating the firm's assets to be comparatively low).

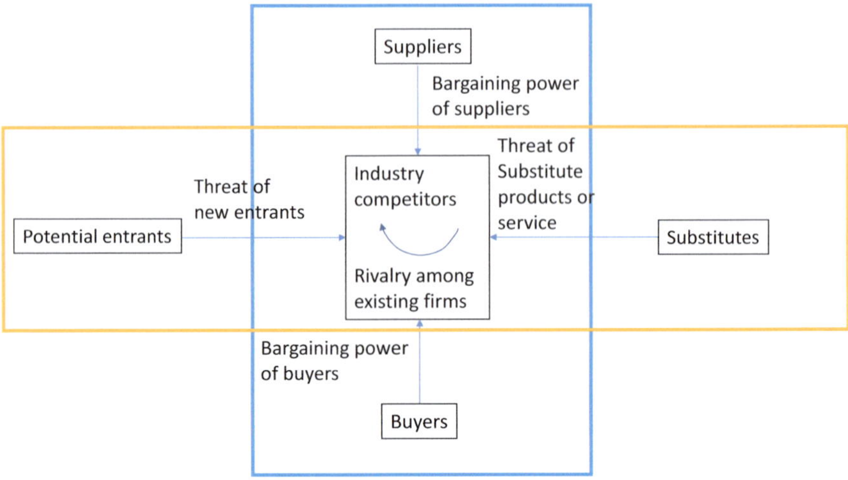

Fig. 6.5 Porter's five forces

- **suppliers** can observe the profitability of their customer firms and try to share some of this economic value added by increasing prices. The intensity of this force has to do with vertical integration choices and with the bargaining power of the parties involved.
- **customers** too can observe the profitability of their suppliers and, especially if competition and bargaining power allows it, they will try to reduce the price for the goods they buy.
- **new entrants** will observe the industry and, if they consider profitability high enough to overcome existing *barriers to entry*, might decide to enter, therefore increasing competition.
- **other industries** might provide products that are – to some extent – substitutes to the ones being produced by the focal industry. Therefore, they too exert a force that is similar to the one being applied by competitors.

Finally, Porter suggested that competitive advantage can be achieved by following one of four main *generic strategies*, summarized in Table 6.1. In Porter's view, a firm must make a clear choice among these strategies. Trying to pursue more than one of them at the same time would lead the firm into being stuck in an awkward position and not really being effective in any of them.

6.3.2 Competitive Advantage and Innovation Strategy

The competitive advantage approach goes quite deep into analyzing and defining firms' strategic behavior and the connection with innovation strategy is strong, even though it may not be so obvious.

Table 6.1 Porter's generic strategies

Strategy	Industry-wide cost leadership	Cost leadership in different market segments	Differentiation	Focused differentiation
Description	Being the lowest-cost producer in the industry	Being able to serve many horizontally differentiated market segments at a low cost	Being able to serve many vertically differentiated segments and maximizing margins gained in each	Specializing in order to serve a single customer segment or niche
How can it be achieved in the product	Reducing direct cost Exploiting economies of scale by spreading R&D investment over a large volume	Being able to design products that are 'just enough' for the needs of each segment Exploiting economies of scope and spreading the same R&D investment over products in different segments	Being able to design products that maximize customer utility at each segment, at the lowest-cost possible Exploiting economies of scope by spreading the same R&D investment over products in different segments	Focusing design effort so that customers in the segment perceive the firm's offering to be the best concerning quality and cost
How can it be achieved in the process	Exploiting economies of scale by spreading production investment over a large volume Exploiting learning economies	Developing a low-cost process that is 'fine-tuned' to segment needs Exploiting economies of scope and spreading production investment over products in different segments	Exploiting economies of scope and flexibility	Developing 'fine-tuned' process that exactly meets the required performance, at the lowest-cost possible

The first point of contact has to do with the concept of 'entry', which is made up of an initial decision, followed by investment in assets and the organization of these same assets. Innovation is often relevant to this concept, since the firm may have to change its products or processes in order to successfully address the new market. For instance, a firm selling in a new geographic region may have to adapt its offering to local needs. Similarly, a firm may have to develop new products in order to make them attractive to a particular market segment.

The second point of contact between Porter's framework and innovation strategy is related to the role that innovation plays in determining the intensity of the five forces and the way these same forces determine the degree of *appropriability** of the rents deriving from the innovation (Cohen and Walsh 2001; Teece 2000; Teece and Pisano 1998; Levin et al. 1987).

If a firm is actively innovating products and processes, the forces being exerted by its competitors, suppliers and customers will be abated. Especially if innovations are based on specific and scarce resources, tacit knowledge and unobservable routines, competitors may find it hard to imitate the firm and match the quality of its offering and its prices. Other factors that may influence the appropriability of above-normal returns are legal protection (see the later section in this chapter), lock-in effects and rapidity. Suppliers who are not directly involved in developing these innovations will not have much bargaining power, since the focal firm may easily switch to other suppliers, or simply threaten to do so. On the other side, customers, unable to find comparable offerings at other firms (in terms of technical features, innovative content and quality of the product or service) will not be successful when asking for lower prices.[7] Instead, a firm that is not innovating at all, that is simply incorporating innovations proposed its suppliers, or whose innovations are characterized by weak appropriability, will be subject to a strong force coming from competitors (who can obviously use the same suppliers) and from the suppliers themselves.

When looking at the force coming from potential entrants, innovation can play an ambiguous role. A long stream of investment in incremental innovation, typical of the mature stage of an s-curve, can determine a strong barrier to entry, since it would be difficult for any entrant to build competencies comparable to incumbents' and reach the same level of technological proficiency. Conversely, radical innovation often destroys barriers to entry. In fact, radical innovations often go in the direction of providing technical means that make entry cheap and easy for about any player.[8] Moreover, radical innovation tends to occur during technological revolutions, when new technologies—or technologies that were originally alien to the industry—become relevant. When this occurs, the way is paved for firms that happen to have the newly required competencies and wish to enter.[9] Finally,

[7] This discussion is quite straightforward for incremental innovation, but can become more nuanced in the case radical innovation. Forces exerted by customers, suppliers and competitors will certainly be much weaker if a firm introduces innovations that do not significantly disrupt an existing paradigm. Conversely, a radical innovation that leads to a new paradigm will see a complete reshuffling of the value chain, which will lead—at least initially—to a somewhat unpredictable definition of the 'five forces'.

[8] For instance, cloud computing dramatically lowered entry barriers in the Internet industry, since it allowed any startup to set up a highly scalable business, without having to invest in equipment, but simply paying for the capacity used.

[9] Examples are abundant. A number of Chinese manufacturers of electric cars come from the battery manufacturing industry. The shift that led from analogue to digital photography allowed the successful entry of firms who operated in consumer electronics and had the right competencies

innovation also changes the degree with which products can be substitutes to one another, and therefore influences the force applied by other industries.[10]

A third point of contact between innovation strategy and Porter's approach is related to the role played by innovation strategy in each of the four generic strategies outlined in Table 6.1. Specifically, the strategic role of R&D and innovation effort can be related to both its outputs and to its cost. Concerning output, innovative activity can lead to improved products and processes that are well-suited to support the generic strategy that the firm intends to follow. For instance, a firm pursuing a cost leadership strategy may benefit by continuously innovating its production processes in search of greater efficiency. Conversely, a carmaker that wants to focus on the sports cars segment will be likely to benefit if it directs its innovation strategy to enhance the features that customers in that segment are likely to value the most. In many industries, the cost that a company incurs in when working on research, development and industrialization of its new products is quite significant and always represents a fixed cost.[11] A company may therefore try to improve profitability by sharing development costs over the highest production volume possible. This can be done by pursuing either economies of scale or economies of scope. In the former case, the *same* product may be offered to a wider audience, for instance by operating on multiple geographic markets. In the latter case, the same R&D investment may be shared across a number of *similar* products, targeted to different market segments. This latter option leads to the concept of platform product development that will be analyzed further in Chap. 9 and then, more extensively, in Chap. 14.

6.4 Intellectual Property Rights and Competitive Advantage

When a firm bases at least part of its competitive advantage on technology, it quickly realizes that the sustainability of this advantage is critical, since competitors will readily try to imitate its products and processes. Not having an adequate

in technology and in operating the related business model. Similarly, new types of glass based on wood fibers are being currently being developed and the main actors are not glass-making companies, but firms who operate in the wood processing industry.

[10] The air transport and railway transport industries were traditionally separate from one another, with the former representing the fast and expensive way to move between two locations and the latter the cheaper and slower one. The roughly simultaneous emergence of low-cost carriers (a business model innovation) and of high-speed trains (a process innovation from the perspective of railway companies) have made the two industries close competitors for short-range trips. In many cases, air transport can be cheaper than train transport and even slower, if one considers the trip from cities to airports and back, together with check-in, security clearance and boarding times.

[11] In order to provide some reference figures, the development and industrialization of a traditional motor car can cost approximately 1 B€. In the case of a commercial airplane, the investment can be in the order of 15 B€. When divided by expected production runs, one can notice that the R&D cost per unit sold often is of the same order of magnitude as contribution margins. So, if a firm is able to spread the same investment over a higher production run, profitability can be substantially increased. Conversely, if demand falls short of forecasts, profitability will be destroyed.

defense strategy against imitation would represent a big loss to the firm. In fact, competitors would not only be able to compete head-on with it, but would do so with the advantage of not having to recoup the investment that the firm has made when developing new competencies and new technology.

The management of a firms' *intellectual property* (IP) therefore is a key element of innovation strategy. This section will attempt to very briefly outline the main aspects and current challenges in this vast and quickly evolving field.

In general, intellectual property has to do with *patents*, *copyright** and *trademarks** in general. Patents are associated to technical inventions and are therefore central to technological innovation. Copyright is associated to the various forms of artistic expression, with the notable inclusion of a strongly technical subject, such as software. Trademarks are associated to corporate or product identity and to Marketing. Nowadays, further alternatives for the protection of outcomes of intellectual activities are emerging, such as *Open Source** and *Public Copyright Licensing** (among them, *creative commons** (CC) licenses). Furthermore, recent evolutions in the field of copyright have led to the use of *Not Fungible Tokens** (NFT), an application of blockchain and distributed ledger technology.

Given the perspective adopted in this book, the focus of the following discussion will be on patents.

6.4.1 Patents

The mainstream approach with which a firm that has developed a new invention can try to avoid imitation is to use a legal mechanism called a *patent*. A patent is a legal document with which a government agency called a *Patent Office*, having successfully examined a *patent application* describing the invention, grants monopolistic rights to the *applicant*[12] to produce and sell the invention for a predetermined period of time (usually 20 years). The patent application contains a detailed account of relevant *prior art*, in order to highlight the novelty of the invention, a *technical description* of the invention, and a set of *claims*. The claims are the heart of the patent, since they represent the legal boundaries of what exactly will be covered by monopolistic rights, if the patent is eventually granted. Therefore, elements that are contained in approved claims will be subject to monopoly, while elements that are not can be freely imitated by competitors.[13]

With some variations across countries, patent applications are confidential until the patent is awarded. At this point, the patent is published, so that competitors may be made aware of what they can and what they cannot legally produce and sell.

[12] For the sake of simplicity, we will only talk about applicants, though patent law makes a distinction between *inventors* (who are always natural persons) and *applicants* (which may either be natural persons, or a legal entity, such as inventors' employers).

[13] One can draw an analogy between intellectual property and land, with patents being the equivalent of property deeds, and the claims the equivalent of the topographic description of the land. The analogy between intellectual property and land is not perfect and is often criticized by scholars, but is quite useful to allow a quick understanding of the subject.

Since the patent is published, competitors are also made aware of the new state of the art and may try to improve the invention or circumvent it. In other words, they may try to achieve better or similar outcomes in a way that is 'sufficiently different' from what is spelt out in the claims, eventually achieving a new invention and applying for a new patent.

For a patent to be awarded by the Patent Office, the invention must satisfy a set of criteria, which can slightly vary from country to country. The most common criteria include the following:

- *patentability of subject matter*. Governments usually make an up-front decision of which subjects can and cannot be patented. For instance, while an inventor can patent a complex mechanical device, a scientist is not allowed to patent a law of nature he has discovered. As we will see in the next section, patentability of subject matter usually raises hot debates, especially for emerging fields of technology;
- *novelty*. For a patent to be issued there must be an invention. Therefore, it is not possible to patent something that is already part of prior art. A patent will obviously not be awarded to an applicant if someone else has already commercialized it, or described it in a paper, in a previous patent or in some other form of public demonstration. What is less evident is that the patent cannot be awarded even if the same applicant or inventor has made it public through any means.[14] Therefore, successful patenting requires keeping a high degree of confidentiality during the development of the invention. When an inventor must interact with another party before filing a patent application, he/she will always ask to sign a *Non-Disclosure Agreement*. These legal documents are used not only to bind the other party to confidentiality, but to constitute evidence that the nature of the knowledge exchange is not public and may not be considered as an impediment for a future patenting activity;
- *non-obviousness*. In order to be patented, the invention must not only be new, but must represent a sufficiently marked improvement over prior art. The general principle is that the technical solution must not be obvious 'to a person skilled in the art'.[15] This requirement avoids the patenting of trivial inventions or minor improvements that routinely occur in any product design and development project;
- *utility*. This principle varies across countries, but is generally aimed at avoiding useless or fantastic inventions. The applicant must therefore demonstrate that the invention is useful to solve some kind of problem and that it can work in practice. In the European Union, the requirement is stronger and calls for the invention to be susceptible of industrial application.

[14] Some countries, such as the USA, grant a *grace period* allowing an applicant to file a patent application within a given interval (12 months) after a public disclosure event.

[15] In European patent law, this criterion is substituted by the concept of *inventive step*. This concept requires the patent application to identify a problem that is not adequately solved by prior art, that the invention allows to solve the problem, and that it does so by using means that are not obvious to a person skilled in the art.

From a more general perspective, one can wonder about the societal role of patents, and this often gives rise to strong debates. After all, governments usually dislike monopolies and even set up antitrust agencies in order to fight them. Why should governments set up a patent office with the mandate of authorizing 20-year-old monopolies?

The main explanation lies in the fact that—if patents did not exist—there would be no incentive for individuals and for firms to invest in technological development. In fact, it would not be rational to pursue any R&D effort knowing that, after having sustained the related costs and risks, competitors could simply copy the invention and take out the firm's competitive advantage, without even having to recover the investment made.

Besides the purely economic reason, one might also argue that allowing this kind of behavior would not be fair. A government granting a patent therefore allows the inventor to become a monopolist and 'tax the citizens, as a reward for the investment made to develop the invention.[16] At least in principle, the patenting mechanism should be self-regulating with respect to the importance of the invention. An invention with little market appeal will be rewarded with meager monopoly profits, while a highly useful invention will have wide commercial success and lead to abundant profits. Moreover, a commercially attractive, but technically marginal, invention will attract competitors, who will readily improve or circumvent it. In such a case, the original 20-year monopoly would effectively end in a short time. Conversely, a highly creative invention will last much longer before it is made obsolete and the inventor will therefore benefit from a much longer effective monopoly.

Besides providing incentives to innovators, patenting also provides two additional benefits. First, if patents did not exist, the only weapon available to investors would be secrecy. This would be an inefficient way for society to pursue innovation, since it would force would-be imitators to 'start from scratch' instead of allowing them to learn prior art by reading previous patents and then trying to improve them (i.e., 'standing on other inventors' shoulders'). Finally, patenting creates a system of legally defined property rights that allows the commercial trade of intellectual property, which can lead to an efficient societal use of the outcome of inventive activity.[17] If the inventor does not feel able to personally exploit the

---

[16] The exact definition of the rights granted to patent assignees requires a careful balancing between the private interests of the inventor and of society at large, especially when dealing with sensitive matters like health care. One typical issue is the righteousness of allowing patent holders to sell drugs at monopoly prices, which may be prohibitively expensive for poor patients in developing countries. However, while it is reasonable to think of adapting the patent system to these cases, it would be quite questionable to scrap it altogether. It is in fact likely that, if patents did not exist, we would not have 'new drugs for all', but simply 'no drugs for anyone'. For a deeper discussion on this topic, one can refer to Schacht and Thomas (2005) and Boldrin and Levine (2013).

[17] In general, a precondition for economic development is the existence of a legal framework defining property rights over assets (Claessens et al. 2003). In the case of real assets, such a legal framework allows economic development in traditional economies by providing an incentive for saving and investing, reducing the need to spend on security, and allowing trade and financial

monopolistic rights tied to the patent, he/she can sell or license the patent. When considering whether to finance a new venture, a bank or an equity investor will consider a patent holder favorably, since the patent shows that the invention does not breach other inventors' rights (i.e., it enjoys what is called *freedom to operate*) and it grants monopolistic rights. This makes the new venture more likely to be profitable and—if something should go wrong—it provides the chance of selling the patent in order to recoup at least part of the investment.

6.4.2 A Strategy for Intellectual Property

When an invention has been developed, a firm must define a strategy for managing this intellectual property and for sustaining the competitive advantage that might derive from it. This subsection will discuss both patenting and other possible approaches.

6.4.2.1 Patenting Decisions

The most obvious approach to manage intellectual property is to apply for a patent. If the patent is awarded, the patent holder can use it in a number of ways. First of all, it can directly *exploit its monopolistic rights* by producing a product based on the patent, provided it has the organizational and financial resources that are needed to successfully develop, produce, distribute and service the product. In many cases, the firm might lack these resources, or might consider that it will not be able to fully exploit the market potential of this invention. For instance, it might feel unable to develop products quickly enough, to come out with a wide enough range of products covering a number of market segments, or to expand internationally. In such cases, it might decide to transfer the monopolistic rights to another party by either selling the patent or by licensing it. In the case of a sale, the firm completely ceases its involvement with the invention. In general, the uncertainty surrounding the commercial value of the invention will lead the buying firm—who will have to bear all future risks—to offer a limited amount of money. As a matter of fact, patent sales often occur in bulk, when the transaction is associated to an entire patent portfolio or to the acquisition of a firm.

Instead of selling the patent, the firm may opt to *license* the monopolistic rights to a *licensee*, in exchange for *royalty* payments. Depending on the agreement reached, the license may be unlimited or bear limitations in time, geography, destination market and sublicensing rights. The royalty fee is usually specified as a percentage on the revenues accruing from the sale of products based on the patent, so that risks are split between the two firms. The licensor may also cooperate with the licensee's product development activities in order to facilitate the *technology transfer* process, in which case additional payments may be required in order to

transactions. Coming back to the analogy between land and intellectual property, one can therefore consider a framework defining intellectual property as a necessary element of economic development in the contemporary knowledge economy.

cover the related effort. Finally, the firm may decide to do nothing and simply use the patent to prevent other firms from developing products based on the technology it has invented.

In many instances, the technology landscape can become so highly fragmented that commercial products can be developed only by having simultaneous rights to use the intellectual property owned by multiple firms. In these cases, and in order to simplify the management of IP and avoid costly lawsuits, firms may decide to *pool their patents* and *cross-license* their patent portfolios, thus giving each other a license to use their respective IP. In case of asymmetric patent portfolios, cross-licensing can be accompanied by transfer of payments to the firm with the largest or most significant portfolio. Hence, both to gain a stronger bargaining position in cross-licensing deals and to be well-positioned in case of lawsuits, firms have a strong incentive to build large patent portfolios, sometimes without too many regards concerning the actual value of each patent.

From a financial perspective, the value of a patent can be accounted for by considering the costs incurred to develop the technology. A better valuation can be made by estimating the net present value of future cash flows that might derive from its commercial exploitation. Building on this concept, an emerging area in IP Management is associated to using intangible assets, such as patents, in corporate finance (Lev 2001). Patents may be appraised and used as collateral in order to secure loans from banks or enter commercial agreements with other parties. Similarly, patents and related royalty streams may be securitized,[18] thus allowing the innovating firm to quickly raise financial resources.

IP Management is becoming a very complex field, requiring skills and competencies that go beyond the capabilities of most firms. Because of this, IP has spawned an industry with a number of intermediaries acting as advisors, brokers, appraisers, licensing and enforcement agents, etc.

Together with undeniable advantages, the choice of patenting an invention also has a number of drawbacks, which must be carefully considered before engaging in this process.

First of all, patenting is quite expensive, especially to individuals and small firms. An initial cost comes from administrative fees due for filing the patent application and for renewing the validity of the patent over time (*maintenance fees*). An additional cost comes from hiring a good patent attorney, who will have to spend time and effort in order to write a strong patent application. In fact, the application must have carefully worded claims that might make it through the examination procedure and—most of all—hold up well in court, should someone challenge the validity of the patent and/or should the inventor decide to sue infringers. While it is possible for an individual to write a patent application personally, the risk of coming up with a badly written, weak and economically

[18] Securitization of IP means generating and selling a financial instrument based on a given pool of IP assets. The buyer of the security pays the asset holder a given sum up front, in exchange for a stream of payments that depends on the revenues generated by the assets themselves (Cohen 2005).

worthless patent is very high. Filing costs may escalate quite steeply in the case of complex inventions, in which a strong protection can be achieved not by filing a single patent, but a number of related ones, thus developing a *patent thicket*.

A second cost is related to the geographic coverage of the patent. Unfortunately, there is nothing like a worldwide patent, and inventors must rely on a system of national patents that cooperate with each other under the auspices of a United Nations agency, named World Intellectual Property Organization (WIPO). Therefore, in order to gain effective protection and maximize the economic value of the invention, it is wise to obtain patent protection in a sufficient number of countries, starting from the ones in which the market for the invention might be larger.[19] This leads to a multiplication of expenses, especially considering that the economic weight of developed countries is shrinking, and that market potential will therefore be spread over a high number of countries.

A third cost is associated to the *enforcement* of the patent. A patent holder must spend considerable resources to observe the relevant industry and make sure that no imitator is infringing on her IP. This is relatively uncomplicated for simple products that can be easily reverse-engineered, but it can become prohibitively difficult for complex products, and especially for process inventions. If infringement is suspected, the patent holder must then start a legal procedure that can have uncertain outcomes, requires considerable time and implies significant costs in order to pay lawyers and expert witnesses.

Aside from its costs, patenting exposes inventors to a number of risks. Having to disclose the invention allows competitors to gain know-how and use it to improve the invention or circumvent it. Competitors may also notice that the commercial use of the invention requires the use of some other technology, and come up with one or more *blocking patents,* thus reducing its economic value. Finally, competitors may simply wait for the patent to expire and then legally start commercializing the product themselves.

6.4.2.2 Alternatives to Patenting

Given the significant problems connected to patenting, a firm may keep the intellectual property as a *trade secret*. This strategy would obviously not make sense in the case of products that are easy to reverse-engineer. Conversely, it can be the best solution in the case of process inventions and it is commonly followed in continuous process industries, such as food and basic materials. When deciding to keep an invention as a trade secret, the firm avoids disclosure and the related risks of circumvention. Competitive advantage will last as long as the secret is kept or until a competitor independently comes up with the same—or a similar—technology,

[19] In practical terms, worldwide coverage is obtained by using a procedure defined by the so-called Patent Cooperation Treaty (PCT). In a nutshell, within 12 months of the first filing in the initial country, the applicant has the opportunity to file a *PCT application* expressing interest in international extension of the patent. The PCT application is at first examined at international level and—if the response is positive—the applicant can within 30 months move into the so-called *national phase*, in which he/she must decide in which countries the patent should be extended and start the related procedures.

and this may well occur beyond the 20-year duration that a patent would have granted. At the same time, keeping the secret involves significant costs in security management and in ensuring the loyalty of the employees who share critical know-how. Should a competitor manage to illegitimately gain access to this knowledge and copy the invention, it may be difficult to bring forth a successful lawsuit. In most jurisdictions, it will in fact be up to the original inventor to prove that the imitator has illegally accessed its own proprietary knowledge.

In some cases, and especially in the case of small firms in fast-moving industries, inventors often decide not to patent, but *to rely on the speed* with which they think they will be able to continuously improve the technology. By following this option, imitators are legally free to imitate the inventor's products, but the inventor expects them to take longer than he/she needs to improve the technology further. This strategy brings forward two risks. The first one is associated to making a wrong estimate on the speeds of development and imitation. The second risk is that competitors may attempt to patent the technology they have imitated. Such a patent would of course be invalid since—once commercialized—the original product constitutes prior art. However, the patent office might ignore this, especially if the inventor has gained a very small access to the market. In this case, the inventor might bring a lawsuit against the imitator and ask for its patent to be invalidated. In order to do so, he/she would have to produce evidence of the timing at which the original product had been disclosed to the public.

In order to avoid such a costly and time-consuming process, firms who do not wish to patent—but who want to avoid being 'patented out of their own inventions'—may choose to make an *invention disclosure*. In other terms, the firm publishes a description of the invention in such a way that the timing is officially recorded and that there might be no doubts about priority. Sometimes, specialized disclosure newsletters or bulletins are used to this purpose, which have the advantage of being regularly consulted by patent examiners when assessing prior art.

6.4.3 Patent Scanning and Industry Landscaping

Patents are not only a means to protect technological knowledge, but they are also evidence of the technological knowledge possessed by a company and an industry. In this sense, they can become a strategic tool for assessing and positioning the knowledge possessed by competitors, or for mapping the knowledge represented in a certain industry and/or geographical region (Trippe 2015).

*Patent Scanning** is a set of methods that allow a high-level examination of pending or current patents to reveal business and technology trends, but not to the extent of grasping their content in depth.

The first general information that can be provided is a general description of the technological domain and of its trends. For instance, Fig. 6.6 shows a technological and geographical description of how patenting activity evolved in the high-pressure laminates industry following specific events (e.g., the Grenfell Tower fire at London

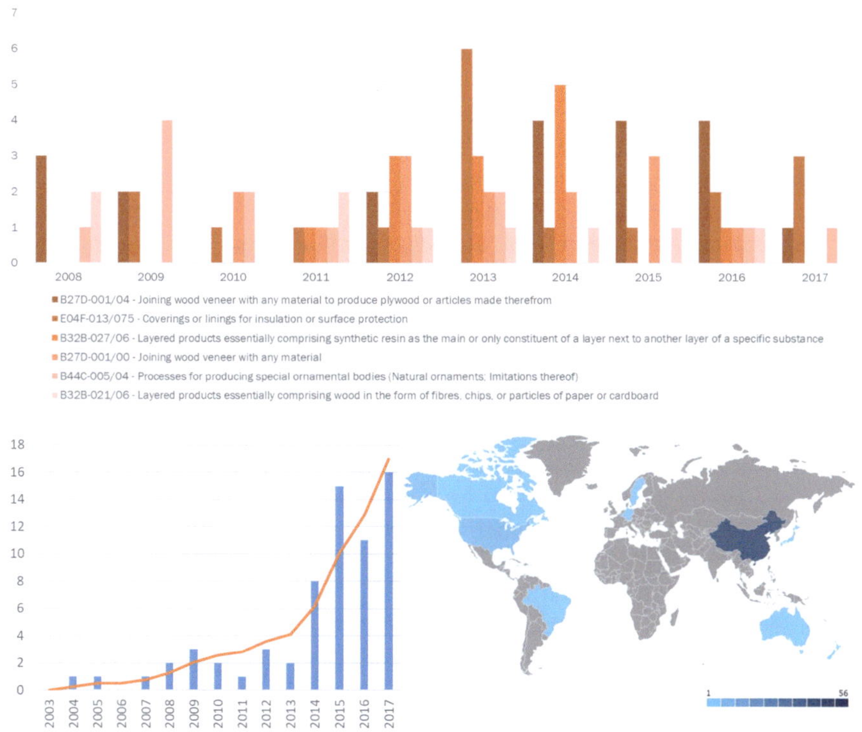

Fig. 6.6 Industry trends from 2008 to 2017 in the high-pressure laminates industry (top); the 67 patents on high-pressure laminates with ecological cores (bottom)

in 2017 led to an increasing interest in fire resistance) or specific new requirements (new sustainability requirements are leading to patenting activity in bio-based resins).

Patent scanning also provides information about competitors in the industry, along with their innovation strategy, or to spot possible entrants from other regions or industries that might become key players. This is particularly important to identify technologies and players that might bring disruptive innovation to the industry. Furthermore, one can spot top patent assignees on a specific patent class and where they are located, or can observe what specialization is being pursued by competitors in a specific geographical area.

All this information can help a firm address the many possible technological choices and define its innovation strategy.

6.4.4 Emerging Issues in Intellectual Property Management

The growing complexity of intellectual property management has recently led to a debate on the adequacy of current rules and institutions. This debate is of direct interest to policymakers, and also to practicing managers, who must constantly grapple with problems for which solutions do not appear to be perfectly defined.

6.4.4.1 Patentability of a 'Critical' Technology

One first issue is concerned with the patentability of inventions that are of growing importance in modern society. For instance, there is uncertainty on whether life forms, software, algorithms and business methods should be patentable or not.

The debate on patentability of life forms deriving from genetic engineering techniques is highly specialized and is also laden with ethical aspects. This prevents us from discussing it in this short chapter section; interested readers may refer to specialized literature (Knoppers 1999). Moving on to 'soft' inventions (i.e., software, algorithms and business methods), the debate is highly relevant to the services industry, which nowadays makes up a significant part of the economy of advanced countries. For firms operating in services, the availability of patenting as a tool for protecting inventions can have a significant impact on their competitive advantage and on the incentive they have to engage in research and development (Spulber 2011). At the same time, critics of patentability of 'soft' inventions highlight the risk of granting protection to trivial or highly abstract innovations that might have nothing to do with technology.

The debate is still ongoing, with a strong discussion on the essence of technology in contemporary society and with significantly different outcomes across jurisdictions. For instance, the USA is relatively open to patentability of these artifacts and attention paid by its judicial system seems to be shifting from arguing about principles to the more pragmatic problem of avoiding the issuance of bad or weak patents. The European Union has instead taken a different stance and has in principle ruled out patentability of software, algorithms and business methods. However, software can still be patented in the EU if it can be framed as a *computer-implementable invention** solving a technical—but not a business—problem.

6.4.4.2 The Fragmentation of the Technological Landscape

Another important issue is connected to the growing complexity of intellectual property. In some industries, such as telecommunications, IP is becoming highly fragmented, with many narrow patents being assigned to different firms. This makes it very easy to inadvertently infringe other actors' IP and creates situations in which a single denial to license could block other firms' activity, thus paving the way for costly judicial action or *patent wars*. In many cases, and especially when dealing with IP that is essential to operate a standard, governments can force patent holders to sign mandatory licensing agreements. These agreements often go under the acronym FRAND, that summarizes the requirements of being *fair* (e.g., the licensor does not extract benefits that are independent of the license), *reasonable* (e.g., in terms of pricing) and *non-discriminatory* (i.e., all licensees are treated in the same way).

The fragmentation also creates opportunities for so-called Non-Performing Entities, or *patent trolls**, to buy patents in bulk and then opportunistically look for licensing revenues by threatening legal action against manufacturers who might be infringing on their IP. The legitimacy of this activity is often questioned, since patent trolls do not carry out R&D, do not have a real business interest in the commercial exploitation of the inventions, and do not represent the interests of inventors. However, courts find it hard to reject their cases, since legally acquired property rights are always valid, regardless of the rights-holders' belonging to the group that the law originally wished to encourage by creating the rights themselves.

A fragmented technological landscape leads to another perverse mechanism that can impact the credibility of the patent system. When companies file numerous patent applications for relatively narrow and potentially worthless inventions, a patent office can be overwhelmed by the workload. In order to solve this problem, it is not thinkable to discourage excessive patenting by increasing administrative fees, since this would be especially damaging to individual inventors and small firms. Another alternative would be to increase the staff of patent examiners, but the cost to taxpayers would be quite high, not counting that it would be difficult to recruit and train highly skilled professionals in bulk. So, if one does not want to delay the lead time with which patent applications are examined, which would leave applicants in an uncertain state for an unacceptably long time, there will be a temptation to examine patents less thoroughly (Shapiro 2004). Patent examiners know that only a minority of patents will actually lead to products and have any economic value. Of these, only a small portion will lead to litigation. Therefore, there is an implicit incentive to examine patent applications somewhat less thoroughly and leave the final decision on their validity to the court, for the ones that will eventually be challenged.[20] While this effect can be rational from an administrative perspective, it has the fault of leaving patent holders in a state of uncertainty on the very existence of their IP rights and on their related value.

6.4.5 Alternative Policies to Patenting

The evident limitations of the patenting system have led critics to propose alternative approaches for dealing with IP and creating other incentive mechanisms to inventors. These mechanisms are often linked to public policies and therefore are partly associated to the topics covered in Chap. 1.

One of the possible approaches is *subsidization*. With patenting, private firms invest in R&D *ex-ante* and, in exchange, the government grants them the right to 'tax citizens' with monopolistic rents *ex-post*. One could imagine a system in which the government raises an innovation tax *ex-ante* and funds companies' R&D under the requirement that the results be openly shared.

Critics to this idea claim that the system would not work, since project selection and resource allocation would be performed quite badly by a government

[20] In 2012, the US Patent Office even introduced an internal Appeals Board to prevent patent validity cases from going directly to judicial courts. This has led to a strong debate that focused on the administrative nature of this entity, on its way of operating, and on the opportunity given to large firms to carry out opportunistic challenges to smaller competitors.

committee, compared to the market-based interaction between multiple corporate shareholders, managers and customers. Moreover, the incentive to work hard for completing a government-funded project that simply covers costs would not be as strong as the one coming from the opportunity to gain a monopolistic status under private funding. However, there are cases in which market forces might not be able to select the projects that maximize social utility. For instance, pharmaceutical companies are often criticized for working on *lifestyle drugs* addressed to rich countries and not dedicating enough effort to develop potentially life-saving medicines for emerging countries or for *orphan diseases**, where the financial rewards might be lower.

Another popular alternative approach to patenting is *Open Source*, which is currently extending from the domain of software to other fields, such as electronic hardware. The main concept behind Open Source is that inventors should allow others to freely use and add to their developments, since this can greatly increase the speed with which innovation evolves. If a wide community of technologists—instead of a single company—works on the same problem and collectively generates a standardized solution, both technical performance and diffusion may be accelerated (Raymond 2001).

Of course, if one does away with monopolistic rights associated to patenting, some kind of incentive for performing R&D must be restored somehow (Lerner and Tirole 2005). According to proponents of Open Source, this individual incentive might be relatively small, since R&D effort is shared by a wide community. Moreover, even though they cannot use the technology as monopolists, developers will be able to profit from their effort by selling related services, such as customization, training and complementary goods. The incentive mechanism works on the principle that the greater an individual's contribution to the technology, the greater the specific knowledge gained and the public reputation, and the more valuable these services will be on the market.

Despite some initial skepticism due to a somewhat strong ideological rhetoric, Open Source strategies have proved themselves quite successful in a number of fields. Most Internet companies use Open Source as the underlying technology, since their economic value comes from the services that they run, the way users are attracted, and the profitability of the business model. Furthermore, many established firms in the software industry have developed strategies that mix Open Source elements with proprietary software, in order to exploit the benefits coming from both models of IP management.

Prizes and competitions are further possible approaches that may stimulate technological development. During the early twentieth century, significant advances in the development of motor cars and airplanes came from contests and from attempts to breach previous performance records. Even today, many advances in the field of motor cars derive from the efforts spent in Formula 1 car racing.

This approach might not on its own create sufficient incentives for significant R&D investment and technological development. Moreover, being based on a tournament-like mechanism, in which multiple participants have to invest and only one wins the prize, the magnitude of the investment will tend to be quite small.

However, despite these drawbacks, prizes can be considered a useful complement to other means for providing incentives to innovation.

6.5 Shaping and Blue Ocean Strategies

When attempting radical and potential disruptive innovation, companies may profit from taking bold strategic moves, trying to generate and become leaders in new markets and industries, rather than fighting for market share in existing ones. This idea is sometimes termed *Blue Ocean Strategy**, from Kim and Mauborgne's (2005) popular book. The rationale is that, by challenging incumbents on their home turf, entrants are likely to be at a strong disadvantage. The old technology may have generated lock-in, while incumbents enjoy strong competitive advantage thanks to complementary assets and reputation. Even more, the way customers conceive products and their relationships with producers might be well-suited to the old technology, but might not fit well with the new one. Therefore, entrants may enjoy greater likelihood of success if they propose a 'new game', in which they have a clear technological advantage, knowing that incumbents will find it difficult to adapt to it, instead of playing the old one.[21]

The same argument can apply to firms that are currently operating in highly competitive industries (or 'Red Oceans'), who should carefully consider whether to continue competing in the traditional way, or propose radically innovative strategies. This school of thought is a clear departure from Porter's 'structuralist' model, based on preexisting industries and markets, and has been defined 'reconstructionist', since it aims at their redefinition (Kim and Mauborgne 2009).

When following a Blue Ocean approach, strategy must be based around a set of key elements. The first is the concept of *Value Innovation**, which requires aligning the economic value of the firm's offering (which broadly includes products, services and channels) to customer segments, defining target prices that might make the offering attractive to customers and profitable for the firm.

*Shaping strategies** are generally not limited to product innovations, but are usually extended to the entire business model. So, firms should not simply think of designing new products, but design a business around an aggregate of available technology and unmet user needs. Given the relevance of business model innovation, the following Chap. 7 will be dedicated to examining this topic in greater depth.

Authors such as Martin (2009) have advocated making use of creativity instead of simple analytical skills, when dealing with strategy. To this purpose, he has suggested using *Design Thinking** (Brown 2009; Cross 2011), which is a general problem-solving approach that has emerged from the Design community and has

[21] As an example, one can observe the introduction of the Wii by Nintendo in 2006. Nintendo aimed at broadening the user base in the games-console industry by making games easier, natural and fun to play. This went against the prevailing trend followed by the industry, which focused on computing power and visual performance, in order to cater to the specific needs of a 'core' of passionate game players.

Fig. 6.7 Foundational elements common to Shaping Strategies (after Hagel et al. 2008)

been applying to problems ranging from the design of complex systems to policy-making.

Other elements typical of shaping strategies include dropping Porter's dichotomy between cost leadership and differentiation strategies, and trying to pursue both of them at the same time, while striving to maximize speed of adoption of the innovation.

Following Hagel et al. (2008), shaping strategies usually start (as in Fig. 6.7) with the definition of a 'vision' for value creation. In order to implement the vision, the firm must decide which elements it wants to be responsible for, and which should instead come from users, customers, stakeholders and complementors. At the core of shaping strategies is the realization that it would be very difficult for a single firm to provide all the necessary elements on its own. Furthermore, it can be much more efficient to partner with existing actors and their established competencies, rather than trying to replicate them from scratch and maybe even competing against them.

One key enabler of shaping strategies, which is quite typical of Internet companies such as online marketplaces and social networks, consists in the development of 'platforms' that may easily create and aggregate stakeholders and support the rapid expansion of the firm.

Thanks to the platform and to the actors and resources that have been attracted to it, firms can devise strategies to quickly grow and scale up their activity, both to guarantee returns on the investment made and to achieve critical mass before any competitor does.

6.6 Innovation and the Resource-Based View

6.6.1 The Basics of the Resource-Based View in Corporate Strategy

The approach to strategic management that is closest to innovation management probably is the so-called *resource-based view* (or RBV in short). From a theoretical perspective, the RBV is based on the same Evolutionary Theory of the firm we introduced in Chap. 2. The firm and its evolution are investigated by observing its inside, rather than its external behavior. Moreover, the firm and its organization are defined as a collection of assets that operate together thanks to bundles of routines, engaged in an endless game of path-dependent adaptation to an unstable environment. Therefore, while Porter's approach places rational decisions taken by managers at the center of strategy, the RBV considers the differences between firms and managers' ability to determine and exploit them as the main source of competitive advantage.

In the RBV jargon, routines that involve resources are termed *organizational competencies*. Among the many competencies that any firm possesses, a subset of them will have the potential of leading to strategic differences and creating competitive advantage. These can be termed *core competencies** (Prahalad and Hamel 1990). Consequently, according to the RBV, strategic decision-making consists in understanding which competencies are core, ensuring their continuous growth, together with that of the underlying resources, and then using them to enter markets in which above-normal profits may be achieved. In RBV terminology, a *capability** is a competence whose strategic value has been understood and is therefore used for strategy setting. Finally, *dynamic capabilities** are higher-level competencies that a company uses in order to adapt, develop and reconfigure the existing portfolio of resources and competencies.

The RBV leads to the idea of managing strategy through core competencies (Hagel and Singer 1999). The main idea behind this approach is that—instead of recognizing attractive markets and then structuring itself internally in order to serve them—the firm must recognize its core competencies and then find markets in which they may be profitably exploited. More specifically, core competencies can generally be recognized as a company-specific *blending* of generic competencies. If one considers technological core competencies, generic competencies would be represented by established technological domains, such as the various fields of engineering. The core competence would therefore come from a company-specific *technological fusion* of these fields, so that one firm may specialize in 'optics and electronics' and another one in 'mass producing products in which functional materials are applied to substrates'.

Having acknowledged the core competence of the firm, managers can think of entering industries, serving markets and developing products that might allow at that particular time the most profitable exploitation of these competencies and exiting when appropriate.

The firm will witness a continuous strengthening of its core competencies, due to the experience gained in many different fields, and will progressively enter and exit industries, depending on their coherence with core competencies, attractiveness and stage of maturity.[22] Upon entry in a given industry, the firm will have to develop

[22] A 'core competency' must not to be confused with the 'core products' offered by the firm. On the contrary, firms that define their strategy around core competencies tend to shift quite readily from mature products and markets to emerging ones, in which they feel such competencies might lead to competitive advantage. Not many firms are close followers of this strategic management approach. Well-known examples are Canon ('blending optics and electronics') and Corning ('amorphous materials technology', such as glass and ceramics). Both of these companies have

additional *integrative competencies* that are specific to the development of products in that particular field and *market competencies* that are specific to interacting with that market.

This process of entry and exit leads to diversification in a way that may superficially resemble the one followed by firms using the product portfolio approach. However, while the latters' entry and exit decisions are fundamentally based on economic and financial considerations, the former will use coherence with core competencies as the main criterion. When the firm is not sure about this coherence, or when it feels that the integrative and market competencies that must be developed in order to support entry might lead it too far away from its core, it will probably avoid this decision, even if the market is financially attractive. It might therefore look for alternative strategies, such as technology licensing or vertical disintegration (i.e., selling key components), in order to profit from its competencies but without defocusing from them.[23]

6.6.2 Bridging the Core Competencies and Competitive Advantage Approaches

In many environments, and especially when competition if very intense and industry dynamics very fast, neither the resource-based nor Porter's competitive advantage approaches are able to provide an adequate support to strategic management. The former tends to look at strategy from a rather long-term perspective, in which firms tend to remain faithful to their core competencies and rather gradually shift their attention from one industry to another one. The latter too tends to have a long-term perspective, since it hypothesizes that competitive advantage can be sustained over time. So, firms will tend to remain grounded in the industries where—sometimes after significant investment and fighting—they have become able to extract above-normal returns.

In a fast-moving environment, both approaches can fail to support managers in proactively defining the firm's future direction. Managing in a turbulent environment requires a continuous process of maneuvering, learning, and discussing with downstream markets and with providers of production factors, including

progressively entered and exited industries, depending on the degree with which their core competencies were felt to be relevant.

If one broadens the concept of competency to the 'capability to serve a market', one can also consider firms such as Sony ('electronics for entertainment', which is the intersection between a market and a technology) and IBM ('technology for supporting business activities'). By observing the history of these firms, one can notice that these companies tend to have a long lifetime and pass nearly unscathed through the business cycles and the s-curves of the products they produce.

[23] For instance, in the '90s, Canon licensed its technology for laser printers and digital photocopiers to competitors, such as Xerox and HP, possibly giving up the possibility of becoming a near-monopolist player in that market. Interestingly enough, it also performed a strategy of partial forward integration, effectively competing against its two customers. This allowed Canon to develop an adequate understanding of the market and, even more, to avoid that double marginalization could lead to an excessive reduction of potential demand.

technology. Some authors have termed this perspective as *theory of temporary advantage* (D'Aveni 1994; D'Aveni et al. 2010). They stress the idea that competitive advantage is seldom sustainable, and it is more appropriate to depict it as a prolonged sequence of temporary advantages. In such a context, the firm should not only think about defending its current competitive advantage, but also constantly look at how to establish its next one. The problem is by no means trivial, since the underlying decisions and tradeoffs are quite difficult to frame and implement.

One first issue is related to allowing the *cannibalization* of a currently profitable business activity by a future one, which might be the foundation of the next temporary advantage. In general terms, it is usually quite obvious that the switch will have to be made at some point in time, but the crucial point is to understand when.[24] An early switch might determine first-mover advantage in the new paradigm, but also limit funding from the prematurely interrupted old one, and giving up the potential benefits of performing 'sailing ship effect' investments. Moreover, a fast mover risks entering an s-curve which might take many years before it goes through the incubation phase and starts providing adequate returns.

A second issue is associated with *path dependency and relatedness*. The firm will have a strong competitive position if it tackles technologies and markets that are somewhat close to its current ones. However, a tradeoff often arises if—by moving a little bit further away from them—it might address a field with larger business potential, albeit accepting to have a weaker competitive advantage.

A third problem has to do with the possibility that, in a rapidly changing environment, *uncertainty* may be so high that the firm may decide not to conceive any strategy at all, regardless of the internal inertia with which it would be able to execute it. This kind of situation leads to a tradeoff between the risk of not doing anything, thus missing out a possible trend, and of making costly and aimless moves before the time is ripe to take any clear action.

In order to frame the strategic management process within such a context, one can consider that a firm's core competencies will not be cast in stone. Competencies will evolve in time, together with the corresponding sequence of market entry and exit decisions. Following Helfat and Raubitschek (2000), this mechanism can be represented as in Fig. 6.8.

The model separates the firm into three elements. The first is the 'knowledge system', representing the firm's portfolio of competencies, as given by its resources and routines. The second element can be termed the 'activity system' and represents the aggregate portfolio of product and market development projects that describes the firm's activities. This project portfolio can be viewed as the core element of the firm's business plan. The third element is the 'learning system', which feeds back into the knowledge system and is concerned with updating the firm's competencies.

Learning can occur in two ways. One is incremental and is due to the continuous use of core competencies in different industries. This implies substantial *experiential learning* (or learning-by-doing) and knowledge *spillovers*, when results that

[24] As already discussed, Netflix was able to perform a timely and successful cannibalization of its original business model, when it shifted from mail delivery of DVDs to streaming.

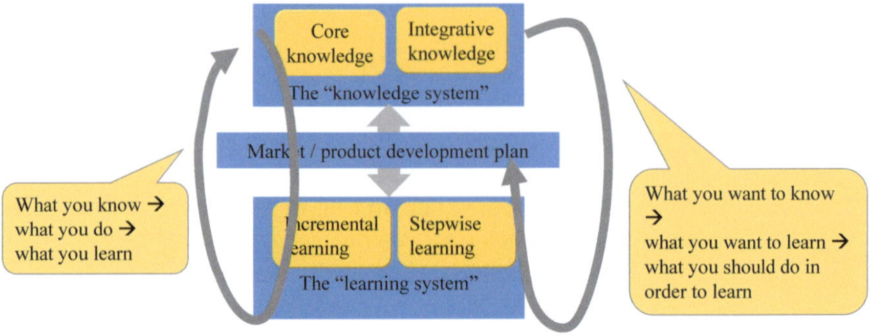

Fig. 6.8 Evolution of corporate competencies

emerge from the work done in one industry turn out to be valuable in another one. A second mechanism is the *stepwise learning* that occurs when the firm wishes to make radical changes to its knowledge base. Stepwise learning generally happens when the technological and/or market environment is about to undergo significant modifications and the company wants to be ready for, or react quickly to, this change.[25]

Based on this representation, it is possible to conceive the formulation of innovation strategy according to two main avenues. One can be viewed as a somewhat passive *competence exploitation* path, in which the firm defines a portfolio of product and market development projects, based on the competencies it possesses. By executing this plan, and thanks to incremental learning and spillovers, the firm will find itself with an updated knowledge base that will eventually become the basis for a new plan. In other terms, what the firm does is based on what it knows, and then learns by doing. A second path consists in a proactive *exploration* of new fields, in which the firm focuses on the competencies it feels will be needed in the future. The firm may do so by both enacting stepwise learning activities, without an immediate impact on the product and market development plan, and/or by purposely deciding to work on products and in markets for which it currently lacks competencies, with the very objective of developing them. By following this path, the formulation of an innovation strategy consists in defining a desired portfolio of competencies and a plan for developing them. In other terms, the firm focuses on learning what it wants to learn and does so through all the learning mechanisms available.

The two paths are complementary and not necessarily alternative to one another, since a firm may follow both exploitation and exploration strategies at the same time. Though in different ways, both paths operate on the two 'levers' of formulating and planning two portfolios: a portfolio of competencies and a portfolio of product and market development projects. We can consider these two portfolios and

[25] As an example, one can think of Kodak in the '80s. By envisioning that digitalization might revolutionize the field of photography and make film obsolete, the firm decided to invest significant resources in the development of the new technology.

their reciprocal interactions to be at the heart of innovation strategy and Chaps. 8 and 9 will deal with them in depth.

6.7 The Formulation of an Innovation Strategy

Before closing the chapter, it is appropriate to remember that formulating strategy is not the application of a recipe but—rather—a complex decision-making process carried out in a complex and uncertain environment.[26] As such, even though it may be aided by frameworks that aid the effort, an adequate attention must be paid to the cognitive processes and biases with which managers operate.

Bingham and Eisenhardt (2011) have proposed a classification of strategic decision-making areas, highlighting the fact that managers mostly rely on heuristics that are the fruit of personal traits, training and professional experience (and that—therefore—are subject to path dependency). The main four heuristics are *of selection* (i.e., which opportunities should be pursued, and which should be dropped), *procedural* (i.e., what actions should be carried out to pursue the selected opportunities), *priority* (i.e., creating a ranking of opportunities) and *temporal* (i.e., specifying when to act and at what pace).

Concerning the type of heuristics, managers operating in relatively stable environments will tend to use an evolutionary approach, in which strategy is progressively formulated as response to the observation of the changing environment. However, this approach has clear limitations in the case of significant modifications to the environment and/or radical strategic initiatives. Martins et al. (2019) have therefore proposed alternative approaches for formulating strategy, which are summarized in Table 6.2. Needless to say, these approaches will require additional effort, in order to break loose from established ways of operating.

Finally, a successful strategy does not only require defining a smart route to follow, but also making sure that it fits with the available resources and ensuring that it will be executed properly. In turn, this creates a connection between the dynamic capabilities of the firm and the cognitive and managerial capabilities of the executives, who must be able to lead the drafting and the execution of the strategy. An interesting framework that brings these concepts together has been proposed by Helfat and Peteraf (2015). The underlying idea is that any strategic initiative must be recognized as an opportunity, defined, and then enacted. This implies, as depicted in Fig. 6.9, the availability of a tryad of managerial capabilities, of *sensing the opportunity*, *configuring/seizing the corresponding action* and *transforming/executing it*. These managerial capabilities have to do with the cognitive abilities to

[26] It is nowadays quite common to address this type of situations by using the acronym VUCA, which means *Volatile* (implying the speed with which phenomena evolve), *Uncertain* (which has to do with the difficulty of capturing reliable data and making forecasts of phenomena and scenarios), *Complex* (which is associated to the number of factors and interdependencies that might shape the future) and *Ambiguous* (tied to the difficulty of understanding the way the observed system works).

Table 6.2 Alternative approaches to firm strategy

Approach	Determinant of change	Ideation	Design
Evolutionary	Exogenous change leads to uncertainty	An initial hypothesis arises somehow	Trial and error until satisfactory solution arises
Rational positioning	Exogenous shocks force the firm to change	Rational economic analysis suggests a solution	The solution is detailed and implemented
Cognitive—analogical reasoning	Intuition or exogenous change determines a redefinition of mental schema	A solution arises by applying an analogy from another industry («we are the X of Y»)	The solution is defined by selectively replicating/adapting elements from the source model to the target
Cognitive—conceptual combination		A solution arises by incorporating substantially new features (modifier) in the current model	The solution is defined by identifying the modifier elements and *integrating* them in the current model

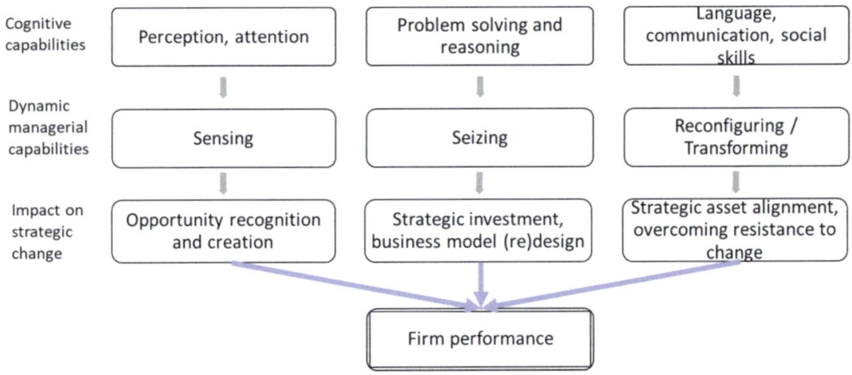

Fig. 6.9 Cognitive and managerial capabilities for strategy

perceive opportunities (which implies being able to 'read through the lines' and 'capture weak signals' in data and events), to carry out a rational and quantitative analysis and design of the actions to be carried out, finally, communicating and mobilizing the organization in the strategic direction.

The framework shows the difficulty that any firm will encounter in structuring and executing an innovation strategy, unless the cognitive capabilities that are available in its top management team are not sufficiently rich and balanced. The model also has interesting implications on managerial education and on managerial selection processes, who must make sure that the above abilities are effectively present in the people who will have to run these activities (Fig. 6.9).

References

Bain JS (1959) Industrial organization: a treatise. Wiley, New York

Boldrin M, Levine DK (2013) The case against patents. J Econ Perspect 27(1):3–22

Bingham CB, Eisenhardt KM (2011) Rational heuristics: the 'simple rules' that strategists learn from process experience. Strateg Manag J 32(13):1437–1464

Brown T (2009) Change by design: How design thinking transforms organizations and inspires innovation. Harper Collins Publisher, New York

Claessens S, Laeven L (2003) Financial development, property rights, and growth. J Financ 58 (6):2401–2436

Cohen WM (2005) Patents and appropriation: concerns and evidence. J Technol Transfer 30(1–2):57–71

Cohen WM, Walsh JP (2001) R&D spillovers, appropriability and R&D intensity: a survey-based approach. In Spivack RN (ed) Advanced technology program's international conferences on the economic evaluation of technological change, NIST Special Publication 952. USGPO, Washington, pp 22–29

Cross N (2011) Design thinking: understanding how designers think and work, Berg, Oxford, New York

D'Aveni RA (1994) Hypercompetition: managing the dynamics of strategic maneuvering. Free Press, NewYork

D'Aveni RA, Dagnino GB, Smith KG (2010) The age of temporary advantage. Strateg Manag J 31:1371–1385

Faulkner DO, Campbell A (2006) The Oxford handbook of strategy: a strategy overview and competitive strategy. Oxford University Press, Oxford

Grahovac J, Miller DJ (2009) Competitive advantage and performance: the impact of value creation and costliness of imitation. Strateg Manag J 30(11):1192–1212

Guidehouse Insights (2021) Guidehouse insights leaderboard: automated driving systems: assessment of strategy and execution for 15 companies developing automated driving systems. https://guidehouseinsights.com/reports/guidehouse-insights-leaderboard-automated-driving-systems, Retrieved 23 Oct 2022

Hagel J, Singer M (1999) Unbundling the corporation. Harward Bus Rev 77(2):133–141

Hagel J, Brown JS, Davison L (2008) Shaping strategy in a world of constant disruption. Harv Bus Rev 86(10):80–89

Helfat CE, Raubitschek RS (2000) Product sequencing: co-evolution of knowledge, capabilities and products. Strateg Manag J 21:961–979

Helfat CE, Peteraf MA (2015) Managerial cognitive capabilities and the microfoundation of dynamic capabilities. Strateg Manag J 36(6):831–850

Kim WC, Mauborgne R (2009) How strategy shapes structure. Harward Bus Rev 87(9):72–80

Kim WC, Mauborgne R (2005) Blue ocean strategy: from theory to practice. Calif Manage Rev 47 (3):105–121

Knoppers BM (1999) Status, sale and patenting of human genetic material: an international survey. Nat Genet 22:23–26

Kopp BS (1968) Conglomerates in portfolio management. Finan Anal J 24(2):145–148

Lerner J, Tirole J (2005) The economics of technology sharing: open source and beyond. J Econ Perspect 19(2):99–120

Lev B (2001) Intangibles: management, measurement and reporting, The Brooking Institution Press, Washington DC

Levin R, Kievorick A, Nelson RR, Winter SG (1987) Appropriating the returns from industrial R&D. Brook Pap Econ Act 3:783–820

Martin R (2009) The design of business: why design thinking is the next competitive advantage. Harvard Business School Press, Boston

Martins LL, Rindova VP, Greenbaum BE (2019) Unlocking the hidden value of concepts: a cognitive approach to business model innovation. Strateg Manage J 9(1). Special Issue: Business Models, Demil B, Xavier Lecocq X, Ricart JE, Zott C (eds), pp 99–117

Miles RE, Snow CC (1978) Organizational strategy, structure and process. McGraw Hill, New York

Miles RE, Snow CC (1984) Designing strategic human resources systems. Organ Dyn 13(11):36–52

Porter ME (1979) The structure within industries and companies' performance. Rev Econ Stat 61 (2):214–227

Porter ME (1985) Competitive advantage: creating and sustaining superior performance. The Free Press, New York

Prahalad CK, Hamel G (1990) The core competence of the corporation. Harv Bus Rev 68(3):79–91

Raymond L (2001) Determinants of web site implementation in small businesses. Internet Res 11 (5):411–424

Schacht WH, Thomas JR (2005) Patent law and its application to the pharmaceutical industry: an examination of the drug price competition and Patent Term Restoration Act of 1984, CRS Report for Congress

Shapiro C (2004) Patent system reform: economic analysis and critique. Berkeley Technol Law J 19(3):1017–1047

Spulber DF (2011) Should business method inventions be patentable? J Legal Anal 3(1):265–340

Teece DJ (2000) Managing intellectual capital. Oxford University Press, Oxford

Teece DJ, Pisano G (1998) The dynamic capabilities of firms: an introduction. In: Dosi G, Teece DJ, Chytry J (eds) Technology, organization, and competitiveness: perspectives on industrial and corporate change. Oxford University Press Inc, New York

Trippe AJ (2015) Guidelines for preparing patent landscape reports. World Intellectual Property Organization, Geneva

Business Model Innovation

7

Abstract

This chapter focuses on the concept of the business model of a firm and on business model innovation, which is at the heart of a number of innovation strategies. The chapter focuses on the amply diffused Business Model Canvas and on its practical use within both established and startup firms. The discussion also presents the e3value modeling methodology and presents recent literature on the process by which business models can be designed.

This chapter focuses on the emerging topic of business model innovation. The concept has already been mentioned in the previous chapters, and specifically in Chap. 1, during a preliminary discussion on different types of innovations, and then in Chap. 6, when introducing the concept of 'Blue Ocean' strategy. Nevertheless, given its growing relevance in the field of innovation strategy, it is appropriate to dedicate a separate chapter to the topic. The subject will be discussed at first by providing some basic information and definitions, and then by exploring in depth two approaches that can be practically used to analyze and design business models.

7.1 What is a Business Model

The concept of '*business model*' is relatively recent, since the term came in use only at the beginning of the twenty-first century, when Internet companies started to emerge (Mahadevan 2000). For these firms, it was quite apparent that their activity differed markedly from the 'traditional' operational model of buying, transforming and selling goods or services and on the financial model of profiting from the difference between value of sales and the cost of goods sold. Search engines, auction sites, social networks and music streaming companies provide services to—

M. Cantamessa and F. Montagna, *Management of Innovation and Product Development*, https://doi.org/10.1007/978-1-4471-7531-5_7

and generate their revenue from—multiple parties in ways that are often quite asymmetrical and difficult to understand.[1]

When researchers started giving a closer look at the phenomenon, it became apparent that business models and *business model innovation* mattered for firms operating in other industries as well. First, many firms were introducing business model innovations that were bringing disruptive change to their respective industries. As a typical example, one can think of low-cost airlines, which represented a significant innovation in transportation, though the major change they introduced was clearly associated to the business model and not to the technical content of flight. Moreover, it became recognized (Chesbrough 2010) that business models represent the 'format' with which firms can bring technological innovations to the market and are therefore key to determining success and failure of the same innovations.[2]

Given the importance of business models and of business model innovation, researchers started working on precise definitions (Lee 2001; Shafer et al. 2005), on methods for supporting their study and on the relationships between business models and strategy (Magretta 2002).

As highlighted by Teece (2010), a business model represents a "*conceptual, rather than financial, model of a business*" and is therefore aimed at representing the constituent elements of a business and their coherence, rather than its profitability.

With some adaptation from Shafer et al. (2005), the key elements of a business model are represented in Fig. 7.1. These elements cover the main strategic choices that define a business, the resources which enable the firm to create value, the positioning of the firm in its value network and a high-level definition of cost and revenue structures. At the same time, the elements in the business model cover both sides of supply (i.e., what is being offered and how it is going to be produced—by the firm and by its suppliers and partners) and demand (i.e., who is the customer and how it can be reached).

Once a business model has been defined, it is possible to 'put flesh on the bones' and use it to generate strategies, business plans, profit and loss and financial projections. Conversely, strategies and business plans cannot really be articulated if not by referring to an underlying business model. Therefore, a firm's business model can be viewed as a more generic concept than its business strategy and as a precursor to it. At the same time, the former provides a high-level description of 'what' the firm will do, but stops short of the latter's aim of stating how this 'what' can be effectively reached, in order to generate and maintain competitive advantage.

[1] For instance, a service like Google may provide a free Internet search service to users, but make substantial profit by selling highly targeted advertising services to other firms.

[2] This is by no means a recent phenomenon. For example, Thomas Edison* (1847–1931) realized that an isolated firm selling incandescent light bulbs would not be sustainable in a world where electricity was not commercialized. So, he developed a business model around the idea of selling complete electrical direct current systems, including local power generation and lighting. Eventually, this business model lost against George Westinghouse's, which was based on alternate current systems that allowed remote generation of electricity and its long-distance transportation.

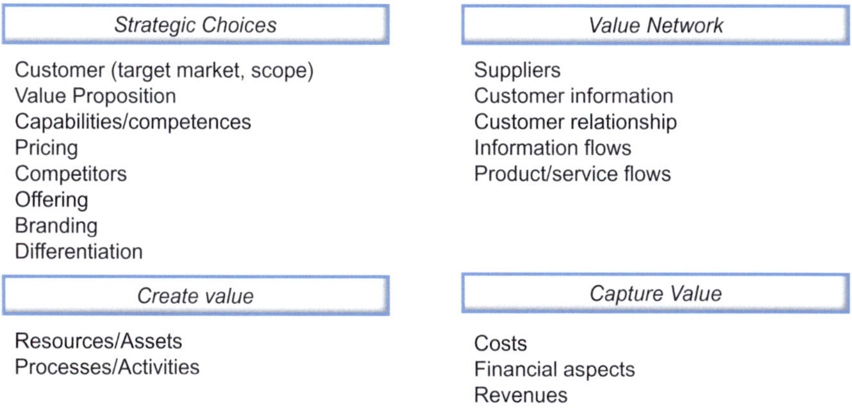

Fig. 7.1 Main elements of a business model

The literature has come up with a few approaches for representing business models. The main difference lies in the focus, which can be on an individual firm or on a complete ecosystem. Two well-known corresponding approaches are discussed in the next two sections.

7.2 The Business Model Canvas

The 'Business Model Canvas'—or '*Canvas*' in short— (Osterwalder and Pigneur 2010) is a qualitative and informal tool for structuring business models. The Canvas came out of the authors' initial research on the constituent elements of business models and their ontology (Osterwalder 2004) and—being very simple and intuitive—it has become a de-facto standard in both professional practice and academia. Its simplicity can be misleading, however, and its underlying concepts have to be correctly understood, if one wants to obtain sound results and not naïve ones. The following subsection will provide a description of the Canvas, while the following one will discuss its use.

7.2.1 The Main Elements of the Business Model Canvas

At its core, and with reference to Fig. 7.1, the Canvas can be considered a concise, structured and static map of a subset of the key elements that make up the business model. However, it does not fully explicit the relationships between elements, nor does it represent the dynamic behavior of the firm's business model within its broader value system.

As shown in Fig. 7.2, the Canvas brings together four main areas of the business model: the product and the infrastructure (which cover the 'offer' side of the business

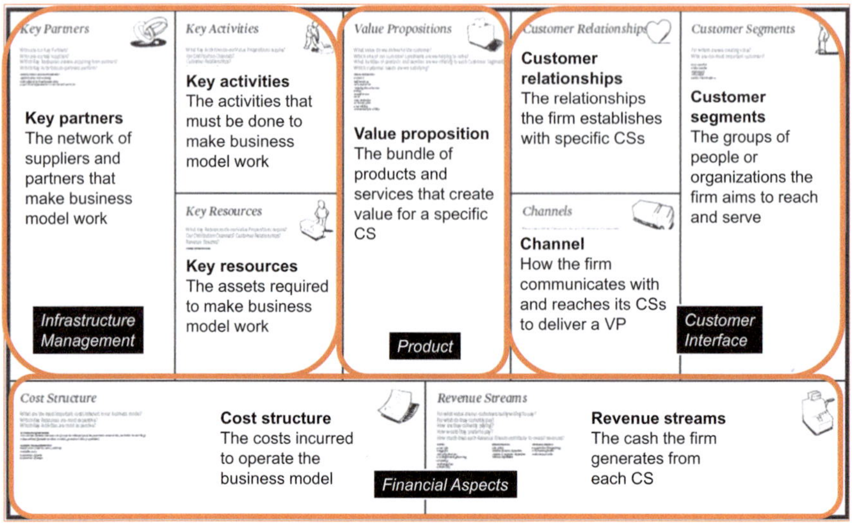

Fig. 7.2 Business Model Canvas

model), the customer interface (which covers the 'demand' side) and the associated financial aspects, based on the costs that characterize the side of 'offer' and the revenues deriving from the side of 'demand'. In turn, these major areas are split in nine sections, which will be described in detail in the following. The order with which we start the discussion is meaningful for teaching purposes and does not necessarily represent the order with which the Canvas should be tackled. As a matter of fact, it is quite likely that the compilation of a Canvas will undergo many iterations and back-and-forth jumps between sections, in order to achieve coherence.

7.2.1.1 Customer Segments (CS)

Customers are key actors of a business model. In fact, a business is sustainable only if it offers some*thing* that is of value to some*one* who is willing to pay and cover the costs. The CS section of the Canvas specifies the specific groupings of customers the firm intends to serve and provide value to (and, conversely, it implies which CSs will *not* be served). The definition of a CS should be narrow enough to provide a clear indication of the strategic choices that are being made (e.g., for a magazine, 'readers' is too generic as descriptor, while 'teenage female readers' might be more adequate).

The CS section can list a number of different segments. Differences may be due to the variety of business relationships (e.g., readers and advertisers are both customers to a newspaper, but their nature is obviously not the same), needs (e.g., a family and a 30-year-old single may not have the same needs when it comes to buying a laundry machine), willingness to pay (e.g., spectators willing to go to a premiere and people waiting for a later and cheaper show) and distribution channels (e.g., online shoppers vs. customers who prefer a traditional retailing experience).

Customer segments should not only be identified and described, but also quantitatively estimated. This information will be of great value when dealing with revenue streams later on.

Together with listing the CSs, it is important to generate a sufficiently deep understanding of customers' problems, needs and wants, so that they can later be associated to the firm's value propositions. Methods for eliciting and structuring customer/user needs will be amply discussed in Chaps. 13 and 14 and—among them—the Empathy Map may prove to be particularly suitable.

7.2.1.2 Value Propositions (VP)

At the heart of a business model and directly in face of customer segments is the concept of 'value proposition', which describes the objects of the transactions that will be carried out with customers. A value proposition is not simply a product or a service being offered, but a descriptor of the value that underlies the product, makes it attractive and might give it some kind of competitive edge. It represents the way with which a customer need is met, or a problem is solved, in a way that is possibly better, faster or cheaper than with competitors. In Osterwalder's terms, it is "*a bundle of benefits that a company offers customers*".[3]

The concept of value proposition is not an absolute, but is specific to the CSs that have been previously identified. A strong VP for one customer segment may be quite uninteresting to another segment. Therefore, providing a precise description of the value content of a VP in relation to its target CS is a key exercise in defining a business model. So, it is quite common to iterate back and forth a number of times in order to come up with a coherent set of VPs and CSs that map well into each other.

7.2.1.3 Channels (C)

Once the team has defined CSs and the corresponding VPs, one can define the channels that allow the delivery of the latter to the former. Examples of channels can be physical vs. virtual (e.g., stores vs. websites), direct versus indirect. The role of channels is not simply to sell products and services, but to follow the customer over her/his purchasing process, from the initial raising of awareness and evaluation of the products, and all the way to ensuring effective delivery and after-sales support.

Of course, channels too must be matched with the specific VPs and CSs that make up the business model, and coherence must be sought for all of them.

[3] For instance, while "a portable and Internet-connected electrocardiograph" is a good descriptor for a product, a better wording for its value proposition could be "a professional-grade ECG device supporting onsite telemedicine processes". The accent on 'professional grade' can be useful, because it may set the device apart from other portable but lower-quality competitors, as well as the accent on the 'telemedicine processes' shifts attention from a purely technical feature to the corresponding outcome, which can be of value to a potential customer segment.

7.2.1.4 Customer Relationships (CR)

Customer relationships add another twist to creating a connection between VPs and CSs. While channels have to do with 'institutional' or 'physical' entities that relate the firm to the customer, CR provides a description of the complete set of connections and bonds that engage the customer within these channels. For instance, the CR section may list elements such as personal assistance, automated services, user communities.

CRs cover the time of the initial purchasing decision (i.e., customer acquisition) and also the subsequent phases in which the firm still deals with existing customers, for the purpose of retaining them and increasing sales.

When coming up with CRs, coherence must be ensured with VPs, CSs and Channels.

7.2.1.5 Revenue Streams (RS)

The revenue streams section closes the portion of the Canvas that covers the demand side of the business model. Given the previously developed blocks, the firm can come up with a description of its revenue mix and a preliminary evaluation of its entity.

Revenues are of course closely related to VPs, since the value that is delivered to a CS determines—together with existing competition—each customer segment's willingness to pay. When defining the revenue stream that can be expected to accrue from each CS, the firm may also consider the possibility of allowing cross-subsidization between CSs.[4]

When defining revenue streams, the firm will have to decide on its pricing structure. The list of alternatives is quite endless, but a few main decisions can be considered to be the following:

- **Transactions versus recurring**. In the case of transaction-based pricing, the customer is charged each time she/he adopts the VP (e.g., you pay when you buy your newspaper). Conversely, in a recurring scheme, the customer is charged ex-ante for being granted the availability of the VP (e.g., you pay an annual subscription and get all the newspapers, even if you won't read all of them). In some cases, pricing can be a mixture of the two (e.g., in many cities, bike sharing services charge a fixed subscription fee, plus a usage fee when bike rides exceed 30 min).

[4] *Subsidization** means that one CS does not cover the costs it generates, but the overall business model stands on its feet because margins from some other CSs cover the difference. This typically happens in multisided markets, in which one side subsidizes the other. For instance, the readers of newspapers are subsidized by advertisers. Subsidization also occurs in the case of vertically differentiated offerings. For instance, a passenger flying economy class with a cheap ticket is subsidized by business class passengers paying full fare. However, given the cost structure of an airline flight, which has high fixed costs and negligible variable costs, it's more profitable to carry an additional customer who doesn't fully cover his costs, rather than let a seat to be empty. Similarly, in 'freemium' business models used in digital services, paying users are given a bundle of functions that make up a VP of higher value, and they subsidize the limited features being enjoyed by the non-paying users.

- **Fixed versus subject to discounts**. Prices can be fixed in a way that is independent from CSs, volume and product features. Conversely, the firm may engage in some form of price discrimination, offer volume discounts and so on.
- **Fixed versus proportional to value**. This decision is relevant for products or services that are complements to other products or services. For instance, at-home delivery can be charged a flat fee of 2 $ or can be computed proportionally to the value of goods being delivered.
- **Sale versus rental**. In the former case, the customer gains ownership over the good, while in the latter it simply purchases the right to use it. A typical example is the selling vs. the leasing of a car.
- **Static versus dynamic pricing**. In the former case, prices are fixed and do not change as a function of supply and demand at the time of consumption, and vice versa. For instance, taxi fares are subject to static pricing, while online limousine services (Uber is a well-known example) allow prices to vary according to supply and demand.

Choosing an appropriate pricing structure is not trivial, since it can have significant impact on customer behavior. For instance, a recurring fee can be off-putting at first but—once a customer has been acquired—it creates a much stronger incentive to consume than a per-transaction pricing scheme would. Dynamic pricing can be an efficient way to match supply and demand, but it can also be confusing and disturbing to unsophisticated customers and deter them from consumption.

Once a pricing structure has been defined, the firm can multiply the expected revenues per user by the size of each CS and come up with a preliminary estimation of its overall revenues and their composition by CS. This is not enough to assess the profitability of the business model, since this will also depend on costs. However, having a ballpark figure of the expected revenues can provide the decision-maker with a preliminary idea of the size of the business opportunity that is being developed.

7.2.1.6 Key Activities (KA)

Once the VP and the demand side of the business model have been developed, the attention can now move toward the supply side, which is responsible of making the production of the VPs and their delivery to customers possible. By analyzing the VP, C, CR sections of the business model, it is possible to come up with a list of activities that are most needed to support them.

When developing a Canvas, one will not waste time with activities that are generic and not particularly critical to the business model, but will focus on the ones whose execution is essential. Therefore, identifying KAs does not only serve the purpose of completing the Business Model Canvas, but also allows managers to understand which parts of business operations will have to receive adequate attention both when formulating strategy and when executing it.

7.2.1.7 Key Resources (KR)

The key activities that have been listed in the previous panel are made possible by key resources, whose identification is critical for understanding the business model and—later on—for making strategic decisions on their ownership. These resources can be physical (e.g., facilities, equipment, access to scarce raw materials, etc.), intangible (e.g., brands, patents, etc.) or human (e.g., skills and competencies). At this point, it is enough to list the resources, without actually defining how these same will be generated or used, either through direct investment or by accessing resources that are available on the market.

In the KR section one can usually omit those resources that are not specifically crucial to the business model and are common to any company. For instance, any firm will need good banking services, but 'access to credit' should be considered as a KR only for business models that require significant capital investment or working capital.

While in the KR section one can include both long-term and current assets. Long-term assets are the fruit of investment, and the value they produce goes beyond the initial outlay (e.g., machinery, personnel's skills, etc.), while the latter assets are going to be quickly converted to VPs (e.g., stocks of raw materials, consumables, etc.).

7.2.1.8 Key Partnerships (KP)

In the KP section of the Canvas, the firm reflects on vertical integration decisions with respect to the key resources it has previously listed. Specifically, the firm can decide whether each KR should be owned by the firm ('make'), made available by using suppliers ('buy') or by associating with complementors and strategic partners ('ally').

This decision will depend on the nature of each KR, as we will elaborate it in Chap. 8. If the KR is absolutely *core* to the business model and coherent to the firm's core competencies, it is likely that 'make' will be the most appropriate choice. Conversely, if the KR is *generic* and not particularly linked to the firm's core competencies, contracting at arm's length with suppliers will in general make sense.

The decision might be harder in the case of *non-core, but co-specialized* KRs, which must be developed in a way that is specific to the firm and its business model. In this latter case, the firm will probably be unable to ensure the availability of these competencies by simple arms-length contracting with generic suppliers. These suppliers may not exist at all, since these KRs are firm-specific. Moreover, suppliers who could develop them might be unwilling to make the required investment, unless they have sufficient guarantees and a long-term commitment from the focal firm. In these cases, 'ally' might be the most advantageous choice.

Making the right decisions at this point is quite critical to ensure the correct execution and profitability of the business model. Using resources coming from parties who are ineffective or inefficient, neglecting potential economies of scale and not considering risk, might in fact the wreck the most beautiful business model one could conceive.

7.2.1.9 Cost Structure (CS)

In general, the allocation of KRs to either the firm, to market-based transactions or to long-term partnerships, allows the definition of the business model's cost structure, which is the last section of the Canvas. This allocation allows the firm to understand and perform a preliminary evaluation of the distribution of its costs among the main cost categories (i.e., fixed, semi-fixed and variable). Moreover, it will allow a preliminary estimation of economies of scale and economies of learning that might arise when operating the business model.

The evaluation of the business model's cost structure is a crucial point, since it is now possible to compare the CS with the RS and evaluate the degree to which the model is profitable or not. In most cases, profitability is not to be considered statically, since business models will usually be unprofitable at the beginning. So, a dynamic view over multiple time periods will allow the firm to consider the evolution of demand, revenue and cost, so that—hopefully—the initial investment may be recovered.

7.2.2 Using the Canvas in Practice

From the short description given above, it is quite apparent that the Canvas is first of all a powerful discussion and communication tool, since it allows a synthetic description of a complex system such as a firm's business model. As such, the Canvas can be used in multiple ways and with quite some degree of flexibility. Among the many instances of use, it might make sense to focus on the four that are of greater relevance to innovation strategy.

7.2.2.1 Using the Canvas in New and in Existing Firms

The Canvas can be used either for designing a business model from scratch or for redesigning the business model of an existing firm.

In the former case, the team that works on the project will simply develop a single target business model or, better still, a set of alternatives to be compared with one another. In the case of existing firms, the team should start with a Canvas that faithfully depicts the current (or 'As-Is') business model, without yet trying to correct its possible shortcomings. Then, the 'As-Is' Canvas can be critically analyzed, so that inconsistencies can be highlighted and improvements proposed, thus leading to the drafting of a new ('To-Be') business model.

In order to make the migration from the starting point to the target business model possible, the team may also come up with a roadmap characterized by a number of intermediate business models. For new firms, this may first of all be needed to ensure a sustainable growth path for the enterprise. Moreover, the intermediate stages can be useful for making 'field experiments' on the evolving business model. Intermediate business models can have the same role for established businesses, as well. In fact, no company would ever give a 'go ahead' to a strategy that would require a deep change to its current business model and a blind investment in a future, but obviously quite uncertain one. When the business model

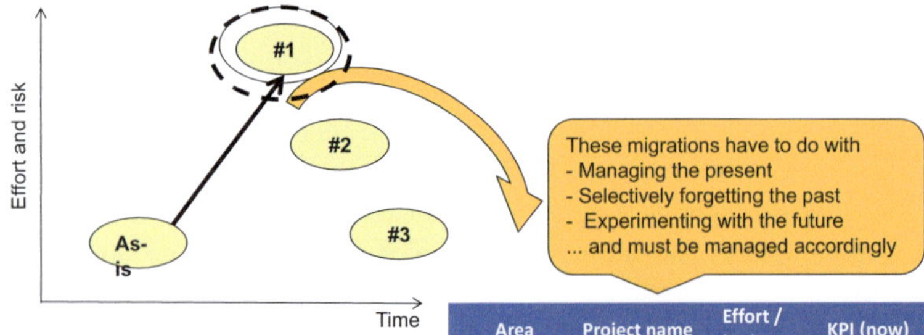

Area	Project name	Effort / budget	KPI (now)	KPI (target)
Value Proposition	Project 1			
	Project 2			
Channels	Project 3			
...				

Fig. 7.3 Developing a migration plan for business model innovation

will be progressively implemented, results coming from the market will provide evidence on its actual attractiveness and sustainability. This evidence may suggest revisions—or even radical changes—to the planned business model.

Introducing substantial changes to an intended business model that is not living up to expectations is called pivoting. *Pivoting** should not be considered as a failure, but simply as a part of the learning experience associated to the development of the business model itself.

As shown in Fig. 7.3, each move from one business model to the next can be expressed in terms of the projects that will have to be carried out for developing the new components highlighted in the Canvas. Each project will be characterized by the time and effort required and by performance objectives in key performance indicators (KPIs), so that a complete and detailed migration plan can be developed, along with the required budget.

When working on a business model Canvas, in both cases of existing and new firms, the team will need to go through a number of iterations before it comes up with a coherent and satisfying result. It is quite likely that, throughout these iterations, a subset of the elements in the business model will not change. Should this happen, the firm is recognizing such elements to be central to its business model. Identifying these central elements is important, because the firm can use them as the 'pivots' around which a set of candidate business models may be generated and progressively adapted.

Which elements are going to be recognized as stable obviously depends on the circumstances in which the business model is being generated. Pivotal elements are the ones that the company considers to be at its 'core' and capable of generating competitive advantage.

In principle, it is possible to imagine three main starting points. If the firm wants to develop a business model and a strategy around what it recognizes to be its core assets and competencies, the KA and KR sections of the Canvas will be the first

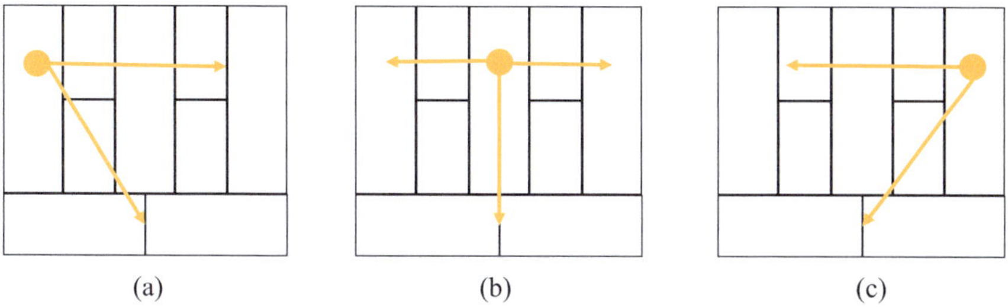

Fig. 7.4 Pivoting around the recognized 'core' of the firm

ones to be compiled and will remain stable during the development of the business model (Fig. 7.4, panel a). In other instances, a company may have developed a technology, a product or a range of products and endeavor to build a business model around them. In this case (Fig. 7.4, panel b), the stable reference point will be found in the VP section of the Canvas. Finally, a firm may consider its core strength to lie in the relationships it has with a reference market. In this case, the development of the business model will revolve around the elements in the CS, C and CR sections of the Canvas (Fig. 7.4, panel c).

7.2.2.2 Unbundling a Business Model

When working on the business model, and specifically on its supply side, one must take a number of 'make' vs. 'buy' vs. 'ally' decisions, which have a bearing on the vertical integration of the firm.

In some occasions, a firm may wish to consider a more radical perspective, which consists in the '*unbundling*' of its business model. Following Hagel and Singer (1999), businesses usually require the operation of three main processes. These are the development of new products (in the authors' original terminology, 'product innovation'), sales (and/or 'customer relationship management') and production (and/or 'infrastructure management'). The authors argue that a firm will usually excel at one—or at most two—of these processes. So, companies should try to recognize their strengths and weaknesses and 'unbundle' the processes that are not coherent with their core competencies. Unbundling can lead to decisions such as outsourcing them, spinning them off or running them in partnership with other organizations.

The Canvas can be a very helpful tool for studying a possible unbundling solution. To this purpose, one can tag elements in the KA and KR sections in order to associate them to the three processes mentioned above. Unbundling would then consist in making sure that the KA and KR elements pertaining to the processes that are recognized as non-core are associated to a key partner in the relevant section. Of course, such a change will affect both the cost structure and the revenue streams and will therefore have to be evaluated with respect to its profitability.

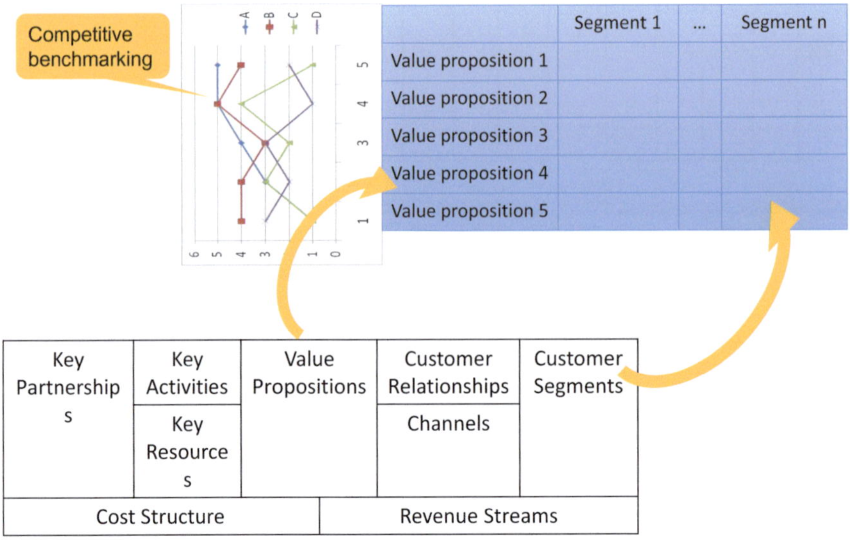

Fig. 7.5 Analyzing and benchmarking value propositions against customer segments

7.2.2.3 Competitive Benchmarking

One of the areas of the Canvas where analysts must ensure coherence is the relationship between VPs and CSs. It is in fact very easy to conflate VPs with a description of product quality that is biased by the firm's inner perception. In order to avoid this distortion, it could be in some cases advisable to perform an analysis at a deeper level. At first, the relevance of each VP to each CS should be analyzed in a rigorous way, making sure that no VP and no CS remain isolated. In visual terms, this can be easily done by using an incidence matrix, such as the one shown on the right-hand side of Fig. 7.5.

This preliminary assessment can be supplemented by performing competitive benchmarking, in order to evaluate how the firm's actual and/or prospective products are able to deliver the VPs, and how they stack up when compared to competitors' offerings. This should be preferably done by using actual customer data, but—that missing—self assessment can also be used. The comparison can be performed by using the profile chart on the left-hand side of Fig. 7.5.

7.2.2.4 Scenario Analysis

Business modeling is usually a long-term and strategic endeavor that points quite far into the future. In the likely case that future and uncertain events might have a positive or negative impact on the inner coherence, or on the profitability of the business model being studied, it is advisable to perform a simple scenario analysis.

The first step consists in identifying a limited number of significant dimensions of uncertainty and their potential outcomes. Dimensions of uncertainty can be associated to *Policy, Economic, Social and Technology (PEST) analysis**.

Examples of uncertainty dimensions can be the emergence of an enabling technology, the continuation of macroeconomic phenomena or the adoption of a given policy decision. For simplicity, outcomes are defined according to discrete levels (i.e., outcome *X* may either happen or not; *Y* may lead to levels that are low, medium or high, etc.).

The combination of these outcomes allows the definition of a discrete number of potential scenarios (for instance, three uncertainty dimensions with Boolean outcomes will generate eight scenarios) which can be listed, so that the ones that are not considered feasible can be pruned out (i.e., outcome *X* may not be possible if outcome *Y* does not occur).

Once the feasible scenarios have been enumerated, the team may develop a Canvas for a number of alternative business models, which may be more or less applicable or attractive in each scenario. Once this has been done, it is possible to compile a comparison table mapping alternative business models with the possible scenarios and use it to select the business model to pursue.

7.2.3 Understanding the Coherence of the Business Model

Throughout the previous discussion, it has been stressed that a business model is a conceptual description that has the aim of evaluating the coherence of its constituent elements. The validation of this coherence can be studied under different perspectives.

One first perspective for validation is simply semantic. The team that is performing the business modeling exercise can check that the elements that are being placed in the Canvas fit together according to business logic and common sense. Performing a thorough check at this semantic level is by no means a trivial task, since it is quite likely that—at first sight—the team will not be able to spot incoherencies. This is often the case, when the elements are not carefully and sharply worded.[5]

A second perspective concerns economic and managerial aspects, and consists in working at sufficient depth with the bottom sections of the Canvas, in order to make sure that expected revenues may cover costs. Of course, this is just a preliminary evaluation and not a full-blown business plan. However, if a rough-cut evaluation does not lead to a positive outcome, it is quite reasonable for the analysts to try improving the business model at this level, rather than going deeper in the financials and hoping that things might improve by adding detail. The bottom section of the

[5] For instance, suppose that a fast food chain is trying to make a stab at the teen obesity problem by proposing a VP of 'healthy and competitively priced fast food' to a CS of 'poor and uneducated teenagers'. A coherence check would make sure that the channel is appropriate to the VP and CS, and it would probably be better to have 'neighborhood outlets', rather than 'premium outlets in fashionable streets'. Moreover, the business model would be incoherent if the firm did not add some element in the customer relations section aimed at making sure that the CS will acquire a greater appreciation of this particular VP. Note that this quite important issue could easily be overlooked, if the CS had not been precisely labeled with the adjective 'uneducated'.

Canvas requires the team to operate on a quantitative—rather than purely qualitative—stance. This is only possible if some quantitative element has been introduced in the upper sections of the Canvas.

Specifically, an estimation of revenues will require the definition of the following elements:

- An estimate of the **size of each customer segment**, starting from a *Total Addressable Market* (TAM*, i.e., the universe of potentially interested customers in that segment), which must be whittled down to a *Served Available Market* (SAM)* that considers the existence of competitors and the inability of the firm to make its offering universally known and available. Then, this will be further reduced to a *target market* made up of the most likely buyers, given the channels and the effort that the firm intends to spend in customer acquisition.
- An attempt to set **price levels** according to the pricing scheme that has been decided. In turn, this requires a careful evaluation of the VPs in the eyes of the specific CSs. The value of the offering leads to a willingness to pay, which depends on the type of good or service that is being discussed. In some cases, value will have to be derived by comparison with other goods (e.g., a high-end watch may be compared to other watches and to jewelry). In other cases, it may require to imagine being in the customers' shoes and make a return on investment or payback calculation from their perspective (e.g., an energy saving device can be priced in order to make sure that the investment is paid back in a given time). Of course, firms should not forget to include sales taxes or VAT, which might significantly impact final consumers' choices.
- The **intermediation costs associated to the channels** that are being chosen. So, indirect sales will be dominated by the markups that are added at each tier, leading to a significant decrease in the revenue available for the focal firm. Conversely, direct sales will be characterized by 'customer acquisition costs' (CAC) that will not impact the revenues, but will certainly increase the costs.[6]

On the side of costs, the analysis will be carried out by considering:

- Key activities and key assets that the firm has decided to perform internally and own and which will lead to a set of **fixed costs** (due to **capital expenditure** needed to buy the assets) and **variable costs** (due to **operational expenses** in running them). These costs will also depend on the capacity that is required, which is directly tied to the sales estimates that have been made on the demand side of the Canvas.

[6] When performing initial analysis, CAC is generally compared to the lifetime value (LTV) of each customer, to ensure that CAC > LTV. LTV is computed as the financial perpetuity of the annual margins deriving from each customer, given by the difference between an annual revenue (or ARPU – Annual Revenue per User) and an annual variable cost (or CTS – Cost to Serve). More precise estimates will also include an estimate of churn rates (i.e., the likelihood that a customer abandons the firm).

- Key partnerships will also lead to a set of fixed and variable costs, which will depend on the level of specificity of the activities that will be performed by the partners. In general, the higher the degree of specificity, the higher the investment that the focal firm will be required to make.
- **General and administrative expenses** that are not being explicitly considered by the business model, but that will nonetheless impact the firm's profit and loss statement.

A third and slightly more complex approach to checking coherence in the business model is based on studying *causal connections* between its elements (Casadesus-Masanell and Ricart 2010). This can be done in a structured way by using *Causal Loop Diagrams* (CLD)*. These diagrams consist in a graphical and qualitative representation of the relationships between variables in a complex system. Variables are represented as nodes in a graph, and arrows represent causal relations between variables, with the + or − sign indicating the direction of the relationship. A positive link from variable A to variable B means that if A increases (decreases), B will increase (decrease). Conversely, a negative link will indicate that if A increases (decreases), B will decrease (increase). Figure 7.6 shows a very simple example of a CLD that represents a few elements that could make up a business model for a low-cost airline.

CLDs are generally used as a preliminary step for structuring a complete and quantitative analysis under the *Systems Dynamics** approach (Sterman 2000). However, they can be also used in order to have a qualitative understanding of

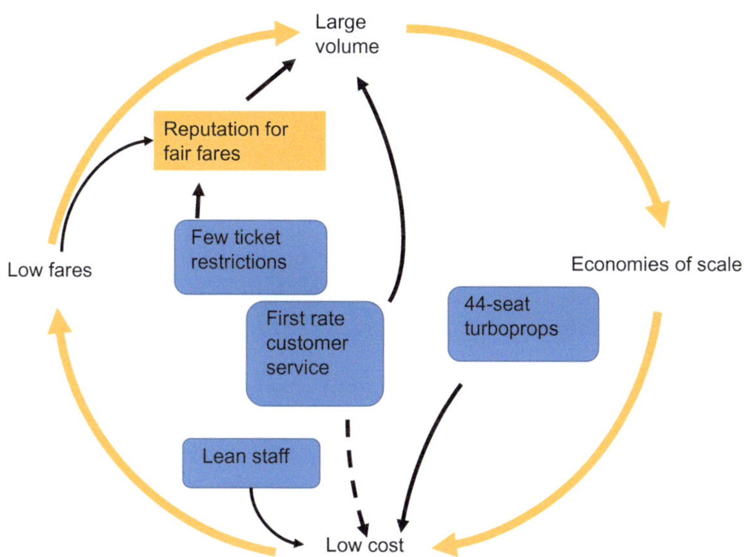

Fig. 7.6 Using a Causal Loop Diagram to study the coherence of a business model (case of a low-cost airline)

system behavior. Specifically, observing the loops in a CLD, it is possible to identify whether a phenomenon is likely to lead to self-reinforcement or to a balanced equilibrium. The former occurs when the number of negative links in the loop is even (including zero), since no causal effect will stop the variables from moving in the same direction. Conversely, a balanced equilibrium will arise if the number of negative links in the loop is odd.

When using CLDs to represent business models, it may be helpful to distinguish between *rigid* and *flexible* causal connections, in order to have a better appreciation of the velocity of the dynamics that is involved. In a rigid link, the impact of A on B is quick, whereas a flexible link will lead to a slower impact. For instance, it is likely that price changes may impact demand quite rapidly, while product quality will allow a much slower development of reputation. When analyzing a business model CLD, it is particularly interesting to identify virtuous circles, or self-reinforcing loops that lead to the creation of greater value. These loops can be viewed as the main sources of competitive advantage of the business model, since they allow the firm to progressively improve its profitability and generate barriers to entry, since imitating firms will find it difficult to catch up on the same path. At the same time, the elements that make up the virtuous circle are to be viewed as particularly critical, since any breakdown that might occur would likely lead to the demise of this competitive advantage.

7.3 Representing a Business Model from the Perspective of a Value System: e3value

Up to now, the discussion has dealt with representing business models from the perspective of a focal firm. Nonetheless, the Canvas described in the previous section has made it clear that not all activities pertaining to the business model will actually be carried out by a single company, and that key partnerships will have to be defined in order to make the model operational. In fact, Zott and Amit (2010) have stressed that "*business models are boundary spanning systems of activities centered on a focal firm*".

So, given that business models cannot be fully understood if not by looking beyond the boundaries of the focal firm, the same firm must make sure that the relevant activities and partners are not only listed, but that they can actually work as a coherent whole.

In order to achieve this result, it is possible to use business modeling tools that broaden the perspective to the entire '*value network*' within which the firm is operating in. The term 'value network' is purposely used as an extension to the familiar concept of 'value chain', since the relationships between the participating entities may not necessarily be linear, as the term 'chain' would imply.

Among these modeling tools, it is possible to introduce *e3value* (Gordijn and Akkermans 2001), a tool that strikes an interesting balance between modeling power and ease of use. The e3value can be viewed as a dialect of UML (Huemer et al. 2008), a System Modeling Language that is widely used in the field of Software Engineering. With respect to UML, e3value proposes a simplified notation and approach. Moreover, e3value focuses on describing the multiple actors involved, the main value-creating activities they perform and the related exchanges of goods, services or financial resources. It does not consider the minute description of the activities being performed, since the aim is to suggest strategies to business managers, rather than to provide technical specifications to system developers. The language has been used in many different applications, and a survey of which can be found in Pijpers et al. (2012).

The main idea behind the e3value approach is that complex business models arise out of the cooperation between *multiple actors*, which are jointly engaged in providing products and services to one another. This allows value creation through two avenues. One is the 'traditional' approach of adding value to inputs and selling the outputs, while the other is the exploitation of complementarity between activities and the associated outputs.

Especially when working with the latter form of value creation, it is critical to make sure that activities and their pricing are structured in a way that all actors are left with sufficient economic margins, thus making it profitable to take part in the overall value network. For instance, consider the managers of a magazine, who must decide on the pricing of space given to advertisers, the number of pages assigned and the type of advertisers accepted. These three elements must be structured in a way that will provide advertisers a return on their marketing investment, at the same time making sure that the readers will not be annoyed by commercial content and therefore reduce their willingness to pay the cover price.

In this type of context, it is likely that a focal firm will have a central role in coordinating the business model, but this will not amount to having formal authority over its partners. Therefore, it is only through the attractiveness of the business proposition offered to each actor that partners will be drawn to the business model and take part in the value creation process. Therefore, when considering a business model that straddles multiple firms, it is not enough to work on a purely conceptual point of view (i.e., making sure that relevant parties are listed and designing their functional interactions). Conversely, one must consider—at least at a high level of abstraction—financial flows as well. This means moving from a conceptual business model to a preliminary business plan, which must be drafted in order to highlight the profitability for each individual partner and for the value system as a whole.

The e3value provides a graphical notation and, through its software embodiments, a tool that allows covering these requirements. Table 7.1 provides a synthetic overview of the notation, while Fig. 7.7 provides a sample representation of a business model.

According to this, in Fig. 7.7 it is possible to see a sample e3value model that depicts the business model of a shopping mall and which exhibits a number of the constructs discussed above. When using the e3value editing tools, analysts can not only create a graphical description of the values system, but also provide basic information leading to preliminary business plans. Specifically, one can input

Table 7.1 Main elements in the e3value notation

e3value construct	Definition	Example
Actor	An economically independent entity that acts within the value system. An actor can be a consumer, a firm or a business unit of a firm	
Value object	An entity that has economic value to an actor and is exchanged within the value system. Value objects can be physical or not (e.g., immaterial services). Money is a value object	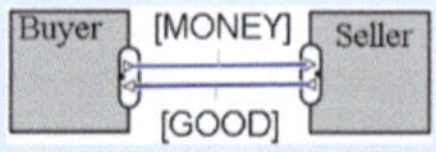
Value port	Represents the way a value object goes in (i.e., is consumed) or out (i.e., is produced) by actors	
Value offering	A value offering is a grouping of value ports that are equally directed. They definitively represents a bundle	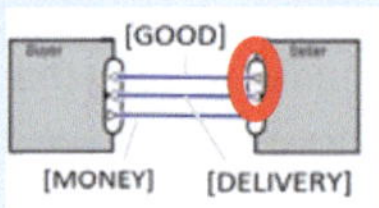
Value interface	In each actor, the value interface represents the grouping of incoming and outgoing value offerings. It represents reciprocal exchange among actors	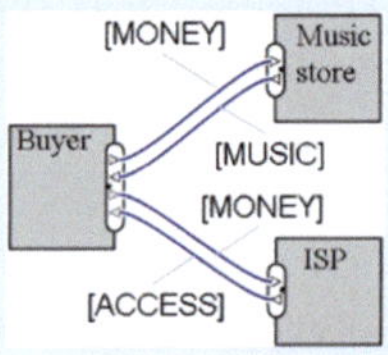
Value transfer	Connects two value ports together and represents trade between actors	
Market segment	Is a grouping of actors who exhibit strong similarity in the way they assign economic value to objects	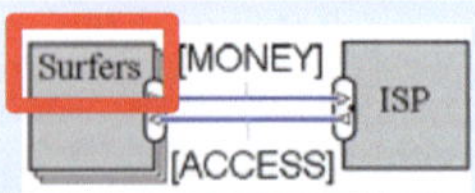
Value activity	Represents activities being carried out within each actor. It allows to separate the	

(continued)

Table 7.1 (continued)

e3value construct	Definition	Example
	'who' (i.e., the actor) from the 'what' (i.e., the activity) and to fully describe authors that carry out multiple activities	
Actor compositions	Represent groupings of actors that partner together tightly. For instance, an actor composition might expose a single value offering to customers, in a way that the latter are not aware of which actor provides what value object	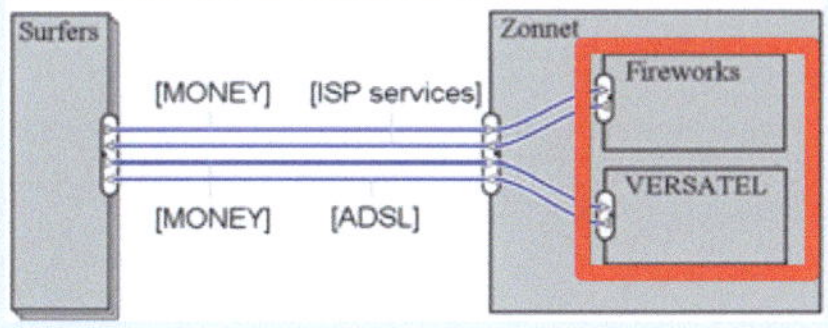
Dependency path	In the case one wishes to add detail, a dependency path shows the order with which value transfers occur. The path has a starting point in which a stimulus is originated and has an end. The path progresses from the origin to the end by replying to the question "how can this happen?" Conversely, one can trace the path from the end to the stimulus by replying to the question "why is this happening?" The path may branch through OR gates (when a stimulus is satisfied through one mean or another one) as well as through AND gates (when a stimulus is satisfied through one mean and another one)	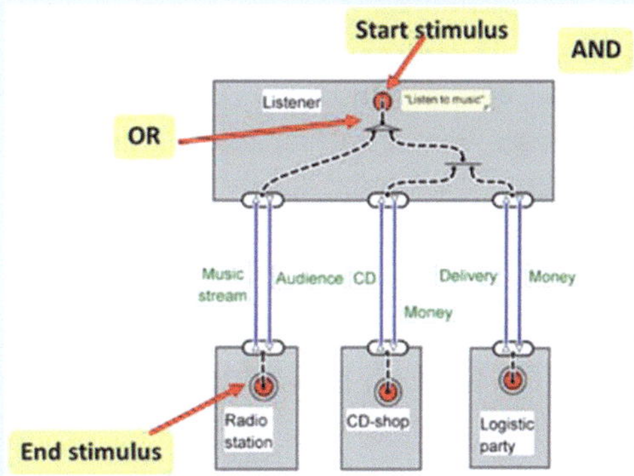

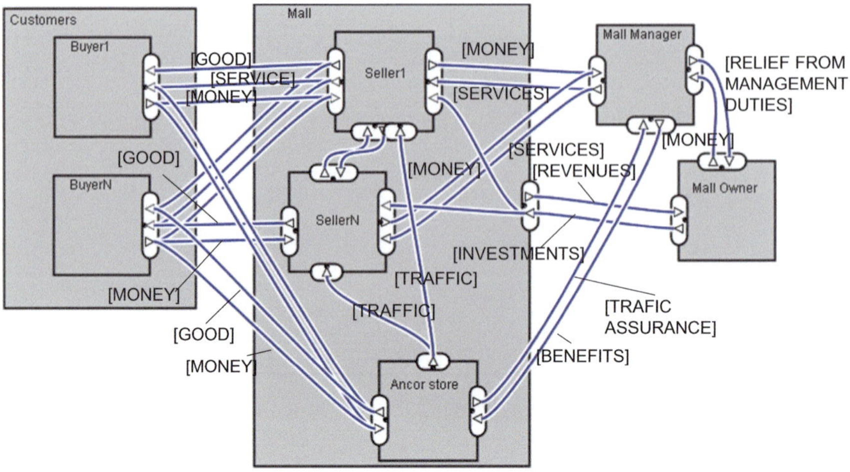

Fig. 7.7 Sample e3value model

information on flows (i.e., demand that a market segment will generate for a given value object) and value (i.e., cost/price data associated to value objects, as well as fixed and semi-fixed cost data associated to activities). This can lead to the automatic generation of financial spreadsheets representing the basic financial flows occurring to each actor which—as mentioned before—can provide a preliminary evaluation of the viability and attractiveness of the business model.

7.4 The Design of a Business Model

The importance of business model innovation has led a number of scholars to focus on the approach that can be followed to develop successful business models, rather than on the individual content of the same.

Zott and Amit (2003, 2007, 2010, 2011) have developed an interesting theory of *Business Model Design*, framed as a boundary spanning (i.e., which includes multiple firms) organizational design effort. Their approach frames the design effort around three phases, summarized in the following Fig. 7.8.

The first phase has to do with a thorough analysis of the situation, trying to understand the existing constraints, the needs and the activities of potential stakeholders, competitors' actions and the ways with which value could be created in such an environment.

The second phase deals with designing mechanisms for value creation within the new business model, and then making a critical analysis of the potential value proposition being generated, through an analysis that encompasses different perspectives that go under the acronym NICE (i.e., novelty, lock-in, complementarity, efficiency). In doing so, it becomes critical to have a clear understanding of the

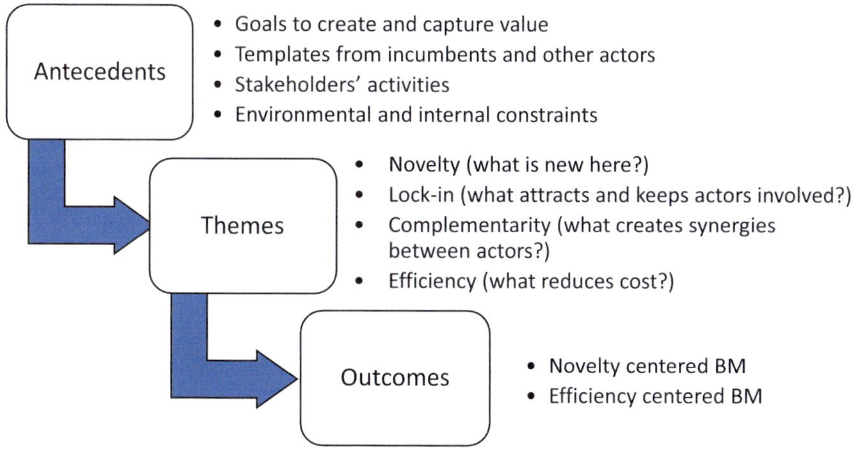

Fig. 7.8 Three phases of Business Model Design

activities to be carried out ('activity content') of the way with which they are enacted ('activity structure') and of which actor will be in charge of performing them and under which procedures and/or incentive mechanisms ('activity governance').

Finally, the last phase has to do with summarizing the impact of the new business model, trying to understand whether its innovative content is more focused on the value proposition ('novelty-centered') or on the efficiency with which value is being created ('efficiency-centered').

References

Casadesus-Masanell R, Ricart JR (2010) From strategy to business models and onto tactics. Long Range Plan 43(2–3):195–215

Chesbrough H (2010) Business model innovation: opportunities and barriers. Long Range Plan 43:354–363

Gordijn J, Akkermans H (2001) Designing and evaluating e-business models. IEEE Intell Syst 16 (4):11–17

Hagel J, Singer M (1999) Unbundling the corporation. Harward Bus Rev 77(2):133–141

Huemer C, Lieg P, Schuster R, Zapletal M (2008) A 3-level e-business registry meta model. In: IEEE international conference on services computing, Hawaii, USA

Lee CS (2001) An analytical framework for evaluating e-commerce business models and strategies. Internet Res: Electronic Netw Appl Policy 11(4):349–359

Magretta J (2002) Why business models matter. Harv Bus Rev 80(5):86–92

Mahadevan B (2000) Business models for internet-based e-commerce: an anatomy. Calif Manage Rev 42(4):55–69

Osterwalder A, Pigneur Y (2010) Business model generation: a handbook for visionaries, game changers, and challengers. Wiley, Hoboken

Osterwalder A (2004) The business model ontology: a proposition in a design science approach. Ph.D. thesis, University of Lausanne, Ecole des Hautes Etudes Commerciales HEC, 173

Pijpers V, de Leenheer P, Gordijn J, Akkermans H (2012) Using conceptual models to explore business-ICT alignment in networked value constellations. Requirements Eng 17(3):203–226
Shafer SM, Smith HJ, Linder J (2005) The power of business model. Bus Horiz 48(3):199–207
Sterman JD (2000) Business dynamics: systems thinking and modeling for a complex world. Irwin/McGraw-Hill, Boston
Teece DJ (2010) Business model, business strategy and innovation. Long Range Plan 43(2–3):172–194
Zott C (2003) Dynamic capabilities and the emergence of intra-industry differential firm performance: insights from a simulation study. Strateg Manag J 24(2):97–125
Zott C, Amit R (2007) Business model design and the performance of entrepreneurial firms. Organ Sci 18(2):181–199
Zott C, Amit R (2010) Business model design: an activity system perspective. Long Range Plann Special Issue Bus Models 43(2–3):216–226
Zott C, Amit R, Massa L (2011) The business model: recent developments and future research. J Manag 37:1019–1042

Innovation Strategy as the Management of Competencies

8

Abstract

As discussed in Chap. 6, a core component of innovation strategy has to do with the proactive management of corporate competencies. This topic is discussed at first from a methodological perspective, followed by a detailed discussion on a variety of approaches that can be used for this purpose, such as internal R&D, Technology Acquisitions, Corporate Venturing, Hiring, Equity and Non-Equity Strategic Alliances, Co-Development, Sourcing R&D activity from other parties, Licensing and Complete Outsourcing. The chapter ends with a discussion on the Open Innovation paradigm, which encompasses many of the previously mentioned approaches, and on issues pertaining to its implementation in a corporate setting.

Chapter 6 ended with a comprehensive model of innovation strategy, which based this element of corporate strategy on two main pillars. The first pillar was the definition of a desired portfolio of competencies and the planning of their development. The other pillar was the outlining of a portfolio of product and market development activities. This chapter will cover the former pillar, while Chap. 9 will tackle the latter.

8.1 Mapping and Planning a Competency Portfolio

In order to manage and develop the competencies portfolio of a firm, the first issue to be solved has to do with classifying and assessing competencies. There is no fixed rule in performing this exercise, and each firm will be likely to develop its own method.

Competencies have been defined as the organizational routines that involve resources, but it is obviously difficult to objectively classify and measure routines. It is therefore much easier to work on the resources and assets that are being used by

M. Cantamessa and F. Montagna, *Management of Innovation and Product Development*, https://doi.org/10.1007/978-1-4471-7531-5_8

the company. The firm can therefore define a directory of relevant competencies (e.g., for an IT company, one could think of a list of technologies and programming languages that it uses) and then make an inventory of the *resources pool* that is associated to each. A firm may therefore consider its human resources and use the headcount of employees who share the same competence. In the case of competencies requiring capital-intensive assets, the company could also use fixed assets (e.g., the book value of laboratory equipment associated to each competency). Alternatively, in the case of industries where intellectual property is highly relevant, the firm could list competencies by using patent classifications and then measure them by counting its patents.

Just as in product portfolio management, visual tools such as bubble charts can be quite useful to support the representation of competencies. Management will identify two axes for locating each competence and the size of the related resources pool will be represented by the diameter of the bubble (see Fig. 8.1 for an example). The choice of the axes depends on the objective of the analysis; candidates include technological maturity, competitive impact of the competence, relatedness to the firm's core competencies, perceived risk, etc.

By looking at the bubble chart representing the current competencies portfolio ('as-is'), it will be relatively easy for management to identify which competencies have to be strengthened, developed, or progressively decreased or decommissioned, thus outlining a desired ('to-be') competence portfolio. Of course, the chart is only a visual tool, and the analysis will have to rely on an adequate understanding of innovation and competitive dynamics in the specific industry, together with a general idea of strategy at corporate level.

From an organizational perspective, the development of competencies is a process that requires relatively long time constants. It must be therefore undertaken on time, since it will probably be too late to start acting when the firm perceives its competitive advantage to be faltering. Moreover, managing competencies implies

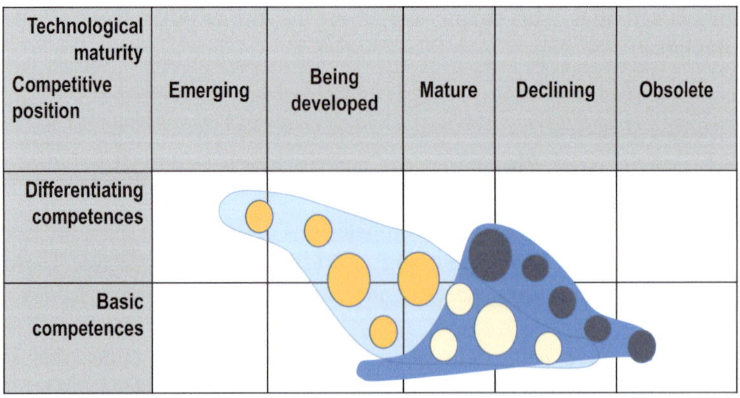

Fig. 8.1 An example bubble chart representing an 'as-is' portfolio of corporate competencies (area to the right) and a 'to-be' one (area stretching top left to bottom right)

operating at a deep level within the firm, and considerable inertia and resistance to change are likely to emerge as a reaction. Firms therefore might think of staffing their top management teams with a balance of executives representing the firm's past (who are likely to have been promoted from within the firm) along with managers representing future needs and directions (who may be recruited from the outside, or be chosen among insiders who have followed somewhat less orthodox career paths).

When managing competencies at a strategic level, one may wonder what constitutes a 'good' portfolio of competencies and resources with respect to size, scope and experience. The academic debate on this issue can be quite confusing, but researchers seem to agree that economies of scale do exist (Macher and Boerner 2006). In other terms, firms with larger pools of competency-related assets are likely to achieve better results, because of greater variety among resources of the same kind and because critical resources can be used with higher efficiency. When looking at scope, i.e., the number of different competencies listed in the portfolio, two elements must be considered. Greater diversification leads the firm to lose focus, which has a detrimental effect. However, a wider scope also provides room for spillovers between different competencies and, when uncertainty is very high, increases the likelihood of having 'the right competencies at the right time'.[1] Following Macher and Boerner, the net impact of these two opposing effects depends on experience, since the ability to capture opportunities due to spillovers and chance will be greater for firms that have built substantial experience in the related fields. It comes as a corollary that startup firms ought to be more focused, while mature firms will benefit by progressively expanding their competency portfolio.

The bubble chart representing a portfolio of competencies can be developed by looking only at the firm being analyzed. However, it can also be used for competitive benchmarking, by placing it side by side with a similar chart representing—possibly with some approximation due to lack of information—selected competitors.

Managers might also decide to map the competencies of the entire value network the firm is part of. This allows to work on '*knowledge partitioning*' within the value chain and to support a broader innovation strategy that may also involve key players operating in its ecosystem. Working at this level will make it possible to identify whether competencies have to be developed internally by the firm ('make'), by using selected suppliers ('buy') or by associating with complementors and customers ('ally').

This decision will depend on the strategic stance that the firm has with respect to each competence. Clear decisions can be taken if the competence is recognized to be *core* (in-house development will obviously be recommended in this case) or *non-core and generic* (contracting with generic suppliers will in general make sense, unless cost analysis suggests differently). The decision might be somewhat more involved in the case of *non-core and co-specialized* competencies, which

[1] Leiponen and Helfat (2010) use the analogy of buying tickets to a lottery.

must be developed in a way that is specific to the firm. In this latter case, the firm will probably be unable to ensure the availability of these competencies by simple arms-length contracting with generic suppliers. These suppliers may not exist at all, since the competencies are firm-specific. Moreover, suppliers who might be in the position of developing these firm-specific competencies might be unwilling to make the required investment, unless they have sufficient guarantees and long-term commitment from their customer. Therefore, the firm will have to decide between internal development or building long-term agreements with other actors.

It comes without saying that failing to identify the need for co-specialized competencies can be one of the most damaging mistakes a company might make in drafting its strategy.

Once the desired competencies have been identified, the firm must decide on the best way to develop them. A variety of approaches is available and will be described in detail in the following sections of this chapter. The approach to choose will have to be decided case by case. In general, firms may work on two main tradeoffs. The first tradeoff is between the *time required* to develop the competencies, versus the *economic appropriability* of the related benefits (Fig. 8.2, right side). A second tradeoff is between the *level of understanding that is possessed* by the firm and by external parties (Fig. 8.2, left side). In this latter case, the larger the difference, the more the firm will favor an internal development of competencies.

Aside from these considerations, a last and crucial aspect that has to be considered is the firm-specific *capability of executing* a given strategy. In fact, it does not make any sense for a company to pursue a strategy that—in principle—is the most appropriate, if the firm does not have the experience or the skills required to implement it. For instance, should an acquisition appear to be more suitable than a strategic partnership in developing a desired competency, managers should ask themselves whether the firm has any experience at all in running such a complex process. If not, it might be advisable to choose the second-best option.

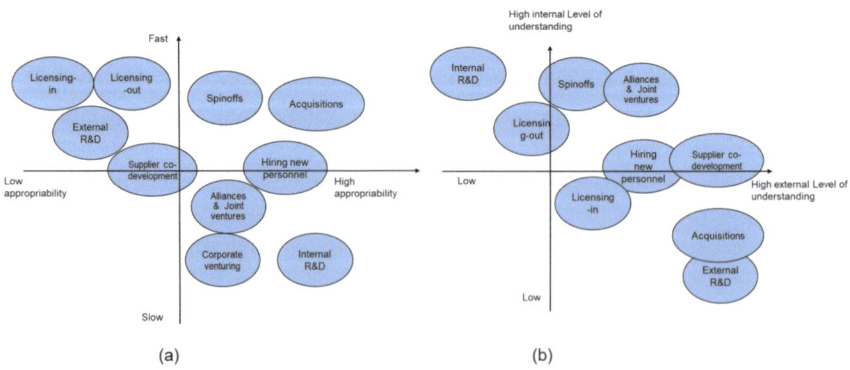

Fig. 8.2 The main tradeoffs in choosing an approach for competency development

8.2 Internal Research and Development

The traditional and mainstream way to grow a desired competency was to develop it internally by allocating resources—human, material and financial—and having them execute research and development projects. Nowadays, this still is a common and powerful approach, but must constantly be assessed against the other means described in this chapter.

8.2.1 The Uncertain Relationship Between R&D Expenditure and Corporate Performance

In Business and in Economics, it is still customary to use R&D expenditure as a proxy for measuring the degree with which a specific firm or an entire economy dedicates itself to the growth of its knowledge base. In the former case, one will commonly use the ratio between R&D expenditure and sales,[2] while in the latter, the ratio between R&D expenditure and GDP. These ratios are often used as indicators for the innovativeness of firms and economies, which is of course misleading, since they conflate an input (investment) and an output (innovative performance). When considering R&D activity at corporate level, one could wonder whether this effort is causally linked to corporate performance—which seems to be a highly popular view—or not. Research on this question has always concluded that the link between R&D intensity and corporate performance is highly ambiguous.

The traditional explanation justifying this connection is that R&D activity provides both *direct* and *indirect benefits*. Direct effects consist in the development of innovative technologies and products, potentially leading to competitive advantage. Indirect effects consist in increasing the firm's absorptive capacity, which makes it more adept at using externally developed knowledge and/or in imitating competitors.[3]

An alternative position (Knott 2008) considers the firm's *organizational intelligence* (i.e., its capability of changing and adapting its resources and routines) as an important mediating factor. A firm with high organizational intelligence will be able to capture opportunities coming from the knowledge generated by R&D activity and it will be therefore appropriate for such a firm to make this kind of investment. Conversely, a firm with low organizational intelligence will not be able to extract value out of R&D and should therefore avoid committing too many resources into it.

[2] In the case of small firms, this ratio is sometimes meaningless and it might therefore be advisable to use the frequency of R&D activity, rather than its intensity (Cuervo-Cazurra and Un 2010).

[3] One might also put forward a somewhat more cynical argument, stating that the correlation does exist, but with a reversed causal link. The claim would therefore be that profitable firms have ample financial resources and tend to overinvest in R&D. Thanks to this effort, they occasionally get some significant results that allow to increase an already existing competitive advantage and, when they don't, the investment made is usually not so large to cause a significant decrease in performance.

Other authors (Li and Hwang 2011) have identified *financial leverage* (i.e., the level of indebtedness) and *demand fluctuations* as another important mediating factor. The argument is that R&D is likely to deliver positive results only if projects are assured stable financing throughout their duration. Now, a highly indebted firm will risk not being in the position to provide a stable R&D budget since, should demand turn down because of the business cycle, the need to service debt will force managers into reducing non-essential expenditure, among which R&D. Should this occur, the firm would have to terminate projects before they are finished, without receiving any benefits and having wasted the money spent to start them. Moreover, a firm with high leverage may use debt to finance projects related to material assets that can be put up as collateral—such as buildings or generic production equipment—but not so much to immaterial assets, such as the ones originating from R&D.

Finally, *firm size* also can play a role in determining the R&D intensity of a company. R&D lends itself to significant economies of scale, since the higher the sales, the more the investment will be spread over each unit sold. Moreover, given the risk associated to R&D projects, a large firm with a large portfolio of projects will face less aggregate risk than a small firm working on a few projects only. It is therefore obvious for small companies to invest less R&D money than large ones. Exceptions can be found for small firms that enjoy substantial contribution margins or that anticipate a significant growth path as a consequence of this investment, and have investors willing to face the associated risks. Furthermore, many R&D activities are internally characterized by economies of scale and by minimum efficient scales. For instance, a given activity may be carried out only by purchasing costly lab equipment and/or by hiring a group of researchers with distinct specializations. It is obvious that only a large firm would be in the position to support such an investment from both financial and economic perspectives. A smaller firm, able to spend less than the required sum, would not have a competitive laboratory and would not reap results from it.

8.2.2 Some Typical Features of R&D Activity

When evaluating the appropriateness of using internal R&D projects to build new competencies, a few elements must be considered. As a starting point, the main advantage of internal R&D is the high degree of appropriability of the competencies generated. This must however be compared against a number of other critical aspects.

One first issue is defining an R&D budgeted for the corporation or the business unit. In general, it is not trivial to identify and compute the financial returns coming from R&D investment, which makes it difficult to define an *'optimum' amount* to be spent. Companies therefore tend to rely on ballpark figures that associate the R&D budget to other variables (e.g., R&D/sales or R&D / profit) and make comparisons with their competitors in the same industry. Therefore, it has become customary to work on 'standard' and industry-specific R&D/sales ratios, such as the 11–12% used in R&D-intensive industries like software and pharma, 8% in electronics and 4–5% in automotive and aerospace.

Internal R&D is not only a matter of the amount of money that is being spent, but also of *the way it is spent*. Specifically, path dependency can limit the capability of the firm to develop new competencies that are far from its existing knowledge base, or to do so in a reasonable time. Especially if top management is not knowledgeable in technology (Cummings and Knott 2018), internal resources will be in a strong position allowing them to effectively block efforts that imply a significant deviation from established competencies ("you know, we've tried it, but it doesn't work. Do you really want us to try again?"). The so-called *Not-Invented-Here Syndrome** is not a representation of intellectual laziness, but is simply due to the fact that internal resources will find it easier and be more productive to continue with the old competencies and technology, rather than going for new ones. Consequently, unless there is a clear mandate with strong incentives and suitable organizational arrangements to pursue the new competencies, they will tend to fall back on the old ones. In conclusion, unless it is managed very carefully, R&D is best suited for long-term endeavors and/or for achieving competencies that are not too distant from the firm's current ones.

When considering R&D projects, the issue of *worker productivity* and *incentives* is of capital importance. In fact, when dealing with intellectual work in general, it is difficult to measure productivity. The difference between a brilliant and highly motivated employee and a mediocre and dispirited one can be very high, and the firm should consider whether its R&D personnel are delivering as much value as their salaries are worth. Concerning incentives, financial rewards and career opportunities may matter, though research tells us that R&D personnel are motivated by other aspects as well, such as intellectual challenges and independence (Sauermann and Cohen 2010).

At the same time, management should consider that R&D results are correlated to *quality of effort*, rather than to quantity, and this calls for sophisticated approaches for measuring worker productivity. Unfortunately, the type of results that firms and their R&D personnel aim to achieve may also differ. As noted by Xiaoying et al. (2009), firms will push to develop highly specific know-how, since this increases the sustainability of its competitive advantage and reduces the market value of human resources at the center of these competencies.[4] Conversely, the employees involved in R&D will tend to generate broader and more generic competencies. The firm will therefore constantly risk losing key personnel, who might leave for a competitor or start their own business right at the time at which results of R&D investment may be reaped.[5]

[4] If a firm develops competencies that have the same value to any firm in the industry, employees will have a strong bargaining position, since they could command a much higher salary by shopping around for a job and trying to extract all the economic value of their know-how, or simply threatening to do so. This would not be possible if the value of their know-how is specific to their current employer.

[5] Firms may introduce *Non-Compete Clauses* in employees' contracts. However, these clauses are generally quite difficult to enforce since—if they are too strong—courts might simply reject them as invalid.

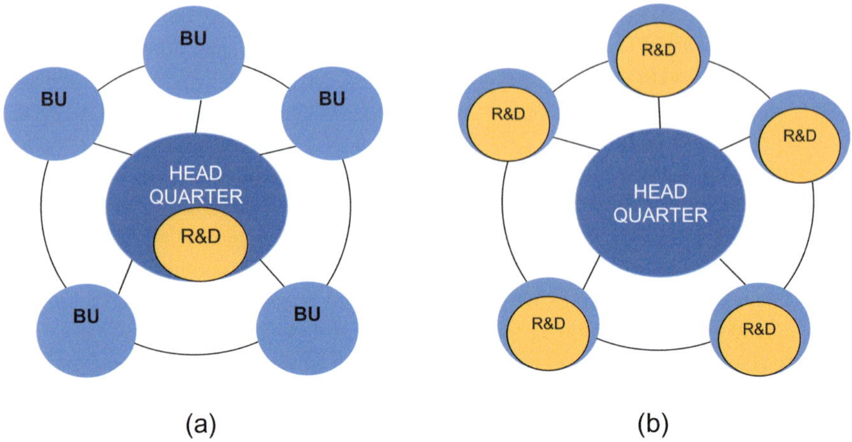

Fig. 8.3 Centralized (**a**) and decentralized (**b**) R&D in large firms

8.2.3 Positioning R&D Activity in Large Firms

In the case of large and differentiated firms organized around a number of Business Units (BUs), management must take a strategic decision concerning the organizational positioning of R&D units. Two solutions are traditionally available, with either a *centralized* R&D unit serving all BUs (Fig. 8.3, left), or having *decentralized* R&D units in each BU (Fig. 8.3, right). Deciding on which alternative to follow implies finding a tradeoff between advantages and disadvantages of each.

The main advantages coming from centralization of R&D are related to *economies of scale*. A single unit with a critical mass can make better use of costly resources, create strong links to the scientific community and provide interesting career incentives to researchers. Moreover, centralized R&D activity leads to greater chances of *spillovers* (i.e., sharing the results of a project among other projects and among BUs), at the same time avoiding duplication of effort. In decentralized R&D, the latter is a constant problem, since it is very common for BUs to develop similar projects at the same time without each BU being aware of the others' activity. Finally, with centralized R&D, *strategic guidance* and *budgets* are defined by the firm's top management, which ensures greater financial stability, a longer-term view and alignment with corporate strategy.

Against these clear advantages, centralized R&D units also have disadvantages. A first drawback is connected to the *geographical centralization* of an R&D unit, which may limit the possibility of accessing talented researchers that live elsewhere in the world.[6]

[6] Access to talent, both in terms of recruiting employees and in establishing connections to academic centers of excellence, is a key problem in industrial R&D. It therefore follows that large companies will tend to locate their R&D centers in areas where talented individuals can be found or attracted, because of either academic quality and/or quality of life.

A second disadvantage is associated to the risk of falling into the so-called *ivory tower syndrome*. This occurs when researchers work on highly challenging and potentially promising projects, but fail to make an effective connection with BUs and their needs, so that few results ever make it into product development.[7] In order to avoid this ivory tower effect, large firms have devised with a number of organizational structures and mechanisms.

First, a constant flow of information between the R&D center and BUs can be promoted using both IT tools and corporate events, similar to trade fairs, for matching research results to business needs. Co-financing mechanisms can also be used to ensure that R&D projects are meaningful to BUs. For instance, the R&D budget may be split between the research center and the BUs, and the former can be allowed to start a project only if it convinces one or more BUs to use their budgets to co-finance it. Organizational mechanisms can also help, for instance by favoring mobility of BU and R&D personnel and managers. Another approach could be to introduce a matrix organization in the research center, so that managers with BU experience are in charge of budgets and project portfolio definition, while managers with scientific background are responsible for human resources and running projects.

Finally, very large firms may introduce a hybrid organizational structure for R&D, with both centralized and decentralized R&D centers. The former centers focus on individual technologies and work at a more fundamental level. The latter ones are localized within each BU and work on projects at a higher TRL that are closer to business needs, acting as a bridge to product development.

Firms can also complement their R&D structures with other organizational mechanisms that can provide additional incentives to personnel that belong to the R&D and/or to operational functions. Some firms allow their staff to use some 10–20% of their time to pursue personal projects. This is usually connected to *intrapreneurship* schemes, in which employees whose personal projects show promising results can act as 'internal entrepreneurs' and are given the chance to bring them toward actual development, thus managing progressively higher budgets and responsibilities.

In other cases, company may promote *corporate spinoffs**. Spinoffs provide a powerful incentive to employees who may be considering starting their own business, since the originating firm can be a very useful shareholder, providing finance and the complementary assets that a startup would typically lack. Spinning off a new company can allow the further development of projects that are non-core to the firm and would probably not go into development if kept inside, because of strategic misalignment and organizational inertia, or because they would require the

[7] A well-known example is Xerox's Palo Alto Research Center (PARC), which the multinational purposely created close to Stanford University and far away from its corporate offices in New York State in 1970. PARC researchers came up with many groundbreaking inventions in the field of computing, such as Graphical User Interfaces, text editors, and Ethernet networking technology. All these inventions could have given Xerox a strong lead in the IT industry, but none of them were exploited. Through licensing deals, they contributed to Apple Computer's initial technology.

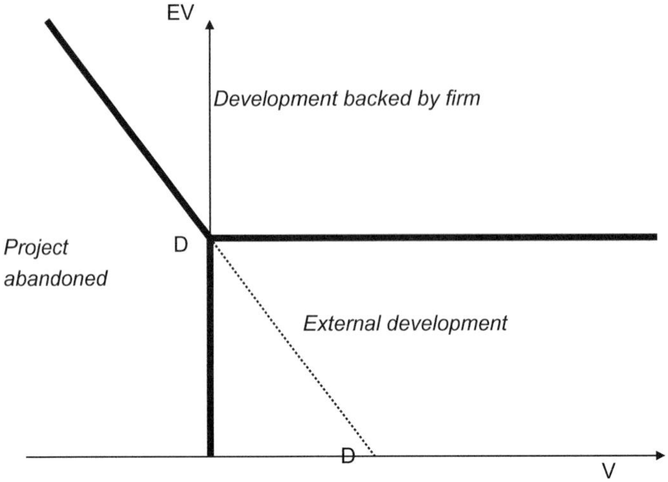

Fig. 8.4 Tradeoffs in external versus internal development of projects

participation of other financial or industrial partners. Besides, the firm avoids completely losing the competencies that an employee might bring with her/him if she/he decided to start an independent company.

The decision on whether to develop a new project internally or externally can be analyzed by considering three main factors (Cassiman and Ueda 2006). The first factor is the intrinsic value of the project V (i.e., the economic value to the employee deciding to leave the firm and founding an independent startup). The second factor is the extra value EV which can be achieved if the firm backs the project and allows the use of its assets. The third factor is the discounting D with which the firm will consider the project when evaluating it. This discounting can be due to the opportunity cost of committing resources to this particular project, the cannibalization of the existing business and the incapability to fully understand the value of the new project. The firm will decide not to back development if $V + \mathrm{EV} - D < 0$ while the researcher will not start a venture if $V < 0$. As shown in Fig. 8.4, the combination of these two decision rules leads to three possible outcomes.

If $V > 0$ and $\mathrm{EV} > D$, the project will be backed by the firm while, if $\mathrm{EV} < D$, it will be developed externally. If $V < 0$ and $V + \mathrm{EV} - D > 0$, it will be backed by the firm while if $V + \mathrm{EV} - D < 0$, it will be abandoned by both parties.

8.3 Technology Acquisitions

Acquisitions are a common event in corporate history and can occur for many different reasons, such as gaining access to a market or a customer portfolio, securing physical assets (e.g., plants and machinery) or obtaining desired competencies. The latter case is often referred to as a *technology acquisition*.

Acquiring an existing firm represents a very quick way for an innovation-oriented firm to obtain desired competencies and to do so with a high degree of appropriability (Zhao 2009). In general, acquisitions can create value in both cases of the acquiring and the target firm having similar knowledge assets, or different and complementary ones. The former case typically supports incremental innovation, while the latter can be an avenue for radical innovation (Makri et al. 2010).

Acquisitions are characterized by a complex process that involves significant *transaction costs**.[8] The process usually entails four steps. At first, the acquirer must *scout* for target firms that might have the desired competencies and be open to be acquired. After a preliminary negotiation, it must perform technical, business and legal *due diligence* in order to estimate the value of the target firm and uncover any elements that may suggest that the acquisition entails an excessive price or risk. Finally, it must *negotiate* the terms of the acquisition and after the *closing* of the deal, it must *integrate* the acquired firm.

Integration means blending the two organizations together so that they perform together as a single entity. Following the evolutionary theory of the firm we are familiar with, the organizational routines that define both companies will have to be modified so that the acquired firm's resources act together, and within a new set of routines, with the acquiring firm's ones. In the case of technology acquisitions, integration is a somewhat paradoxical process. The acquiring firm buys another firm because of its competencies, but cannot fully benefit from them until it destroys them in order to merge them with its own. Because of path dependency, the integration process is complex and may yield uncertain outcomes. Especially when the acquired company is large and, as it often happens when an incumbent attempts to deal with a technological discontinuity, the new organization will have to deal with considerable inertia coming from both organizations, thus potentially impairing the potential synergies or delaying them significantly.

Following Sears and Hoetker (2014), a successful integration will have to generate adequate absorptive capacity on the side of the acquiring entity and ensure a sound management of the potential conflicts that might emerge on both sides. Both aspects depend on the degree of overlap between the two firms' competence base, following the scheme in Table 8.1. The framework shows that the situation that is easier to manage (on the bottom left) is also the one that is of lesser value, in terms of broadening the competencies base.

Moreover, integration may fail if *resource retention* strategies are not correctly defined and if key employees of the acquired firm—either because of uncertainty, or because they don't like the idea of working for the new firm—decide to leave. While full integration may require time in order to fully provide the economic value sought, the acquired firm nonetheless remains an operational entity in its own right. Therefore, the acquiring firm will be able to reap some benefits while it is still performing the integration process.

[8] In economics, a transaction cost is the cost incurred to take part and to execute a business transaction, on top of the price paid.

Table 8.1 Degree of competence overlap between the acquirer target firms

Target overlap	Acquirer overlap	
	Low (target knows little of what acquirer knows)	High (target knows much of what acquirer knows)
Low (acquirer knows little of what target knows)	• Low absorptive capacity • High diversity • Low conflict	• Low absorptive capacity • High diversity • High conflict
High (acquirer knows much of what target knows)	• High absorptive capacity • Low diversity • Low conflict	• High absorptive capacity • Low diversity • High conflict

In many cases, technology acquisitions are subject to the so-called *winners curse**.[9] Following Ransbotham and Mitra (2010), when a firm is being considered for an acquisition, its shareholders will value it at an intrinsic enterprise value EV. A prospective acquirer i will value it at a reservation price $RP_i = EV + VS_i - CI_i$, where VS_i is the value of the acquirer-specific synergies or growth options that the acquisition might lead to after integration is performed, and CI_i is the cost of the integration process. For a technology acquisition, the asset being sought is a set of competencies and its value-creating potential after integration and, especially if the acquirer is much larger than the target firm, it is possible for RP_i to be orders of magnitude greater than EV. The parties will negotiate an acquisition value AV, which will lie somewhere between EV and RP_i. If there is only one potential acquirer, it is likely that AV will be higher than, but nonetheless quite close to EV, since the target firm will not have many levers available to negotiate a higher price. The scenario might change significantly if the target firm manages to have more than one potential acquirer enter the scene, thus starting an auction-like process. In this case, the acquisition value will become marginally higher than the second-highest reservation price $RP_{i'}$. If the value of the target firm's competencies is not specific to the highest bidder, the two suitors' reservation prices will not differ substantially, and the acquisition value can become much higher than EV. This brings potentially huge benefits to the shareholders of the acquired firm, and a real risk to the winner that the benefits of the acquisition might not materialize as foreseen.[10]

[9] When an asset is sold at an auction, each candidate buyer will participate knowing her *reservation price*, which is the maximum price she/he is willing to pay for the good. Reservation prices are participant-specific, since each potential buyer will value the asset differently to the others. The auction will end when a bid is made, that is marginally higher than the second-highest reservation price. Therefore, when someone wins an auction, he discovers that he was the participant willing to the pay the most and that he ended up paying something above the second-highest reservation price. The term '*winner's curse*' represents the likelihood that the asset has been overpaid. This probability will be lower if the value of the asset is highly specific to the winner, since he will secure the asset at a bid that is substantially lower than the reservation price. The opposite holds for an asset whose subjective value is very similar for the highest bidders.

[10] Bidders may also increase their reservation prices by an additional amount that considers the competitive value of excluding a competitor from acquiring the target firm. The phenomenon is similar to what happens in professional team sports, when well-funded clubs pay outstanding sums

Many acquisitions in the high-tech industries follow this process and lead to outstanding prices, such as eBay's acquisition of Skype in 2005 or Facebook's acquisition of Oculus Rift in 2014. Of course, acquirers will try to avoid the auction-like process by imposing a short timeframe to conclude negotiations and asking the target firm to grant exclusive negotiations (a *no-shop agreement*). Therefore, firms that feel they could become interesting acquisition targets will try to preempt this kind of request by starting as many informal talks as possible with potential acquirers, before being formally approached by any one of them.

In summary, acquisitions are a powerful, quick, but somewhat risky way to gain access to desired competencies, which itself requires a capability to execute them correctly and profitably.[11] From the perspective of target firms, the popularity of acquisitions as a way to gain competencies and technological assets and the potentially high valuations at which these transactions can occur has become a very good match with the venture capital financing process which, as already discussed, is aimed at securing exit events.

8.4 Corporate Venturing

In corporate venturing, the focal firm acts as a venture capital fund and takes equity stakes in young technology-based companies, either directly, or by setting up an independent *corporate venture capital* arm. Investment targets for corporate venturing include startups that are developing technology of interest to the focal firm, which becomes well positioned to become a privileged customer, distributor or acquirer. In other cases, the focal firm may use a corporate venturing unit to manage the stakes it owns in its own corporate spinoffs. Finally, the focal firm may take stakes in firms that are developing technology or products that are complementary to the firm's core competency. This latter strategy is termed *ecosystem venturing* and follows the rationale that, by supporting the growth of these companies, demand for the focal firm's products will increase.[12]

From the perspective of the focal firm, corporate venturing allows to create a portfolio of promising technologies and 'let them grow freely in its backyard'. An outright acquisition would in fact be financially riskier, and the need to integrate the acquired companies could hinder a technology development process that is still

for high-performing athletes even if their roster is already competitive ("it's better for us if *X* sits on our bench than have him on the pitch, but playing against us").

[11] Cisco is one quite well-known example of *serial acquirer*, who has acquired and successfully integrated more than 150 companies between 1993 and 2013. When a firm's innovation strategy leads to acquiring an average of more than one firm each month, the acquisition process itself becomes a routine (or dynamic capability, following the definition given in Chap. 6).

[12] As an example, Intel, the PC microprocessor maker, has over the years made a number of investments in computer game publishers, since complex games that stress computer hardware are a driver of demand for the replacement of personal computers.

ongoing. Corporate venturing typically occurs in industries where technological change is significant and where complementary assets have an important role. Of course, firms using this approach must have good absorptive capacity and sufficiently high cash flow (Dushnitsky and Lenox 2005).

From the perspective of the investee company, having a corporate investor or a corporate venture capital fund as a shareholder can be an interesting proposition. Besides the equity finance provided, the focal firm will probably be quite open to do business, and having a powerful firm as investor and business partner can increase the growth prospects of a fledgling startup. However, the investee could also risk becoming a *captive supplier* to the focal firm, thus reducing the returns to its other shareholders and even making it difficult to do business with other customers, who are competitors to the firm making the investment. In addition, given these pros and cons, it is possible that the best entrepreneurs will be wary of corporate investments, while the weaker ones will be more open to such an opportunity, thus leading to the risk of adverse selection.

Therefore, firms that want to be active corporate investors must define and follow clear policies for treating investees fairly, without making an exploitative use of their dual role of shareholder and potential business partner. By failing to do so, they would risk incurring in the paradox of being unable to close deals with the firms having the highest growth potential or being forced to pay a price premium when making the investment (Dushnitsky and Shaver 2009). Setting up an independent corporate venturing arm, staffed with general partners who are not employees of the firm and whose compensation is, just like in normal VC firms, tied to the value of the portfolio companies, may help reduce this risk. Additionally, a corporate venturing entity might decide to co-invest in syndicated deals, with a purely financial VC fund taking the lead, in order to reassure entrepreneurs of their future behavior.

8.5 Hiring Human Resources

A firm that lacks competencies in a given area may resort to hiring employees that already possess the underlying knowledge, either as fresh graduates from a relevant academic curriculum or as professionals who already have some work experience in the field.

The hiring of multiple individuals has *significant transaction costs*, which can ultimately be comparable to those required to acquire a company of the same size. The hiring process includes searching for candidates, assessing their competencies, interviewing them and negotiating employment contracts. Moreover, just like an acquiring firm has to perform a *complex integration process* of the target company, a hiring firm must carry out multiple and small integration processes in order to fully exploit the knowledge of its new employees. This is by no means an easy task, since managers will have to spread their attention to all new recruits, taking care that the way with which they enter the organization strikes the right balance between conforming to established norms and procedures, at the same time effectively introducing those elements of novelty that justified their hiring in the first place.[13] Moreover, while an acquired firm usually continues its operations throughout the integration process, the productivity of new hires is usually low, until their integration is completed.

Compared to an acquisition, hiring generally entails a smaller risk of running in the 'winner's curse', since job markets are usually more liquid than the market for companies, and this makes it easier to identify a going price for a professional with given skills. However, this might not be true in the case of competencies that are just emerging. The scarcity of experts might allow them to command very high salaries, at least in the short term, and until more people gain these skills and rebalance the job market.

In any case, the hiring process is always *quite risky*, since it will be difficult for an employer to foresee the true value of a new employee. A person may have been an excellent performer in a previous firm, but the different set of complementary resources and routines and the different corporate culture will not necessarily lead to the same value when joining a new firm. Academics would say that individual performance is *institutionally specific* (Toole and Czarnitzki 2009).

Recently, firms have also started using acquisitions as an alternative approach to mass hiring, a practice termed '*acqui-hiring*'. In these cases, the target company is not chosen because of the organizational competencies and assets it has developed, but simply because of the individual competencies that are included in its workforce. In this case, the integration of the target company will be carried out differently than with a standard acquisition and have a greater focus on individuals and teams to be associated to the existing workforce, rather than on departments and business units.

8.6 Non-equity Strategic Alliances

A second family of methods for gaining access to competencies is involved with creating links with external entities, thus accepting to give up part of the appropriability of the related economic value. The basic way to do so is to create a *strategic alliance* with another firm and pursue a joint program of activities, which may include R&D and product development.

[13] Trying to integrate human resources with competencies that are significantly different from the ones that already exist in the organization is generally quite difficult. Management must explicitly support the change and make the use of these competencies mandatory in projects. If they fail to do so, organizational inertia will make it very difficult for a new employee to change established routines with her new knowledge. Social pressure will tend to make her conform to the old routines, effectively making the hiring useless from a strategic perspective.

Alliances can create value when participating firms have complementary competencies, and when competitors cooperate in order to pursue economies of scale and/or to define common standards.[14] Research also suggests (Hoang and Rothaermel 2010) that successful alliances usually have the objective of exploiting existing knowledge assets and exploring their potential synergies, rather than for outright exploration of new fields in which participating entities have no prior background.

For classification purposes, under the term alliance we will only consider 'non-equity' relationships that are based on contracts, which include neither the exchange of shareholdings, nor the establishment of a joint venture (which will be the subject of the next subsection).

When firms are bound by contractual, and not by institutional links, the outcome of the relationship will depend on the strength that the contract has in specifying and directing the actions to be taken by each party. Unfortunately, contracts are known by economists to be *incomplete* and *difficult to enforce*[15] and this provides room for opportunistic behavior and *free riding**,[16] thus leading to poor results. In practice, the opportunistic behavior may sometimes come from a managerial decision not to allocate the required resources to the common projects that make up the alliance. In other cases, top management may have allocated resources, but failed to fully motivate the middle managers and the employees involved, explaining the importance of the alliance projects and providing adequate incentives. When an employee is pressured and must choose whether to work on an internal project or on an alliance-related one, it will then be likely that he will opt for the former. This is especially true if work on alliance projects involves unusual organizational responsibilities, frequent or extended trips abroad, etc.

A further element that might restrain participating entities from fully committing to the alliance is the risk of *exposing valuable intellectual property* to the other firms (Giarratana and Mariani 2014). This might occur when running the common

[14] For instance, in the late '90s, a group of competing mobile phone makers decided to develop a common operating system for next-generation smartphones. They therefore joined forces among themselves and with Psion, a maker of handheld computers, who contributed with its competencies in software. The result was the Symbian OS, which for a few years was a contender in the nascent industry.

[15] Regardless of the effort spent by managers and their legal teams, no contract will be able to contemplate all the possible circumstances that might occur and the eventual emergence of unforeseen events will give leeway for the parties to act opportunistically. Moreover, if a party does not fulfill contractual obligations, the other party can only start legal action and ask a court to restore its rights. However, this usually is an expensive and lengthy procedure with an uncertain outcome. Even if a court should decide in favor of the claimant, it is likely that damages paid will not be able to compensate for the broken relationship and the related opportunity costs. The greater is unenforceability, the more room is created for opportunistic behavior.

[16] '*Free-riding*'* means that one party exerts less effort than it should, in the expectation that some other party will compensate for this behavior. The greater the number of participants, the greater will be the incentive to free-ride and—therefore—the likelihood that it might occur. Alternatively, a party may fear that other parties want to free-ride and will wait for a sign of commitment before actually fully committing itself. Of course, if each party behaves in this same way, activities will never fully start.

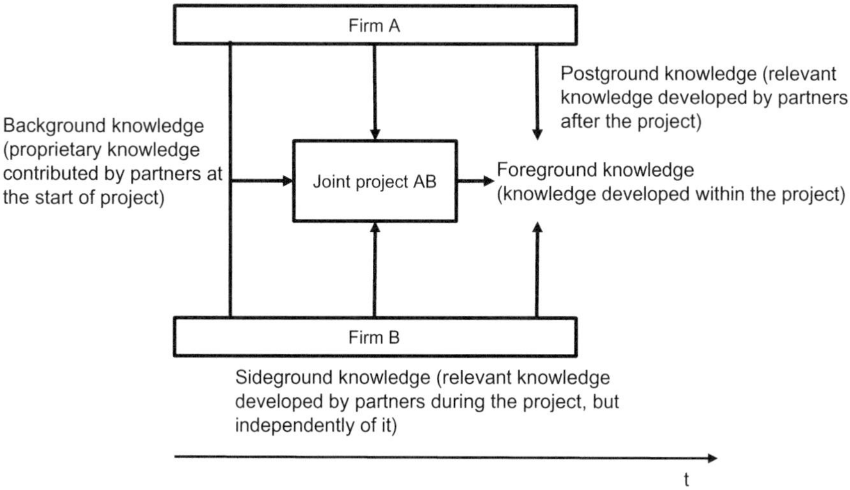

Fig. 8.5 Different forms of knowledge relevant to an alliance

project (e.g., a firm that tries to convince a competitor-alliance partner to use one of its proprietary technologies in the common project will have to expose some critical elements in order to justify this choice), or as an unwanted and unrelated side effect (e.g., a representative from a competitor that is visiting for a joint meeting might overhear a discussion at the company canteen during lunchtime).

Managing IP in the context of strategic alliances is quite critical and requires a sound management of the different forms of knowledge that may be relevant to the alliance. Specifically, the participating firms will have to define clear rules for dealing with 'background', 'sideground', 'foreground' and 'postground' knowledge (see Fig. 8.5 for an explanation).

Given the complexity of alliance management, it comes as an obvious conclusion that purely formal mechanisms based on written contracts will often not be able to ensure a full and successful unfolding of activities. Relational mechanisms and cultural norms will therefore be needed to compensate for this (Hoetker and Mellewigt 2009; Devarakonda and Reuer 2018).

8.7 Equity-Based Alliances and Joint Ventures

Equity-based alliances and joint ventures have similar objectives to non-equity alliances, but are based on a deeper institutional linkage between participating entities. These entities may exchange shareholdings directly or—more commonly—in an alliance-specific and newly incorporated legal entity called *joint venture*.

At least in principle, the institutional nature of the linkage creates less ambiguity than a purely contractual one. Co-ownership of the joint venture defines an incentive to pursue its success and, should this not be strong enough for the partnering institutions, the incentive should work at least for the management team and the employees who are formally assigned to the venture. Moreover, in case of unforeseen circumstances and disagreements over each partner's duties, the governance structure of the joint venture can become an important mediator, attempting to solve disputes before having to take legal action.

These advantages over purely contractual alliances might however not be fully achieved if the partnering institutions are very different in size and if they tend to take an intrusive role in the joint venture's governance.

Benefits of a joint venture must also be balanced against its drawbacks. The first one is related to the cost of establishing and maintaining a new legal entity, which requires separate accounting, a board of directors, dedicated management, and so on. These fixed costs will discourage from recurring to this solution, if not for projects of significant current or prospective value. From the organizational perspective, the personnel that are assigned to the joint venture or are directly hired by it are no longer detached members of the parent organizations. This means that both formal and informal knowledge transfer from the participating firms to the joint venture and vice versa will be quite hard to achieve. As a consequence, the joint venture will tend to lose access to potentially valuable competencies that reside in its parent companies, while it will not be easy for the same parent companies to absorb competencies generated by the joint venture.

8.8 Co-development

Co-development often occurs when a company needs to develop components that require highly specific competencies, but does not wish, or is unable, to build these competencies in-house, and therefore has to ask a supplier not just to produce the component, but to develop it too. In this context, specificity means that these competencies and the ensuing components will be of little or no value, if offered to competitors of the customer company.[17]

In principle, a basic solution to this problem could consist in asking the supplier to make an investment I in an R&D project and then promising to pay a comprehensive unit price, P, based on expected volume, V, variable production cost, VPC, and an acceptable contribution margin, MC. This price P would amount to $P = (\mathrm{VPC} + I/V)(1 + \mathrm{MC})$. However, this *revenue-sharing* co-development

[17] In this context, the supplier is not required to act as a consultant or as a provider of R&D services, but is asked to provide both the R&D and the production of the firm-specific components that are being developed. This specificity can come from peculiar technological choices, from the desire to serve a particular market niche, or from the fact that the technology is just emerging and competitors have not yet shown interest in it.

arrangement would probably not be accepted by the supplier, since multiple risks would fall on his shoulders. The first source of uncertainty is *technological* (or *development*) *risk*. The project, which ultimately comes from the customer's highly specific wishes and cannot rely on established competencies, may not lead to the expected results. A second element of uncertainty is *market risk*, since sales volume of the end item that the component will be part of may be less than the expected value *V*. The supplier has little control over *V* (which will probably depend on many factors other than the quality of that component) and might not have enough market knowledge to assess whether *V* is a reasonable or an overly optimistic forecast.

In principle, these risks could be attenuated with contractual clauses granting minimum volume purchases, and/or adjustments to *P* depending on volume. However, the supplier would still incur the much greater risk of *post-contractual hold-up*. Since the new competencies are specific to the customer and cannot be used to supply other firms, the investment *I* becomes a sunk cost. This creates a very strong incentive for the customer to wait until the project is ended and then renegotiate the terms of the contract, asking the supplier to decrease the price, theoretically all the way to $P' \cong \mathrm{VPC}$, and regardless of volume *V*. At this point, the supplier would have no alternative, but accept or go to court, which would entail a score of other costs and risks.[18]

Due to these risks and to the likelihood of post-contractual hold-up, the parties may define a co-development agreement based on *investment sharing*. The customer would then finance the development work by paying the investment *I* up front and, since the supplier now bears no more risk, price will be set close to marginal cost $P' \cong \mathrm{VPC}$. Such an agreement will usually grant the customer exclusive rights to the competencies generated, including ownership of, or exclusive licensing rights to, any intellectual property that might be developed under the project it has financed. This will prevent the supplier from trying to use these competencies to serve other customers that might eventually show some interest.

An investment-sharing co-development agreement is not without risks, however. The customer may have problems monitoring the development work being performed by the supplier, as in a typical *principal-agent**[19] situation. For instance, the supplier might not use the financing exclusively for the project or might try to steer its technical contents toward objectives that fit with its own strategy, rather than the customer's. Furthermore, the appropriability of the competencies will be weak, since the effective owner will be the supplier, who will gain higher bargaining power, especially if the exclusivity clauses mentioned above are difficult to enforce. The customer might therefore decide to enter an *innovation-sharing* co-development agreement and also share part of the R&D work with the supplier.

---

[18] In some industries, a supplier would not even dare sue a customer over a supply contract of this kind, in fear of losing future contracts with the same counterpart, as well as with other customers.

[19] In Economics, a '*principal-agent problem*'* occurs when a principal asks an agent to perform an activity, but the decisions taken by the latter might be self-interested and therefore not aligned to the interest of the former.

Major source of uncertainty / Innovation type	Uncertainty in translating R&D to a product	Uncertainty in translating R&D and on timing	Uncertainty on timing
New to the world	Innovation sharing	Innovation sharing or investment sharing	Investment Sharing
Incremental	Revenue sharing	Revenue sharing or Investment sharing	Investment Sharing

Fig. 8.6 Choosing among the different types of co-development

Following Bhaskaran and Krishnan (2009), the suitability of the three co-development alternatives essentially depends on the type of innovation and the major source of uncertainty (Fig. 8.6).

In the case of incremental and competence-enhancing innovations, and if the main source of uncertainty is technical, the supplier will probably be in the position to make a sound assessment of risks and accept the basic revenue-sharing contract. Conversely, if uncertainty has to do with marketing and timing, an investment-sharing agreement will relieve the supplier of those risks that it has no way of assessing and controlling. In the case of radical and competence-destroying innovations, investment-sharing might still be used if the main source of uncertainty is related to the market. Instead, if uncertainty is also related to translating the technology into a product, an innovation-sharing agreement may prove to be beneficial, since it will allow the two entities to work together in overcoming technical difficulties, and at the same time allow the alignment of their incentives and limit the principal-agent issue.

From the side of the supplier firm, entering a co-development agreement with another firm can also be considered as a strategy for developing new competencies. This is especially true if the customer firm is technologically advanced (Alcacer and Oxley 2015) and willing to entrust not only production, but also development activities.

8.9 Sourcing R&D Activity from Other Parties

An alternative approach to developing competencies consists in paying external parties to carry out the corresponding R&D work and then defining some kind of a technology transfer mechanism to absorb the results. In this case, the firm will have to identify the external entity that is best suited to carry out the work (which involves balancing technical competencies, price and transaction costs) and defining routines and processes that may ensure an efficient engagement with them. Among the most common candidate entities, firms can consider:

- **Competitors' R&D units**. Competitors might accept an offer to cooperate on specific projects, in order to profit from temporarily unused capacity, or to increase returns from competencies they have developed. While a competitor might be willing to provide technical results, it is highly unlikely that it will accept to transfer competencies and know-how.
- **For-profit R&D firms**. In a number of industries, it is possible to identify firms whose main business is to perform outsourced R&D activity by leveraging on low cost or specific competencies. The business term for addressing these firms varies by industry. The pharmaceutical industry refers to *Contract Research Organizations*, while other industries may use the term *Engineering Services Provider*. In this case too, suppliers will be unlikely to allow an actual transfer of competencies and knowledge and will limit themselves to delivering contractually specified results.
- **Universities and Non-profit R&D organizations**. As already mentioned in Chap. 1, academic entities have a key knowledge-generating role in the innovation process. Therefore, they are interesting candidates to engage in, especially when dealing with competencies and activities that are related to the upstream (with the lower TRL) phases of the innovation process. Universities often call this process *technology transfer**, which can lead to a number of approaches for cooperation with industry, including licensing of intellectual property, performing contract-based research and undertaking government-funded research projects jointly with firms. Moreover, academic organizations may foster the creation of *academic spinoffs** based on internally developed technology or know-how, with the purpose of facilitating relationships with industry.

  Cooperation with universities can be an attractive proposition for corporations because of the high level of their competencies, and because an academic organization is not usually as keen on preserving its proprietary know-how as a for-profit firm. In case contract-based research may lead to patents, the firm that provides finance to the project will usually gain ownership to the patent, or at least exclusive licensing rights. Moreover, it also may be relatively easy to transfer tacit knowledge, since the firm can easily hire junior academic staff, such as Ph.D. students, that have been involved in research projects. Moreover, the firm may negotiate the *secondment* of some of its own R&D personnel, who may stay with the academic lab throughout the project.

  Partnering with universities is however not exempt from problems. One first issue is related to the organizational *fragmentation* of universities around departments and academic disciplines, which contrasts with the multidisciplinary nature of industrial problems. Moreover, universities and companies operate at somewhat different timescales and with differing objectives. A university lecturer will divide her/his time between a host of research projects, teaching activities and administrative duties. Moreover, she/he will usually focus her/his efforts to achieve research results that are publishable, which usually implies a development stage that is earlier than the results that a firm would expect.

In order to avoid this potential mismatch, firms and universities nowadays tend to forge medium- to long-term agreements that involve multiple departments (sometimes called *University Technology Partnerships*), thus allowing inter-disciplinary work and allowing a longer time frame to accommodate these different objectives. Some universities may also move to a 'matrix-like' organization, setting up inter-disciplinary and temporary *research centers,* which draw scientists coming from departments around a common problem relevant to industry.

Furthermore, in order to make dialogue with industry easier, they can set up *Technology Transfer Offices* staffed by professional and not academic personnel.

Another issue in university partnerships is related to the *confidentiality* of results. As mentioned, academic researchers tend to disseminate the knowledge they generate through teaching and publishing. Publishing in important journals and being highly cited by their colleagues is the main avenue for achieving promotion. Firms must therefore make sure that the researchers they finance do not inadvertently publish important elements of knowledge that might leak to competitors, or preempt patenting activity. Cooperation contracts usually specify that manuscripts have to be signed off by the firm before submission to journals and conferences, but enforcement of these clauses is not always easy.[20]

Given the inconveniences that firms might experiment in attempting direct interaction with universities, a number of countries have promoted *applied research centers*, such as the well-known Fraunhofer Society in Germany. These centers act as bridging institutions between academia and industry, with a strong focus on applied research and industrial problem-solving, engaging their own staff together with academic personnel.

- **small firms and individual inventors.** Many inventions and technologies are developed by small firms, who are often unable or unwilling to fully exploit it. The same holds for individuals who have developed specific competencies or inventions and are not in a position to create a business around them. A firm that is interested in a given technology may therefore find it quite easy and inexpensive to cooperate with the original inventor, rather than to try to replicate it internally. At the same time, small firms may be wary of value appropriation by the large firm, and this may lead to an adverse selection mechanism (Diestre and Rajagopalan 2012).

Outsourcing the generation and development of innovative competencies and solutions appears to be a very attractive concept, allowing firms to tap into the possibly unlimited richness of talent and ingenuity that can be found in the world.

[20] Confidentiality clauses in industry-sponsored research also raise a number of issues related to ethics and public welfare. Firms are likely to allow dissemination of research leading to positive results, but to deny it in case of negative ones. Aside from ethical issues, it may be socially inefficient to be prevented from warning other researchers that a given solution does not work.

However, the main problem is connected to the *transaction costs* a firm seeking for a solution will encounter in finding the right 'solver' and creating an agreement that will be attractive for both parties.

A number of mechanisms have emerged over the years. Following Pisano and Verganti (2008), they can be classified according to two axes. The first is the *openness of the system*, which can range from closed invitation-based mechanisms, which require ex-ante knowledge of potential solvers, and open mechanisms where virtually any potential solver is free to apply, and which require the capability to screen and select them ex-post. The second axis is the *governance mechanism*, which can range from strictly hierarchical (i.e., the seeker firm decides which solution is preferable) to flat, in which decision-making is shared between the seeker firm and the group of solvers. The former requires a well-defined innovation strategy while, in the latter, the firm uses the community not only as a source of competencies, but also as a contributor for defining innovation strategy. Table 8.2 shows the resulting four categories.

*Crowdsourcing** marketplaces for innovation are a special type of the 'innovation mall' model and are becoming quite popular for implementing open strategies of this type. With crowdsourcing, problems and challenges are posted on dedicated Internet marketplaces, and prizes and/or grants are awarded to the solvers that come up with the best solutions or demonstrate they have the best competencies for solving them. Well-known examples of such services are Innocentive, originally developed by the pharmaceutical company Eli Lilly and that now acts for many other players in the same industry, and Ninesigma, that works on many other industries, from aerospace to consumer products.

From an economic perspective (Boudreau et al. 2011), crowdsourcing works as a 'contest', in which many actors respond and only one of them wins. So, a tradeoff must be found between the investment required to respond to the challenge, the size of the prize and the probability of winning it, which depends on the number and qualification of competitors. If a participant expects a high number of competitors, she/he may not find a strong enough incentive to invest time and effort to come up

Table 8.2 Four different types of interaction models

Governance participation	Hierarchical	Flat
Open	*Innovation mall*—anyone can participate, but the seeker firm decides the winner (e.g., crowdsourcing marketplaces)	*Innovation community*—anyone can propose solutions and the community decides on the winners (e.g., Open Source software projects)
Close	*Elite circle*—the seeker firm invites participants and decides the winner (e.g., firms in the furniture and household goods industry with designers)	*Consortium*—the seeker firm invites participants and shares technical decisions with them (industry consortia)

with a strong bid and will therefore apply only for those problems for which a solution has already been developed or is close to being so. For this reason, innovation contests are often operated in stages. The initial challenge has the objective of qualifying the best solvers and does not require candidates to make substantial investment to place a bid. Once these candidates have been identified, further developments will be run through a contest run among a small pool of competitors, sometimes with a small financial contribution.

Another aspect to be considered is the difficulty of ensuring that problem owners and problem-solvers may engage in a fruitful and efficient discussion. In many fields, the tacit nature of technological knowledge makes it difficult and time-consuming to define a common language and interaction process, which may add to the costs and the difficulties of this approach. In fact, cooperation platforms sometimes provide brokerage and consultancy services aimed at overcoming such problems.

8.10 Licensing

Firms can also decide to license other entities' intellectual property, in order to get a foothold in a technological area of their interest, and progressively start developing competencies around it. Many patent portfolios can be targeted for this purpose, since they are essentially unused either in absolute terms or in specific regional or industrial areas.

In the case of large firms, the patent might not have been exploited because it is related to a non-core market, or because the firm decides to pursue different technologies. Large firms, such as IBM, often market their intellectual property quite actively, sometimes allowing very easy access to small firms and startups. Universities and research centers too might have portfolios with inactive patents, since this kind of organization cannot directly exploit inventions and must rely on licenses to do so. Therefore, it is quite common for academic institutions to file patent applications without a clear exploitation strategy, which creates an interesting opportunity for potential licensees. Moreover, universities may be quite glad to support the transfer of knowledge that is relevant to the patent via contractual research, consultancy activities or facilitating the recruitment of skilled personnel. Finally, SMEs might be financially or organizationally constrained to fully exploit their IP and be quite happy to derive some additional revenue from it.

Entering a technological field via licensing can be an interesting approach when a firm wants to catch up lost ground and expects to be able to build additional competencies by working on the technology. In any case, IP contracts should be drafted with great care, to ensure that the firm will effectively be free to operate independently. For instance, grant-back clauses may force the licensee to transfer rights on improvements back to the licensor and may be highly limiting.

Opposite to acquiring patents or licenses to patents, a firm may also decide to sell its IP or act as a licensor for patents that are not being used. This may become not only a source of revenue from an unused asset but, even more, may allow the exploration of strategically relevant activities, in partnership with the firms that are interested in this IP.

8.11 Complete Outsourcing

In this context, complete outsourcing means that the firm recognizes that a given technological area is non-core, and therefore gives up any further attempt to develop it either internally or through partnerships. From this moment on, the outcomes of this technology will be seen as 'black box' components to the firm's products, and the firm will limit itself to specifying its needs, negotiating purchasing agreements and checking compliance with specifications.

In doing so, the firm may accept 'off the shelf' technology and components, which will not enable any sort of differentiation with respect to competitors. Should the firm wish to differentiate itself, the only way would be to enter a co-development agreement with the supplier and finance a firm-specific project under the provision of exclusivity.

Complete outsourcing may lead to issues and problems that usually become visible only after some time. When a firm decides to disengage from development in a given technological area, it progressively disperses its competencies. At first, some of the employees that were working in that area with technical responsibilities may be given purchasing roles and will obviously become very attentive and technically capable buyers. As the years go by, these employees either migrate to other functions or retire, and the firm will substitute them with personnel that does not have the same technical expertise and understanding of the technology. It therefore can become difficult for the company to effectively specify, buy and verify the components it needs.

Moreover, when all competitors in an industry share the same decision of outsourcing a given technology or component, this might lead to the emergence of monopolistic suppliers. Should this happen, the industry bears the risk that this supplier will become complacent and stop innovating, thus damaging the prospects for its customers. This is often at the origin of many trend reversals in industry, which often oscillates between outsourcing and vertical integration choices.[21]

[21] For instance, a number of medium–high end Swiss watch manufacturers have recently reversed the trend of buying mechanisms from specialist suppliers and started developing and producing them in-house, as was done by the traditional 'manufactures', in order to avoid the commoditization of their products.

8.12 The Framework of Open Innovation

We close this chapter by introducing the concept of *Open Innovation** (or OI), which can be viewed as the 'umbrella' term that encompasses a number of approaches for accessing and developing competencies outside the boundaries of the firm. Originally, the term OI has been made popular by Chesbrough's seminal work (2003), which identified an emerging practice by a number of large corporations, among which Procter and Gamble probably is the best-known example.

Large firms traditionally ran their development activities as 'closed funnels' (left-hand side of Fig. 8.7). They would start many R&D projects to build a technological base, only a minority of which would actually lead to products with some market return. When evaluating the productivity of their R&D expenditure, firms realized that substantial resources were being spent on reproducing competencies and/or results that already existed in the world and—at the same time—they were not able to find alternative applications for 'lost' projects that did not have a clear internal exploitation path in the firm's existing markets. Realizing this inefficiency, firms started working on practices that were later labeled as 'Open Innovation'. First, they decided that a given percentage of their R&D budgets had to be spent on projects run with external entities or to acquire technology that had already been developed by other parties. Furthermore, R&D units were directed to find revenues from lost projects by using licensing or spinoffs or to use them strategically to enter new markets. The resulting 'open funnel' model is shown on the right-hand side of Fig. 8.7.

By investigating such phenomenon, Chesbrough (2003) described OI as "*A paradigm that assumes that firms can and should use external ideas, as well as internal ideas, and internal and external paths to market, as the firms look to advance their technology*". More recently, Chesbrough and Bogers (2014) have provided an updated definition that shifts the focus away from Technology toward

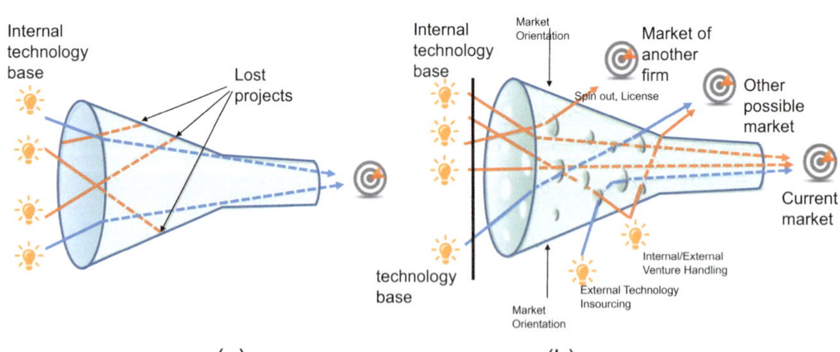

Fig. 8.7 **a** The traditional (closed funnel) approach to innovation; **b** the Open Innovation (open funnel) approach

Knowledge "*A distributed innovation process based on purposively managed knowledge flows across organizational boundaries, using pecuniary and non-pecuniary mechanisms in line with the organization's business model*".

One might therefore consider all the approaches discussed in this chapter as forms of OI, with the single exception of the first one (i.e., 'internal R&D'). This is true in principle, but readers should be warned that many firms, when thinking of OI, tend to restrict the concept to the contents of Sect. 8.9 (i.e., 'external sourcing of R&D').

In any case, running an Open Innovation program requires setting up a dedicated organizational unit, processes and a budget, and ensuring adequate governance mechanisms. Many OI initiatives have had limited success, and research has indeed tried to understand what might be the main critical success factors in this field (e.g., Monteiro and Birkinshaw 2017).

One main element consists in the definition of a *strategic charter for OI*, which must include a clear orientation in the weight to be given to 'technology push' and 'demand pull'. The former will lead to OI activities that are quite independent of the firms' business units, with a high propensity toward risk, and will try to come up with innovative solutions that may be useful in a future perspective, but not necessarily be of immediate relevance. The obvious risk, in such a case, is that OI will not have a real connection to the firm's innovation processes and be perceived as a relatively useless 'ivory tower'. The latter will instead ensure that OI is carried out with a constant connection to BUs, trying to find solutions to explicit and immediate needs, and will be less risk prone. In this case, the risk is that OI will not deliver solutions capable of transformative change on the organization.

A second important strategic element consists in defining the *objective of OI* and the associated targets to scout for. Among them, one can list:

- **providers of enabling technologies**, in order to facilitate the development and incorporation of their solutions in the firm's products and processes, eventually leading to corporate venturing and/or co-design activities;
- **developers of innovative products**, that might allow a broadening of the firm's offering, with a potentially win–win solution on both sides (i.e., the innovator can access the firm's complementary assets in distribution, while the firm avoids having to invest in product development), eventually leading to strategic alliances, joint ventures and corporate venturing;
- **competencies pools**, leading to externalized R&D and, eventually, to corporate venturing and acqui-hiring;
- **potentially disruptive competitors**, leading to a better understanding of new technology and business models and, eventually, to alliances, joint ventures and corporate venturing;
- **opportunities for learning and developing absorptive capacity**. Even though this might seem like a somewhat limited objective, it is to be viewed as a very important potential outcome. A firm operating OI will in fact acquire a privileged observatory on the leading of edge of technology and innovation, increasing its ability of innovating both through OI and via internal developments.

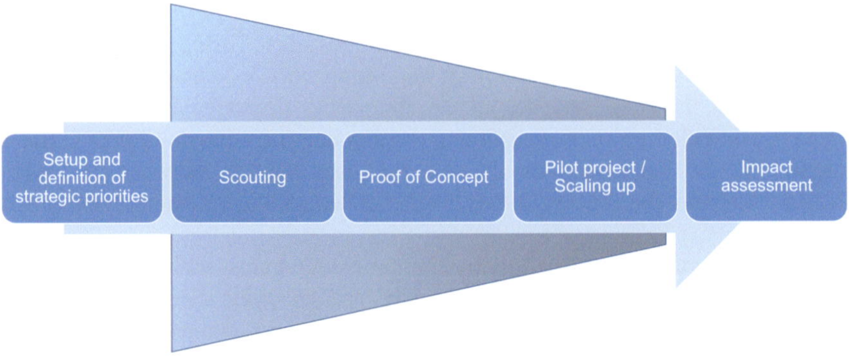

Fig. 8.8 The funnel-shaped OI process

After having defined the main objectives for OI, it will be necessary to define *a clear process for scouting, testing, scaling up the innovative solutions and, finally, assessing their impact*. The process can be shaped like a funnel (see Fig. 8.8, and stage-gate systems in Chap. 9) since, given the high risk involved, only a fraction of the proposals being scouted will be actually implemented at scale.

In general, proposals deriving from scouting initiatives will be screened, and a fraction will lead to '*proof of concept*' (PoC)* projects, generally carried out in cooperation with the firm's BUs. PoCs have the objective of ensuring that the innovative proposal can work properly both in general, and in the specific context of the firm's operations. The outcome of a PoC is generally quite limited (a 'demonstration', or *demo*, in short), but can provide enough confidence to make the further investment to scale up and bring the innovation in real production. This can be done either outright or, quite often, going through a pilot stage, in which the innovation is experimented in real operations, but on a limited scale.

All these activities require a *budget* and it is critical to decide the absolute amount, the fraction to be allocated to each phase, as well as the decision-making mechanisms on spending it (e.g., when deciding on a PoC, what is the decision-making role of the OI unit and of the BUs? How are the costs split among them?).

Managing the OI process is fraught with difficulties, since it has to do with pushing externally sourced innovations, usually coming from smaller organizational entities, in the BUs' innovation processes.

The focal firm must therefore learn how to deal with actors that are organizationally less structured than ordinary partners and suppliers, have a weaker bargaining power and are likely to value timeliness of business decisions more than the margins they can extract from a contract. The *partnering process* between a large firm and a small player engaged in the former's OI activities can therefore lead to numerous breakpoints that the OI department must manage with great care.[22]

Moreover, the firm may experience very *high resistance to change* and a variety of reactions, which are commonly grouped into the 'Not Invented Here' syndrome (Antons and Piller 2015). BU managers may view the external innovation as something that will take time and effort away from the priorities of the business. They may perceive it as a threat to their reputation (i.e., think of a small startup that has developed a solution that the internal R&D department has not been able to) and they may miscalculate the cost/risk/benefit balance of the proposed solution, especially by mistrusting a solution that does not come from a reputed and well-known supplier. Reactions arising from NIH can vary, from simply delaying actions until the relationship is severed, to opportunistic raising doubts over risks and disadvantages while downplaying benefits, and may even reach the point of active tampering and sabotaging pilot activities.

All these difficulties point in the direction of making sure that OI is not simply an impromptu initiative, but becomes a well-structured dynamic capability for the firm. A discussion on possible managerial practices and frameworks will be presented in Chap. 10.

As a final comment to this discussion, it should be quite clear that a firm's ability to innovate by enacting all the alternative approaches discussed in this chapter in a significant and effective way is subject to very significant economies of scale and will therefore be highly dependent on its size. In fact, each of these approaches requires the availability of specialized professionals, whose presence in a small firm is unlikely to be sustainable.

Moreover, the risks involved in each initiative will be sustainable only if one can rely on a sufficiently broad portfolio of alternatives. Finally, most of the approaches that have been discussed are reliant on the attractiveness of the firm as a potential business partner, employer, or acquirer, which depends on its reputation and on its size.

It might therefore be hypothesized (though current research has not reached conclusive results yet on this topic) that the variety of contemporary approaches to innovation is increasing the role of economies of scale, to the clear advantage of large firms. At the same time, this model of innovation creates interesting opportunities for smaller firms, provided they abandon a traditional and 'isolated' approach to innovation, and engage with large companies within their open innovation ecosystems.

[22] Examples of such issues can be both organizational and technical. Among the former, excessive lead times in moving from an initial contact to the signing of a contract, or the time and effort to be spent in achieving a 'go-ahead' from corporate functions, from legal to purchasing, in the new relationship. Among the latter, misperceptions in the degree of maturity of the proposed technical solutions, or the need to involve further partners, such as system integrators, in order to integrate a very 'narrow' solution in a more complex system.

References

Alcacer J, Oxley J (2015) Learning by supplying. Strateg Manag J 35(2):204–223

Antons D, Piller FT (2015) Opening the black box of "not invented here": attitudes, decision biases, and behavioral consequences. Acad Manag Perspect 29(2):193–217

Bhaskaran SR, Krishnan V (2009) Effort, revenue, and cost sharing mechanisms for collaborative new product development. Manage Sci 55(7):1152–1169

Boudreau K, Nicola L, Lakhani K (2011) Incentives and problem uncertainty in innovation contests: an empirical analysis. Manage Sci 57(5):843–863

Cassiman B, Ueda M (2006) Optimal project rejection and new firm start-ups. Manage Sci 52 (2):262–275

Chesbrough H (2003) Open innovation: the new imperative for creating and profiting from technology. Harvard Business School Press, Boston

Chesbrough H, Bogers M (2014) Explicating open innovation: clarifying an emerging paradigm for understanding innovation. In: Chesbrough H, Vanhaverbeke W, West J (eds) New frontiers in open innovation. Oxford University Press, Oxford

Cuervo-Cazurra A, Un CA (2010) Why some firms never invest in formal R&D. Strateg Manag J 31(7):759–779

Cumming T, Knott AM (2018) Outside CEOs and innovation. Strateg Manag J 39(8):2095–2119

Devarakonda VD, Reuer JJ (2018) Knowledge sharing and safeguarding in R&D collaborations: the role of steering committees in biotechnology alliances. Strateg Manag J 39(7):1912–1934

Diestre L, Rajagopalan N (2012) Are all 'sharks' dangerous? New biotechnology ventures and partner selection in R&D alliances. Strateg Manag J 33(10):1115–1134

Dushnitky G, Shaver JM (2009) Limitations to inter-organizational knowledge acquisitions: the paradox of corporate venture capital. Strateg Manag J 30(10):1045–1064

Dushnitsky G, Lenox MJ (2005) When do firms undertake R&D by investing in new ventures? Strateg Manag J 26(10):947–965

Giarratana MS, Mariani M (2014) The relationship between knowledge sourcing and fear of imitation. Strateg Manag J 35(8):1144–1163

Hoang H, Rothaermel FT (2010) Leveraging internal and external experience: exploration, exploitation, and R&D project performance. Strateg Manag J 31(7):734–758

Hoetker G, Mellewigt T (2009) Choice and performance of governance mechanisms: matching alliance governance to asset type. Strateg Manag J 30(10):1025–1044

Knott AM (2008) R&D returns causality: absorptive capacity or organizational IQ. Manage Sci 54 (12):2054–2067

Leiponen A, Helfat CE (2010) Innovation objectives, knowledge sources, and the benefits of breadth. Strateg Manag J 31(2):224–236

Li M-YL, Hwang N-CR (2011) Effects of firm size, financial leverage and R&D expenditures on firm earnings: an analysis using quantile regression approach. Abacus 47(2):182–204

Macher JT, Boemer CS (2006) Experience and scale and scope economies: trade-offs and performance in development. Strateg Manag J 27(9):845–865

Makri M, Hitt MA, Lane PJ (2010) Complementary technologies, knowledge relatedness, and invention outcomes in high technology mergers and acquisitions. Strateg Manag J 31(6): 602–628

Monteiro F, Birkinshaw J (2017) The external knowledge sourcing process in multinational corporations. Strateg Manag J 38(2):342–362

Pisano GP, Verganti R (2008) Which kind of collaboration is right for you? Harward Bus Rev 86:78–86

Ransbotham S, Mitra S (2010) Target age and the acquisition of innovation in high technology industries. Manage Sci 56(11):2076–2093

Sears J, Hoetker G (2014) Technological overlap, technological capabilities, and resource recombination in technological acquisitions. Strateg Manag J 35(1):48–67

Sauermann H, Cohen WM (2010) What makes them tick? Employee motives and firm innovation. Manage Sci 56(12):2134–2153

Toole AA, Czarnitzki D (2009) Exploring the relationship between scientist human capital and firm performance: the case of biomedical academic entrepreneurs in the SBIR Program. Manage Sci 55(1):101–114

Xiaoying L, Wang J, Xiaming L (2009) Can locally-recruited R&D personnel significantly contribute to multinational subsidiary innovation in an emerging economy? Int Bus Rev 22(4): 639–651

Zhao X (2009) Technological innovation and acquisitions. Manage Sci 55(7):1170–1183

Innovation Strategy as Project Portfolio Management

9

Abstract

Chapter 6 showed how project portfolio management has a key role within the formulation of an innovation strategy. In fact, it allows a rigorous management of initiatives aimed at both advancing and exploiting corporate competencies. The chapter discusses the main elements relevant to the implementation of a project portfolio management system, which includes the classification of project types, the setting up of a Stage and Gate evaluation process and the different quantitative and qualitative methods that can be used to decide on project initiation and termination, along with their typical difficulties and pitfalls.

This chapter provides a complementary view of innovation strategy to the one that has been developed in Chap. 8. Still following the resource-based view of the firm that was discussed in Chap. 6, we will concentrate on the decisions pertaining to the selection of the projects and activities that should be performed, given the competencies the firm currently possesses and the ones that it wishes to develop. Since most activities related to innovation are project based, it is possible to borrow concepts, methods and tools from the field of *Project Portfolio Management** (or PPM).[1]

[1] Project Portfolio Management is a key area of the Project Management literature. This literature historically has had a bottom-up approach at projects. *Project management strictu* sensu deals with individual projects at operational level, while *program management* deals with programs, or clusters of related projects, at a tactical level. Finally, *project portfolio management* copes with the strategic decision on which projects and programs should be initiated. In the context of Innovation Management, one can assume a top-down approach in which the firm must first decide what to do ('doing the right things'), and then focuses on execution (doing things right').

M. Cantamessa and F. Montagna, *Management of Innovation and Product Development*, https://doi.org/10.1007/978-1-4471-7531-5_9

9.1 Project Portfolio Management in a Strategic Perspective

PPM is one of the areas of the broader discipline of Project Management, and focuses on the selection of the projects that must be started, managed and terminated, for the purpose of implementing the strategy of an organization and achieving the consequent objectives. Tools used in PPM are a relatively recent addition to the toolbox of practicing managers. The recent success of PPM is due to the recognition of its strategic relevance, and to the diffusion of IT-based tools that make it easier to collect and represent information on multiple projects.

Due to its strategic relevance, firms that do not make a correct use of PPM run the risk of being unable of formulating and enacting a sound innovation strategy. In general, lack of discipline in PPM leads firms to start too many R&D and innovation projects (in the remainder of this chapter, we will simply call this the 'R&D portfolio'), without a clear strategic focus and/or without a strong understanding of their alignment with the type and quantity of resources, either available or to be developed. It is always easy to give a go-ahead to a potentially interesting project, but attention should be paid to the strategic reasons that justify it (which, in turn, determine what specific outcomes should be expected from it) and to the resources that are required in order to ensure success with respect to timing and results.

It is not uncommon for companies to be quite unaware of their R&D portfolio. Examples of this inability are not knowing how many and which R&D projects are active at a given time, being uninformed of the ratio between the overall workload and the available employees (which often turns out to be greater than 100%), and assigning personnel to an excessively high number of simultaneous projects. The immediate outcome of not having enough resources assigned to projects, and allowing them to work inefficiently, is that projects are doomed to be delayed. When this occurs, project managers will scramble to get hold of more resources, thus initiating a highly political wrangle, with endless meetings and backdoor dealings, which leads to high influence costs and even greater inefficiency.

A firm that does not use PPM correctly will also find it difficult to terminate unsuccessful projects when it becomes clear that the desired results are not emerging, and to do so before costs become too high. When dealing with R&D and innovation, not being able to terminate projects is a problem indeed. It means either that the firm is unable to manage its project portfolio and keeps on running unsuccessful projects or—conversely—that its innovation strategy is so conservative that it concentrates on projects whose probability of success is close to one. Keeping bad projects alive also deprives good projects from accessing resources and achieving their potential results.

Not having a sufficiently developed PPM system, firms will also find it difficult to explicitly relate projects to innovation strategy. As a first result, the selection process will be weak and highly subject to emotion, political influence and cognitive bias (Hammond et al. 2002). Furthermore, if projects have fuzzy objectives, they might be overloaded with incoherent goals. For instance, one may start a project aimed at adapting a product for a specific market segment, characterized by a limited budget and requiring a quick completion time, and hastily accept a supplier's proposal to experiment with a new technical solution at the same time. The likely outcome will be ending up with a failed project, missing both objectives.

Formulating and executing an innovation strategy therefore requires a robust PPM system, which can be conceived as being based on three pillars. These three pillars will be discussed in the following sections and are categorizing projects, defining a process for evaluating projects, and using tools for decision-making.

9.1.1 Categorizing Projects and Defining Roadmaps

When dealing with R&D and innovation activities, the first element to be recognized is that projects are not all the same. These differences should be recognized and managed through an appropriate categorization, thus avoiding comparisons between projects of a completely different nature. These differences can be identified along a number of dimensions, such as risk, project size and project scope, as discussed in Chap. 6. Concerning the latter, projects may be characterized by different degrees of innovative content with respect to the product (or specific subsystems), the process, etc. However, the main difference to be identified is related to distance that the project's results will have from the market, as discussed in Chap. 1. According to this classification, one can develop finer differences, as in the following list:

- **Basic research projects**. The aim of basic research projects is to develop knowledge in a given technological domain, generally without having identified a clearly defined application yet. In profit-making firms, projects of this kind are not very common, but are not impossible to find. A firm may pursue them because it considers the knowledge that could be generated useful for either building further developments, or for increasing its absorptive capacity.
- **Applied research and technology development projects**. The aim of an applied research project is to generate knowledge, and/or to develop and demonstrate the suitability of a given technology in solving a problem that is related to the business. The result usually consists of a demonstrator and is not yet directly applicable to the firm's products and processes, since the validation performed is related to the technology *per se*, without having fully worked on its integration with other subsystems that would make up a real product.[2]

[2] When operating in industry, the concepts of basic and applied research must be related to the specific industry, and do not necessarily imply activities that have, in absolute terms, a high scientific or inventive profile, or that generate leading edge technology. For instance, a motorcycle manufacturer experimenting and testing a new commercially available component, such as LED headlights, is carrying out what can be considered to be applied research. In fact, this activity is aimed at creating knowledge ("Let's try fitting these newfangled LED headlights on a couple of our bikes and see if they work") and is not yet directed to a specific product (in which case the objective would be "Let's design our next bike with LED headlights").

When dealing with basic or applied research projects, Scholefield (1993) suggested a classification that is associated to the objectives of research projects. On the one side, one can have 'maintenance' R&D objectives, with the purpose of ensuring the survival of the firm and keeping up its competitive advantage. Conversely, one can have 'growth' R&D objectives, aiming at upgrading the firm's competencies by mastering a new technology or looking for technological breakthroughs.

Concerning development projects, one can use the following classification:

- **Next-generation and/or platform development projects**. In this kind of project, the focus is not on a technology, but on the integration of different technologies and solutions in order to create the foundation for future products, processes and services. The result usually is a prototype which engineers can rely on, in order to proceed with a product development project that will lead to a commercial launch. Alternatively, the result can be a *platform**,[3] defined as "*a collection of technological assets that are proven to work together*", and might serve as the basis for developing a *family* of multiple *derivative products.*
- **Product development projects**. These projects aim at launching a specific product addressed to a given market segment. Depending on the existence or not of a prior platform development, a product development project can be a more or less complex activity. This topic will be deepened in Chap. 16.
- **Customization projects**. Finally, customization projects have the objective of adapting an existing product to the needs of a specific customer (e.g., "this delivery company wants a special version of our vans for its fleet") or market niche (e.g., "a version of car model X, adapted to the specific regulations in force in country Y").

9.1.2 Platform Product and System Development

The platform concept has become a key element of innovation strategy in the nineties (Meyer and Lehnerd 1997) and has found a renewed interest thanks to digitalization (Hermann et al. 2016), which multiplies the opportunities of developing systems with open architectures, capable of interfacing with different modules, which can change over time.

Platform-based product development consists in structuring the system development pipeline in a two-tier structure. Platform projects establish a technological basis that, until obsolete, allows the firm to come up with a set of smaller (in terms

[3] The term 'platform' is often used, but with different meanings that might lead to confusion. In general, a platform may be intended as a set of assets that enable future actions. In this context, we refer to the meaning of 'platform' used in the technical literature, as a set of subsystems that are interoperable and that enable future developments or customizations. This should not be conflated with the use of the same term that refers to IT-enabled services business model that enable interactions between multiple parties (e.g., two-sided markets).

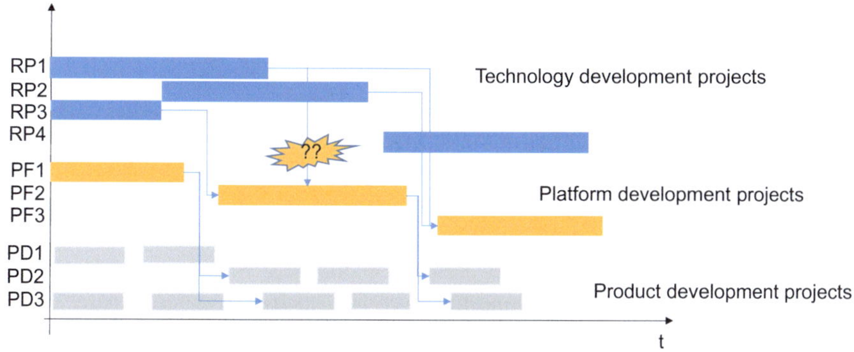

Fig. 9.1 The role of platform development projects in innovation strategy

of cost and development time) derivative development projects.[4] This strategy exhibits a number of advantages (Cantamessa 2005).

One first advantage is the potential sharing of components among product versions, which can lead to significant economies of scale in manufacturing and purchasing. Furthermore, a properly configured platform strategy allows the overall development cost of the platform projects and of its derivatives to be less than what would have been spent with an equivalent number of independent projects.

Moreover, since the duration of derivative projects is short, this allows a quick reaction to changes in the market.

As a final and very important advantage, a platform-based product development strategy tends to keep more innovative activities apart from the less innovative ones. As shown in Fig. 9.1, platform projects are used to transfer results from research and technology development into product development. This approach gives research projects a clear objective ("we must finish project X by month K, so to feed its results into platform project Y"). It also allows the testing of new and risky technologies in the context of a large-scale project that—not being directly connected to a product launch—is not generally subject to an exceedingly tight schedule. Platform projects can therefore be used to validate a set of new technologies—individually and with respect to their interoperability—and to create the know-how needed to deploy them into derivative product development. At the same time, by taking the more innovative design tasks out of derivative product development, engineers are discouraged from overdesigning individual products ("why don't we try technology X in this new product Z?"), which generally leads to increased cost and lead time, and often with dubious benefits. For instance, Fig. 9.1

[4] In quantitative terms, a firm may define the degree with which derivatives are to be dependent on the platform by expressing the percentages of components (in number or in value) that must be common and based on the platform, contrasted to the ones that can be specific to each derivate product. Similarly, the degree of innovation of each platform with respect to its predecessor can be represented by the percentage of new components (again, in number or value), contrasted to the ones that are to be *carried over* from the previous one.

shows one incoherent proposal of 'inserting' a new technology in the midst of platform development project.

Platform-based product development also has a few risks and drawbacks. Aside from the greater complexity in the portfolio management process, the firm may mismanage the strategy and end up with greater overall development costs.

Moreover, it may be forced to delay the launch of new products until the platform project is ended or find itself locked in the platform's technology for a long time. Finally, component sharing is known to be a double-edged sword (Krishnan and Gupta 2001). If not carefully planned, it may lead to insufficient differentiation between products and therefore lead to loss of market share.

Due to the importance of the subject, Chap. 16 will explore the issue in depth and propose a number of practical approaches for a sound definition of platforms and product families.

9.1.3 Technology Road Mapping

The idea behind technology road mapping is to create a high-level view of the set of projects of different types that the firm is considering as candidates for its portfolio, together with their mutual relationships and their placement in time. In general, a *technology roadmap** provides a comprehensive view of the project portfolio in order to ensure its intrinsic coherence and does not cope with its economic or financial sustainability. Specifically, a technology roadmap allows having a compact view of the portfolio and of the way with which results from 'upstream' activities will feed into 'downstream' projects. In general, the following elements will be included:

- **Triggers**. Triggers are events that are exogenous but relevant to the innovation strategy of the firm. Examples of triggers are the time at which a governmental policy or a standard comes in force, the release of a new technology by a key vendor, etc.
- **Markets**. The market section of a technology roadmap depicts the markets in which the company wishes to operate and the timing of entry and exit.
- **Products**. The product section represents the timeline of development and introduction of individual products in the markets identified above. The relationship between products and markets can be simple, with each product being related to a single market. However, it can also be defined by a matrix, with each product being sold in different markets or market segments.
- **Platforms**. In case products are organized in product families related to platforms, this section represents the development of the platforms themselves.
- **Technology**. This section represents the development of individual technologies that the firm considers to be enabling and relevant with respect to the downstream development of platforms and products.
- **Capabilities**. Finally, this section represents the projects used to build the competencies that are needed in order to sustain the downstream R&D activity, which can include actions associated to all the approaches discussed in Chap. 8.

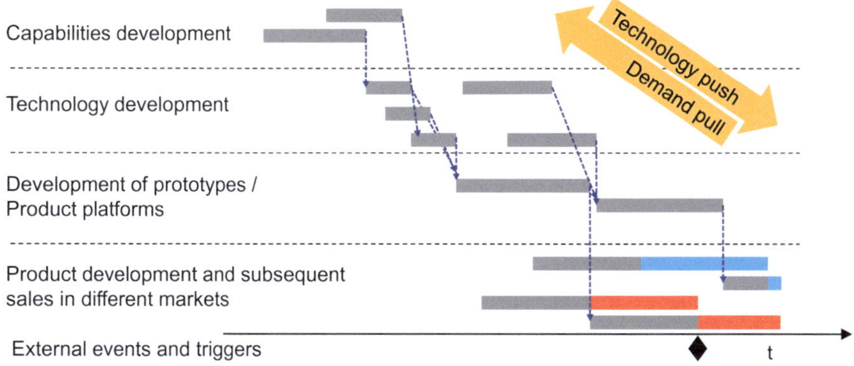

Fig. 9.2 Example of a technology roadmap

A fictitious example of technology roadmap is shown in Fig. 9.2. The graphical layout of this roadmap is based on layers, with development work being represented in gray and sales in white. Depending on the case, alternative layouts can be used, such as trees and flowcharts.

The coherence of the technology roadmap can be studied by progressing top-down, from left to right, following a 'technology-push' approach, which shows how and when upstream results can be used by downstream activities. Conversely, one can move bottom up, from right to left, following a 'demand-pull' perspective. This implies checking that needs that emerge from the market are appropriately addressed by the upstream activities and at the right time. In practice, road mapping will be carried out by iterating the two approaches until a satisfactory proposal is reached.

For a deeper analysis, the connections between technologies, products and market needs can be studied with more sophisticated methods. In general, these methods use *incidence matrices*, in which features at different upstream and downstream levels are mapped onto each other in order to translate market needs into technological priorities.

Figure 9.3 represents a simple example, in which future customer needs and their importance (usually coming from market research and rated on a 1–10 scale) are at first mapped on system performance indicators through a qualitative rating of relevance where 1 implies 'very weak', 3 'weak' and 9 'strong'. The mapping is performed through a simple mechanism of weighted sums.[5] Then, performance indicators are mapped on enabling technologies in order to prioritize them.

[5] The approach closely resembles *Quality Function Deployment**, which a widely used methodology in product development (see Chap. 14 for further details). The weighted-sum mechanism is very coarse, since importance and relevance indicators are qualitative concepts, and their numerical representation is based on an ordinal, and not a cardinal, scale. Strictly speaking, ordinary algebraic operations should not be applied to this measurement system, but the approach is nonetheless commonly practiced, under the condition that outcomes are not taken as precise measures but as rough indicators, and that the number of rows and columns in the matrices is sufficiently high, thus suggesting that major errors should be 'evened out'.

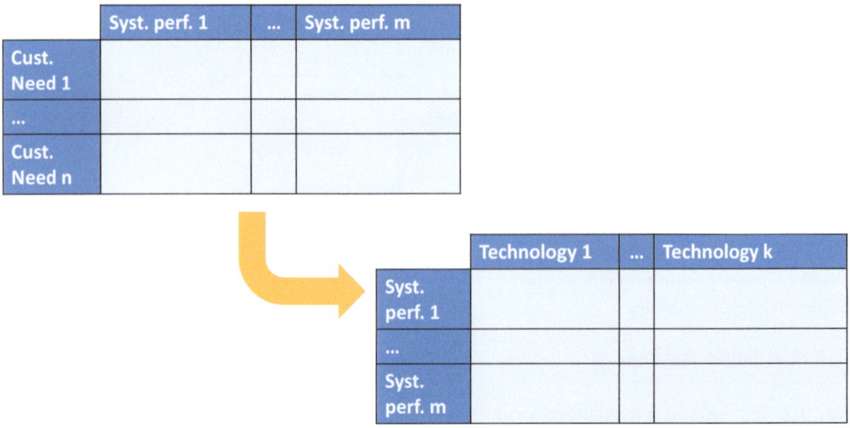

Fig. 9.3 Using incidence matrices to align technological development to market needs

More elaborate methods have also been proposed, such as CFTP (or *Customer-Focused Technology Planning** (Paap 1994), in which the analysis is enriched by operating by markets and customer segments and by adding comparisons with competitors.

9.2 Defining a Process for Project Evaluation

All projects in an organization follow a lifecycle that goes from the initial inception, through its planning and to its execution—usually in a number of major phases—to the exploitation of its results within the organization. When the number of projects is high, firms will tend to reduce the specificity with which each project is organized and managed, and try to fit projects within a relatively stable and routinely framework. Therefore, while the contents of each project will be unique, the way with which it is structured and managed will tend to be similar to other projects, or at least to the ones belonging to the same category. It is therefore not a contradiction to state that, when dealing with R&D and innovation, projects will be subject to a relatively standardized process.

Understanding and structuring this process is strategically necessary, first of all because the firm must be able to analyze its strengths and weaknesses (Hansen and Birkinshaw 2007) in the various stages of development. Secondly, the process that represents the project lifecycle provides an important opportunity for selecting projects during their progress. In general, innovation projects exhibit a significant uncertainty with respect to technology and market risk, which makes it very difficult to make an *ex-ante* assessment of their viability. If forced to make an up-front selection, firms would probably only accept projects with very low uncertainty, but this would imply a very strong bias in favor of incremental innovation. On the opposite, firms should accept a high number of projects, but be ready to terminate them as soon as they turn out to be unprofitable.

In some industries, uncertainty is so high that an *ex-ante* selection would be impossible to make. The most extreme example probably comes from the pharmaceutical industry, where it is customary (Fig. 9.4) to start from an order of magnitude of 10,000 candidate chemical compounds and make an in silico (i.e., through simulation) or in vitro (i.e., in labs) screening in order to come up with a set of 200–300 candidates. These are then tested in vivo (i.e., on animals) until a set of 5–6 candidate compounds are left. What follows are three rounds of *clinical trials* on humans, aimed at assessing safety, dosage, efficacy and short- and long-term side effects on a progressively higher number of volunteers. At the end, the pharmaceutical company may hope to end up with one drug that can be launched on the market, so that margins from sales may cover the development cost for the entire pipeline of candidates on which the firm has worked.

Less extreme cases can be found in the automotive industry, where a firm may typically examine a dozen or so concepts, draft a few alternative specifications for the most promising ones and then give a go-ahead for the development of a single vehicle.

Managing a pipeline—or a funnel—of projects with high individual uncertainty requires a careful understanding of the relationship between project cost and evolution of uncertainty. In fact, the economic foundation of a system in which many projects are started and progressively weeded out, and only a few make it to the market, is related to the way cost and uncertainty evolve over time (Fig. 9.5).

In general, when a project progresses from its initial to the final phases, project costs escalate over time, leading to an s-curve. Hence, the initial phases of projects are not very expensive, until a given moment after which substantially greater investments are required. In the pharmaceutical industry, this generally occurs for clinical trials. In manufacturing industries, it usually coincides with the detailed design phase, which requires substantial amounts of engineering person-hours. At the same time, it is common for uncertainty on technical feasibility and market acceptance to decrease while the project progresses, generally quite quickly during

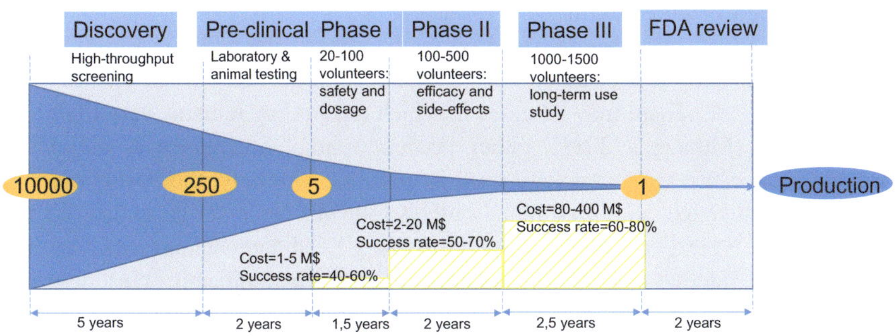

Fig. 9.4 The development pipeline in the pharmaceutical industry (from Burrill and Co. 2009)

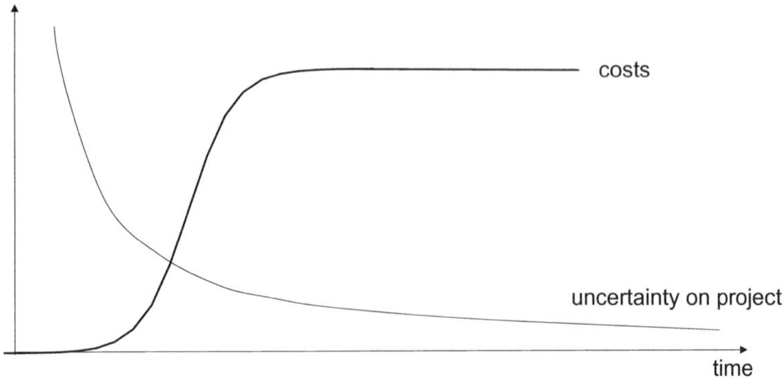

Fig. 9.5 A typical evolution of cost and uncertainty during project evolution

the initial phases, and then more slowly. This is especially true if the later and more expensive phases are aimed at perfecting the product, rather than making sure that it will work.

In such a situation, the early phases have a strong '*lever effect*', since a limited investment can lead to strong returns with respect to uncertainty reduction. In this case, firms can afford to make a preliminary investment in a high number of projects who are likely to fail, since the return coming from screening the good ones outweighs the loss coming from the failed ones. Of course, the firm will have to be highly selective at the point of the s-curve where the slope of the cost curve becomes larger.

In order to formalize a procedure of this kind, Edgett and Cooper (Cooper 1990; Cooper and Edgett 1996; Cooper et al. 2002a, b I-II) have developed the widely used concept of a *Stage-Gate®* system. In a Stage-Gate system, projects are divided into *stages*, whose deliverables are to be examined at *gates* in which continuation of the project to the next stage or termination will be decided. In case managers decide to continue, a budget is allocated so that the project team can work on the next stage, but not beyond. At the subsequent gate a further decision will be taken, based on the new information that will be developed during the stage.

Stage-Gate systems are widely used, though sometimes leading to unsatisfactory results (Schmidt et al. 2009). This may be due to a variety of reasons that have to do with the design of the system.

First, Stage-Gate systems are based on the hypothesis described in Fig. 9.5, with relatively cheap early phases that lead to a significant reduction in risk. This may not always be the case. There may be cases in which uncertainty remains very high until the very end (Montagna 2011), either because market acceptance is difficult to evaluate, or because the technology used is unstable. Similarly, the cost curve may have a different shape, with high initial outlays followed by lower costs due to small adaptations. A typical case comes from when using Lean approaches (see Chap. 11), such as in the Internet industry, in which firms make a significant investment in a software platform, and then perform continuous experimentation and fine-tuning of new products, services and solutions, until they reach the desired results.

Secondly, the Stage-Gate system must be structured in a way that is appropriate to the project category. When doing so, one must balance the budget allocated to each stage with the contents of the deliverables that will be asked for at the next gate, and with an expected rejection rate. In the end, one cannot achieve more than what one spends for. If a firm requires an amount of uncertainty reduction during the early phases that is not supported by an adequate budget, there is a significant risk that potentially successful projects may be terminated. The number of stages is also important, since a low number will make the selection process ineffective, while a high number will make it excessively bureaucratic. For conservative and risk-averse firms, it may also be difficult to 'shape' the system in a way that submission of projects is actively encouraged at the entrance of the funnel and enforcing strict 'go-no go' decisions in the subsequent phases, so that only a few projects make it to the end.[6]

Finally, the way the Stage-Gate system is run is also quite critical. The panels in charge of 'go-no go' decisions must be carefully composed, with a good balance of managers and specialists, internal and external experts. Moreover, decisions must be implemented without deviations. If news spread that a terminated project can be salvaged through internal politics, the Stage-Gate system will be viewed as a useless procedure, adding to internal bureaucracy, that can easily be sidestepped. Finally, firms should not 'bury' terminated projects, but try to extract as much knowledge as possible from them, and structure and disseminate these 'lessons learned' throughout the organization.

9.3 Project Selection

Once a firm has defined distinct project categories (i.e., 'what' is being decided upon) and defined a process for selecting projects (i.e., 'when' should decisions be made), there still is a key issue to be solved. This is related to the criteria and to the methods that ought to be used for performing this selection (i.e., 'how' should decisions be made).

This problem exhibits substantial similarities to the portfolio selection problem that is found in finance, which suggests that some methods might be borrowed from this latter field. A financial portfolio manager must optimally allocate a finite fund to a number of investments in financial securities by taking investment and divestment decisions. Each security will be characterized by an inherent expected return and level of risk, with some degree of correlation existing among the securities. In a similar way, a project portfolio manager will have to take initiation

[6] In fact, there may be a striking and difficult-to-reconcile difference between the organizational culture required by early phases, in which openness and risk-taking should be promoted, and the subsequent ones, in which correct execution and detailed analysis ought to prevail. Some firms have decided to split their Stage-Gate systems in two sections, and to assign them to different organizational units (e.g., the pharmaceutical company Eli Lilly has decided to set up a separate 'discovery' organization dealing with the early phases of the innovation funnel).

and termination decisions on projects and allocate a finite budget made up of financial, human and physical resources among them. Projects too may be characterized by expected returns and levels of risk, and positive correlation may be present if project outcomes are complementary, while negative correlation may be due to substitute outcomes (e.g., two similar products) or to projects that compete over scarce resources.

The conceptual similarities between financial and project portfolio management should not make us neglect some substantial differences between the two fields. First, *projects are not traded on the market*, which means that financial valuations and investment/divestment decisions are substantially more difficult to make and execute. Moreover, the decision of buying and selling financial securities is relatively fine-grained, while initiating or terminating projects is not, and leads to a Boolean type of decision (e.g., you can easily sell 484 shares out of a 1452 holding, but a project can't be closed down by 1/3, and will have to be either continued or terminated). Secondly, project portfolio management has a clear *strategic role* within the firm, and managers will righteously wish to include non-financial issues in the decision-making process, which may at times include tacit and even emotional aspects.

Because of this, managers will tend to reject 'black box' decision-support methods that prescriptively give solutions without providing a clear view of the underlying reasons. Conversely, managers may accept methods that help them rationalize a process they can understand and feel as their own.

9.3.1 Top-Down Versus Bottom-Up Selection

In general, the project portfolio selection process can be viewed as either a *bottom-up* process or a *top-down* one.

In the bottom-up approach, projects are proposed by corporate functions and selected according to their inherent viability and compliance to strategy. In the top-down approach, strategic directions are translated into a budget allocation decision, which assigns resources at first to predefined project categories and objectives (e.g., *X* million dollars to radical innovation projects in domain *Y*), and then to individual projects proposed by corporate functions.

The bottom-up approach seems easier to implement and leaves more *creative freedom* to the lower ranks of the organization, which are closer to the needs of business operations and to the market. At the same time, it requires very rigorous decisions and might make the implementation of a given strategy difficult, should proposals coming from the bottom not go in the desired direction. Conversely, the top-down approach requires more *ex-ante* activity in the budgeting process and grants less freedom to the organization. At the same time, it makes it easier to enforce a strategy that has been defined at corporate level.

In practice, firms will always adopt some kind of hybrid between bottom-up and top-down portfolio selection. The degree with which either should prevail and the way they should be managed depends on corporate culture and on the type of innovations the organization is trying to pursue.

9.3.2 Financial Methods for Project Selection

Any innovation project involves the investment of financial resources in the expectance of future returns, which will generally be in terms of additional margins from sales. Therefore, project selection must consider financial aspects, which can be evaluated according to the methods described in the following.

9.3.2.1 Net Present Value

According to finance theory, the key indicator that ought to be used when assessing a project is the *Net Present Value** (NPV) of future marginal cash flows. As in (1), the analyst should compute the cash flows that the project would lead to (negative during the investment phase, positive when enjoying its results) and compare them to the cash flows one would have by not initiating the project. These marginal cash flows at each time period, CF_t, ought to be discounted according to a given interest rate i, since the economic value of future cash flows will be reduced because of their inherent uncertainty. In algebraic terms,

$$\mathrm{NPV} = \sum_{t=0}^{T} \frac{\mathrm{CF}_t}{(1+i)^t} \tag{9.1}$$

and the firm will accept projects for which $\mathrm{NPV} > 0$.

This seemingly straightforward approach underlies a number of quite significant issues. The first one is related to the use of NPV itself, since many firms tend to neglect it, and use the much simpler *Payback Time** (PBT) criterion instead. Following PBT, the firm defines a maximum time for recovering investments, T, and will accept projects only if the future cash flows comply with this threshold. In mathematical terms, if

$$\sum_{t=0}^{T} \mathrm{CF}_t \geq 0 \tag{9.2}$$

From a theoretical perspective, the PBT method is inherently flawed, since it does not discount cash flows, and because it purposely neglects any event that might occur after T. Therefore, a project with short PBT, but low long-term returns (dashed line in Fig. 9.6) will be preferred to a project with a longer PBT and substantial long-term returns (continuous line in Fig. 9.6). As such, the use of PBT leads to a strong bias toward incremental innovations.

Despite these shortcomings, many firms use PBT not only because of its simplicity, but also because of particular situations that may at least partially justify its use. In first instance, there might be cases in which cash flows beyond T are irrelevant, because of technological obsolescence or rapidly changing demand. Secondly, liquidity is generally a major concern for firms, and PBT aims at a quick recovery of the cash employed in projects. This is especially important when operating in very turbulent environments and the firm does not have adequate

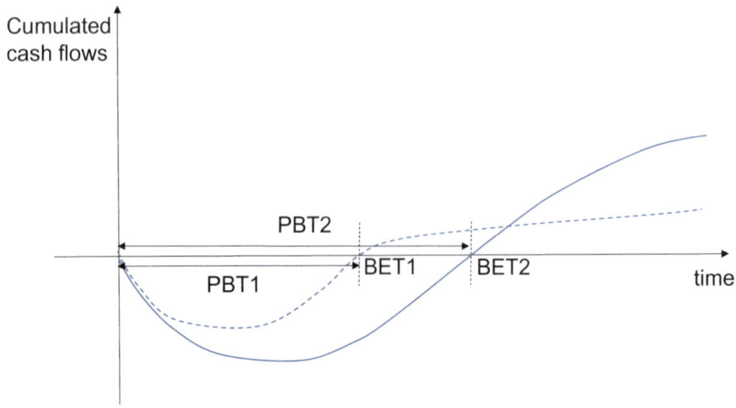

Fig. 9.6 The bias introduced by Payback Time in project selection

visibility of future opportunities. Committing to a number of profitable, but long-term projects, might be dangerous for the firm since—should a much more profitable opportunity arise midway in the progress of these projects—the necessary financial resources would be unavailable.[7]

Aside from the choice of the criterion to use, it is generally quite difficult to make a proper use of financial methods. One first issue is related to the concept of *relevant costs* (refer to Chap. 14 for a classification). When evaluating marginal cash flows, one should not consider sunk costs, i.e., costs that have already been incurred and that would not be recovered by any of the decisions that are currently at stake. Therefore, past expenditure should never be considered when deciding on a new project, or on the continuation of an existing one. For instance, a sentence like "we have already spent 2 million on this project, so we should consider that…" is unacceptable.

Subtler is the problem of potentially sunk costs that are associated to existing resources, human or capital. For instance, the cost of technical personnel is often a sunk cost, since by deciding for or against the project, it may be unlikely for the firm to hire or fire anyone, and salaries will have to be paid regardless of the actual use of these resources. As a general rule, a cost associated to a resource should be included if one of the two following conditions is true; (i) the number of resources does depend on the initiation or termination of the project (e.g., if personnel can be hired or fired) or (ii) the number of resources cannot been modified, but the number of competing projects is such that, should the project not be approved, these resources would nonetheless be saturated by alternative uses (e.g., if personnel can

[7] In a perfect market, this would not seem to be an issue, since the firm would be able to raise new financial resources in order to deal with an emerging and highly attractive opportunity. However, perfect markets are more an abstraction than reality. In any case, even if financing means were available, it is likely that the firm would find key resources tied up in its existing portfolio of long-term projects and be unable to expand its organization efficiently in order to work on the new projects.

always be allocated to other projects). In this latter case, the cost of resources is technically 'sunk', but is considered as an opportunity cost.[8]

When considering human resources, condition (ii) is generally true, since managers usually have the problem of selecting among more projects than they can cope with. However, they should be wary of exceptions that might occur, which might lead to killing projects that would be marginally profitable, and have to pay the salaries just the same.

Besides personnel, another critical example may come from the use of highly specialized physical assets, such as laboratory equipment, test benches and so forth. In these cases, it is very common for firms to fall into a 'sunk cost trap' and charge hourly costs for the use of this equipment, derived from annual amortization rates and a budgeted expected use, and end up with not enough projects and underused resources.

Table 9.1 shows a typical example of this mistake, with a firm whose projects require specific costs, together with the use of a laboratory. The lab has an annual amortization rate equal to 1.28 M€/yr. and an expected usage of 1600 h/yr, leading to an hourly cost rate of 800 €. The table clearly shows that there are not enough candidate projects to saturate the lab and that, by including hourly cost, the firm will mistakenly not start projects C and D, leading to lower profits and an even more underused laboratory.

It is generally quite difficult to spot these mistakes analytically or by going over accounting reports. The best approach is to walk around facilities, looking for idle resources and asking what is happening. Quite often, these resources will have been idled by a project selection process that has erroneously accounted for sunk costs associated to the original investment. In other cases, these assets should not be there, and the company should consider disposing of them (but, as long as they are available, they should be used). Following the case described above, the easiest way to notice the problem would be to walk to the lab, notice that it is working much below capacity (480 out of 1600 h/yr) and start asking questions.

A second problem in the financial evaluation of projects is associated to the differential nature of the cash flows that should be evaluated in NPV, since the cash flows due to the project—if activated—should be compared against the cash flows that would emerge by *not* starting the project.

In general, it is not straightforward to foresee what would happen by not having the project, and it is tempting to assume that 'nothing' would happen (Fig. 9.7). Such an assumption might be misleading, since the decision not to start a project could lead to the deterioration of the firm's competitive position, the loss of market share of current products, and so on, which implies progressively negative cash flows with respect to the current status.

A final issue is associated to the definition of the discount rate in NPV, which must consider the risk and uncertainty that is specific to the project being evaluated. Following financial theory, one should be able to define such rate by analogy with

[8] Of course, not considering sunk costs when taking a managerial decision on the use of resources does not mean that these costs should not be considered for cost accounting.

Table 9.1 An example of a mistaken decision due to sunk costs

Project	Project cost (k €)	Use of lab (hrs.)	Use of lab (k€)	Total accounting costs (k€)	Expected margins after the project ends (k€)	Accounting profits (k€)	Decision	Actual use of lab (hrs.)	Actual profits (without non-relevant costs) (k€)	Actual profits (without non-relevant costs) *by accepting all projects* (k€)
A	B	C	D	B + D	E	E−(B + D)	F	C'	E−B	E−B
P1	100	200	160	260	300	40	Y	160	200	200
P2	200	100	80	280	300	20	Y	80	100	100
P3	300	300	240	540	450	−90	N	0	0	150
P4	400	400	320	720	600	−120	N	0	0	200
P5	200	300	240	440	550	110	Y	240	350	350
Totals		1300	1040					480	650	1000

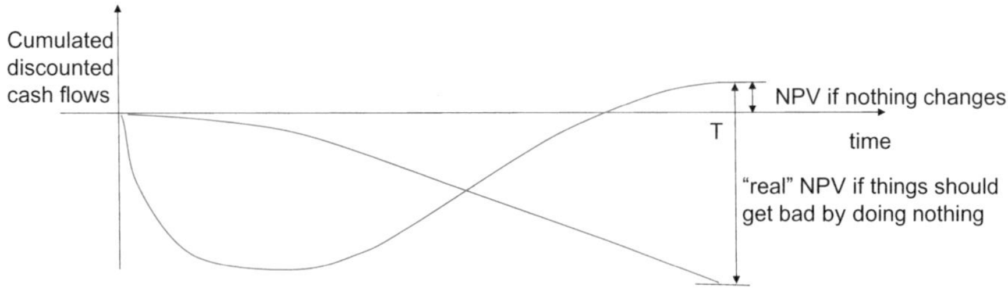

Fig. 9.7 Comparing cash flows resulting from launching and not launching a project

tradable securities having similar risk profiles, using the *Capital Asset Pricing Model** method. Unfortunately, this method requires the identification of one or more firms with publicly traded shares and whose risk profile is the same as one of the single projects being considered. This is possible only in the few cases in which one can find such public companies focused on individual projects. Alternatively, firms can define discount rates that are specific to project categories, following the classification that was discussed in the previous section of this chapter, though few firms have sufficient knowledge to make a proper assessment of this kind. In the end, it is likely for firms to simply use the *Weighted Average Cost of Capital** (WACC) which, on the one side, is the average interest rate that is required by shareholders and debtholders of the firm while, on the other side, it represents the risk associated to the overall activity portfolio of the firm. The use of WACC can be misleading, because it does not represent project-specific risk. So, using WACC will lead to a bias that favors projects that are riskier than the average, and goes against the less risky ones.

The complexity of computing risk-adjusted and project-specific rates of return suggests the use of simpler, albeit somewhat rougher, approaches. One such approach, termed *Expected Commercial Value** (ECV), consists in accounting for project-specific technological risk, P_t, and commercial risk, P_c, separately from the interest rate, and considering the sequential nature with which these two risks manifest themselves. ECV is based on a very simple decision tree (Fig. 9.8) where DC represents costs related to technical development and CP the costs associated to production and launch. ECV can readily be computed as follows, provided expert opinion can be used to provide estimates for P_t and P_c.

$$\mathrm{ECV} = (\mathrm{NPV} * P_C - \mathrm{CP}) * P_t - \mathrm{DC} \tag{9.3}$$

Since development activity is usually limited by a given budget, it can be interesting to select projects that do not only have a high ECV, but also provide a high level of ECV per dollar invested. This leads to the so-called Productivity Index (PI), which is often used in conjunction with ECV.

$$\mathrm{PI} = \frac{\mathrm{ECV}}{\mathrm{DC}} \tag{9.4}$$

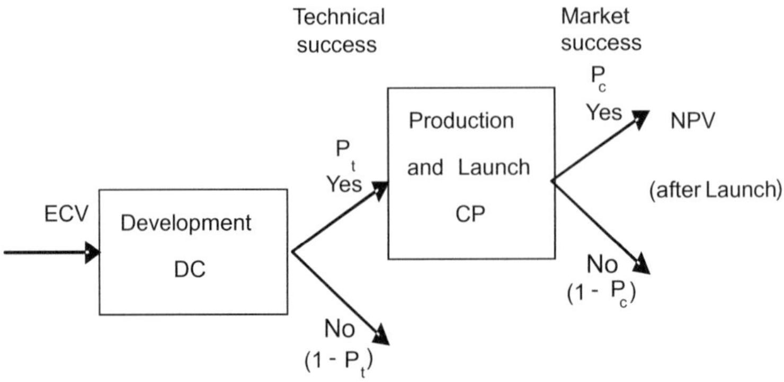

Fig. 9.8 The decision tree representing the ECV method

9.3.2.2 The Option Value of Projects

In the previous section, we discussed the opportunity of supporting project portfolio decisions with Stage and Gate processes. Even if such a process is not formalized, it is only logical for firms to consider investment in new projects on a step-by-step basis, so that further investment is decided for only if the results of previous phases are sufficiently promising.

From a financial perspective, this kind of behavior is very interesting, since decisions to fund the *i*-th stage of a project will not depend on the intrinsic returns deriving from stage *i*. Rather, the decision will be based on the consideration that—should stage *i* be successful—the firm will be able to invest in stage *i+1* (something that could not be done if stage *i* were not funded). In turn, the value of stage *i+1* is associated to the future possibility of operating stage *i+2*, and so forth, until the *n*-th and final stage of the project, whose value will effectively be associated to the profits that will accrue because of the completion of the overall project. In financial terms, stage *i* of a project has both an *intrinsic value*, due to the NPV of the cash flows it will generate independently, and an *option value*. The latter value is associated to the fact that, by engaging in stage *i*, one acquires the option (but not the obligation) to engage in a future activity that might bear further economic value.

This mechanism is similar to financial options traded on stock exchanges. In financial markets, an investor can buy a *European call option* on a security (the 'underlying asset') which gives her the right, but not the obligation, to buy that security at a given date (the 'exercise date') and at a given price (the 'exercise price'). At the exercise date, should the security's value be greater than the exercise price, the investor will exercise the option and profit from the difference. Conversely, should the share's value be less than the exercise price, the option will be worthless. The price of the option can be calculated as its expected value, which depends on the difference between the current price of the underlying asset and the exercise price, the time before the exercise date comes, and the volatility of the underlying asset's price.

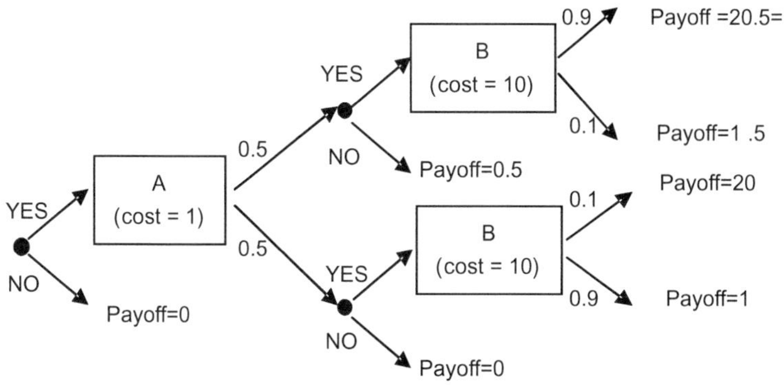

Fig. 9.9 Using decision trees for a first-approximation valuation of project-based real options

A project in two stages, *A* and *B*, bears a strong analogy to a call option, with the former representing the option and the latter the underlying asset ("if you invest in *A* you will be able, if things go right, to invest in *B*"). Cost of project *A* is equivalent to the price of the option, its duration to the exercise date, the cost of project *B* to the exercise price, and the uncertainty in the value of project *B* to volatility. However, not being associated to a traded security, we will speak of a *real option** (i.e., associated to real activities) instead of a financial one.

The valuation of projects as real options is not an easy problem to solve. Valuation methods used for financial options (e.g., the *Black and Scholes** formula) are generally not applicable, since they require the use of parameters that are not available for non-traded assets. Specific methods have therefore been developed, such as the replicating portfolio (Copeland and Antikarov 2001). Such methods are relatively complicated and require specialist knowledge, which has somewhat limited their diffusion and prevent us from discussing them in depth. For a rough-cut valuation, it is possible and intuitive to represent real options as decision trees, with the awareness that the approach might however not be very precise.

For example (Fig. 9.9), suppose that a firm must consider investment in a two-stage project with stages *A* and *B*, whose costs, respectively, are 1 and 10. *A* has a 50% chance of being successful, and this determines the success probabilities of *B* and the related outcomes (if *A* is successful, *B*'s likelihood of success is high, and vice versa).

If one does not consider the option value of stage *A*, the decision-maker must choose among the following alternatives (9.5)–(9.7) and will rationally choose doing nothing.

$$\text{NO(don't start any project)} : \text{payoff} = 0\,\text{M EUR} \tag{9.5}$$

$$\begin{aligned} \text{YES-NO(start project}\,A\,\text{only)} : \text{payoff} &= (0.5 \times 0.5 + 0 \times 0.5) - 1 \\ &= -0.75\,\text{M EUR} \end{aligned} \tag{9.6}$$

$$\begin{aligned} \text{YES-YES (start } A \text{ and } B\text{)}: \text{ payoff} &= [8.6 \times 0.5 + (-7.1 \times 0.5)] - 1 \\ &= -0.25\,\text{M EUR} \end{aligned} \tag{9.7}$$

Instead, if one considers an alternative with options (start *A* and go on with *B* only if *A* is successful), the payoff will be as in (9.8):

$$\begin{aligned} &\text{YES-YES(only if A is successful)}: \text{payoff} = (8. \times 0.5 + 0 \times 0.5) - 1 \\ &= 3.3\,\text{M EUR} \end{aligned} \tag{9.8}$$

Phase *A* therefore has an intrinsic value equal to −0.75 M€ (as in (9.6)). However, if one also considers it as an option with respect to *B*, its value climbs to 3.3 M€ (as in (9.8)). Therefore, its option value with respect to *B* will be the difference between the two and equal to (3.3 + 0.75) = 4.05 M€.

9.3.3 Optimization Methods

Financial criteria provide information on profitability of individual projects, but fail to compare projects against each other when resources are limited and when projects are interdependent. To this purpose, one can use optimization methods.

In this case, project selection can be modeled as a *mixed-integer optimization** problem in which the firm attempts to maximize the combined economic value of the selected projects, at the same time complying with constraints in the use of resources. A simple optimization model is the following:

Sets:

$i = 1 \ldots m$ is the set of resource types;

$j = 1 \ldots n$ is the set of candidate projects;

$t = 1 \ldots T$ is the set of planning 'time buckets' on which planning is performed, up to a horizon *T*.

Parameters:

r_{ijt} is the use of resources type i by project j at time t;

v_j is the NPV of project j;

d_{it} is the availability of resource type i at time t.

Decision variables

$xj = 1$ if project j is started, $= 0$ if it is not.

$$\max \sum_j v_j * x_j$$

subject to

$$\sum_{j} r_{ijt} * x_j \leq d_{it} \quad \forall i, t$$

$$x_j \in \{0, 1\} \quad \forall j$$

The model is based on the standard *knapsack problem** often used in Operations Research, extended over multiple time units. The model can be developed further, in order to include inter-project constraints. For instance, if projects j' and j'' are mutually exclusive, one would add a constraint such as $x_{j'} + x_{j''} \leq 1$. Or, if project j' can be started only if project j'' is also started, the constraint to be added would be $x_{j'} \leq x_{j''}$.

An optimization model such as the one outlined above is characterized by significant computational complexity. In the case of small project portfolios, the model is straightforward to develop, implement and solve with general-purpose spreadsheets such as Microsoft Excel® and their optimization functions (i.e., Excel Solver®). For larger portfolios, computational complexity can be too much for simple optimization tools. Therefore, dedicated solver engines ought to be used, which of course leads to a limitation on the applicability of such methods.

When using optimization models, decision-makers usually tend to be skeptical of 'black box' algorithms that dictate which projects should be started and which should not. Consequently, it is generally best to use optimization models in an iterative way, by performing 'what if' analysis and by 'playing with constraints'. Decision-makers can start from the initial solution and think of deviations they feel might work well or have strategic relevance, force them by adding constraints, and run the algorithm again to see the impact on the objective function.

Subject to these additional constraints, the objective function will worsen (at the limit, it might remain the same) and the difference can be interpreted as the immediate opportunity cost that the managers should be willing to pay in order to enact their 'strategic' deviation from the solution that the model suggests as optimal.

9.3.4 Multicriteria Methods

Financial and optimization methods are based on quantitative information, such as financial flows and resource availability and effort. Given the strategic nature of project portfolio management, the problem decision-makers have to face is *not fully structured*.[9] Therefore, optimization methods could result less appropriate, or even

[9] Operations Research defines a degree of *problem structuring**, which depends on the degree with which the constituent elements of the problem (i.e., the objectives and the alternative choices) are defined. If objectives and alternatives are completely defined, the problem is structured; if one of the alternatives and objectives is missing, the problem is partially structured and, if both are undefined, the problem is completely unstructured. The degree of structuring of a problem is an essential feature of the problem and determines the methods that can be used to solve it. Typically, optimization models are appropriate for fully structured problems, while multi-criteria models are used for partially structured or completely unstructured problems. For the latter, their application requires the use of structuring tools before the solving phase.

misleading, because of attempting to force objectives that are not fully defined or subject to uncertainty into mathematical formalisms.

For instance, decision-makers may have to consider problem dimensions that can have both quantitative and qualitative character, such as:

- economic and financial aspects (e.g., NPV, PBT, etc.);
- operational aspects (e.g., effort required, budgeted cost, time required, etc.);
- uncertainty and difficulties (e.g., market and/or technology risk, project complexity, ease of further implementation, etc.);
- relation with the firm (e.g., coherence with existing or desired corporate competencies, strategic alignment, relevance for competitive positioning, contribution to corporate or brand image, etc.);
- target (e.g., market, customer segment);
- product attractiveness (e.g., product features, ease of use, etc.).

Some of these qualitative dimensions listed above could be included in the frame of quantitative methods with some ease. For instance, risk can be captured by using discount rates. However, it has already been highlighted that this 'translation' can result more complex than expected and subject to mistakes. For other dimensions, such as 'strategic alignment', the translation would even inherently result in quite arbitrary. Therefore, given the impossibility and the potential incorrectness of translating a set of heterogeneous criteria in quantitative and monetary terms, it is conceptually and practically advisable to use specific *Decision Aiding** methods.

Among the many multicriteria methods available in the literature, the *Analytical Hierarchy Process** (AHP) appears to be the most suitable, since it is a powerful tool in structuring objectives around a strategic direction before supporting the specific phase of choice. This is done by structuring a hierarchy of criteria and identifying the values of relative weights, so that they may faithfully mirror a previously defined strategy. From a technical point of view, AHP performs this weighting of criteria and evaluation of alternatives through pairwise comparisons, which are then followed by a validation of the coherence in the overall results.

Despite the appropriateness of AHP to PPM (Lockett et al. 1986; Levine 2006), its diffusion in this context is relatively limited, due to some controversial aspects of the method itself and the risk of making mistakes in its application.[10] On the contrary the mainstream—though methodologically questionable—approach is represented by simpler comparison tables.

Comparison tables consist in defining a set of criteria, in assigning weights to each criterion based on a qualitative scale (e.g., 1–5), and then evaluating each project against each criterion, again using a qualitative scale. Additionally, a threshold score might be assigned to each criterion, so that projects will be considered only if all thresholds are met. Each project whose scores are above the

[10] For going deeper in multicriteria methods and specifically to investigate concerns about the use of AHP, readers are referred to specific contributions in the literature (Saaty 1980; Roy 1996; Belton and Steward 2002).

Weight	*w1*	...	*wn*		
Criteria	*c1*	...	*cn*		
Threshold	**	...	medium	Above?	Score
Project1	***	...	low	x	x
Project2	***	...	medium	v	21
...	**	...	high	...	
ProjectN	**	...	medium	v	19

Fig. 9.10 An example of comparison table for multicriteria evaluation of a project portfolio

thresholds will then be given an overall score that is computed as a weighted sum, as in the example provided in Fig. 9.10.

The firm's strategic intent is expressed *ex-ante* through the choice of criteria, their weights and the related thresholds. The method is very appealing because of its simplicity and because of its transparency, since the reasons leading to the decision of selecting or excluding a project are evident (in fact, the method is often used by public agencies when awarding procurement contracts or research grants).

However, its quantitative fundamentals are quite weak. Performing weighted sums when both scores and weights are based on cardinal scales is one first and obvious issue. Another potential problem comes from the use of a 'flat' set of criteria, a number of which may be highly related to one another, and this may lead to inadvertently overweighting them. Therefore, caution should be taken when using a method of this kind. One can be confident of the outcome if it is simply used to create a first ranking and then a broad grouping of projects (e.g., 'highly attractive', 'moderately attractive', 'unattractive'). It may be somewhat more critical to use it for discriminating among individual projects, especially if differences between project scores are narrow.

9.3.5 Mapping Methods

Another widely used approach for supporting the project selection process is based on graphical representations of the project portfolio. *Mapping methods* can be contextualized in the field of *data visualization**,[11] i.e., the graphic representation of data and information, in this case, data and information about projects.

The approach is attractive and widely used, since maps provide an intuitive way to represent and study the 'big picture' of a complex project portfolio and are fairly easy to implement with spreadsheets. The only caveat to be remembered is that they provide a visual representation of the problem, but not a guidance of what to do with it. Even more, a visual representation leads to an inherent bias to developing a balanced solution, which may not necessarily be the optimal one.

[11] The field of data and information visualization has emerged from research in *Computer Graphics** and *Data Mining** and is being increasingly applied in a variety of applications, from scientific research to business, in the latter case for financial data analysis, market studies, manufacturing production control, etc.

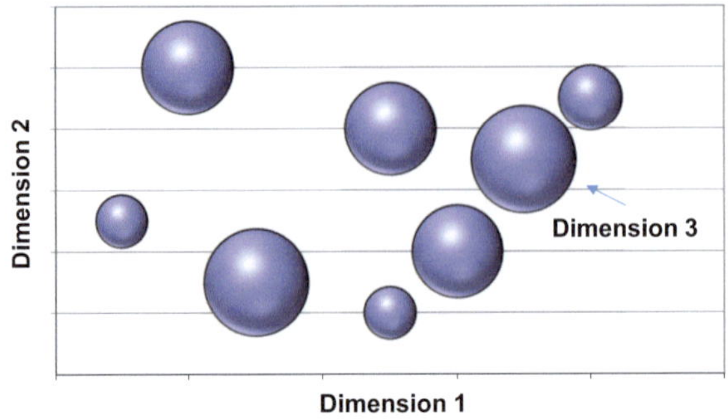

Fig. 9.11 A bubble chart representing a project portfolio

A project portfolio map is usually based on a selection of quantitative and qualitative dimensions, which can be drawn from the list provided in the previous section. One popular graphical representation is the bubble chart (Fig. 9.11), in which the *x*- and *y*-axis represent two dimensions that the company considers to be relevant to its strategy and decision-making effort. Each project is then located on the *xy* plane, with a third dimension being the size of the bubble, and representing the 'size' of the project.

Alternatively, project portfolio maps can represent the division of budgets or available workforce among the individual projects, or among project categories. This can be done either with pie charts (Fig. 9.12) or, by adding the time dimension, with area charts (Fig. 9.13).

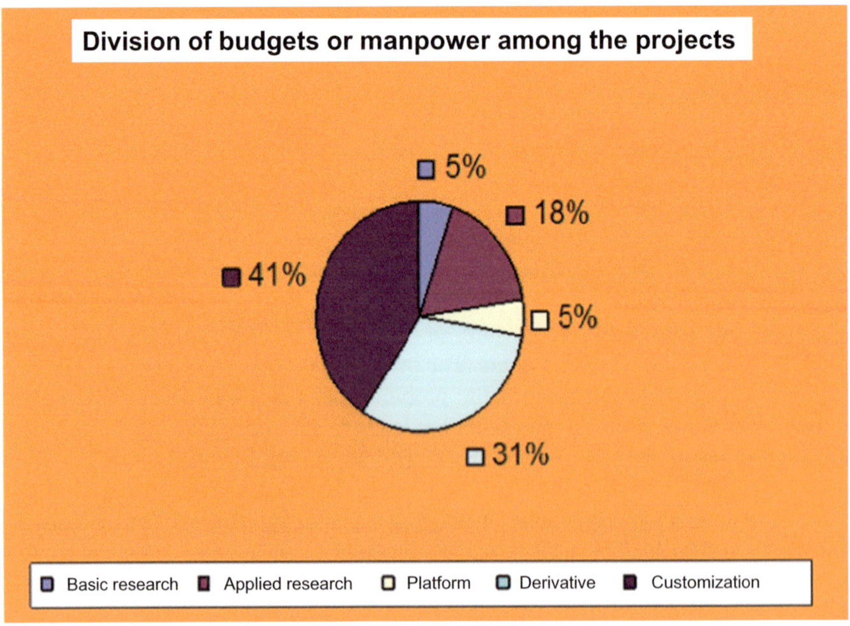

Fig. 9.12 A pie chart representing a project portfolio

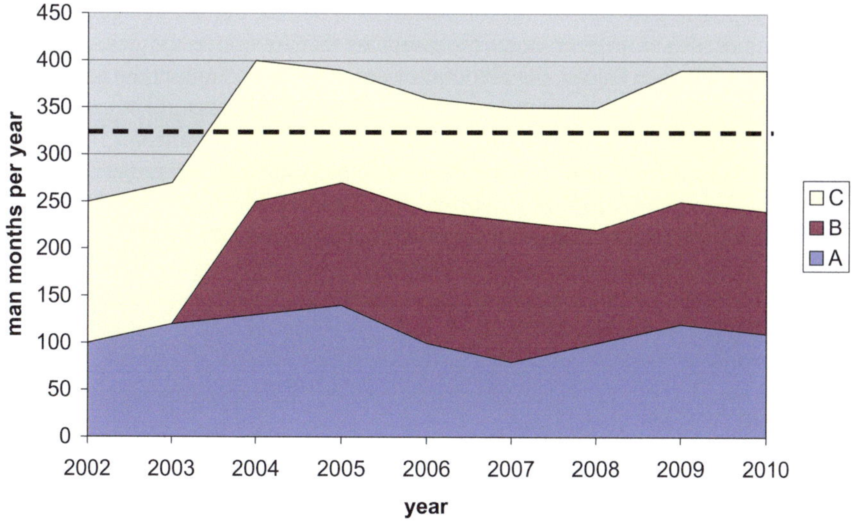

Fig. 9.13 An area chart representing a project portfolio

9.4 Best Practice in Project Portfolio Management

The previous discussion has shown that project portfolio management is a quite demanding process, which requires significant managerial attention. Empirical research on firms that have engaged in project portfolio management (Cooper et al. 1999) shows that satisfaction with PPM is associated to the quality of the process and to its alignment to the strategic needs of the firm.

Specifically, one first element of best practice is associated to the degree with which the PPM system is made explicit and clear and then rigorously applied. A PPM system with unclear rules, and which does not effectively determine the fate of projects will be next to useless, since all actors will quickly learn ways to get around the system and keep their favorite projects in the pipeline.

Moreover, best practice suggests combining 'hard' financial elements together with 'soft' and qualitative ones. This insight suggests making a combined use of methods, such as financial and multicriteria ones. For instance, financial methods could be used to perform a preliminary screening of candidate projects. Multicriteria methods could be used to refine the selection by assuming a broader perspective. Then, managers could go back to financial methods in order to suggest changes to the portfolio and/or to individual projects in order to refine the solution further.

Finally, projects should not be considered in isolation, but in aggregate. This is important because projects compete for resources that are scarce and generally hard to reproduce. For this reason, the real cost of project X is not simply the accounting cost involved, but the opportunity cost of the greater value that could be gained if X were not activated, thus allowing freed resources to work on project Y. This suggestion leads to the obvious conclusion that comparisons between projects must be performed. This can be done in a rigorous way with optimization methods, but also in a more intuitive and visual way by using mapping methods.

References

Belton V, Steward TJ (2002) Multiple criteria decision analysis, an integration approach. Kluwer Academic Publishers, Norwell, MA

Burrill GS (2009) Analysis for PhRMA based on publicly available data. In: Report Burrill and Company, 2009

Cantamessa M (2005) Fundamentals of product portfolio management. In: Clarkson J, Ekert C (eds), Design process improvement. Springer-Verlag, London

Copeland T, Antikarov V (2001) Real options—a practitioner's guide. Texere, New York, NY

Cooper RG (1990) Stage-gate systems: a new tool for managing new products. Bus Horiz 33 (3):44–54

Cooper RG, Edgett SJ (1996) Critical success factors for new financial services. Mark Manag 5 (3):26–37

Cooper RG, Edgett SJ, Kleinschmidt EJ (1999) New product portfolio management: practices and performance. J Prod Innov Manag 16(4):333–351

Cooper RG, Edgett SJ, Kleinschmidt EJ (2002a) Optimizing the stage-gate process: what best-practice companies do—I. Res-Technol 45(5):21–27

Cooper RG, Edgett SJ, Kleinschmidt EJ (2002b) Optimizing the stage-gate process: what best-practice companies do-II. Res-Technol 45(6):43–49

Hammond JS, Keeney RL, Raiffa H (2002) Smart choices: a practical guide to making better decisions. Harvard Business School Press, Boston

Hansen MT, Birkinshaw J (2007) The innovation value chain. Harv Bus Rev 85(6):121–130

Hermann M, Pentek T, Otto B (2016) Design principles for industry 4.0 scenarios. In: Proceeding of the IEEE 49th Hawaii international conference on system sciences, Koloa, HI, USA, Jan 5–8, pp 3928–3937

Krishnan V, Gupta S (2001) Appropriateness and impact of platform-based product development. Manag Sci 47(1):52–68

Levine HA (2006) Project portfolio management: a practical guide to selecting projects. Wiley, San Francisco

Lockett G, Hetherington B, Yallup P, Stratford M, Cox B (1986) Modelling a research portfolio using AHP: a group decision process. R&D Manag 16:151–160

Meyer MH, Lehnerd AP (1997) The power of product platforms. Free Press, New York

Montagna F (2011) Decision-aiding tools in innovative product development contexts. Res Eng Des 22(2):63–86

Paap JE (1994) Technology management and competitive intelligence: new techniques for a changing world. Competitive Intell Rev 5(1):2–4

Roy B (1996) Multicriteria methodology for decision aiding. Kluwer, Dordrecht

Saaty TL (1980) The analytic hierarchy process. McGraw-Hill, New York
Schmidt JB, Sarangee K, Montoya MM (2009) Exploring new product development project review practices. J Prod Innov Manag 26(5):520–535
Scholefield JH (1993) The development of an R&D planning model at ICP. R&D Manag 23 (4):20–30

Part III
Making innovation happen—The Product Development Process

Organizing Product Development Activities

10

Abstract

This chapter deals with organizational issues pertaining to the management of innovation and technical development activities within firms. The discussion is based on the classic—and mostly still valid—results deriving from Tom Allen's research, which include the use of codified knowledge, the role of gatekeepers and the influence of physical space. The discussion is enriched with recent research findings, which consider the role of innovators, the impact of contemporary information and communication technology. Finally, the chapter discusses issues pertaining to organizational design of development teams, along with their staffing in an international setting.

With this chapter, we depart from the high-level and strategic perspective that has characterized the discussion up to the previous chapter and start the third part of this book delving deeper into the operational aspects of managing innovation and product development activities. In doing so, we will consider the organization of applied research and product development projects together, not in the ignorance of the differences existing between these two fields, but because the managerial challenges that characterize them make them quite close to one another. This chapter will deal with organizational aspects and focus on human resources, while the following Chap. 11 will view such activities as business processes.

The first main statement that can be made in discussing the organization of innovation activities is that they generally require an ad hoc organization, regardless of the overall organization of the firm that hosts them. The reason is that innovation activities are very different from other typical corporate processes, with respect to knowledge intensity, involvement of actors from different corporate functions and duration. In order to provide a broader discussion on similarities and differences between innovation and other corporate activities, one can refer to the following Table 10.1.

M. Cantamessa and F. Montagna, *Management of Innovation and Product Development*, https://doi.org/10.1007/978-1-4471-7531-5_10

Table 10.1 Similarities and differences between innovation and other corporate activities

	Similarities with innovation activities	Differences from innovation activities
Research	Objects being discussed are technically similar, as are the language and methods being used Both activities are highly knowledge-intensive	The purpose is different (i.e., producing abstract vs. commercially useful knowledge) Research is based on explicit and formalized protocols and knowledge and tends to specialization, while innovation tends toward multidisciplinarity Both use abstract languages, such as mathematics, but innovation makes a more extensive use of representations of concrete artifacts (e.g., technical drawings, prototypes, etc.) Most researchers work for non-profit institutions that grant freedom to perform research within a wider scientific community. These communities (or 'invisible colleges') set research agendas and use conferences and journals to promote the field and to appraise the scientific merit research results and individuals. Conversely, innovation is not as free and tends to occur within boundaries of hierarchical and for-profit organizations
Administrative and business-support activities	Both are highly information-intensive. Incremental innovation activities are often significantly repetitive	In administrative activities, knowledge is incorporated into codified processes, and actors are supposed to simply follow them Innovation activities are knowledge-intensive and more unpredictable than administrative ones, with respect to duration and outcome
Production activities	These activities share the focus on obtaining concrete artifacts	Innovation activities have the goal of producing knowledge allowing the production of goods and services. Conversely, production activities can be seen as 'users' of such knowledge

As a consequence of these peculiarities, it is quite customary for firms to define an organizational structure dedicated to innovative activities that leans on, but is somewhat different from, the overall organization. In defining these ad hoc structures, two opposing criteria of organizational design must be managed. One criterion adopts a process view and focuses on *communication flows*, stipulating that organizational proximity between actors should depend on the strength of the information flow between them. Therefore, if Mr. A and Mrs. B have to exchange a substantial amount of information, we will tend to locate them into a same unit, regardless of the fact that Mr. A might be a software engineer and Mrs. B a specialist in Logistics. The other criterion adopts a '*functional view*' and stresses that resources with similar skills and jobs should be located in the same organizational unit, since this will lead to greater specialization and efficiency in the allocation of work. Hence, if we locate software engineers Mr. A and Mr. C in the same unit, the firm will benefit because this unit will be highly focused and potentially more proficient. Moreover, whenever Mrs. B will need to work with a software engineer, management will have the choice of making Mr. A or Mr. B available, depending on their workload.

As it will be discussed in the following, *organizational design** is mostly concerned with striking the right tradeoff between these two criteria, given the kind of projects the organization is trying to work on.

Before tackling this discussion, a word of caution ought to be spent on the limitations of organizational design, especially when focusing on communication flows. When one attempts to design an organization that is suitable for dealing with a given type of innovation problems, she/he is assuming that the kind of problems to solve, the skills and competencies required and the communication flows that are likely to occur are all known. In other words, it is assumed that the problem is well defined, 'problem setting' is completed and that resources must simply be arranged for effectiveness and efficiency. When dealing with incremental and competence-enhancing innovations, this is likely to be the case. However, if a firm intends to work on architectural or radical and competence-destroying innovations, this would be a somewhat lofty assumption. Moreover, a focus on communication invariably leads to concentrating on explicit information flows that can be traced and documented, but not on the many micro-level communications that naturally occur in teams, not to mention the conversations that individuals have with themselves when solving problems.

As a conclusion, most of the following discussion will be applicable to incremental innovation occurring in medium to large companies and for projects with a relatively long duration. Conversely, the discussion will have to be considered with some caution in case of radical innovations or of very small firms and short projects.

10.1 What We Know of Organizations Engaged in Innovation

R&D and innovation activities have been studied from an organizational perspective since the '70s, when a researcher at MIT, Tom Allen* (1931–2020), decided to concentrate on this topic and published his results in a seminal text (Allen 1977) that remains a key reference in the field of *Organization Theory**. Allen's studies stemmed from the intuition that development and innovation activities were different from Research, since engineers and designers—despite some common traits—acted in markedly different ways from scientists.

It is striking that the validity of Allen's findings has been repeatedly confirmed by researchers for decades and until now. This despite modern organizations can rely on an ICT 'weaponry' that was unimaginable in Allen's times, when information was stored in books and technical documents and when communication could only occur on paper or verbally, with face-to-face contacts or over telephone lines.

In these last years, moreover, thanks to further advances in ICT and the acceleration caused by emergency measures during the COVID-19 pandemic, the situation has further evolved, with a broader and more frequent shift to virtual and dispersed organizational structures. Many firms are now operating in a rather experimental way, some of them fully embracing virtualization while other ones trying to 'bring people back to the office'.[1] Researchers will therefore have significant amounts of data available to study these phenomena and try to understand whether IT and novel organizational forms lead to a real departure from the established body of knowledge, or simply represent a different implementation of these same principles.

Broadly speaking, Allen's findings continue to demonstrate their validity, and it is not clear whether this is due to the intrinsic nature of innovation activities, or whether because technology has not yet evolved to the point of causing significant change. This is the reason why a large part of this section will be dedicated to presenting and commenting these seminal results, in the consciousness that—at some point in time in the future and in specific industries—its validity may become weaker.

10.1.1 The Role of Literature and Formalized Knowledge

People who are involved in research and innovation may be engaged in different activities, but generally share a common educational background in science and technology. They may have attended the same universities, are likely to have used the same textbooks during their undergraduate studies and to have read and written similarly structured research papers and thesis during their graduate studies. One might therefore expect that, once engaged in their professional lives, they will have a similar attitude toward literature and make similar use of it.

In fact, researchers' entire profession revolves around literature. A researcher will cite previous literature in order to express the current state of the art in his/her field and to support the relevance of the research challenges he/she is working on.

[1] In the post-pandemic environment, firms are making quite different choices: Google is making significant real estate investments and has stated that it wants its employees back in the office, while Ford is activating remote employment contracts for certain types of job tasks.

He/she will use literature to define research hypothesis, to explain and justify the validity and appropriateness of the methods and protocols used, while the findings will be compared to hypothesis and to other results found in literature. Once published, his/her findings will enrich the body of literature, thus allowing other researchers to cite the paper, to build on it and, in doing so, granting the researcher recognition in the community and a quick career.

Somewhat surprisingly, when observing engineers and designers involved in development work, the picture changes significantly. As Allen discovered, literature has a very limited role as a basis for action and as a source of knowledge. When observing a sample of projects, he found that nearly 70% of solutions to technical problems were sourced from personal contacts, either from peers and colleagues or from customers. Only around 20% of the solutions were derived from literature, with the rest undefined.

On a more detailed analysis, which consisted in comparing similar projects performed by different organizations,[2] Allen found that the use of literature as an information source was negatively correlated with the quality of the solution. By delving deeper in this curious finding, he realized that this was not due to the use of literature *per se*, but to the fact that inexperienced personnel would make a greater use of literature, even if without getting much out of it. Conversely, experienced developers would achieve good results without having to use literature, but relying on their own knowledge or on interpersonal communications.

Following this finding, one can conclude that using literature in technical work is not harmful, but that it is not enough to make up for lack of experience. The reasons can be traced to the relationship between knowledge and innovation. As it has already been amply discussed, innovation activities mainly rely on knowledge that is tacit, organizationally embedded in people and routines and contextually situated in the specific technical and business setting the firm is operating in.

Conversely, the literature tends to incorporate knowledge that is codifiable, abstract and general, and aimed at a much broader audience. Therefore, its applicability in a specific setting will be somewhat limited, and will depend on the availability of some actor able to identify the most appropriate knowledge elements, translate them into terms that are appropriate to the problem under hand, and combine them with the existing knowledge base. Consequently, and somewhat paradoxically, the literature seems to be of limited use to the inexperienced engineers, who would need it most, while it could be useful to the most experienced ones, who however generally do quite well without using it.

Not unsurprisingly, Allen also found that, the more a written source is close to the company's specific setting and context, the more likely will it be used. Therefore, trade journals will be used more than academic ones and internal technical reports more than conference papers. Moreover, people will search for

[2] Allen had the chance of studying dual-supply projects, in which a customer awarded similar development contracts to two different suppliers, in order to minimize risk. This was an ideal research setting, since it allowed him to compare the activities being performed by different organizations, using the same project content.

written sources of information by following a 'pecking-order' approach, starting from sources they are most familiar with (i.e., books and/or reports available in their own personal archive) and progressively widening the scope of their search (e.g., borrowing from a colleague) until they find a satisfactory answer. This approach is not due to indolence, but to the fact that a source will be close by and familiar if it has been useful in the past and—exception made for radical innovation—it is quite likely that it will still be for the problem at hand too. Moreover, having used a written source before implies that finding the right information and making good use of it will probably not require much effort.

In Allen's times, the managerial implication of this finding was that firms ought to concentrate on the tacit knowledge element, on interpersonal communication and on the way individuals communicated and accessed their 'local' knowledge stores, rather than on building large libraries.

Nowadays, a similar lesson can apply as well, warning firms to be careful in designing the ICT infrastructure that is supposed to support innovation activities. While a communication component enabling interaction between different actors is undoubtedly important, the idea of building large repositories of codified knowledge risks being ineffective, since each individual will still face the problem of understanding what can, and ought be, used and how.

10.1.2 The Role of Interpersonal Communication

One key element of Allen's research was the documentation and analysis of '*communication acts*' between people involved in innovation activities. When this research was originally performed, personnel engaged in development work were asked to fill in survey forms reporting on "with whom did I speak today". Today, similar research is often performed by tracing electronic communications (e.g., metadata from e-mails, messaging systems, voice or video calls), which allows collecting massive amounts of data in a much easier way, and then studying the resulting communication networks.

The main idea behind this study was the recognition that interpersonal communication had a key role in suggesting solutions to technical problems. This led to the obvious research question of understanding what kind of communication could ensure the best results.

Allen relied on the availability of a sample of 'twin projects' in order to provide an answer to this question. Quite surprisingly, he discovered that the intensity of communication within the team (i.e., the number and frequency of communication acts) did not correlate with project success. A similar result emerged when trying to correlate quality of the outcome with the 'diversity' of communication within the team (i.e., the number of people to whom the communication acts were directed to). Success therefore did not seem to be an issue of communicating with immediate colleagues. Instead, Allen discovered that the variable that was most correlated to project quality was 'communication diversity' with people external to the team and external to the technical function the project belonged to. As it appears from

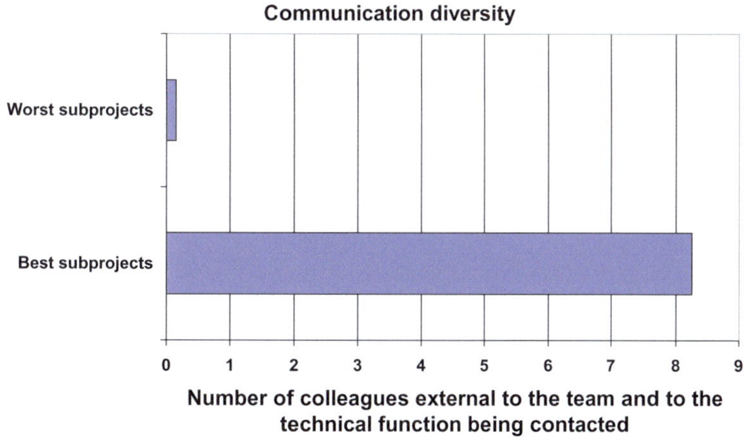

Fig. 10.1 Impact of external communication on project success

Fig. 10.1, the worst projects were the ones in which no outside contact was made, while the best ones were characterized by a significant attempt to 'reach out'.

The implication of this finding is that '*consultancy networks*' are critical to project success and firms therefore ought to nurture them. However, this kind of knowledge sharing activity is not easy to perform. Employees in need of support, and especially the more inexperienced ones, might not know who should be contacted and might feel uneasy in asking for help, since this would imply admitting some degree of ignorance. On the other side, the person who could provide such knowledge might be overloaded with work and be somewhat wary of spending time on someone else's project, especially if this person is inexperienced and therefore unlikely to be able to return the favor. On the opposite, internal consultancy works fairly well among experienced personnel. These are likely to know 'who knows what', have such a reputation that they do not feel ashamed of asking for help (especially if in the name of a junior co-worker) and are likely to be heeded to, since they may return the favor in the future.

10.2 Gatekeepers and Innovators

10.2.1 The Role of Technological Gatekeepers

By studying these dynamics, Allen discovered that technical organizations naturally tend to develop *hub and spoke** communication networks, in which a small fraction of actors become hubs and are responsible for most of 'knowledge brokering' both within the firm and beyond. These hubs are given the name of *technological gatekeepers** and are a key element of technical activities in all industries. As shown in Fig. 10.2, John might have a problem and ask Bill, a 'gatekeeper', for

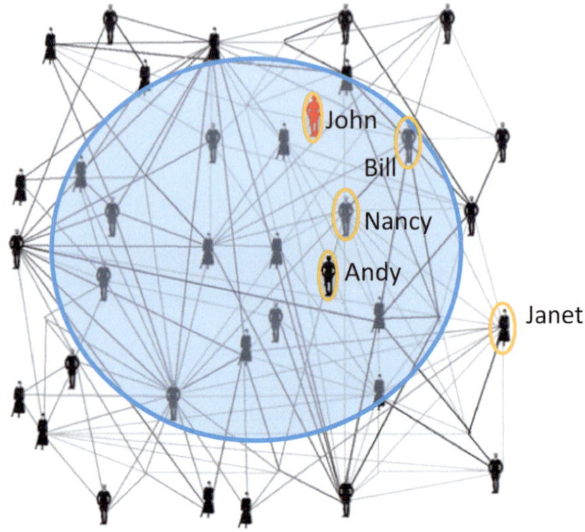

Fig. 10.2 Gatekeepers' role in technical problem-solving

help. Bill might not have the answer, but thinks that Nancy, another gatekeeper, might. In turn, Nancy might direct to problem to a young co-worker, Andy, or to Janet, a friend she met at a recent conference and who works for another company in the same area. John will therefore be able to contact Andy or Janet and, should he find it difficult to apply their knowledge to the problem, Nancy and Bill will probably assist him. It is clear that, without these two gatekeepers, John's problem would have never found a solution.

Gatekeepers are a very peculiar organizational actor. With some exceptions that will be discussed in the following, gatekeeping is not a job but, rather, a *role* whose recognition is purely informal. It is therefore quite surprising to discover that technical organizations include a few people (usually around 5–8%) who are highly critical to the functioning of innovation processes, but that their very existence and their activity occurs in an informal way.

Becoming a gatekeeper depends on a particular blend of technical expertise and personal traits that can be found in a minority of people. Gatekeepers usually are technically very competent and proficient, which gives them a reputation for professional excellence and, sometimes, the promotion to a supervisory position, such as team leader, which reinforces their visibility. At the same time, they exhibit a high degree of curiosity and eagerness to learn new things and to keep in contact with like-minded people. In short, gatekeepers belong to that minority of technical personnel who routinely read trade journals and are very happy to visit trade fairs, where they spend time discussing technical issues with exhibitors. Moreover, they often spend their own money to become members of professional associations and then read newsletters, attend conferences and take part to the activities of local chapters. Their role stems from the fact that, by being familiar with the 'outside world' and by being at the same time fully immersed in firm's context and its current knowledge base, they are particularly able to tap into the most appropriate source of knowledge or expertise and facilitate its translation and application to firm-specific problems.

At the same time, the informal nature of the gatekeeping role makes it quite fragile, difficult to manage and subject to a number of threats. In some cases, a gatekeeper's role might not be recognized at all. Being a highly effective professional, a gatekeeper may be overloaded with work, preventing him/her from spending some time in knowledge brokering. Missed recognition can also lead to conflicts with both line managers and project managers. In the former case, a gatekeeper might inadvertently challenge their technical authority.[3] In the latter case, internal consultancy might be seen by project managers as a waste of time.[4]

In other cases, a gatekeeper's role may be recognized, but mismanaged. For instance, a firm might recognize a gatekeeper and suggest he/she spends all of his/her time in knowledge brokering. However, not having to personally perform technical activities any more, this will gradually lead him/her to lose that deep contact with the technical context that provided the roots for the gatekeeping role at the beginning. Quite often, in recognition of his/her competence, a gatekeeper might also be promoted to a managerial position. At this point, the nature of his/her work changes quite dramatically, since technical contents will give way to administrative ones and this too will hinder the gatekeeping activity.[5]

In order to avoid these problems, firms have tried to come up with organizational solutions for supporting gatekeepers and providing them with incentives. One of these solutions is the so-called *dual-ladder* career path. In such a system, talented technical personnel are assessed for their specific capabilities and directed to either a traditional managerial career, or to a technical one. In this latter case, the person is progressively relieved of operational work, allowed to become a 'knowledge manager' in a given domain and given a widening set of responsibilities on gatekeeping and on coordinating colleagues who are involved in the same career path at a more junior level. In the most structured of these organizational settings (Hirsch 2006), these responsibilities may include:

[3] Think of a junior employee telling his boss "*Mr. Jones, you are telling me that my solution is wrong and will not work. But I asked David, whom we all know is a great gatekeeper, and he told me it's surely going to work, since he knows company X has successfully applied a similar approach*".

[4] Imagine a project manager saying "*Alice, we all know you are a great developer and you come out with great solutions, but you spent all yesterday helping out with those other two projects. Remember you are full time allocated on my project and we are running late for next week's review meeting. Please don't make me say this again*".

[5] Promotions are also well known to be subject to the so-called *Peter effect**, which states that personnel will tend to be promoted to their 'level of incompetence'. In other words, people are promoted based on their performance in their current job, which can be a bad predictor of the performance they might have in their next career step. After promotion, in case this latter performance is poor, the person will not be promoted any further and will therefore be stuck into a job he is not capable of performing well.

- the creation and development of *communities of practice** facilitating knowledge sharing among personnel in the same domain, which is especially important in large and dispersed organizations;
- the support to the human resources function, for activities such as recruiting, assessment, professional development and resource retention;
- the support to technical functions, by working on the assessment and standardization of technology, support methods and tools.

Dual-ladder career path systems are very appealing, but their implementation is by no means trivial and can run into many problems. For instance, it is not easy to find the right balance between technical and gatekeeping activities, as well as between internal knowledge sharing and external networking. Moreover, salary issues can be quite complex to solve, since a credible dual-ladder career path would require having comparable salaries on both 'ladders'. This can sometimes be hard to justify from an economic perspective and to implement on a contractual basis. Moreover, it can lead to jealousy between people belonging to the two ladders, which end up in professional lives that are very different from one another.

If one attempts to compare initial and recent research findings on gatekeepers, not much has changed, despite the role of ICT and data and its impact on organizations. In fact, ICT has made it easier and cheaper to codify knowledge and to share and access it, while AI algorithms are making it easier to extract information from data. This has led to a huge increase in the amount of available information and knowledge. Now, all this knowledge requires higher-level and tacit knowledge in order to correctly filter and interpret it, understand its implications and limitations and apply it. Similarly, the analysis of big data and the interpretation of the resulting evidences requires high-level domain knowledge (Agostini and Filippini 2019).

So, instead of making gatekeepers obsolete, ICT and big data seem to simply have increased the need of knowledge integration roles and broadened the scope of gatekeeping activity, which nowadays involves helping others find the right information in digital repositories, or supporting data analysts in their analyses, thanks the use of their situational knowledge.

Moreover, research also shows that knowledge sharing networks, despite the use of ICT, will nonetheless exhibit 'structural holes' that require the intervention of traditional gatekeepers (Tortoriello 2015). According to the so-called 'variance hypothesis' (Dahlander et al. 2016), innovation performance is greater if organizations perform broad searches for new knowledge, which is unlikely to be performed successfully without gatekeepers. To this purpose, some authors (Singh and Wassenaar 2016) suggest that gatekeepers' activities should be separated among the search for new knowledge (which requires a stronger focus on indirect contacts) and combinatorial knowledge (which instead requires a stronger focus on tacit knowledge).

Concerning the sources used for technical problem-solving, research still tells us (Dodgson et al. 2007) that personal communication is still dominant over codified knowledge. At this moment, there is no way to know whether this is an inevitable feature of technical problems, or whether it is due to the fact that ICT systems are still not able to be an effective substitute as far as knowledge management is concerned. However, ICT is having a key impact on innovation activities with the diffusion of simulation and virtual prototyping systems. Thanks to these systems, technical solutions can be searched for extensively and somewhat cheaply through trial and error, instead of relying on prior knowledge. It is therefore possible—though this hypothesis still must be verified—that the role of gatekeepers might be progressively reduced, because simulation and virtual prototyping tools will enable relatively inexperienced people to experiment and innovate, even without having an extensive communication network feeding them with hints and insights on potential solutions.

10.2.2 The Role of Innovators

Gatekeepers, as described above, are technical problem-solvers who therefore are key actors in innovation processes. However, they are not, *stricto sensu*, innovators themselves. If we consider an innovator to be a person capable of matching a market need and opportunity with a technical solution, and to support the ensuing change at an organizational level, this latter role appears to be quite wider and more strategic than that of the gatekeeper.

Firms do not only need gatekeepers, but they also need innovators to guide the innovation process with an entrepreneurial mindset (Cohn et al. 2008). Framing and managing the role of the 'innovator' within established firms is a relatively recent topic on which literature is still quite scarce. Given their strategic role, innovators must of course be included in the top management team. Due to organizational inertia and path dependency it would be deceptive to think of a separation of roles in which innovators propose ideas, the board decides on which should be followed, and functional managers look after execution. However, it is not clear how these 'innovators' could be identified, based on a somewhat elusive talent, follow a specific career path—in which criteria for promotion are definitely different from the mainstream ones—and end up joining the top management team.

Some firms have a clear dualistic leadership that brings together innovators and execution-minded talent, but this can usually be traced back to the original founding team (e.g., Bill Hewlett and David Packard in HP). Moreover, quite seldom does this lead to a permanent organizational and leadership structure (the main exception being firms in the field of luxury and fashion, which usually have distinct managerial and artistic directorships).

Given the difficulty of understanding what innovation talent is, research on this topic has mostly been carried out by comparing the approaches to problem-solving taken by widely acclaimed 'innovators' with respect to 'ordinary' executives. Following Dyer et al. (2009), innovators seem to score much higher on cognitive capabilities such as making associations between seemingly unrelated elements, observing situations and placing deep questions, experimenting and networking (Fig. 10.3).

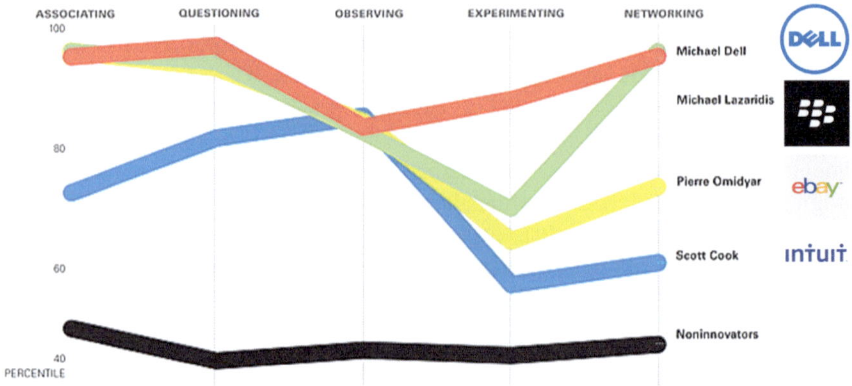

Fig. 10.3 Distinguishing features of well-known innovators compared to non-innovators (modified from Dyer et al. 2009)

A similar approach has also been taken by organizational psychologists, who have recently started dealing with the topic, by identifying personal traits that appear to be prominent in innovators, with the objective of facilitating their identification through assessment tests (Potocnik et al. 2014).

Firms willing to explicitly foster innovation talent often adopt *intrapreneurship schemes* (a concept that was introduced in Chap. 8). For instance, 3M and Google are known to allow employees to spend 10–20% of their time on personal projects. A company like Adobe has instead introduced a number of processes to foster grassroots innovation. Among them, one encourage employees to 'incubate' their ideas in Advanced Technology Labs, under the guidance of 'idea mentors' and of 'entrepreneurs in residence', following a process that mimics the process with which startup firms are supported by accelerators and funded by venture capital funds.

In order to allow an effective innovation activity, and especially when a firm engages in complex open innovation practices, it is necessary to ensure that those who are given this responsibility can operate in close cooperation with top management and with the members of the organization who are in charge of routine activities. To this purpose, it is necessary to develop *appropriate managerial frameworks*.

One possible framework comes from Bower and Burgelman (1983), who initially developed their proposal for intrapreneurship activities. The framework is represented in Fig. 10.4. The rows represent the three main organizational tiers involved. These include the top management of the firm, who must provide strategic direction and oversight; the top management of the innovation initiative (usually made up of a senior manager, reporting to the CEO and a board, with representatives from various functions and BUs); the junior managers and professionals in charge of scouting and managing individual projects. The columns are split in two sections, which represent the focus of the activities carried out.

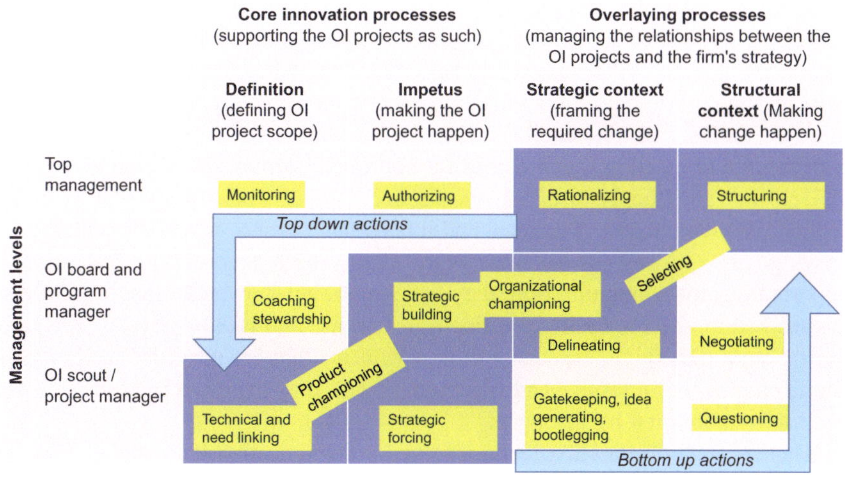

Fig. 10.4 Bower–Burgelman managerial framework (1983)

The left-hand section is concerned with the implementation of the innovation activities as such, while the right-hand section deals with making sure that the innovation activities have a connection to, and an impact on, the firm's strategy. In turn, both sections are made up of a conceptualizing/planning phase (left) and an execution phase (right).

The resulting matrix allows to identify the *responsibilities and actions* that have to be planned for each management tier and in each activity. The yellow boxes shown in the figure are to be intended as examples, but will have to be defined specifically for each individual firm, and then organized in an overall process.

10.3 Physical and Virtual Spaces

While studying communication flows between individuals, Allen discovered that the physical layout of office space had a major impact. Specifically, and quite unsurprisingly, he discovered that the probability of information exchange in a given time interval between any couple of individuals decreased quite dramatically with the distance that separated their respective workplaces. The probability of having one weekly communication act would fall to a meager 10% when this distance would be greater than 15 m, in the case of people having some kind of organizational linkage. The same probability would occur at a distance greater than 5 m for people not tied by such a linkage.

Recent research (Hansen and Lovas 2004) confirmed this finding, showing that knowledge exchange depends on geographic proximity, together with previous acquaintances and relatedness of competencies.

If one couples this finding with the previous one, which stated that project success is associated to speaking to multiple people not belonging to the same project team, a major problem arises. How can a company create the opportunity to have all these 'external' communication acts, without having to squeeze everyone in a 5 m radius from one another? The obvious answer is to design office layouts with the aim of creating opportunities for multiple informal acts of communication.

One first solution is to adopt *open space* (or *open plan**) layouts, in which distance between people is reduced and it becomes easier to notice if a colleague is busy, go up and talk to him/her. While open space layouts do create greater opportunities for communication, they have important drawbacks that anyone who has worked in one can easily confirm, such as noise and lack of privacy. The latter can be quite disturbing, especially in case of firms that are unaccustomed to evaluating employees on results, and where employees may therefore be afraid of being constantly monitored by managers.

Another solution for stimulating informal communication is to create '*nodes of attraction*' in which employees can go and spend some time. The most obvious node of attraction is the coffee area, which can be designed in such a way that employees will find it comfortable enough to stay for a relatively extended break. In this case, conversation will very easily shift from initial general topics and some office gossip, to the discussion of technical problems and to the exchange of views and knowledge pertinent to the projects that participants are engaged in. Other nodes of attraction can be specially designed corridors and atriums, departmental services (such as printers and photocopiers) and so on.

Some firms have adopted more extreme solutions, such as the *non-territorial office*. This kind of layout includes different areas, intended for different types of work (e.g., individual desks, isolated areas for thinking, tables for working in small groups, large conference rooms, etc.). In a non-territorial office, no one has an assigned desk, but people move from one area to the other, depending on the work to be done at that moment. The idea was originally proposed by Allen, but started diffusing only recently, when documentation started moving from paper to cloud-based repositories, thus freeing employees from the constraint of a fixed workplace.

In general, playing with space allows managers to influence behavior and communication flows in a way that formal directives would probably not be able to achieve. For instance, creating a large 'project room' or 'war room' will attract employees working on the same project and automatically decrease the strength of their connection with their functions and line managers.

Thanks to cooperation between corporate managers and architects, a number of R&D centers[6] have been designed with an explicit aim of fostering informal communication between employees, allowing them to easily keep abreast with problems and achievements and exchange knowledge.

[6] For example, most IBM facilities have been designed around the objective of facilitating communication among employees. Other examples are the Renault Technocentre in Guyancourt (focused on interactions between different engineering departments) and the GM development

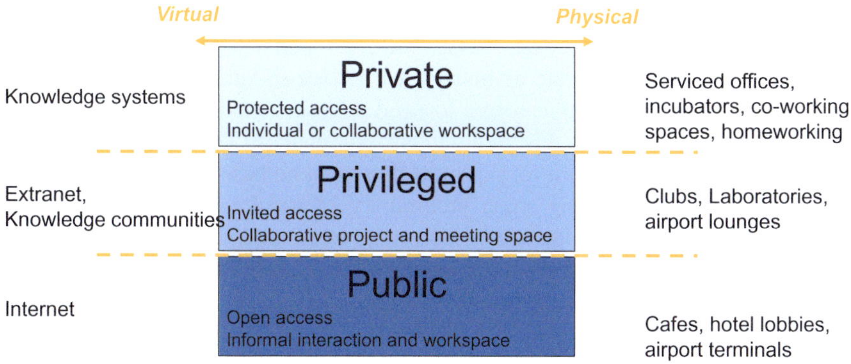

Fig. 10.5 Contemporary variety of physical and virtual workspaces (modified from Hardy et al. (DEGW) 2008)

The concept of workplace is nowadays becoming ever more complex and difficult to manage, since it is no longer limited to the physical office one commutes to every day (see the DEGW report, Hardy et al. 2008). Along with physical workspaces, employees engaged in R&D operate in *virtual workspaces* where they exchange information, store results of their work and are assigned tasks related to their projects. At the same time, knowledge workers can operate in a wide range of workspaces (Fig. 10.5), which includes private facilities (e.g., company premises or their own home), restricted areas (e.g., airport lounges and co-working areas) and public spaces (e.g., cafes and hotel lobbies).

Finally, when discussing these organizational issues, an emerging topic is related to the management of *geographically dispersed and culturally diverse* teams. This dispersion is due to the increasingly global footprint of corporate operations. A firm may have several R&D departments located in different parts of the world. Alternatively, the R&D department in one country may have to jointly design a product with a subsidiary located in another country and with a set of suppliers who operate from other parts of the world.

A central enabler of this globalization of product development activities is the progress of information and communication technology. These tools allow personnel engaged in innovation activities to continuously keep in touch and exchange both structured (e.g., a 3D model of a physical object, a dataset of experimental data) and non-structured information (e.g., a quick exchange of views through a messaging system). People are becoming ever more accustomed to using systems of this kind even when they are physically located close to one another, and this can decrease the difference they perceive when they have to cooperate with colleagues who are located at a high geographic distance.

---

center and plant in Spring Hill (aimed at creating links between engineering and production). Noteworthy is also the case experimented at LTU of a Design Observatory at Lulea University (Sweden), where engineering activities and communication are experimentally stimulated by acting on the built environment and a number of innovative technologies that are incorporated in it.

Literature (Bierly et al. 2009, Choi and Cho 2019) tells us that effective cooperation is based on trust between colleagues. Cooperation through virtual communications does not generate as much trust as face-to-face meetings do. At the same time, virtual cooperation seems to need a lesser amount of reciprocal trust, though it does not cancel this requirement completely. Therefore, it might be helpful to set the groundwork for a project that must be based on extensive virtual communications by planning some preliminary physical meetings or co-located work. Virtual communication also leads to greater independence in individual work, which requires greater clarity of objectives and a more structured approach to project management.

10.4 Organizational Design for Innovation Activities

As discussed at the beginning of the chapter, innovation activities are usually carried out within specific organizational structures. After having introduced the main elements that characterize these activities, it is possible to examine the organizational forms usually adopted by companies and investigate their suitability for innovation projects of different types.

Organizational forms are defined by the types of projects carried out:

- the **existence**, or the **non-existence, of a project manager** responsible for the project, the scope of his responsibility and the authority granted to him with respect to the resources that work in the same project;
- the **degree of separation between resources** working in a project and the corporate functions they belong to;
- the degree with which the project requires **highly specialized resources**;
- the degree with which the project requires close **coordination between corporate functions**. In this context, coordination implies the identification of solutions to technical design tradeoffs that pertain to different corporate functions. For instance, the choice of a material for a product can impact manufacturing costs, technical performance and can have implications on customer service. These tradeoffs will have to be solved both *ex-ante*, when specifying the product, and *ex-post*, when sorting out problems that have been neglected earlier on, therefore managing the resulting design changes.

By blending these four features, four 'standard' organizational forms can be defined. Figure 10.6 portrays these forms, with functions being represented by the vertical bars, managers with black dots, while dashed circles represent the resources involved in the project.

- **Purely functional organizations**. In this case (Fig. 10.6a), work is performed within functions by resources that are not permanently assigned to projects. Information is handed 'over the wall' from one function to another (e.g.,

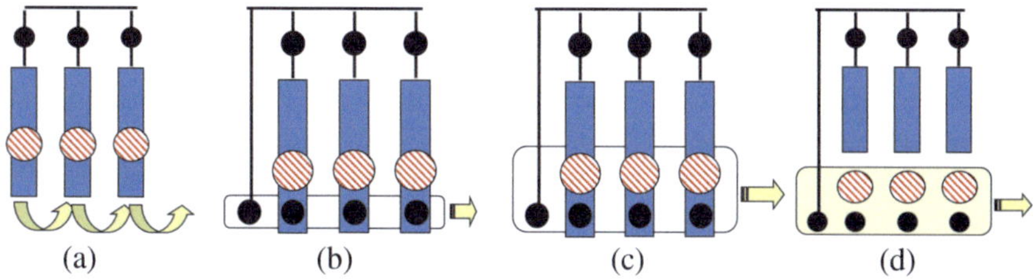

Fig. 10.6 Four standard organizational forms used in innovation activities: functional (a), lightweight inter-functional (b), heavyweight inter-functional (c) and autonomous team (d)

marketing defines a product brief, product engineers then design the product, and manufacturing engineers finally define the process). Coordination is only managed by functional managers, who may not have the time and the knowledge of the specific project to do it effectively. Consequently, this form is suitable only in the case of projects for which inter-functional coordination is not very critical.

This form has advantages, such as the efficiency due to resource pooling within functions and the high degree of technical competence that is allowed by specialization—something that is very useful especially if technological progress is rapid. Another advantage is connected to the simplicity of having resources reporting to a single—albeit possibly ineffective—authority.

- **Lightweight project teams**. This organizational form (Fig. 10.6b) can be considered to be a minor adaptation of the previous one, aimed at solving the coordination problem that was highlighted above.

This form has the advantage of requiring a traditional, functionally oriented firm, to make minimal change to working habits and organizational culture. Work is still carried out within functions by resources who are not permanently assigned to the project. Technical coordination is performed by a lightweight team composed by a project manager and one representative per each function. Neither the project manager, nor the functional representatives have formal authority over the resources that are working on the project. However, thanks to frequent interaction, they are in the position to influence their work and have the opportunity of creating an effective flow of information that may solve most *ex-ante* and *ex-post* coordination issues.

The major drawback of the lightweight team form lies in the ambiguity between the responsibility given to the project manager—which usually consists in delivering the prescribed project results on time and on budget—and his/her lack of formal authority. This ambiguity is often exacerbated by the fact that—since lightweight teams do not directly manage resources—firms will tend to staff them with junior professionals in order to reduce costs. Now, should significant problems arise, it would be quite hard to imagine a junior project manager taking a tough position with a functional manager. This issue can be somewhat abated if the company runs multiple projects under the lightweight team approach. In such case, the firm may appoint a senior program manager to oversee all these projects, have junior project managers reporting to him and assigning him responsibility for interactions with functional managers.

- **Heavyweight project teams**. This organizational form (Fig. 10.6c) corresponds to a matrix organization, in which functions have the responsibility of developing technically competent resources, who are then formally assigned to one or more projects. Resources still work within functions, must report to line managers concerning their technical performance, but report to team leaders and project managers for their work in the project. Inter-functional coordination becomes much stronger, given that resources are assigned to projects and are strongly involved in its evolution. Moreover, project managers are given formal authority over resources and are in the position to solve whatever problems may arise.

  This organizational form is therefore quite suitable for projects that involve architectural change. At the same time, this form has some disadvantages too, the main one being the cost deriving from the senior personnel involved in project management, and from the impossibility of using resource pooling, which may lead to temporarily idle resources. Moreover, the deep involvement of resources in projects can lead to some degree of separation from their technical functions, which in turn reduces their specialization and—potentially—their technical proficiency. A final and major drawback of the heavyweight project team form is related to the complexity of the dual reporting mechanism. In the case of project-based firms (e.g., engineering consultancies, engineer-to-order manufacturers, etc.), this problem is quite manageable, since resources are accustomed to work in a matrix organization, and because there is a clear separation in the duties of functions (i.e., to nurture resources) and of projects (i.e., to use these same resources to create economic value within projects). The problem can instead be quite severe in the case of organizations that are not project-based. In this case, functions might see the assignment of resources to project work as a hindrance to their normal operations, which can lead to tensions or to the assignment of 'second-best' personnel. For instance, a plant manager might not be very happy to know that his best process engineer must spend half of her time for the next three years in a product development team, instead of making sure that the plant runs smoothly and efficiently.

- **Autonomous teams**. In this case (Fig. 10.6d), the firm pulls resources out of their functions and creates a team that is completely autonomous, geographically co-located and under the complete supervision of a project manager. This arrangement leads to a very tight coordination between team elements and project phases. The separation from corporate functions allows greater freedom to experiment but—on the other hand—a lower degree of specialization and capability to tap into existing corporate competencies. Furthermore, the autonomous nature of the team leads to lower efficiency, since capacity must be *ex-ante* allocated to the project team, without relying on other personnel belonging to functions.

  This organizational form is therefore suitable for projects that involve radical and/or competence-destroying innovations. Another critical aspect is associated to the duration of the project. Autonomous teams work well either for projects with very short duration (e.g., a task force set up to solve an urgent problem, or to quickly develop a radically new product concept) or for projects that will effectively never end, because the team becomes the nucleus of a new and permanent corporate division. Conversely, it can be quite difficult to disband such a team at the end of a medium- to long-term project (e.g., 2–3 years), since resources will have to be reintegrated in their original functions after a long absence, during which career paths in the function may have evolved substantially, and while their competencies may have become obsolete.

Following this discussion, companies should to analyze their project portfolio and identify the organizational form that is most suitable per each major project category. The ability of using different organizational forms allows the firm to practice the 'ambidexterity' introduced in Chap. 3, with the coexistence of innovative activities that have different natures and must be managed appropriately.

At the same time, the firm should try to avoid introducing too many differences in its organizational arrangements, so that its personnel do not have to spend too much time understanding the way they are supposed to operate and to whom they are supposed to report within each specific project. Therefore, a firm may decide to work with lightweight teams for the majority of projects and occasionally use autonomous teams in order to introduce a next-generation product.

10.5 Project Staffing

A final aspect of organizational design, which is particularly relevant when dealing with geographically distributed teams, is connected to staffing, which consists in identifying team leaders and individual team components to be assigned to the project. The choice of team members is of course critical to project success. A team made of up of competent professionals who are able to work well together, exploiting their mutual complementarities, is far more likely to achieve the desired results than a team in which individual competencies are lower and/or members are not able to work well together.

Staffing must therefore simultaneously look at individual profiles and at the *degree of diversity* within the team. In general, researchers agree (Adler 2002) that the relationship between team diversity and performance follows an inverted U-shaped curve. The degree of diversity provides richness to the internal debate and to the team's capability of generating new ideas, but up to a given point, beyond which the difficulty in communicating and understanding each other starts prevailing.[7] Among the dimensions of diversity that are relevant to innovation,

[7] Some contributions to the literature suggest the negative impact of diversity might be more significant in hypercompetitive settings and in smaller and younger firms (Hoisl et al. 2016).

researchers have studied gender (Diaz-Garcia et al. 2013; Xie et al. 2020), ethnic background (O'Reilly III et al. 1998) and personality traits (Reilly et al. 2002).

When assigning personnel to a team, therefore, there are a number of dimensions over which individuals can be considered. The main such dimension is obviously the match between the requirements coming from the project's activities and the candidates' 'hard skills' or professional competencies. This first dimension is generally quite easy to appraise, since data on the knowledge and competencies possessed by technical personnel are usually well known and are often codified in 'competencies repertoires'. However, other dimensions may be highly relevant as well, and might be subtler to manage, since companies are not likely to keep profiles of personnel under these perspectives.

- **Personal traits**. Examples include degree of openness, optimism, capability to work in teams and should be considered with some allowance for diversity. A team made of pessimistic members is not likely to explore anything but the most conservative of technical solutions, just like a team made up of optimists might suggest solutions without sufficiently exploring their potential risks.
- **Gender**. The topic is not sufficiently researched, but preliminary evidence suggests that men and women engaged in innovative activity tackle technical problems from different perspectives and following slightly different cognitive approaches. This being the case, diversity in team composition is likely to enrich the innovation process and lead to better results.
- **Experience**. Aside from the obvious statement that experience is valuable, research tells us (Gunther and Ehrlenspiel 1999) that novice and experienced designers tackle technical problems following quite distinct approaches. The former are usually more methodical and use a predefined process as their main problem-solving strategy, possibly in order to make up for their relative lack of experience. The process followed is therefore well structured and well documented, starts with an in-depth clarification of the problem and is followed by the proposal and analysis of potential solutions. Conversely, experienced designers tend to rely on intuition and experience and follow a process that is generally poorly structured and documented, but quicker. Problem clarification, generation of solutions and their evaluation occur simultaneously and iteratively, with a constant focus on the artifact being developed and its subsystems, and not on a predefined process and its phases. Under this perspective, team staffing implies making choices over which kind of problem-solving strategy should be preferred, and on finding ways to manage the conflicts that may arise when personnel who follow completely cognitive approaches have to cooperate.
- **National culture**. Since the '70s, researchers have realized that national culture has an influence on the way with which people operate in organizations and tackle the problems set in front of them by their profession. One of the first and best-known contributions on this topic is Geert Hofstede's research on cultural dimensions performed while he was working at IBM on a huge sample of more than 100,000 individuals (1983). Hofstede identified five '*cultural dimensions*' which define national cultures and namely power distance (i.e., the degree with

which individuals accept inequality in power or wealth), individualism, uncertainty avoidance, masculinity (i.e., the degree with which values such as power, career and money are preferred over caring for others, quality of life, etc.) and long-termism. In 2010, Hofstede added a sixth dimension, dealing with indulgence vs. restraint (the value with which 'enjoying life' is looked upon positively versus with suspicion). Following Hofstede, national culture influences behavior and, therefore, the approach individuals will follow in tackling problems arising in their profession. Moreover, individuals' national culture will lead to diversity within international teams, with the ensuing implications with respect to richness and coordination issues to be faced. Hofstede's findings have often been criticized,[8] especially because differences in national culture explain only a small part of the variance exhibited by individuals, which has to do with subnational divisions within a country and with individuals' upbringing. Therefore, the theory can be used to provide preliminary guidance, but certainly not as a tool for predicting individuals' behavior.

References

Adler NJ (2002) International dimensions of organizational behavior. South-Western Thomson Learning, Cincinnati

Agostini L, Filippini R (2019) Organizational and managerial challenges in the path towards industry 4.0. Eur J Innov Manag 22(3):406–421

Allen TJ (1977) Managing the flow of technology. MIT Press, Cambridge

Bierly PE, Damanpour F, Santoro MD (2009) The application of external knowledge: organizational conditions for exploration and exploitation. J Manag Stud 46:481–509

Burgelman RA (1983) A process model of internal corporate venturing in the diversified major firm. Adm Sci Q 28(2):223–244

Choi O-K, Cho E (2019) The mechanism of trust affecting collaboration in virtual teams and the moderating roles of the culture of autonomy and task complexity. Comput Hum Behav 91:305–315. https://doi.org/10.1016/j.chb.2018.09.032

Cohn J, Katzenbach J, Vlak G (2008) Finding and grooming breakthrough innovators. Harv Bus Rev 86(12):62–69

Dahlander L, O'Mahony S, Gann DM (2016) One foot in, one foot out: how does individuals' external search breath affect innovation outcomes? Strateg Manag J 37(2):280–302

Díaz-García C, González-Moreno A, Sáez-Martínez FS (2013) Gender diversity within R&D teams: its impact on radicalness of innovation. Innov Organ Manag 15(2):149–160

Dodgson M, Gann D, Salter A (2007) The impact of the use of innovation technology on engineering problem solving: lessons from high profile public projects. Technol Anal Strateg Manag 19(3):471–489

Dyer JH, Gregersen HB, Christensen CM (2009) The innovator's DNA: five "discovery skills" separate true innovators from the rest of us. Harv Bus Rev 87(12):60–67

Gunther J, Ehrlenspiel K (1999) Comparing designers from practice and designers with systematic education. Des Stud 20(5):439–452

Hansen MT, Løvås B (2004) How do multinational companies leverage technological competencies? moving from single to interdependent explanations. Strateg Manag J 25(8–9):801–822

[8] Hofstede focused on work-related attitudes, which some scholars believe do not fully capture personal behavior. An alternative framework, focused on the latter, has been proposed by Trompenaars and Hampden-Turner (1994; 2020).

Hardy B, Graham R, Stansall P, White A, Harrison A, Bell A, Hutton L (2008) Working beyond walls: the government workspace as agent of change. OGC and DEGW, London
Hirsch PM (2006) A master reflects on the sociology of work and the discipline. William Form's Legacy, Work Occup 33(1):5–7
Hofstede G (1983) The cultural relativity of organizational practices and theories. J Int Bus Stud 14(2):75–89
Hofstede G (2011) Dimensionalizing cultures: the hofstede model in context. Online Readings Psychol Cult 2(1)
Hoisl K, Gruber M, Conti A (2016) R&D team diversity and performance in hypercompetitive environments. Strateg Manag J 38(7):1455–1477
O'Reilly CA III, Williams KY, Barsade S (1998) Group demography and innovation: does diversity help? In: Gruenfeld DH (ed) Composition. Elsevier Science/JAI Press, pp 183–207
Potocnik K, Anderson N, Latorre F (2014) Selecting for innovation: methods of assessment and the criterion problem. In: Nikolaou I, Oostrom J (eds) Recent and new developments in recruitment and selection, Taylor & Francis
Reilly RR, Lynn GS, Aronson ZH (2002) The role of personality in new product development team performance. J Eng Technol Manag 19:39–58
Singh S, Wassenaar DR (2016) Contextualising the role of the gatekeeper in social science research. S Afr J Bioeth Law 9(1):42–46
Tortoriello M (2015) The social underpinnings of absorptive capacity: the moderating effects of structural holes on innovation generation based on external knowledge. Strateg Manag J 36 (4):586–597
Trompenaars F, Hampden-Turner C (1994) The seven cultures of capitalism: value systems for creating wealth in the United States. Br Jpn Ger Fr Sweden Neth Int Aff 70(3):550–551
Trompenaars F, Hampden-Turner C (2020) Riding the waves of culture: understanding diversity in global business. McGraw Hill, New York
Xie L, Zhou J, Zong Q, Lu Q (2020) Gender diversity in R&D teams and innovation efficiency: role of the innovation context. Res Policy 49(1):103885

The Product Development Process 11

Abstract

The objective of this chapter is to discuss managerial approaches to the management of the product development process. At first, the chapter presents the main phases of product development and the defining features of this process, such as the lever effect, the pervasive presence of iterations and the need to balance tradeoffs associated to project cost, time and outcomes. After presenting classical approaches to the management of product development (i.e., sequential, design reviews, concurrent engineering) and the support that can be derived from information technology systems, the discussion explains the role of flexibility and the emerging approaches that is enabled by flexibility (i.e., Agile and Lean).

This chapter can be viewed as a complement to the previous one, since it focuses on the activities that make up product development, rather than on the organization of the actors that carry them out. This perspective fits well to the contemporary approach to Management, which is strongly focused on processes. It is also closely related to Nelson and Winter's theory of the firm, which is prevalent in the study of Innovation Economics (explored in Chap. 2), as well as to the Resource-based Theory that is popular in the context of corporate strategy (which we discussed in Chap. 6).

However, the process-oriented view of product development calls for a comment and a word of warning that are similar to the one that has also been raised in the previous chapter. If one assumes that there is a predefined process for product development, and that its constituent activities can be ex-ante spelt out and arranged, this implies that the problem to be solved is relatively well structured. In the case of ill-defined problems, one cannot define ex-ante a solution process, since the process will be identified while solving the problem. Therefore, the process-oriented view will be applicable for product development projects with moderate innovative content, in which prior experience can be used as a guide

M. Cantamessa and F. Montagna, *Management of Innovation and Product Development*, https://doi.org/10.1007/978-1-4471-7531-5_11

to allow the formal and detailed definition of a process. Conversely, it may not be so applicable in case of small and informally managed projects and/or very innovative problems, for which no clear solution strategy has been identified and formalized yet.[1]

11.1 The Main Phases of the Product Development

One of the main tenets of the Resource-based Theory is that firms are different and that these differences can be traced to their resources and routines. This is particularly true for the product development process. This process will be significantly different from firm to firm and will depend on the industry, on the products and services being developed, on the upstream and downstream relationships with other firms and, of course, on the individual firm's past history.

Companies need to specify their own specific product development process at some level of detail. This is a mandatory requirement for firms who wish to certify their quality systems around international norms, such as ISO 9001, and is a desirable attribute for companies that are led by project management principles. A well-defined product development process can be used as a template for managing each product development initiative, as a basis for defining workflow on information systems, and to support the engagement with customers and suppliers.

In view of that, despite the uniqueness of product development processes, it is nonetheless useful to define some typical phases, in the knowledge that these represent only a general reference (Fig. 11.1). The following description covers the case of physical, discrete and assembled products, but can be easily adapted to the case of services, remembering that, instead of product architecture, one will have to deal with the architecture of the underlying service process.

The presentation shows a sequence of stages, although this should be considered only as a conceptual organization; due to its own nature, the development process progresses through iterations among these phases, and contemporary management practice suggests performing these phases in parallel, concurrently and/or iteratively.

[1] Furthermore, the very hypothesis that a predefined process can exist clashes against the behavior observed in knowledge-intensive professions by Donald Alan Schon* (1930–1997). Schon (e.g., 1983, 1995) studied a variety of professionals from technical and non-technical fields. He discovered that they seldom follow predefined strategies, but continuously analyze new information arising from the problem and from their attempts to tackle it, and adapt their problem-solving approach as a consequence. In a way, professionals are involved in a continuous 'conversation' with the problem and follow a cognitive approach that can be termed reflection in action. These two visions can however coexist, if one considers 'reflection in action' to be typical of individuals' behavior and the process perspective to represent a product development project in its entirety at organizational level. Chapter 8 will deepen this concept specifically to design activities.

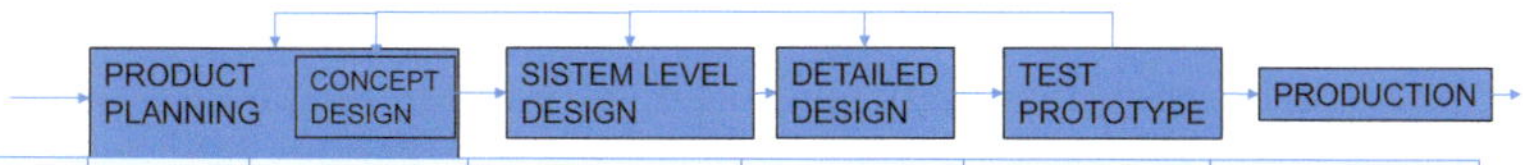

Marketing	X	X				
Purchasing	X	X	X			X
Finance	X					
Technical office	X	X	X	X	X	
Manufacturing		X	X	X	X	X

Fig. 11.1 A reference product development process

- **Product planning** is the initial phase of the product development process. In this phase, the firm has the objective of defining the new product from the joint perspective coming from the market (i.e., "to which customers is it addressed? What needs does it have to tackle?") and from technology (i.e., "what functions and performance should characterize it?"). Product planning is a highly inter-functional and inter-disciplinary phase, in which representatives from most corporate functions, from marketing to customer service, are to be involved.

Product planning generally starts with the dual exploration of tacit or explicit market needs and of technological opportunities. The prevalence of either of the two determinants depends on the type of innovation (i.e., incremental or radical) the company wishes to pursue. Based on this research, the firm will reach a very high-level definition of the product that is often termed *product brief* or *design brief**. The product brief allows the *positioning* of the product with respect to target market segments and to the firm's and to competitors' existing product portfolios and *product pipelines**. Then, the firm will develop a detailed description of customer/user *needs* and *requirements*. Based on the information from Marketing, designers start generating the feasible technical solutions that will be able to fulfill the previously defined customer/user requirements. As it will be discussed in Chap. 15, the development team will collectively select the technical solutions that define a *product concept*. This is the moment in which the firm decides the extent to which it will use an existing product or service concept, or whether to look for a new concept and how different it will be from previous solutions, thus leading to a more radical innovation. In this case, they will generate alternative technical solutions leading to new concepts and select the most suitable one. In parallel to these activities, the cost of the system will be evaluated and a *business case**[2] will be defined, analyzing the returns from expected sales and comparing them against the required investment. This business case will be

[2] A business case is the financial evaluation of an individual initiative and is developed on the background of a corporate business plan (which, instead, represents the financial plan for the entire firm). The underlying assumption of this approach is that the individual project will not be so big as to determine substantial changes to the overall business plan (e.g., it will not require additional funding) and may therefore be evaluated in isolation. In some cases (e.g., an aerospace company that is working on a next generation wide-body jet), the product development project may be so large as to require a review of the entire business plan of a firm.

revised at each of the following steps of the process, and will therefore evolve from a preliminary feasibility study to a fully blown operational plan.

- **System-level design** is the stage at which the firm starts placing further detail on the product concept. First, the firm will translate requirements into detailed system-level *technical specifications*, which will provide designers involved in the subsequent steps with a clear framework of the functions and performance levels that the product or service will have to achieve. Then, the firm will move from the product concept to the definition of a *product architecture* (which—for the time—can be defined as the list of subsystems that make up the product and their reciprocal interfaces) and will make a number of choices concerning the main subsystems and components. Among such choices, *carryover* means deciding on which subsystems or components are to be taken from previous products, and which are to be introduced anew. Percentage of carryover, computed by component count or by component value, is often used as a proxy for the innovative content of the product. *Custom* versus *off-the-shelf* decisions are associated to the choice of which subsystems are to be developed specifically for the new product and which are to be selected among an existing offering coming from either the same firm, or from outside vendors. Finally, *make, develop or buy* decisions dictate the distribution of work between the firm and its suppliers in development and production. 'Make' implies both activities are performed in-house; 'develop' refers to internal development and external production; 'buy' means that both activities are assigned to suppliers.

System-level design is also concerned with defining interfaces between subsystems at functional and geometrical level, and with the specification of subsystems and components. This latter task broadly depends on the nature of the specifications, which can be either *localized* or *general*, respectively, depending on whether they are related to a single or to multiple components.[3] In the case of general specifications, it is necessary to *apportion* specifications to each component. The task is not trivial, since it will generally lead to finding the right balance in a complicated system of tradeoffs.[4] This phase is also concerned with the preliminary definition of all the systems that will interact with the product throughout its lifecycle (e.g., production, maintenance, logistics, etc.).

- The **component-level development** phase is where most of the engineering work is carried out. Materials and components are chosen and designed in detail, while calculations are carried out to dimension them so that the required

[3] For instance, in a notebook computer, mass storage capacity is a localized specification, since this specification at product level (i.e., the gigabytes of mass storage the firm wants the computer to have) can immediately be translated into the same amount at the level of the single subsystem, which is responsible for it (i.e., the hard-disk or solid-state drive). Conversely, cost, mass and battery time are general specifications, since they will be associated to a number of different components. Cost and mass will be given by the simple sum of the same specification for all components, while battery time will be a function of battery capacity and energy usage by other components.

[4] In the notebook computer example, should one provide more battery capacity, which will come at a greater cost and mass, or adopt chipsets that use less energy but are more expensive?

specifications are met. Decisions carried out at the system-design phase will eventually be reviewed.

- The **prototyping and testing** phase can be separated from detailed design for conceptual and historical reasons, though contemporary firms tend to perform them in parallel through simulation. The conceptual distinction is due to the fact that detailed design is concerned with generating or detailing solutions, while prototyping deals with verifying and validating them. Testing can be performed analytically, through simulation, or on physical prototypes. The role of simulation on virtual prototypes is progressively growing, due to its predictive power and low cost. When used, physical prototypes are usually developed in steps and range from preliminary (or *alpha*) prototypes, in which components and manufacturing processes can be significantly different to the final ones, intermediate (or *beta*) in which components and processes are similar to the final ones, and *pilots*, which are built using final components and production processes. Testing on prototypes is often carried out for regulatory purposes, since most products must comply with governmental regulations, and requires the awarding of a certification by the competent authorities before they are commercialized. In most cases, certification requires tests to be carried out on prototypes or early versions of the product itself.
- **Process design** is involved with the detailed design of processes that will enable the product to be produced, distributed and serviced. This also involves designing a large number of the resources required by such processes, such as tools, dies and fixtures for manufacturing, service manuals and tooling for field engineers and so on.
- **Product launch and production** is the final phase in product development. At product *launch*, the firm starts to deploy the resources dedicated to production, in order to build and deliver products to the market. In some cases, production starts on a limited scale by using *pilot* lines, which will later be backed up by other production resources. During *production ramp-up*, production lines are usually started at a slower pace than the nominal one, so to iron out problems and make sure that its production personnel becomes progressively acquainted with the new system.
- **Post-launch activities.** Firms usually allocate a given time after launch, during which the product and the process will be fine-tuned in order to ensure adherence to specifications and market acceptance, after which the development process is officially terminated. In some industries, such as aerospace, the development process is not ended at all and remains active until the product is in use. This allows customers to be continually supported with updates and with the possibility of adapting the product according to emerging needs.

11.2 The Peculiar Managerial Complexity of the Product Development Process

It should be apparent at this point that product development processes are quite distinct from the rest of the processes that are usually performed by organizations.

One first element is associated to the magnitude of the process, in terms of *effort* and *time required*. In some cases, as for instance in or firms characterized with a relatively limited product range, such as aerospace (less it would be in the case of fast-moving consumer goods), the budget required by each instance of the development process has a size that has an observable impact on the annual financial results. As well as, development processes in all industries usually last months or years, thereby spanning multiple reporting periods, and even the career paths of many of the employees involved, so that the team that initiates the process might not be the same that delivers the results at the end.

A second feature of the product development process, as Fig. 1.1 shows, is the *simultaneous involvement of different organizational functions* (e.g., Marketing, Design, Production, etc.) and technical disciplines (e.g., Hardware and Mechanical Engineering) not only across the project in its entirety, but even within each stage. Within such a complex setting, each person involved is asked to contribute not only by applying his or her professional knowledge to execute an activity, but also to creatively propose, elaborate and modify solutions by interacting with colleagues which are characterized by heterogeneous cultural backgrounds and sets of objectives.

In this context, managers have to develop *plans under uncertain conditions*, while designers operate under *significant information asymmetry*, since managers usually do not have a full understanding of specific technical choices (Mihm 2010). This leads to a *principal-agent problem*, in which designers will tend to 'game the system' in a variety of ways. For instance, designers will tend to focus on the specific tasks and components that are assigned to them, while they are relatively unaware of (or care little about) the overall design problem. This naturally leads them to develop 'beauty ideals' that are locally optimal, but that can collectively lead quite far away from the global optimum. Moreover, each managerial approach to product development will define a set of incentives and managerial pressures with respect to timeliness and technical quality of the solutions, which designers may interpret opportunistically (e.g., to make sure that problems will not arise later on, they may to 'over engineer', or if a deadline appears to be extremely important, they might 'cut corners' and propose a technically weak solution).

A final distinctive feature of the product development process is the *presence of iterations*. Most corporate processes are conceived to be performed only once, and any iteration is usually considered as a fault to be avoided. This is not true for the product development process. As in any problem-solving process, and especially due to its creative nature, it is normal to revise choices when designing a new product or service.

Iterations can occur at component level, when a designer makes a technical choice, challenges and tests it, and eventually goes back to revising it in order to improve the solution. Iterations can also occur at system level. For instance, a designer may choose a component and may then be forced to revise his or her decision, because another component developed by another colleague highlights an incompatibility. Similarly, an iteration might occur because a colleague representing field service may point out that a given design choice would make the product difficult to repair. Iterations can also occur between development projects, when choices made for one product may impact another one (for instance, using a component in project A could be economically viable only if the same choice is done for project B). Finally, iterations occur between the firm and the market, whenever it is technically feasible to adapt the product or service to new (or newly discovered) needs.

11.3 Tradeoffs in the Product Development Process

When managing a product development process, moreover, a number of *tradeoffs* have to be dealt with (Smith and Reinerstein 1991). In order to correctly understand this issue, it is possible to analyze the entire product lifecycle from an operational and financial perspective, as shown in Fig. 11.2.

We will now provide a brief explanation of the main tradeoffs; the following subsections will focus on a more detailed description of the variables involved.

Reducing the development time is often a key goal for companies, since it allows a number of strategies useful for entering a market and achieving competitive advantage. At the same time, it is obvious that dedicating less time to development will curb the amount of product quality and performance that the firm will be able to achieve, which will reduce its future attractiveness in the market, thus reducing volumes and/or margins.

Product development also entails significant ex-ante and fixed costs, which will have to be recouped by ensuring that the resulting product quality and performance will allow adequate margins from future sales, so to ensure overall profitability of the project. Finally, reducing time to market could be achieved by increasing

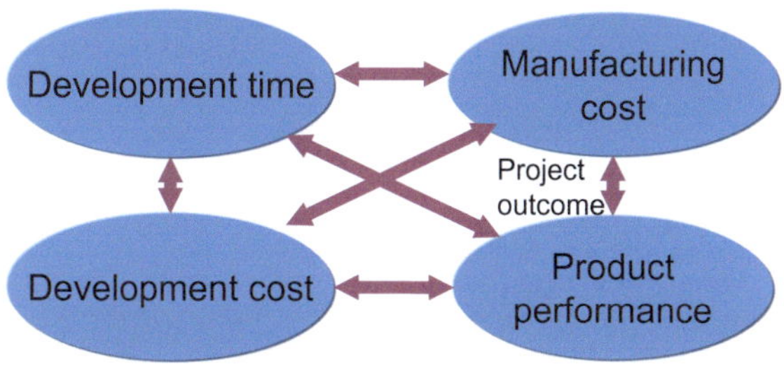

Fig. 11.2 The main tradeoffs in product development process

development costs (e.g., by using external consultants), but the firm will have to evaluate the soundness of similar decisions.

Based on the above, it is unthinkable to manage the product development process keeping these goals as separate, and without considering the consequences that the decisions made to satisfy one can have on the others. Similarly, it is essential to know the intrinsic elements of difficulty that exist for reaching each of them.

11.3.1 The Key Concern of Reducing Time to Market

In order to correctly analyze the variable 'time' in product development, it is necessary to consider the entire product lifecycle from an operational and financial perspective, as shown in Fig. 11.3.

A firm must first of all define a planning horizon *T*, beyond which it expects the product to become obsolete, or it does not feel able to make forecasts. Evaluations will therefore be made up to time *T*, which can be called '*window of opportunity*'. The planning horizon can be split into two phases, the former dedicated to product development, up to *Start of Production* (SoP) and product launch, and the latter to sales.

From a financial viewpoint, the development phase leads to negative cash flow, while the sales phase determines positive cash flows that will eventually allow the firm to recoup the investment made. The slope with which the cumulated cash flow curve recovers after the launch event depends on sales volume (i.e., units sold per time unit) and on the contribution margin (i.e., the difference between unit price and variable cost). *Time to Market** (TTM) is usually defined as the lapse of time

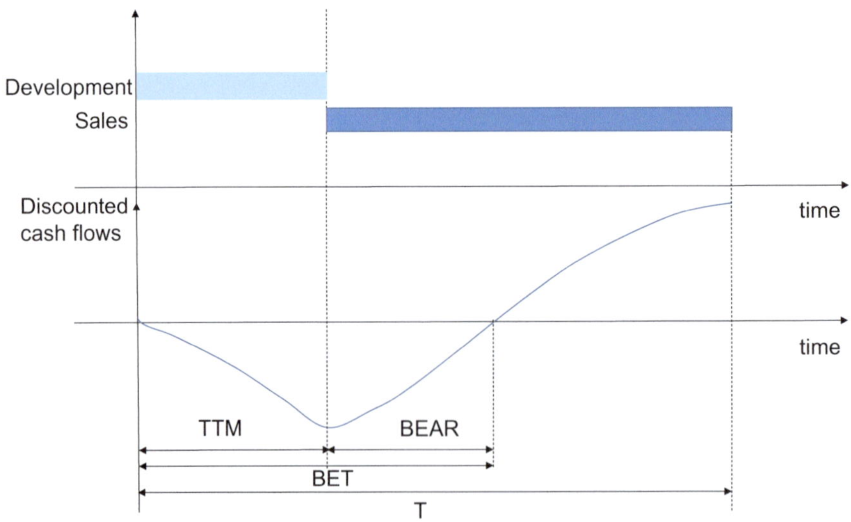

Fig. 11.3 The operational and financial structure of a product development process

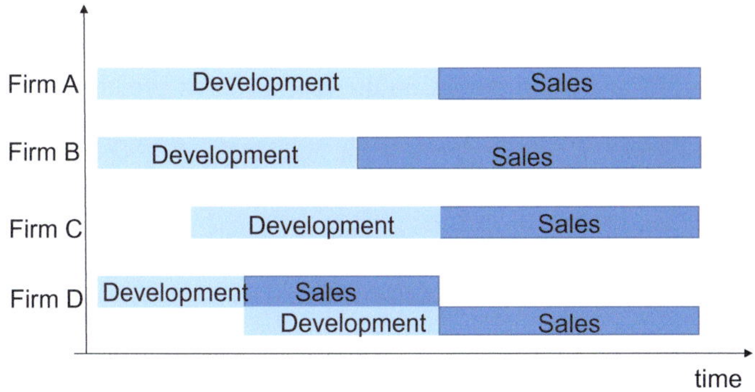

Fig. 11.4 Strategic uses of short time to market

between the start of the project and product launch. *Breakeven Time** (BET) is defined as the interval that goes from the start of the project to the time at which the investment is recovered. *Breakeven after release** (BEAR) is the interval between product launch and the time at which the investment is recovered, and therefore is the difference between BET and TTM.

A number of alternative strategies, as depicted in Fig. 11.4, can be adopted to exploit the key objective of reducing TTM. The most obvious one is to start the project early and to enjoy a longer sales phase (firm B). Alternatively, a firm (C) may postpone the start of the project and keep the same duration of the sales phase. By doing this, the firm will be able to develop a more advanced product, by incorporating up-to-date technology or newly emerging market needs. Finally, a firm (D) may reduce TTM so much that it can launch multiple products within the same sales cycle.

Each of these alternatives has financial implications[5] that the firm must be able to foresee and manage. In view of that, firms usually develop and constantly update a financial model of the product over its lifecycle, using it to perform simulations and what-if analysis. A model of this kind is usually a key part of the business case that is drafted in order to obtain approval of any project before it starts and that must be updated and used throughout the project progress. When working on financial worksheets representing business cases, there are many uncertainties, and the outcomes rest on highly debatable assumptions that might be both endogenous to the firm (e.g., what market share will the product reach?) and exogenous to it (e.g., what will the US Dollar-Japanese Yen exchange rate be in the next five years?). It is therefore advisable not to perform single-point simulations on expected values only, but to envisage a number of scenarios in order to achieve a better understanding of the variance of the outcome and its dependency on key assumptions. Instead of

[5] For instance, a firm must be able to estimate the impact of '*delaying product launch by 2 months in order to incorporate a new component that might allow a 2.30-dollar reduction in unit cost*', or '*allowing a 100.000 € increase to the development budget in order to hire a consulting firm that might help develop a better product, with an expected sales increase of 5–10%*'.

discrete *scenario analysis**, the firm may also decide to use *Monte Carlo analysis** to achieve a broader picture of potential outcomes and their variance.

In the literature, analytical studies on the economics of product lifecycles have also been proposed to aid TTM decisions; however, these approaches are of limited practical applicability, because they rest on very stylized models. It is however noteworthy to mention the model developed by Cohen et al. (1996) who recognized a number of interesting features of optimal decisions in product development, which may be considered as general and relatively practical guidelines. The main outcomes of this model are the following:

- A firm should focus its effort on the phases in which it is able to generate higher added value. Conversely, it should consider outsourcing for other phases.
- If the window of opportunity T is shortened (i.e., in the case of products with fast obsolescence), the firm should intuitively reduce TTM. However, the optimal reduction should be less than proportional (to be precise, it should be proportional to $\sqrt{T}$).
- In the case of large markets and/or high unit margins, a firm should develop high-quality products even if this requires a greater TTM. The resulting higher quality of the product will lead to greater market share, and the related contribution margins will more than repay the added effort and the initial lost sales. Conversely, in the case of small markets and/or low unit margins, the optimal strategy is to go for incremental innovation (i.e., to pursue smaller advances with shorter TTM and limited investments).
- Firms should not use BET as a performance indicator, since this casts an incentive for excessively short and unambitious projects. BEAR is a more balanced indicator, which brings firms closer to the optimal TTM.[6]
- If a firm increases its development productivity (i.e., the increase in product quality per each unit of effort spent), it should carefully consider how to use this advantage. It is usually better to keep the same TTM and substantially improve the product, rather than reducing TTM and targeting the same product quality.

In some cases, firms can dispense with the problem of defining the 'ideal' TTM and leave it to the seasonality that is inherent to the market or is purposely chosen,

[6] A number of firms, among which Hewlett-Packard (House and Price 1991), have adopted performance indicators similar to BEAR. This is not an easy policy to enforce, since BEAR measures the performance of a project not by looking at its execution, but at its future consequences. This runs contrary to the usual training received by project managers, i.e., to deliver given results on time and on budget, since BEAR takes the completely different stance of looking at how quickly the project will recoup the investment made, *whatever* time and budget has been used. From a project manager's perspective, it may seem puzzling that development performance may be calculated by looking at the sales phase. After all, one might argue that what happens during the sales phase depends on functions and managers in charge of sales, production, field service, etc. In reality, this is the very reason why BEAR is a powerful indicator, since it forces managers in charge of development projects to internalize the elaboration of all the processes and assets that will be needed later on and ensure they will work correctly.

in order to provide 'rhythm' to development activities (Brown and Einsenhardt 1998; Souza et al. 2004). When following such a *time-paced* strategy, the firm anchors the launch date for its products at a given period and then decides on the technical contents and the development budget. Launch dates can be determined by demand (e.g., toys must be launched in time for the Christmas selling season), by major trade fairs (e.g., consumer electronics being shown at the Las Vegas CES in January) or by internal traditions (e.g., 'the traditional June salesforce meeting in the Alps'). Time pacing makes project scheduling and management significantly easier, since all actors involved learn by experience what should be accomplished at a given time (e.g., 'each year our customer *X* delivers his prototypes in mid-January and wants durability tests to be completed by the end of February').

In the end, it is important to highlight that having TTM as a key priority can provide evident advantages, such as fast entry in the market (eventually benefiting from first-mover advantage) and the possibility of incorporating superior technologies into new product families, but it leads also to some disadvantages. Given the hurry, the risk of entrapping the project into limited technical objectives could drive incremental innovation processes only, as well as favoring quick solutions versus technically better ones. A number of approaches adopted in the management of the product development process aim at facing these problems and will be analyzed later in the chapter.

11.3.2 Product Performance and Customer Satisfaction

Product features and performance are key to reach customer satisfaction. *Customer satisfaction** has become one of the most relevant goal since the *Total Quality Management** approach has been adopted by the majority of companies worldwide. For people involved in Innovation Management, customer satisfaction is also crucial to ensure that new technologies diffuse. This is particularly true for market-driven and incremental innovations, but it is also relevant for radical and/or technology-push innovations. The comprehension of the variables that drive customer choices, in fact, is focal in those cases in which needs must be envisioned and adoption paths must be built in advance (Montagna and Cantamessa 2019).

A product development perspective devoted to the adoption of innovations, although its valuable rationale, has three major problems.

First, adopters *are not necessarily aware* of their needs. In fact, people can only imagine own needs tracing them back to previous experience. The famous phrase by Henry Ford (1863–1947) "*If I had asked people what they wanted, they would have said faster horses*" suggests that when a radical innovation and a paradigm shift are taking place, past experience cannot provide a solid basis on which future may be anticipated. Building adoption readiness is precisely the reason why companies set up significant communication campaigns to move customers in their convictions unhinging cultural elements, or to build progressive adoption paths so as to allow people to become familiar with technology, to reduce their switching costs and to establish network economies, eventually defining lock-in effects.[7]

The second issue lies in the *difficulty of capturing the entirety of needs*, and we will analyze in detail such issue Chap. 13.

The last issue is related to the *translation of needs into requirements*. Customer satisfaction depends on what will be designed and, consequently, on how the design problem will be posed to designers. This translation depends on the ability of designers to exactly translate into requirements those needs that have been identified, without incurring into hidden traps. One trap lies, for instance, in the very *act of translating* from the customer's mind to that of the marketing employee who is doing the analysis, or from the marketing employee to the designer. Typically, these people do not adopt common perspectives, languages and tools, and what may consequently happen is a misunderstanding, whereby the design intent differs from the original customer need.

In summary, achieving customer satisfaction is not a trivial job, even where the company's effort is very high and all possible actions are taken. Wrong decisions may occur, despite having spent significant time and resources, thus making the previously defined tradeoff decisions even more critical.

11.3.3 Managing Costs in Product Development

One of the first issues that a manager should consider when she/he faces costs in product development is observing their *evolution throughout the project*. At first, one can plot the cumulative costs *incurred* by a firm during the process. This leads to an s-curve (continuous line in Fig. 11.5), since the first phases usually are relatively inexpensive to run, while the majority of costs are incurred in the detailed design, prototyping and production design phases.

At the same time, one can plot a curve of cumulated committed costs (dashed line in Fig. 11.5) that shows the share of overall costs being *determined* by each phase. This curve shows that the most important market and technological choices are taken during the product planning and system-level design phases.[8] So, the later

[7] For the former case, think about when Ikea entered the American market reaching what in the Rogers's diffusion curve would be called 'early majority': Americans were used to expensive, but long-lasting, furniture; the famous communication campaign "*it's just furniture change it*" was aimed at convincing people of the possibility of adopting furniture that did not last a lifetime, because furniture was not linked to indissoluble values (not surprisingly, the campaign visually associated furniture to the buyer's spouse). For the latter case, one can think about Apple and its roadmap that brought consumers from iPods, which made them familiar with portable digital technologies, to iTunes, which introduced significant switching costs for adopters, to iPhones and iPads. As a matter of fact, the basic concept behind the iPad was known to Apple since the 1990s, but was initially a commercial failure, which then later required to develop a strategy to bring it to the masses.

[8] For instance, at the beginning of the development process for a consumer electronics product, the team may decide to adopt a completely new material for the casing of the device. The cost incurred for taking this decision can be very low (i.e., some research on materials properties, a few calls to

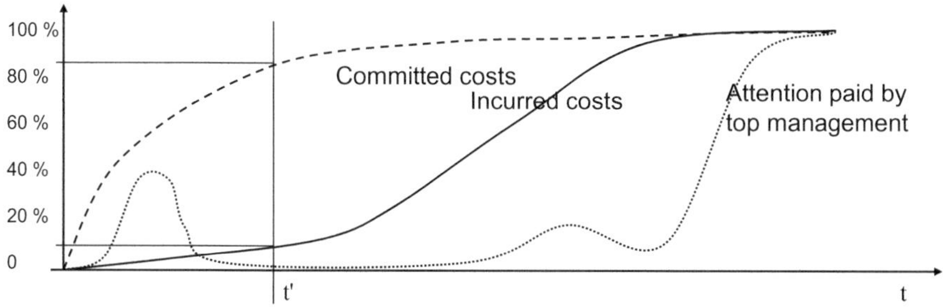

Fig. 11.5 The 'lever effect' in the early phases of the product development process

phases will entail a significant expenditure, but will be unable to make a significant difference in the overall costs, if not by making sure that they are properly executed and major mistakes are avoided.

In summary, while the early phases of the development process have the very important task of 'doing the right things', the later phases of the development process are simply concerned with 'doing things right'. Moreover, while reversing a decision early in the process can be relatively inexpensive, changes occurring during the later phases may have a very strong impact on development cost.

The gap between incurred and determined costs that characterizes the initial phases of the development process leads to a '*lever effect*', and explains why managerial attention ought to be very high at the onset of the process, when costs are estimated, planned and decisions on their commitment are made. In fact, many development processes fail because top management starts paying attention only toward the end, when tangible results show up, but—unfortunately—when it is either too late or too costly to overturn earlier choices (dotted line in Fig. 11.5). The lever effect suggests a product development approach called '*front loading*' which calls for 'shifting the identification and solving of problems to earlier phases of a product development process' (Thomke and Fujimoto 2000). Front loading, however, is based on the hypothesis that problems can be effectively solved early on, by dedicating enough effort. This might not be always true, since the firm may operate in the presence of significant market and technological uncertainty and/or volatility, which would call for an opposite approach. In this latter case, the firm should aim towards a 'late commitment' strategy that consists in keeping options open and—provided this is technically feasible—making modifications until the very last moment before product launch (or even changing the products already in use in the field).

A further problem that managers have to face is the *estimation of development and product costs*. Cost estimation in product development is grounded on

suppliers and a few minutes of debate during a meeting). However, once taken, this decision may determine a substantial portion of the subsequent costs, since the casing built with this new material will require a given amount of engineering work during detailed design, followed by the search for new suppliers, investments in tools and dies, and so on.

historical data, past design experience or manufacturing knowledge. Product costs are built on design decisions that were appropriate with respect to the requirements, the functionalities, the technologies that were characterizing past systems, conceived for past customer needs and markets. Development costs too are estimated on the basis of past processes, that may have been carried out in a different historical moment and competitive environment.

Moreover, the early stages of design processes can't do without making use of qualitative approaches, be them intuitive or analogical techniques, or rely on expert judgments or heuristic rules. This is because, in the early stages, no detailed decisions have been made yet on components, on materials, or on production equipment. More accurate estimations will be provided by quantitative techniques only in the final stages of the development process. Therefore, when managers have to make costing and planning decisions for a product development process, no information is completely reliable, and this will become available when it is too late. Currently, some efforts are being made in the literature to anticipate quantitative estimates in the conceptualization stages (for a review, Altavilla et al. 2018), but these remain isolated efforts.

11.4 Information Technology Support to Product Development Processes

IT has amply impacted the product development process, initially as a support to the activities being carried out and—more recently—by making digital technology a core element of products and services. In this section, we will focus on the former aspect, while the implications of the second will be covered later on.

There are two main ways with which ICT supports product development: the former one being product representation and the latter one management and transfer of information.

Concerning the former, product development has become ever more *virtualized*, in the sense that products and services may be represented in detail with computer models. Three-dimensional representations of physical products do not only allow the visualization of an idea, as in the old technical drawings drafted on drawing boards or on 2D CAD (*Computer-Aided Design**) tools, but also the simulation of the product under multiple perspectives, which collectively fall under the acronym of CA-X. So, for instance, *Computer-Aided Engineering** (CAE) tools allow the simulation of cinematic and dynamic behavior of a product, of stresses when subject to forces and of thermal and fluid flows. Similarly, *Computer-Aided Manufacturing** (CAM) software allows the modeling and simulation of manufacturing and assembly processes, as well as the disassembly processes needed to service the product and disassemble it at the end of its life. Finally, *Computer-Aided Production Engineering** (CAPE) tools enable product developers to represent the production system at a higher level of abstraction (e.g., manufacturing or assembly cell). Analogous simulation tools have also been developed for industries other than manufacturing, such as services (e.g., for simulating business processes), software engineering and so on.

The availability of multiple simulation tools provides a tremendous support to product development, since it allows designers to receive quick and inexpensive feedback on their design choices, without having to engage in costly and time-consuming experimentation on physical prototypes. *Virtual prototyping** has effectively changed the way with which products are conceived, both at the level of the organizational process and the individual cognitive one (Gupta 1996).[9] In the traditional paradigm, designers had to make the best possible choices and then moved to costly experimentation on physical prototypes to validate such choices, in the hope that everything would go well. In the modern paradigm, designers can develop products by trial and error, using virtual experimentation for exploring solutions. This somewhat reduces the need for ex-ante knowledge, though this knowledge does retain value for reducing the time and cost required to converge to a solution. Virtual experimentation also allows product developers to explore innovative solutions that would not have been considered within a traditional product development process, given the time and cost required by experimentation. Simulation also allows front loading of the development process, which means spotting problems and critical issues early in the process. At the same time, and since simulation tools provide sophisticated experimentation capabilities to designers, this calls for a broadening of their skills.

Digitalization moreover has brought up the issue of managing the huge mass of information that is generated and consumed by the product development process. Even in the case of products with moderate complexity, design teams generate a plethora of files for representing the product and its components, which must be multiplied by the number of versions generated at the various project iterations and by the number of different simulations that may be performed on them. Additionally, project documentation may include both structured data (e.g., a database of experimentation results, the Work Breakdown Structure representing project activities, etc.) and unstructured data (e.g., messages exchanged within the development team).

In order to manage this information properly, to provide the appropriate operational support and to offer some means for capturing the tacit knowledge that is used and generated within the development process, information systems have evolved quite significantly. Early *Engineering Data Management** (EDM) systems were little more than file managers for CAD-generated data. These were followed by *Product Data Management** (PDM) systems, which managed all information pertaining to the product during the design phase. These were followed by *Product Lifecycle Management** (or PLM) systems, which manage all information related to

[9] This phenomenon is common to many activities that have 'gone digital'. Whether it is writing text with word processing software, or shooting photographs with digital cameras, users exhibit the tendency to work *by trial and error* and to engage in multiple iterations that would not have been possible with traditional means, such as fountain pens and film-based cameras.

the product throughout its lifecycle.[10] PLM systems are application frameworks that allow the integration of various modules relevant to product development, based on a common infrastructure. Among these modules, the most significant are:

- **Bill of materials manager**, able to represent the hierarchical relationships between components under different perspectives (i.e., engineering, purchasing, manufacturing) and to associate files and documents to relevant elements in the BOMs. These different perspectives stem from the fact that each corporate function may have a different hierarchical view of the same product. For instance, designers may prefer functional analysis as a decomposition criterion; purchasing agents may require a decomposition that is based on suppliers or component categories; manufacturing engineers may instead look at the assembly sequence with which the product is physically built. The role of a BOM manager is to allow these different views, at the same time ensuring consistency among them.
- **Configuration manager**, able to represent variations of the same product, depending on the features that are applied when serving individual customers or market segments. These modules are charged with ensuring that configuration rules are enforced (e.g., 'vehicle X cannot have air conditioning unit Y if engine Z is being used') and must also keep track of engineering changes at product and component level (e.g., 'Dishwasher X sold in market Z used pump Y_1 from date D_1 to D_2. From date D_2 onwards, pump Y_2 must be used') in order to ensure consistency of production and aftersales service.
- **Engineering change manager**, which tracks the evolution of changes to the product by following design iterations before product launch. Additionally, they manage modifications that are applied after the end of the development project, in reply to faults that are discovered and have to be corrected, or to the normal evolution of components.
- **Component and supplier manager**, acting as a centralized repository of components being used within the firms' collection of BOMs, and that facilitates the standardization and reuse of components.
- **Workflow manager**, able to automate the evolution of the project so that—when an activity is completed—the following steps are automatically initiated, and the people responsible for their execution are alerted and provided with the right information.
- **Vaulting system**, which represents the main logical repository of information being stored and modified in the development process. In a vaulting system, files are represented by their inner data and by a set of relevant metadata. Whenever an actor accesses a file in order to modify it, the file is *checked out* and other users are prevented from doing the same, in order to ensure consistency. When the actor ends his work, the file is *checked in*, the vaulting system assigns it a

[10] For a deeper insight on PLM systems, readers can refer to Grieves (2006), Stark (2005), Saaksvuori and Immonen (2002) for a general discussion and to Schuh et al. (2006) and Garetti et al. (2005) for the most popular commercial systems.

new version number, updates the metadata (e.g., who modified it, when and why) and then passes the information to the workflow manager so that other actors who might have to react to this change may be notified.

- **Project and project portfolio manager**, which enable the strategic and operational supervision of development activity.

Modern corporations widely use both PLM and *Enterprise Resource Planning** (ERP) systems. While the former support the product development process, the latter support administrative and production processes. Given the different nature of these processes, it is natural for these two types of systems to follow different philosophies.

ERP systems support repetitive, explicit and finely detailed processes, and are therefore focused on automating transactions. Conversely, PLM systems support processes that are knowledge-intensive, somewhat tacit and relatively unpredictable. They are focused on managing information and tend to support—rather than automate—activities. In the context of PLM systems, the degree with which a firm decides to define processes and activities in a normative way leads to a significant tradeoff between efficiency, predictability and repeatability on the one side, and lack of intellectual freedom and willingness to propose new solutions on the other side. Because of this tradeoff, user involvement in the introduction of PLM systems is probably more critical than for any other kind of business information systems (Neirotti et al. 2012).

11.5 From Traditional Managerial Approaches to Concurrent Engineering

Over the last century or so, firms have developed a number of management techniques for coping with product development and its peculiar features. A first approach, typical of the greater part of the twentieth century, came from *Taylorism**, that was the dominant paradigm in the management of production activities. The product development process was finely split into separate tasks and assigned to specialist engineers. Each design task was performed independently from one another and its completion would lead to an 'over the wall' transfer of information to downstream tasks. In theory, the process did not consider any kind of iterative interaction between design tasks. In practice, iterations would be informally managed by a *chief engineer*, who would engage in a constant dialogue with the designers, and solve any tradeoffs that might arise among the technical solutions they would individually come up with.

Around the 1970s, artifacts became much more complex with respect to the number of components and the heterogeneity of technologies, which made the Tayloristic approach and the 'behind-the-scenes' management of iterations ineffective. Product development therefore became subject to *design reviews**. The project was divided in major phases, at the end of which *milestones* were defined.

At each milestone, a design review would be carried out in order to go over the design choices made in the previous phase and attempt to identify any inconsistency or mistake that might create problems later on. Design reviews would be carried out by managers, customers and peers. The latter could be taken in either as experts in the same technical areas that were involved by the previous development phase, or as representatives from the areas that would be in charge of the downstream phases.

In the 1990s, researchers studying product development in the automotive industry realized that Japanese carmakers were able to perform product development in much shorter timeframes that Western ones, and with a significant reduction in engineering changes after product launch. By studying their managerial approaches, they discovered that this was due to a strong parallelization of design activities that was termed *Concurrent Engineering** (Clark and Fujimoto 1991). In the mainstream interpretation of Concurrent Engineering (Fig. 11.6), design phases are not performed sequentially, with a single large transfer of information on completion of the upstream phase to the downstream one. Instead, activities are performed in parallel, with a continuous and often informal exchange of information.

This parallelization can lead to two main advantages. The first advantage is the decrease in the time required to complete the development process. Although each phase may require more time to be completed, the overlap leads to a shorter overall lead time. The second advantage is related to design quality. In the traditional, sequential approach, mistakes are spotted much later during the process, after substantial time has lapsed and much design work has been carried out. When the mistake is found, it is generally difficult and costly to go back to the original design choice, change it and unwind all subsequent decisions. The overlap makes it possible to find mistakes much earlier, and their correction substantially easier.

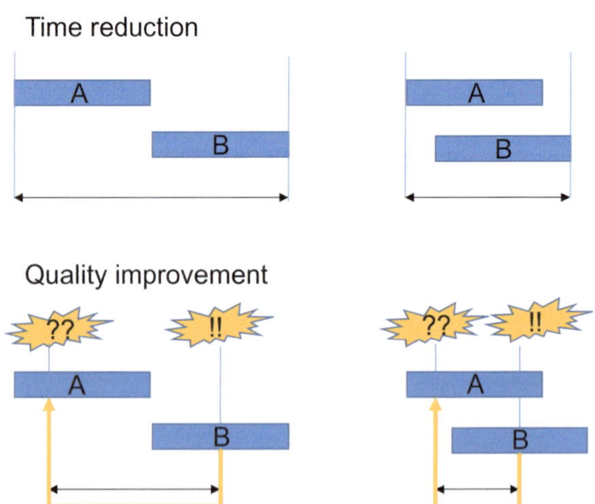

Fig. 11.6 Concurrent engineering and its two advantages

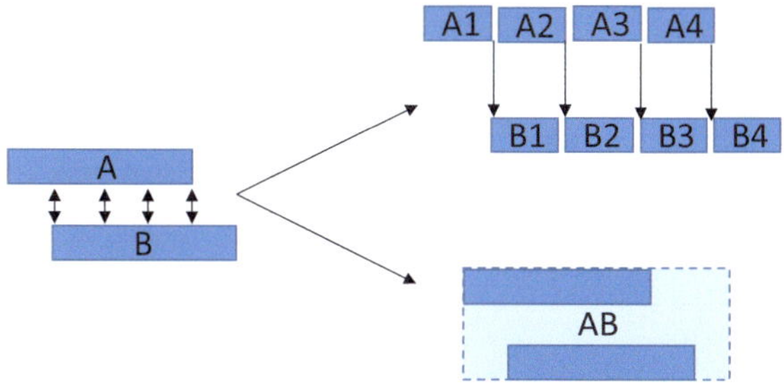

Fig. 11.7 Concurrent engineering viewed from a project management perspective

Alongside these advantages, Concurrent Engineering has been reported to exhibit some drawbacks too. Downstream engineers are forced to operate on imprecise and vague information and must often revise their work when upstream information changes. If the firm does not provide proper incentives, they will be tempted to not perform their activities until the upstream phases are ended and the design is officially *frozen*. This behavior might lead them to provide some words of advice to upstream designers, but not to effectively perform work on the tasks that are assigned to them.

Moreover, managing parallel tasks with a frequent exchange of information is not easy from a project management perspective, in which activities are usually linked by simple end-start constraints. As shown in Fig. 11.7a, one solution might consist in planning a given number of information transfers and fragment activities correspondingly, but this leads to a substantial increase in the number of activities to be managed. Alternatively (Fig. 11.7b), a firm may opt for an informal approach, view the parallelized activities as a single one and leave its management to the inner dynamics of an inter-functional team.[11] This approach works well when the size of the activity and of the associated team is reasonably small, but cannot be applied to larger projects. Specialized techniques for coping with this issue will be introduced in Chap. 12.

Researchers have identified some additional and somewhat less evident pitfalls of Concurrent Engineering (Crawford 1992).

The first one, already cited, is related to the underlying belief that reducing TTM should be a key priority. This may lead development teams to focus on limited technical objectives and incremental innovation. When facing difficult technical problems, time-pressed development teams will be likely to 'cut corners', without even considering the opportunity of making substantial revisions to design choices.

The reliance of Concurrent Engineering on inter-functional teams also makes project success heavily dependent on factors that are quite difficult to control and manage, such as the internal climate, the skills of the project manager and so on. At

[11] These teams are often named with the set of responsibilities that is assigned to them. For instance, a *Design and Build Team* will be tasked with making sure that a subsystem is designed to specifications and its manufacturability is ensured.

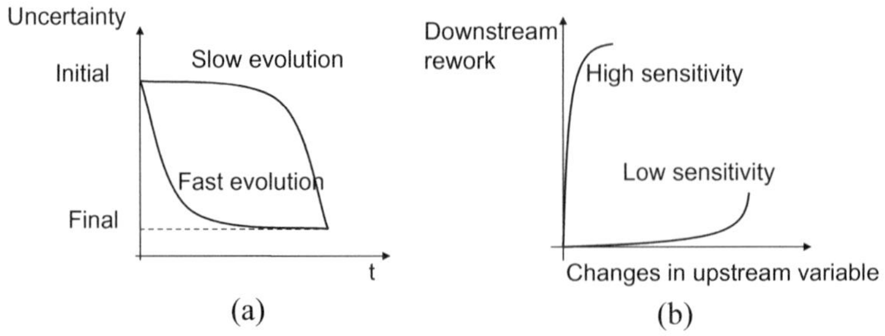

Fig. 11.8 Managing concurrent engineering according to upstream evolution speed and downstream sensitivity

the same time, teamwork requires to spend considerable time exchanging information with colleagues both individually and in meetings. This places a limit on the time available for personal work and, aside from reduced productivity, reinforces the incentive to narrow the technical scope of the project.

When following Concurrent Engineering principles, the amount of task overlap should be defined carefully, since not all activities in a project are suitable. Following Krishnan et al. (1997), the degree with which two phases should be overlapped or kept sequential depends on the evolution speed of the upstream activity, and on the sensitivity of the downstream one.

As shown in Fig. 11.8, an activity has high evolution speed if designers are able to reduce most part of design uncertainty quite early, by making major choices at the beginning of the process, and then spend the rest of their time fine-tuning the solution and performing incremental changes. Conversely, evolution speed will be slow if uncertainty remains high throughout the duration of the task and convergence to the final solution occurs only at the end. It is intuitive to imagine that activities having to do with incremental innovation, proven technologies and well-known markets will have high evolution speed, and vice versa. Sensitivity of a downstream activity to upstream changes may be defined as the relationship between the change in an upstream variable and the corresponding amount of downstream design rework that is required to adapt to this change. Low sensitivity means that limited rework is needed even in the face of significant change, whereas high sensitivity will be present if a lot of rework will be induced even by minor changes. If one combines these two properties, it is apparent that a complete overlap will be possible in the case of fast upstream evolution and low downstream sensitivity, while activities should not be overlapped at all if upstream evolution is slow and downstream sensitivity is high. Intermediate degrees of overlap will then be possible in the remaining cases.

Concurrent Engineering has been introduced as the parallelization of activities that would intuitively be considered to be upstream or downstream of one another, such as the design of the product and the definition of the related manufacturing processes. However, concurrency can also be introduced under another perspective, in which the experimentation of *alternative solutions* is carried out in parallel

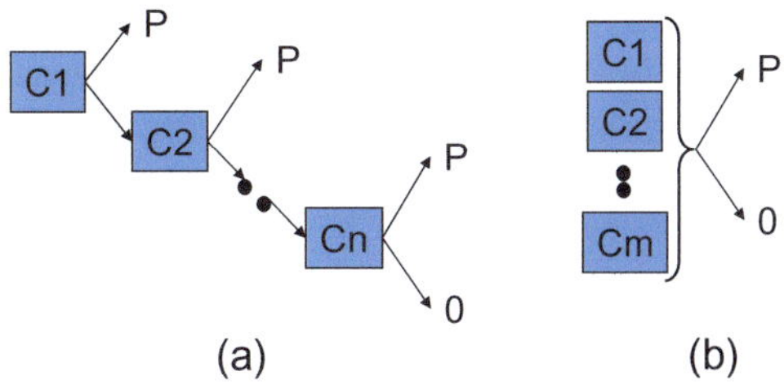

Fig. 11.9 Set-based concurrent engineering

instead of iteratively. This approach is termed *set-based Concurrent Engineering* (Sobek II et al. 1999).

When facing a problem that might be solved by choosing one among a discrete number of technical solutions, companies would generally opt to follow a sequential experimentation, starting from the most promising alternative and moving to the following one in case of failure (Fig. 11.9, left-hand panel). By following this approach, the firm minimizes expected cost, but the expected time required to reach a solution will be high and subject to a high variance. In order to shield the rest of the project from this variance, project managers will assume a high duration of these sequential experiments, and project lead time will therefore become longer. Should the team be lucky and find that the first alternative provides a satisfactory solution, it will generally be impossible to replan the project and anticipate its completion. Moreover, the firm will always remain with doubts with respect to the quality of the adopted solution, since nothing will be known of the alternatives that have not been experimented. This can be highly damaging toward the end of the development process, if the chosen alternative might turn out not to be satisfactory enough. In this case, everyone will start asking "what would have happened if we had chosen *Y* instead of *X*?", though revising this early decision could prove very costly, if not impossible at all.

A firm following the principle of set-based Concurrent Engineering will instead experiment with all alternatives (or a subset of them) in parallel (Fig. 11.9, right-hand side) and choose the most promising at the end. This way, the firm will certainly incur in higher costs, but with no variance. Moreover, it will benefit from a reduced lead time with no variance at all. Finally, having tried all viable alternatives in parallel, the confidence in the one that will be adopted will be much higher.

11.6 From Concurrent Engineering to Approaches Based on Flexibility

Most management strategies for product development processes are aimed at coping with the distinguishing features of such process that were described earlier in this chapter. Most of all, the 'lever effect' that characterizes the early phases of the process has led to developing strategies aimed at ensuring that wrong decisions were not taken, if not early on. Following this perspective, firms have focused on stage and gate systems, feasibility studies and extensive market research. This line of thought appears to be particularly appropriate to a context in which the *marginal cost of a design change is very high* or, in other words, in which *design flexibility* is low (Thomke 1997).

When facing particularly turbulent environments, in which markets and/or technology are rapidly changing, or simply very difficult to predict, it is necessary to adopt a completely different approach to product development. If it is not possible to make sure that wrong decisions are not taken, the alternative approach is to make sure that, whatever happens, reaction is possible and inexpensive. By taking this stance, the key managerial question becomes "*given the tools we have, what is the last responsible moment for taking this decision?*" and to use this delay to capture as much information as possible from the environment.

The extreme case of a such approach is evident in the case of digital services and applications, which are not 'designed and then launched', but 'launched and continuously redesigned'. Such a radically different approach is made possible by the fact that digital technology lends itself very well to redesign cycles, which have been supported by a '*real-time market research*' (Sanchez and Sudharsan 1993). Each cycle is based on capturing data from actual usage (which is cheaper and more precise than running traditional market research), analyzing it through a variety of analytical tools, redesigning new solutions accordingly and experimenting with them, in order to evaluate their impact, before scaling them up.[12]

Then, an important question to be raised is whether this radical shift in product development is confined to the world of 'digital', or whether it can be applied to other industries as well, thanks to the key concept of *flexibility*, as introduced above. While it is fairly obvious that manufactured products will never enjoy the same degree of flexibility as a digital service, recent technology does allow firms to have some degrees of freedom in this respect, and the ability to exploit this opportunity can become a source of competitive advantage.

A first approach with which manufactured products allow flexibility, though this is limited up to the time that the individual product is sold, consists in using appropriate manufacturing processes. *Flexible manufacturing systems**, soft tooling and *additive manufacturing techniques** can be considered to be interesting alternatives to the traditional processes used in mass production.

A second approach consists in using design flexibility as a guiding feature for implementing *platform-based development*. The 'rigid' part of the product will therefore be decoupled from the 'flexible' part, with the latter being assigned to the platform, and the latter to derivative products (e.g., a manufacturer of small kitchen equipment may keep the 'rigid' electromechanical components in the platform and bring annual changes to the shape of the casing in order to match evolving customer needs). While unit costs may be higher, these two types of flexibility grant the firm the freedom to make design changes without incurring significant expenses. The

[12] In digital services, it is customary to perform *A*|*B* testing, in which two groups are subjected to two variants of the same service, and the impact is compared, in order to decide which of the solutions is best.

firm may therefore make last-minute changes to the product and the process before the launch of the new product, by taking into account recent inputs pertaining to technical solutions or coming from the market. Moreover, after launching the new product, the firm may obtain further feedback from the market (which is likely to be more reliable than from prelaunch market research) and redesign the product in order to sell a stream of continuously updated versions of the same.

A third approach that allows to implement flexibility in the case of manufactured products lies in a deep digitization of the product itself. In this case, the product is not conceived as an assembly of predefined parts, each having a given (and rigidly assigned) behavior and function, but as a 'computer with peripherals', whose behavior and functions depend on the software that will be run on the same. The firm will therefore launch a product that is capable of running multiple functions and activities and these will be enabled over time, as technology progresses and/or as customer/user needs arise. A characterizing feature of this approach is that this adaptation can be carried out not only before the product is sold, but also by modifying the products that are already in the field via 'over the air' updates.[13] Thanks to this approach, the firm will also have the strategic option to develop system functions personally, or whether to market the product as an enabling platform and allow the development of an ecosystem of complementors around it.[14]

[13] The obvious example is smartphone development, in which the manufacturer does not think of functions associated to hardware such as 'storing data', 'measuring movements and acceleration', 'localizing the phone', 'recording digital images' in terms of what the user will do with them. The producer will instead consider them as the rigid component of the product, which will enable the second and flexible component, made up of a stream of future applications and services that will use and repackage these features to the benefit of users.

This approach has been spreading in other industries, such as automotive, especially following the example cast by Tesla. Tesla cars are conceived as 'computers with wheels', developed with a significant number of hardware features and peripherals, which are later adapted to emerging conditions and opportunities. Examples of the impact of this approach are abundant and clearly show the radical and potentially disruptive potential of this innovative approach. For instance, over the years, Tesla was able to avoid a number of costly (and reputation damaging) product recalls for safety–critical issues (from increasing road clearance to supervising driver behavior), by simply running 'over the air' fixes that operated on its hardware. Moreover, when Tesla decided to market the Autopilot (i.e., a SAE Level 2 self-driving function), this was not limited to future cars, but included cars that had already been sold, since they already accommodated the required hardware.

[14] It should be evident that, in this latter, case, the purpose of product planning changes dramatically, since the manufacturer will have to reconsider the concept of 'user needs'. In fact, the firm will have to consider only part of end-users' needs, since the rest will be satisfied by complementors (e.g., a smartphone manufacturer will have to consider ergonomics in handling the phone, but not ease of use when ordering pizza, since this 'function' will be covered by the firm that will eventually develop that particular service).

Moreover, the firm will have to consider the needs arising from complementors, to make sure that the product lends itself well to accommodating the future stream of applications and services they will develop (e.g., the smartphone manufacturer will have to decide on the precision of the localization features. In doing so, it will have to consider the needs of service developers, which may include food delivery and taxi hailing, but also someone who might wish to develop an app for taking measurements during home renovations, in the latter case with obviously more stringent requirements).

When operating such deep digitalization of the product, the manufacturer will have to strike the right balance between product cost and flexibility, since a product with built-in functional redundancies will enable a greater amount of flexibility, but will be more expensive to produce and will lengthen its replacement cycle. Conversely, a barebones product will be cheaper and subject to more frequent replacement sales, although it might not be able to accommodate future applications and services, and thus not be attractive for complementors. Such a decision also has an impact on environmental sustainability, which is negatively affected by frequent replacement sales.

11.7 Lean Product Development and Agile Project Management

When uncertainty becomes large, and flexibility allows it, it may become impossible or ineffective to develop a reliable and complete set of product specifications on which to base the development effort. In this case, it is possible to use the concept of *Lean Development** (Ward 2009).

'Lean' marks a significant change with respect to the traditional concept of product development, as shown in Fig. 11.10. In the traditional approach (top panel), the firm is required to make a forecast of the future technological and market scenario, and to 'front load' this information in the product development process. Once the early phases of a stage and gate process have been passed, the product concept will be deemed as viable, the business case will have been approved and the firm will therefore have to develop the new product in an 'open loop' mode, hoping that the baseline on which the final 'go' decision has been made was solid enough. Within the development phase, there will be a number of iterations, but these will not be aimed at reconceiving the product, while—rather—at fine-tuning a myriad of design decisions, in order to come up with a product that best meets the specifications that have been decided upfront.

In the Lean approach (bottom panel), instead, the firm starts with a general vision about the direction to take, and then starts by conceiving, developing and launching a *Minimum Value Product* (MVP) as soon as possible. The MVP can be tested and continuously revised, improved and fine-tuned according to its performance with respect to marketing and financial metrics, in a continuous and deep interaction with different customers' segments.[15] Should this continuous experiment approach failure, the firm should not necessarily abandon the project, but simply make a *pivot*. Pivoting means making substantial and irreversible changes to the product and/or the business model in the attempt to gain the desired results (Pillai et al. 2019). This iterative product and customer development process should

[15] Some authors (Blank and Dorf 2012) talk about '*customer development*' running in parallel to 'product development'.

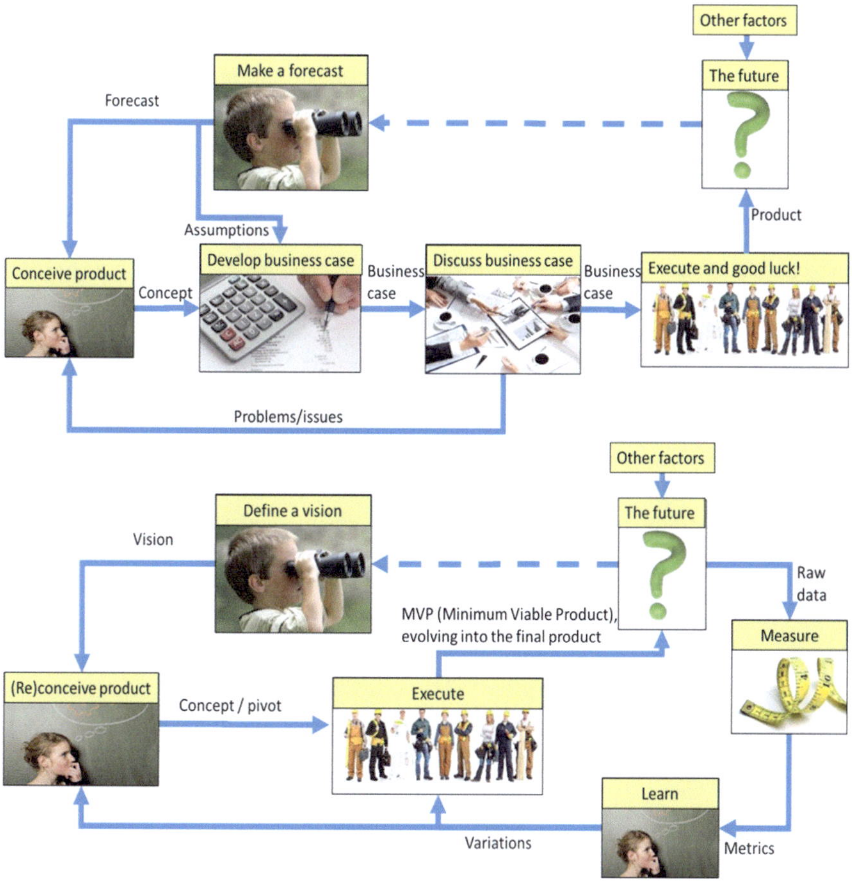

Fig. 11.10 Lean development versus traditional approaches

be carried out in the most objective way possible, which implies the need to define a sound and accepted set of metrics able to effectively highlight what works and what does not work in the product, and guide the firm accordingly in its redesign.

The first point of contact between the firm and the market is therefore the MVP, which can be conceived in different ways. The main choice lies between MVPs that simply represent the future product and functional MVPs. In the former case, the MVP will not be a product, but a faithful representation of what consumers will see during the purchasing process (e.g., mock sales processes, commercial videos, brochures, data sheets, storyboards and customer journeys representing the usage process, landing pages on the web or social media, product packaging). In the latter case, the MVP will be a solution that is usable, since the firm will have implemented a limited set of functionalities, which may or not be based on the final technological solutions.[16] The choice between the two types of MVPs depends on whether the firm wants to reduce the uncertainty associated to the sales process or to the degree with which customers/users will appreciate the product after sales.

It goes without saying that Lean Product Development *is not applicable to all products*. First, it requires significant flexibility. Second, it is clearly aimed at reducing market uncertainty, either in the sales process or in the actual evaluation of the value proposition being conceived. Third, it rests on the assumption that technological uncertainty is minimal or, at the least manageable (in other words, one can safely say "provided customer/users want it, we can deliver it"). These three conditions are usually met in the case of digital technology and services. Conversely, they are not always met in the cases of low flexibility (since each iteration would be very costly) and high technological uncertainty, combined with low market uncertainty.[17]

This discussion seems to draw a line between digital services (to be developed with Lean) and manufactured products (to be developed traditionally). However, the distinction might be quite blurred. For instance, digital services may also have significant technological uncertainty, either because the underlying IT technology is not well developed yet, or because they might rely on mechanisms that are not proven.[18] Secondly, manufactured products may be characterized by the joint presence of rigid components that do not allow lean development,[19] along with digital content or service components that are flexible, and therefore do.[20]

[16] The typical example is represented by 'wizard of OZ' MVPs, in which users interact with IT systems that rely on a back end that is staffed by humans and not yet automated. For instance, a firm wishing to propose investment advice via a chatbot will have to face the uncertainty that users will trust the system. So, before developing the system, it could launch an MVP in which the messaging system is run by human operators, though without telling its first users.

[17] Imagine the paradoxical case of a pharmaceutical company launching an MVP for a cancer treatment under the claim "we do not know if it really works but for the time being, we are just testing whether doctors and patients would be happy to buy it". Similarly, an aircraft manufacturer would never launch a new supersonic jet by saying "technically it might not work, but we are testing the market to see whether people would pay X in order to fly from Paris to New York in 4 hours. And, if we realize that the engines are not feasible, we will try to cope somehow".

[18] An interesting example is the introduction of Universal Basic Income in Italy in 2018. This was conceived since the beginning as a digital service (storyboards and customer journeys included), with the dual objective of providing support to poor people and to act as an active labor market policy to find them a job. The measure failed in this second objective probably because, despite a careful design of the architecture of the digital service, the underlying legal and economic foundations left something to be desired.

[19] Components of manufactured goods can be rigid because of two reasons. First, the manufacturing technology may such that any design change entails excessive costs. For instance, with traditional injection molding technology (but less so in the case of soft tooling), changing the shape of a plastic casing implies having to change expensive injection molding dies. Secondly, once a manufactured product is sold, it is clearly difficult to enact design changes that would require the substitution of physical components (less so if the product architecture has been conceived in such a way as to allow it).

[20] Going back to the previous example, the same pharmaceutical firm might be developing an IoT system to dispense doses of the treatment and ensure therapeutic compliance. Similarly, the same aerospace firm might be developing a new onboard infotainment system for its aircraft. In both

As a final comment, and from a theoretical perspective, it should be noted that Lean Product Development is probably not to be considered a groundbreaking concept. Already in 1994, Rosenberg wrote of 'economic experiments' that firms are forced to carry out whenever the solutions and interdependencies between technology and market needs could not be derived from 'first principles' or market research. The main change therefore lies in the fact that this approach has been become a major and quite diffused organizational innovation, since it is not any more limited to a minority of cases, and implemented rather unwillingly, but can nowadays, and mostly thanks to digitalization, be pursued strategically and in a growing number of cases.

11.8 Agile Product Development

Lean product development, discussed in the previous section, matches quite closely with *Agile Product Development**, to the point that the two are often conflated. The main difference lies in the fact that the former is a strategic approach to developing products and services, while the latter is an operational approach that is focused on the management of these processes.

'Agile' has its roots in the Software Engineering industry which, over the years, has realized that traditional approaches to Software Development were seriously lacking when trying to develop complex IT services, aimed either at the market or at supporting complex organizations. In fact, the sequential ('waterfall') and bureaucratic ('specify and then code') approach to development hindered the possibility of building systems progressively and iteratively, and of interacting with customers/users in order to discover their needs. In face of these challenges, Software Engineering has progressively evolved to the concept of *Agile Software Development**. Many principles of this approach are easily adaptable to completely different technological domains and industries, and this has led to the *Agile Product Development* approach. Among the techniques used by this philosophy, one can list the following:

- **Rolling-wave planning**, in which scheduling is not computed on the basis of a detailed breakdown of all foreseen activities, from the start of the project to its completion. Instead, a detailed schedule is developed for the immediate future (e.g., 4 weeks) only, and a rough plan is drafted for the remainder. The schedule is updated at regular intervals (e.g., every 2 weeks).
- **Loose-tight planning**, in which activities are classified on the basis of their uncertainty and creative content. Tight planning is performed for activities whose duration is relatively certain, while a looser planning allowing for delays is made for the other ones.

cases, technology uncertainty will be quite limited, while the main source of uncertainty will have to do with customer/user acceptance. Therefore, a Lean approach to these components would be certainly advisable.

- **Time-boxing**. The usual practice in project management consists in structuring projects based on an ex-ante detailed definition of activities and on the ensuing identification of key milestones, each of which is defined around a set of expected results. During project execution, project managers must ensure these results are achieved and tend to manage the timing at which this happens with some flexibility, occasionally accepting delays. Time-boxing follows the opposite approach, by breaking down the project into regular intervals, often termed *sprints* (e.g., every 1 or 2 weeks) and then identifying the expected results at the end of each. No delay is allowed during execution and project teams are required to report whatever results have been achieved. This method grants a smoother and more regular management style, while the flexibility with which results can be reported becomes an antidote to the common tendency of 'over engineering' technical solutions.
- **Evolving specifications**. Agile Development usually shuns away from adopting a clear separation between the product specification phase and the ensuing project execution. Instead, specifications are allowed to evolve over time, with a high number of iterations occurring at the release of new solutions, each of which allows the team to gain a deeper understanding of the problem and to fine-tune the specifications by performing experiments, such as A|B tests.
- From an organizational perspective, Agile Development is based on a relatively **informal management style** based on **strongly empowered** and **self-organizing teams**. One main approach of this type is *SCRUM**, which includes a set of techniques and rituals aimed at ensuring consistency in the methodological approach and a common culture in the development team. Among the main elements that characterize these approaches are meetings, monitoring of backlog, and visual aids to task scheduling.

  Meetings are organized both at the beginning of each sprint and at the beginning of each day and are carried out according to a predefined playbook aimed at being quick (e.g., morning meetings aim at defining the agenda of the day and are run by standing up) and at ensuring commitment by the team.

  Project progress is constantly monitored and communicated within the team by using large notice boards and is measured not by focusing on the project (i.e., 'activities performed' in a Gantt chart), but on the product (i.e., 'amount of backlog being ended' over the time axis). Depending on the case, metrics for backlog can include the number of components still to be designed, the number of open issues to be solved, etc.

  Finally, activity scheduling is usually performed visually, by using *Kanban Boards,* in which items from the backlog are placed as post-it notes on a 'to do' area. Team members that are free will go to the board, take charge of a new backlog item and place the note in a 'in progress' area. When ended, they will move the note to the 'done' area and pick up a new item to deal with.

Just like Lean Development, Agile Development applied to manufactured products raises a number of issues and doubts, since such products may not lend themselves perfectly well to this approach. Aside from organizational inertia due to cultural issues and established ways of working, the 'constraint of physicality' (Schmidt et al. 2017) can indeed discourage from undertaking an approach that is based on iterations and acceptance of ambiguity and uncertainty.

Research is still relatively inconclusive in suggesting best practices to be followed in this area. However, the two main solutions that firms may adopt consist in segregating the 'inflexible' from the 'flexible' components of their products, and, respectively, using traditional and agile approaches for each. Alternatively, they might use the so-called *Agile Stage and Gate approach* (Cooper and Sommer 2016) by maintaining a traditional and sequential approach for the overall product development process, but using agile methods within each phase, thus speeding up and reducing the complexity with which design activities are carried out.

References

Altavilla S, Montagna F, Cantamessa M (2018) A multilayer taxonomy of cost estimation techniques, looking at the whole product lifecycle. J Manuf Sci Eng 140(3):030801

Blank S, Dorf B (2012) The startup owner's manual: the step-by-step guide for building a great company. K&S Ranch Inc. Publisher, Pescadero

Brown SL, Eisenhardt KM (1998) Competing on the edge: strategy as structured chaos. Harvard Business School Press, Boston

Clark KB, Fujimoto T (1991) Product development performance: strategy, organization and management in the world auto industry. Harvard Business School Press, Cambridge

Cohen MA, Eliashberg J, Ho T-H (1996) New product development: the performance and time-to-market tradeoff. Manage Sci 42(2):173–186

Cooper RG, Sommer AF (2016) The agile–stage-gate hybrid model: a promising new approach and a new research opportunity. J Prod Innov Manag 33(5):513–526

Crawford CM (1992) The hidden costs of accelerated product development. J Prod Innov Manag 9:188–199

Garetti M, Terzi S, Bertacci N, Brianza M (2005) Organizational change and knowledge management in PLM implementation. Int J Prod Lifecycle Manag 1(1):43–50

Grieves M (2006) Product lifecycle management: driving the next generation of lean thinking. McGraw-Hill, New York

Gupta R (1996) Survey on use of virtual environments in design and manufacturing. In: ASME Design engineering technical conferences (IDETC-CIE), CA, USA, 18–22 Aug

House CH, Price RL (1991) The return map: tracking product teams. Harv Bus Rev 69(1):92–101

Krishnan V, Epiinger SD, Whitney ED (1997) A model-based framework to overlap product development activities. Manag Sci 43(4):437–451

Mihm J (2010) Incentives in new product development projects and the role of target costing. Manag Sci 56(8):1324–1344

Montagna F, Cantamessa M (2019) Unpacking the innovation toolbox for design research and practice. Des Sci 5:e8

Neirotti P, Cantamessa M, Montagna F (2012) Understanding the organizational impact of PLM systems: evidence from an aerospace company. Int J Oper Prod Manag 32(2):191–215

Pillai AS, Goldfarb B, Kirsch DA (2019) The origins of firm strategy: learning by economic experimentation and strategic pivots in the early automobile industry. Strateg Manag J 41 (3):369–399

Rosenberg N (1994) Exploring the black box: technology, economics, and history. Cambridge University Press, Cambridge

Saaksvuori A, Immonen A (2002) Product lifecycle management. Springer, New York

Sanchez R, Sudharsan D (1993) Real-time market research. Mark Intell Plan 11(7):29–38
Schon DA (1983) Reflective practitioner—how professionals think in action. Basic Books Inc., New York
Schon DA (1995) Knowing in action: the new scholarship requires a new epistemology. Change 27(6):27–34
Schmidt TS, Chahin A, Kößler J, Paetzold K (2017) Agile development and the constraints of physicality: a network theory-based cause-and-effect analysis. In: International conference on engineering design, Vancouver, Canada, 21–25 Aug
Schuh G, Assmus D, Zancul E (2006) Product structuring—the core discipline of product lifecycle management. In: 13th CIRP International conference on life cycle engineering (LCE2006), Leuven, Belgium, 31 May–2 June
Smith PG, Reinerstein DG (1991) Developing products in half the time. Wiley, New York
Sobek DK II, Ward AC, Liker JK (1999) Toyota's principles of set-based concurrent engineering. Sloan Manag Rev 40(2):67–83
Souza GC, Bayus BL, Wagner H (2004) New-product strategy and industry clockspeed. Manag Sci 50(4):537–549
Stark J (2005) Product lifecycle management: 21st century paradigm for product realization. Springer, London
Thomke SH (1997) The role of flexibility in the development of new products: an empirical study. Res Policy 26:105–119
Thomke S, Fujimoto T (2000) The effect of "Front-Loading" problem-solving on product development performance. J Prod Innov Manag 17(2):128–142
Ward AC (2009) Lean product and process development. The lean enterprise institute, Cambridge

Project Management for Product Development

12

Abstract

This chapter discusses the application of project management methods in the peculiar context of product development processes. The chapter presumes a basic knowledge of project management, and focuses on the derivation of Work Breakdown Structures from the development process, on project scheduling in both cases of infinite and finite (and single and multiple) resources, and on the management of project iterations.

After having discussed the process with which product or service development is performed, it is immediate to start thinking about ways for managing each individual development project. A firm will likely have defined a standard process that provides a general template. However, each initiative will be characterized by some degree of uniqueness, which suggests the use of tools and methods from the field of *Project Management**.

This chapter cannot attempt to cover such a broad field as project management, which is a discipline with distinct and well-known sources and textbooks. Readers are advised to refer to these, in order to gain at least the basic competence that will enable them to apply methods and tools proficiently in the domain of product development. Therefore, a basic understanding of Project Management is here assumed (specifically, the hierarchical definition of activities in a *Work Breakdown Structure** and the generation of a project schedule, based on network techniques, such as the *Critical Path Method**).

This chapter therefore has the objective of highlighting some peculiar features, typical of the field of product development, which can affect the way Project Management methods can be applied. It is in fact intuitive that development projects are characterized by a degree of uncertainty that is not usually found in other domains, such as construction or event planning, and therefore require a specific approach.

M. Cantamessa and F. Montagna, *Management of Innovation and Product Development*, https://doi.org/10.1007/978-1-4471-7531-5_12

The perspective adopted in this chapter is somewhat traditional and aims at providing a few concepts that can be added to a basic knowledge of project management methods, and that are specific to product development. Our perspective is therefore fundamentally stage-based, i.e., based on the project as a sequence of activities, and aimed at scheduling. Readers should be aware that other contributions to literature, especially from the field of engineering design, follow different perspectives, in which the project is modeled by looking at its activities, at the solution to be achieved, at the problems to be solved, etc. For a review of such approaches, one can refer to Wynn (2007).

12.1 Activities and Resources in Product Development Projects

As it has been outlined in the previous chapter and elsewhere in the text, the product development process is characterized by the central role of information. Especially in contemporary practice, the role of virtual (instead of physical) system models has grown to the extent that product development activities can mostly be considered to be information-producing and information-consuming tasks. It comes as a consequence that the translation of the development process in project management terms is quite straightforward. The development process will likely be represented by a workflow, with activities symbolized by boxes and information flow by arrows connecting the same boxes. *Project activities* will therefore correspond to the same activities outlined in the process. *Project events* will correspond to the ending of an activity and—therefore—to the production of a package of information. Finally, precedence relationships between project activities will be associated to information flows (or, better still, transfers of information). In short, if activity *A* provides a package of information *X* to activity *B*, there will be a precedence relationship that binds *B* not to start before *A* is ended and *X* is produced (Fig. 12.1).

The transfer of information from activity *A* to activity *B* is a critical interface when managing the product development project. In principle, when activity *B* takes ownership of *X* and starts, it sanctions the termination of activity *A* and it provides an implicit acceptance of the validity of information *X* and of the upstream

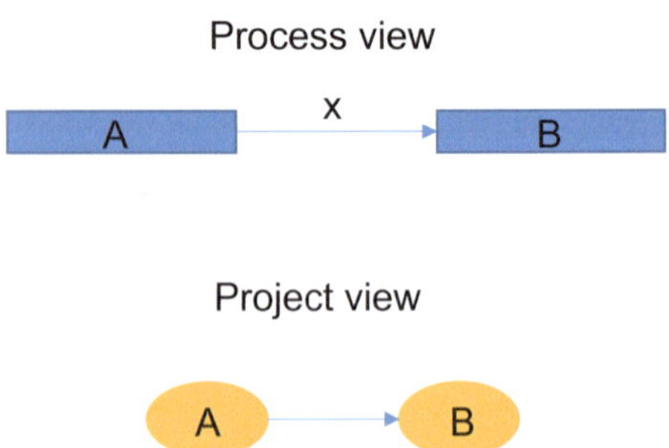

Fig. 12.1 Translating the product development process in a project management structure

activities that led to it. We know that, in product development, this is never completely true and that further iterations are quite likely to occur. Nonetheless, the transfer of ownership of *X* from *A* to *B* implies that both parties do not see any evident problem in this information package, and that activity *B* accepts to take it as an input. This suggests that some degree of formality ought to be introduced at this stage, which can consist in the signing of a document, the minutes of a joint review meeting, and so on.

Moreover, as discussed in the previous chapter, product development is characterized by the possibility of *iterations*, which can force the repetition of an activity, or of a group of activities. This possibility can be modeled and planned; probably, this is the most peculiar aspect of applying Project Management in product development.

Then, when managing the project, the *plan*, the *budget* and the *project organization* must be defined (Fig. 12.2). These are required to estimate the effort, duration and cost required by activities.

In the field of product development, the estimation of project effort, duration and cost is not always an easy task, since many of these activities can be highly specialized and only the managers in charge of each are sufficiently knowledgeable to provide such information. This can make a top-down budget allocation process difficult. Conversely, in having to accept a bottom-up approach, companies bear the risk that managers may behave opportunistically and ask for more time, effort and budget than they really need.

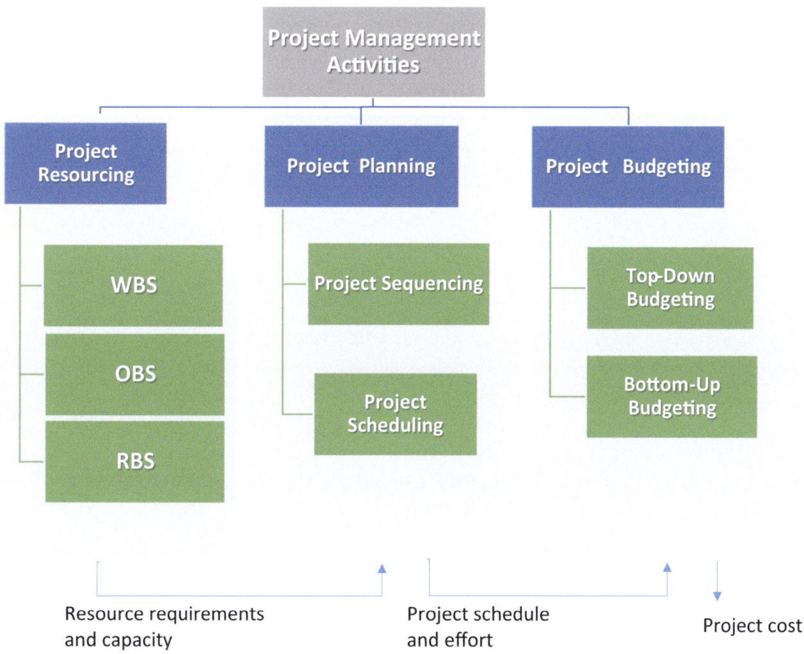

Fig. 12.2 Activities required for planning a project

Leaving opportunistic behavior aside, effort, duration and cost of product development activities can seldom be defined deterministically, since the very nature of the process implies a high degree of uncertainty. This uncertainty can be managed via a number of approaches; overlooking traditional probabilistic approaches, such as PERT, which are nowadays seldom used, Monte Carlo simulation appears to be an interesting alternative (Kroese et al. 2011; Kwak and Ingall 2007; Drexl 1991).

Monte Carlo simulation of projects allows to represent the duration of each activity as a stochastic variable, and to represent also the outcome of the activity in probabilistic terms (i.e., one can assume the probability with which, at the end of an activity, the project will proceed with another activity, or with the repetition of the same activities). In the context of engineering design projects, a further possible approach is *Signposting* (Clarkson and Hamilton 2000), in which activity duration is modeled as a stochastic variable whose uncertainty is tied to the complexity of the project (i.e., number of components, number of relationships among components that could force the propagation of design changes within the project).

A number of project management software tools can support this approach, though their diffusion is still somewhat limited. Therefore, it is common practice to make 'safe' estimates, in the knowledge that the buffer capacity that is consequently built in the project will at some future time be used by some disruption that has not been properly identified at the time of initial planning.

Finally, when dealing with effort, duration and cost of activities, firms should be wary of assuming simplistic relationships between them. In principle, it is possible to start from the assumption that effort (usually measured in units such as person-days or person-months) is given by the product of *Full-Time-Equivalent** resources being used on average throughout the activity, times the duration of the activity itself, or

$$\text{Effort} = \text{FTE} * \text{Duration} \tag{12.1}$$

However, the inverse relationships

$$\text{FTE} = \frac{\text{Effort}}{\text{Duration}} \tag{12.2}$$

$$\text{Duration} = \frac{\text{Effort}}{\text{FTE}} \tag{12.3}$$

(12.2) and (12.3) are not necessarily true, since the resources employed will likely exhibit decreasing marginal productivity.[1]

[1] For instance, if John is able to complete a job in 6 days, this does not imply that John and Anne will be able to complete it in 3 days. This is because John, who was the first choice, is likely to be more productive than Anne. Furthermore, when both people work on the same task, further work may be required to coordinate them. This may be accomplished by having the two actors dedicate some time to this task, or require the contribution of a (costly) third party, i.e., a manager.

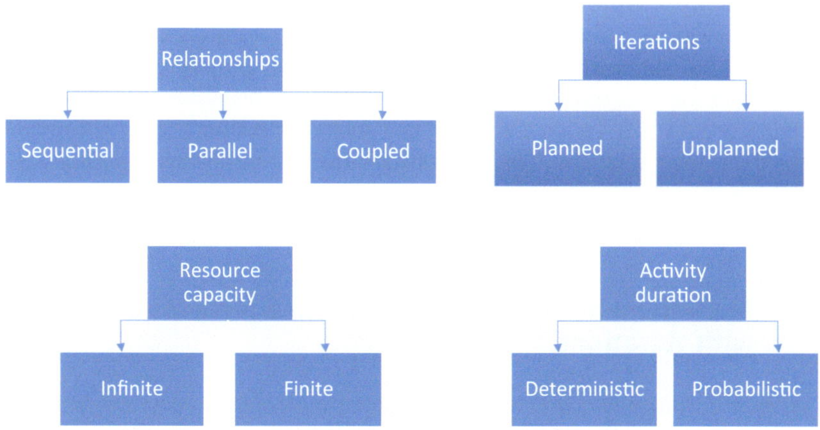

Fig. 12.3 Main descriptors of project planning models

This phenomenon occurs in any economic activity, but can be considered to be especially relevant if specific skills are required, as in the case of product development. Similarly, it might not be correct to state that the cost of an activity is given by the product of effort and unit cost, since the latter might depend on the actual resources being employed (e.g., two employees may have different salaries, or the firm may decide to use external consultants at an extra cost).

In some instances, uncertainty may also lead a firm to formulate the project plan without determining a single triplet of values for duration, effort and cost, per each component activity. Instead, it may estimate two such triplets, one for the normal execution of the activity and another one for an accelerated (or *crashed*) execution, with $T_{\text{normal}} > T_{\text{crashed}}$ and $C_{\text{normal}} < C_{\text{crashed}}$.

In general, resources engaged in projects can be grouped into three main categories. The first is made up by employees that are allocated to the project '*according to use*'. In this case, their cost is associated with the actual time they spend on the project and—if they are not required to work on the project in a given interval—the project does not have to bear the related cost. This reasoning assumes that the firm has other ongoing projects which are able to saturate the time of these resources. If one refers to the organizational forms discussed in Chap. 10, this is common in the case of 'functional' and 'lightweight project team' arrangements. The second category includes employees that are ex-ante assigned to the project for a given proportion of their available time and for a fixed time frame. In this case, which is typical of heavyweight team organizational forms, their cost is charged to the project based on their a-priori assignment, and independently of the actual day-to-day usage. Finally, a third category considers *external resources*, such as consultants. These add flexibility to the pool of internal resources and are generally supposed to be charged on an ad-hoc basis, like the employees belonging to the first category, but at higher hourly rates.

Based on these types of resources and on their amount, project planning can be operated considering 'finite' or 'infinite' resources.

Figure 12.3 summarizes the main descriptors of project planning models: deterministic versus probabilistic, finite versus infinite resources, based on simple precedence networks versus networks that include iterations, and with planned versus unplanned iterations. The last two features will be specifically discussed in the next section.

12.2 Management of Activity Precedence Networks with Iterations

In product development, it is quite possible for the completion of one activity to lead to a new execution of a preceding activity that had already been terminated. These iterations can be 'tight', if the two activities are located close to one another in the project plan, or 'wide', if the two activities belong to phases that are distant from one another.

When planning the development process, potential iterations can be foreseen with some ease by thinking in terms of information flows. Referring to the example in Fig. 12.4, activities *A*, *B* and *C* are *sequential*, since information supplied by *A* is needed to start *B*, and information produced by *B* enables *C*; activities *D* and *E* are *parallel*, since they both use information supplied by *C*; activities *G* and *H* are said to be *coupled*, since both need to use information supplied by the other, and this leads to a clear possibility of iterations occurring between the two. If one looks carefully, coupling is present among the three activities *G*, *H* and *I*, with a possibly complicated pattern of iterations between the three.

If one tries to translate these information flows into precedence constraints, standard project management techniques cannot be used, since coupled activities lead to '*circuits*' in the activity network. These circuits make it impossible to define the starting time of an activity depending on the ending time of the precedent activities, since the latter might depend on the termination of the former. Therefore, if ordinary project management methods are to be used in the field of product development, it is necessary to learn how to identify and manage circuits.

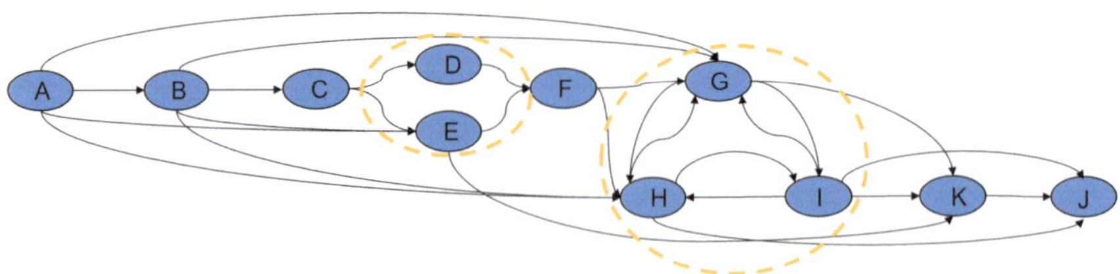

Fig. 12.4 Sequential, parallel and coupled activities in a project

12.2.1 Identifying Circuits and Iterations

The identification of circuits can be performed through direct visual examination of the activity network, if the project is small. When the project is large, formal techniques should instead be used. One possible approach is to represent the activity network in the form of a Boolean *adjacency matrix**, which is called Design Structure Matrix (or DSM, Browning 2001; Eppinger and Browning 2012). In a DSM (Fig. 12.5), activities are listed in rows and columns, and an arc (i.e., an information flow) between activity *A* and *B* is represented by a nonzero value in the cell at the intersection of column *A* and row *B*. In a DSM, one can identify sequential activities by observing strings of '1's that are parallel to the main diagonal and located beneath it. Parallel activities will instead lead to vertical strings of '1's, located under the main diagonal. Finally, coupled activities can be identified by observing nonzero cells above the main diagonal. The position of nonzero values depends on the naming and ordering of activities in the DSM. Therefore, the identification of coupled activities and circuits as nonzero values above the main diagonal can be performed only after having attempted to exchange rows and columns in order to eliminate them. So, referring to Fig. 12.5, it is clear that activities *G*, *H* and *I* are coupled, since any exchange of the corresponding rows and columns would move a nonzero value below the diagonal, but bring a new one above. Conversely, activities *J*, *K* are not coupled, since a simple reversal of rows and columns would bring the corresponding nonzero value below the diagonal.

Fig. 12.5 DSM representation of the project in Fig. 12.4

	A	B	C	D	E	F	G	H	I	J	K
A	A										
B	1	B									
C	1	1	C								
D			1	D							
E	1	1	1		E						
F			1	1	1	F					
G	1	1				1	G	1	1		
H	1	1				1	1	H	1		
I							1	1	I		
J								1	1	J	1
K					1		1		1		K

When using DSMs, circuits can be identified by observing above-diagonal nonzero values, after *partitioning algorithms* (Gebala and Eppinger 1991) have attempted to minimize their number. When these nonzero values are close to the main diagonal, the activities are closely coupled and relatively easy to manage through approaches inspired by concurrent engineering, as it will be discussed in the following. When they are located far away from the main diagonal, this implies the possible occurrence of a large-scale iteration that could evidently be quite disruptive for the project. In this case, project managers ought to anticipate and prevent the iteration from happening, for instance by ensuring an early involvement of representatives from the 'downstream' activity during the execution of the 'upstream' one. Alternatively, they may organize review meetings to make sure that the 'upstream' activity is not terminated until the 'downstream representatives' have checked the results, looking at the implications on their future activities, and provide an approval.

DSMs are very powerful tools for the management of product development processes. Their use is not limited to representing activities and their mutual relationships, but can be extended to other two coordinated matrices. As shown by Eppinger and Browning (2012), it is possible to compile an adjacency matrix for the resources involved in product development, with each matrix value denoting information flow between resources. In this case, the rearrangement of rows and columns allows managers to group resources and form teams based on expected information exchanges. Similarly, and as it will be shown in Chap. 16, a DSM can represent the components or subsystems of the product. Matrix values will represent the existence of inter-component interactions, and the rearrangement of row and columns will allow the definition of the 'chunks' that make up product architecture.

As an alternative approach to circuit identification, one can also consider the 'reachability' information that is inherent in an adjacency matrix. To explain this concept, let us consider a simpler matrix A, as in Fig. 12.6. At first, the matrix can be simplified by iteratively pruning it of activities that do not have either any predecessor or any successor, which of course would be a necessary condition for being part of a circuit. Then, the simplified matrix A' (in Fig. 12.7) can be progressively elevated to powers A'^2, A'^3, …, A'^n to identify circuits of the corresponding order. In fact, if matrix A' contains information on the 1-step paths linking any couple of activities, matrix A'^2 will provide information on the 2-step paths, and so forth. Therefore, if matrix A'^n exhibits a nonzero value on the main diagonal, this indicates that there is an n-step path leading from the corresponding activity to itself or, in other terms, a circuit. Matrices A'^2 and A'^3 in Fig. 12.7 have nonzero values on their main diagonals, which show the existence of 2-step and 3-step circuits involving B, C and D.

This method tells us about the existence of circuits of different length that involve each activity, but does not help in identifying the activities that make up each circuit. This can be done through visual identification.

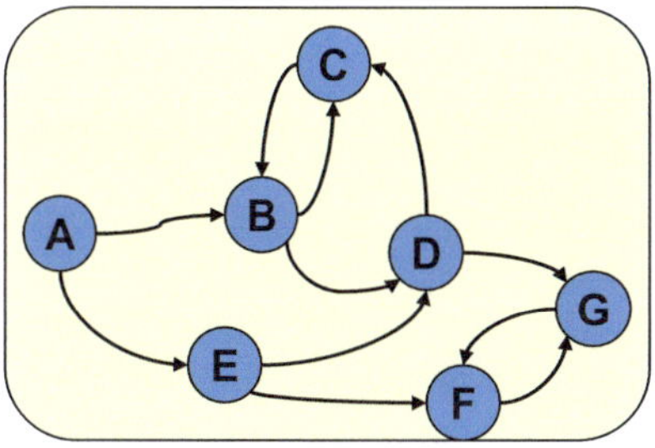

(a)

	A	B	C	D	E	F	G
A	0	0	0	0	0	0	0
B	1	0	1	0	0	0	0
C	0	1	0	1	0	0	0
D	0	1	0	0	1	0	0
E	1	0	0	0	0	0	0
F	0	0	0	0	1	0	1
G	0	0	0	1	0	1	0

(b)

Fig. 12.6 A simple graph (**a**) and DSM matrix (**b**) with circuits to be identified

A=

	A	B	C	D	E	F	G
A	0	0	0	0	0	0	0
B	1	0	1	0	0	0	0
C	0	1	0	1	0	0	0
D	0	1	0	0	1	0	0
E	1	0	0	0	0	0	0
F	0	0	0	0	1	0	1
G	0	0	0	1	0	1	0

$A'=$

	B	C	D	F	G
B	0	1	0	0	0
C	1	0	1	0	0
D	1	0	0	0	0
F	0	0	0	0	1
G	0	0	1	1	0

$A'^2=$

	B	C	D	F	G
B	1	0	1	0	0
C	1	1	0	0	0
D	0	1	0	0	0
F	0	0	1	1	0
G	1	0	0	0	1

$A'^3=$

	B	C	D	F	G
B	1	1	0	0	0
C	1	1	1	0	0
D	1	0	1	0	0
F	1	0	0	0	0
G	0	1	1	1	0

Fig. 12.7 Using the algebraic 'reachability' method to identify circuits

12.2.2 Managing Circuits and Iterations

When circuits have been identified, either visually or algorithmically, managers must cope with them. The case of wide circuits has already been mentioned, and managers will have to find a way to avoid iterations from occurring altogether, since these could severely disrupt project execution. Close circuits can be dealt with by following two approaches that enable the use of ordinary project management techniques.

The former approach is to recognize that close circuits call for a parallel execution of activities, with numerous and informal exchanges of information between the resources that are in charge of the same activities and—eventually—to the formation of single team that will be given the joint responsibility over the entire set of activities.

This 'concurrent engineering' approach suggests that, from a project management perspective, activities in a circuit simply be collapsed into a 'macro-activity', without attempting to go deeper into the details of how it will be managed. When aggregating activities in a circuit, the resulting macro-activity will be likely to have a duration that is longer than the duration of the longest component activity, since interactions and iterations will probably require time. Management may therefore adopt rules of thumb in order to estimate this duration, such as "*we expect twice the time required by the longest component activity*". Concerning the effort required it is safe to hypothesize that, due to the parallelization of activities, the new macro-activity will require the joint presence of all resources originally planned for the component activities and for its entire duration (Fig. 12.8).

Collapsing a circuit into a single macro-activity is quite simple, but cannot always be done. Especially when dealing with activities that require a long time or involve many resources, management is likely to be wary of allowing a significant portion of the project to be self-managed by a large team, without any understanding of what occurs within. In this case, or if the firm wants to keep a tight control over the evolution of the project, another approach is required. This being the case, it is possible to 'break the circuits' and then plan for a given number of iterations to occur. In other words (Fig. 12.9), if activities *B*, *C* and *D* are coupled, the firm may at first decide to start *B* before *D* and *D* before *C* (thus allowing B to start without having information coming from *C*). The second step will consist in planning one or more corrective iterations, in order to adapt the results of *D* and *B* to the new information being generated by *C*. Looking at the project network, this means cutting (or 'suspending') the arc from *C* to *B* in order to allow the sequence $B \rightarrow D \rightarrow C$. Then, this arc must be reapplied, in order to represent the precedence relationships between the first and second iterations.

When following this approach, management will have to make educated guesses on the number of iterations that will be needed, and on the duration of the same activities in the iterations (e.g., "we plan for two iterations to occur, with activity

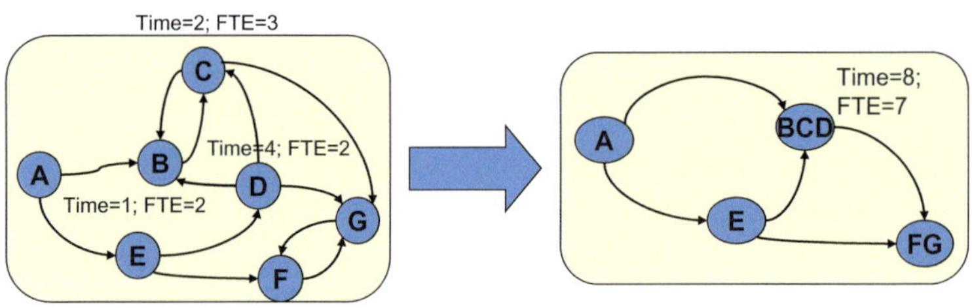

Fig. 12.8 Managing circuits with macro-activities

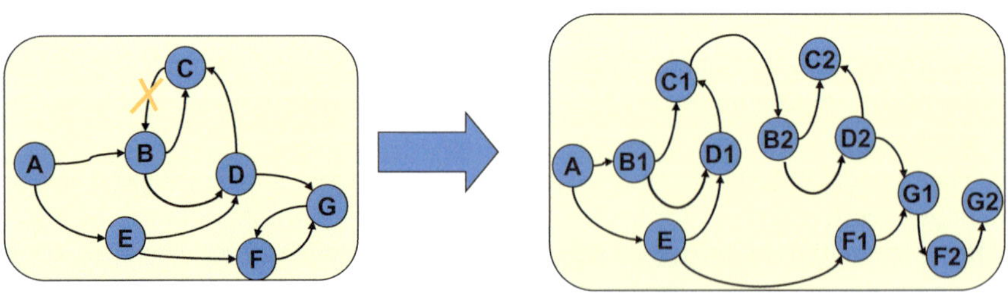

Fig. 12.9 Managing circuits by breaking them, and then reiterating activities

duration being cut by 50% during the second one"). The main decision to be made is then related to the sequence of activities within each iteration, and which arcs should therefore be cut. In principle, one should cut the arcs that minimize the disruption to the activities that—because of that same cut—are forced to start without the benefit of having all the information they would require. One should therefore be able to measure the amount and significance of information in each arc and cut the ones that score less on this measure. In some cases, managers may be able to perform an evaluation of this kind in quite a straightforward way, given their knowledge of the project. Should this not be the case, one can use a heuristic and assume that arcs carry the same information content. Activities can then be ranked depending on the net amount of information they produce. A simple proxy for this measure is to compute the difference between the number of outgoing and incoming arcs connecting each activity to the other activities belonging to the same circuit (i.e., not counting the arcs that connect outside the circuit). The activity with the largest such value will start first and—in order to allow this—its incoming arcs will be cut. Going back to Fig. 12.7, the scores for the three activities B, C and D would have been $B = 2 - 1 = 1$; $C = 1 - 2 = -1$ and $D = 1 - 1 = 0$. So, activity B can start first, and this requires cutting arc $C \to B$.

12.3 Project Scheduling

12.3.1 Scheduling with Infinite Resources

Once the project has been analyzed and circuits solved, activities can be scheduled. In the simplest case, scheduling can be performed under the hypothesis of infinite resources, i.e., assuming that resources can be pulled in to execute the project without any constraint. The *Critical Path Method**, or CPM is the most common approach. Just as a reminder, CPM can be considered to be a simulation of the project from the beginning to the end. It consists in defining an Earliest Starting Time (EST) and an Earliest Ending Time (EET) per each activity, with the difference between the two being equal to the duration of the activity. Each EST is defined as the largest among the EETs of the activities that—according to the activity network—are predecessors to it (in other words, you cannot start an activity until all of its predecessors have ended). The largest EET represents the duration of the entire project. Following Fig. 12.10a, the schedule can be defined by computing values of EST and EET by using a table (Fig. 12.10b) or by plotting a Gantt chart (Fig. 12.10c).

After the project schedule is defined from the beginning to the end, it is possible to backtrack from the end to the beginning, computing Latest Ending Times (LET) and Latest Starting Times (LST) per each activity. The LET of an activity is given by the smallest of the LSTs among the activities to which the same activity is a predecessor (in other words, if you do not want to delay the entire project, you cannot allow an activity to end after any of its successors is—at the latest—due to start).

(a)

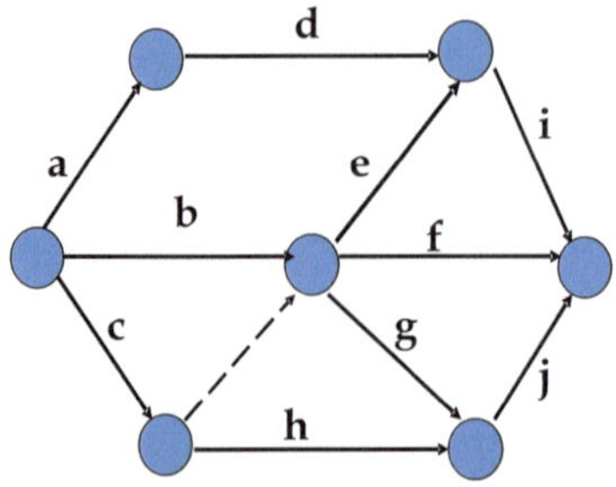

Activity	Predec.	Duration	EST	EFT	LFT	LST	Slack
a	-	2	0	2	9	7	7
b	-	4	0	4	6	2	2
c	-	3	0	3	3	0	0
d	a	2	2	4	11	9	7
e	b, c	5	4	9	11	6	2
f	b, c	1	4	5	13	12	8
g	b, c	2	4	6	8	6	2
h	c	5	3	8	8	3	0
j	g, h	5	8	13	13	8	0
i	d, e	2	9	11	13	11	2

(b)

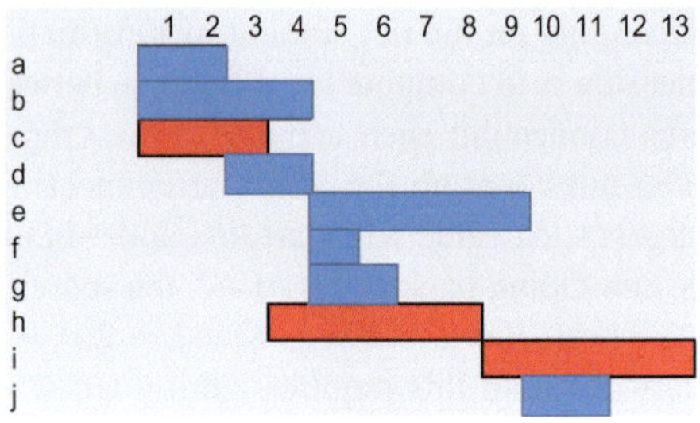

(c)

Fig. 12.10 Using the critical path method (CPM) to schedule a project

The difference $S_j = \mathrm{LST}_j - \mathrm{EST}_j$ is called '*slack*'. A subset of the activities will have $S_j = 0$, which means that the activity cannot be delayed if not by accepting an increase in the duration of the entire project. These activities are defined *critical*. If instead $S_j > 0$, the activity could be delayed by this same amount without any disruption to the project. These activities are called *non-critical*.

It can easily be shown that critical activities are sequential to one another, and therefore identify a '*critical path*' that determines the duration of the project.[2] Activities on the critical path have to be managed carefully, since any delay will have an impact on project duration. Similarly, if one wishes to reduce project duration, actions will have to be taken on the critical activities only. Conversely, delaying a non-critical activity up to the value S_j will not determine any change on the overall project, while reducing its duration will not lead to any benefit.

Along with the Gantt chart, that shows the starting and ending time of each project activity, one can plot a loading chart representing resources usage. In a loading chart, each activity is represented as a rectangle whose base is the duration, and whose height is the number FTEs employed. The associated effort will therefore be given by the surface of the rectangle. In a loading chart, rectangles representing activities are located on the time axis according to the schedule and piled on top of each other (of course, activities can be split in order to fill up any gaps). The envelope of the loading chart therefore shows the profile with which, given the

[2] In general, projects will have one critical path. It is, however, possible to run into 'degenerate' cases with two or more parallel paths of maximum duration, which are therefore both critical.

schedule, resources are being absorbed at any time. Thanks to the loading chart it is possible to compute project cost, for instance in the case in which internal resources are available up to a given level, and external ones have to be called in when needed.

12.3.2 Project Crashing

As mentioned earlier, activities might have been defined according to a standard and to an accelerated—or crashed—level. Project crashing is performed when project duration is not acceptable and managers want to decrease it. The most common way to do so consists in '*crashing*' activities on the critical path one at a time. The critical path will have to be recalculated at each step, since the crashing of an activity may cause the critical path to change.

When determining the activity to be crashed, simple heuristic rules can be used. A popular one consists in using the Cost Change per Time Unit criterion, in which critical activities are ranked according to the following ratio

$$\frac{\text{CC}}{\text{UT}} = \frac{\text{crash cost} - \text{normal cost}}{\text{normal time} - \text{crash time}} \tag{12.4}$$

and then crashing the activity that exhibits the lower value, since it is the one that provides the greater return (in terms of time saved) per cost added.

By following this rule iteratively, a project manager can progressively move on to activities whose crashing is less favorable. The results at each iteration can be plotted on a graph, with time on the horizontal axis and total cost on the vertical one (Fig. 12.11). The resulting tradeoff curve allows choosing the solution that provides the firm with an acceptable combination of cost and time.

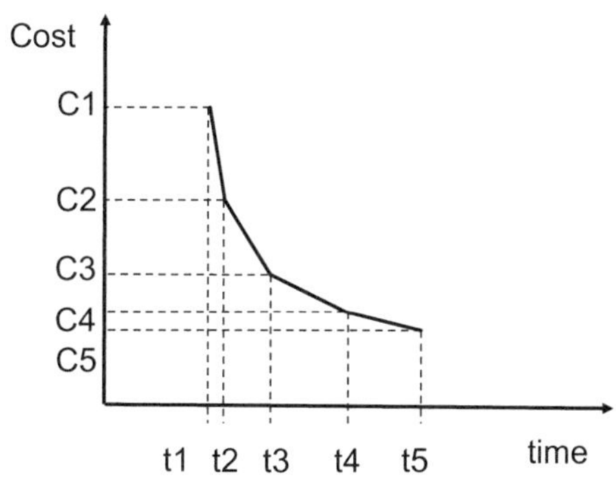

Fig. 12.11 Project crashing and the resulting tradeoff curve

12.3.3 Finite Resource Scheduling

In many cases, resources are not freely available, but are limited. For instance, the project in Fig. 12.10 might have to be scheduled having no more than five FTEs under hand. Scheduling will have to cope with this new constraint, accepting that project duration be lengthened to some extent.

Finite resources scheduling can be performed in a number of ways. The simplest one consists in observing the loading chart that has been drafted in the infinite resources case, and then shifting the activities that correspond to the 'peaks' that violate the resources constraint. This approach is manageable for relatively simple projects, in which constraint violations are not many. When the project becomes more complex, it is advisable to use heuristics based on '*priority rules*' (Elsayed 1982).

Scheduling a project with priority rules implies simulating its evolution from the beginning to the end, drawing the Gantt chart and the loading diagram step by step. At each event, defined as the ending of one activity, one must list the activities that might start because precedence relationships allow them to. If the sum of the FTEs required by these 'candidate activities' is less than the amount of resources available at that time, the constraint is not active and all activities may start without any delay. Note that the amount of resources available at a given time is computed as the difference between the number of FTEs available to the project and the number of FTEs that are engaged by activities that are still under way. Instead, should the amount of available resources be insufficient, there is a conflict between candidate activities. A priority level is then computed per each of these activities (bearing in mind that a candidate activity qualifies if precedence constraints enable it *and* if it the number of FTEs that it requires is not greater than the currently available ones). The candidate activity with the highest priority will start and engage the resources it requires.

The procedure is iterated among the remaining candidate activities until no further activity can start. The scheduling process will then shift to the next event (i.e., to the ending of an activity that is closest in time), and so on. Figure 12.12 provides an example of this process in a simple case.

With finite resources scheduling, the concept of 'critical path' loses its meaning. One might try to identify a critical path by backtracking from the end of the project and progressively looking for the upstream activity that determines the EST of a downstream one. In the case of infinite resources, this relationship would only be due to precedence relationships (i.e., you can start critical activity *X* after all of its predecessors have ended, and the predecessor that ends the last is critical as well), thence the implication that critical activities form a path. In the finite resources case, an activity's EST could also be due to the progressive freeing up of resources.

In other words, with limited availability of resources, you can start activity *X* when predecessors have ended *or* when a completely unrelated activity that was saturating resources ends. Therefore, the activation of one activity at the end of another one does not imply a connection in the activity network, and the two activities therefore might not lie on a same path. Moreover, the connection between

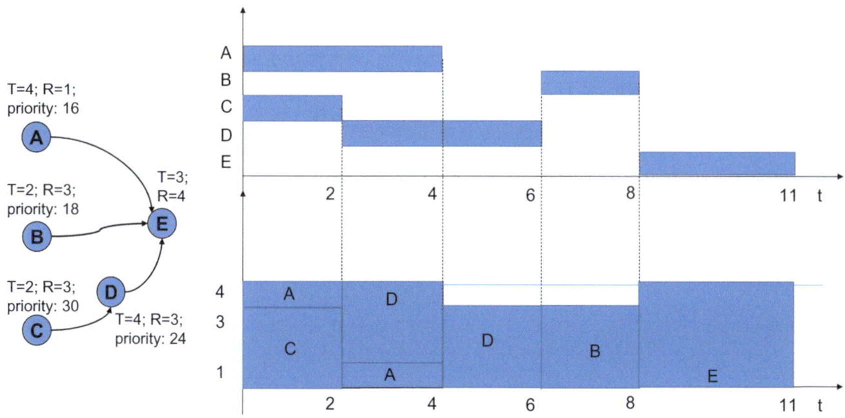

Fig. 12.12 Using priority rules for project scheduling under finite resources

the ending activity that frees up resources and the subsequent one does not depend on the network structure, but on the specific schedule that has been chosen.

Therefore, one cannot formally speak about critical activities and critical paths in finite resources scheduling. From a managerial perspective, it might nonetheless make sense to label the set of activities that are critical because of any of the two reasons outlined above, since these will have to be managed carefully, if delays to the project are to be avoided.

A wide range of priority rules has been developed and the choice depends on the activities that the project manager wants to bring forward. Before describing individual priority rules, it is however important to notice that the priority of an activity is not to be computed by looking at that activity in isolation, but at the impact on the overall project, if the activity is placed at the top of the priority list, thus enabling the launch of its dependent activities. In other words, the priority of an activity should be computed by looking at the paths that are originated from it and—in the common case that more than one path is originated—to the one that maximizes its priority. With this principle in mind, it is now possible to discuss the following main priority rules. Other priority rules, which generally represent linear combinations of the ones listed below, have also been proposed in literature (Elsayed and Nasr 1986).

- ACTIM (Activity Time) is one the simplest rules. Priority of an activity is given by the maximum—computed over all the paths originating from that activity—of the sum of the durations of the activities belonging to the path. If one thinks in terms of the loading diagram, priority is given to the rectangles representing activities that have a wider base and take up more time.

$$\text{ACTIM }(A) = \max_j \sum_{i=1;i\in j}^{n} d_i \tag{12.5}$$

- ACTRES (Activity Resource) states that priority of an activity should be computed by looking at the product of duration and resources usage of the activities in the path. Looking at the loading diagram, priority is given to rectangles that take up a large surface.

$$\mathrm{ACTRES}(A) = \max_j \sum_{i=1;i\in j}^{n} d_i * r_i \tag{12.6}$$

- ROT (Resources Over Time) looks at the ratio between FTEs and duration. On the loading diagram, priority is therefore given to 'tall' rectangles.

$$\mathrm{ROT}(A) = \max_j \sum_{i=1;i\in j}^{n} \frac{d_i}{r_i} \tag{12.7}$$

The example in the following Fig. 12.13 shows the calculation of priorities for activity A of a simple network.

The previous discussion has covered the case of projects with resources of a single type. However, it is quite common for projects to involve resources of different types, which cannot be substituted with one another (e.g., hardware engineers and software developers).

When dealing with multiple resources, it is customary to perform finite resource scheduling for the resource type that is the bottleneck to the project, i.e., for which the ratio between required resources and available ones is the highest. In fact, by scheduling on the bottleneck resource, the project will last longer and activities will tend to spread out in time. Therefore, the non-bottleneck resources will be even less likely to cast a constraint on the project. The easiest way to identify the bottleneck resource is to quickly schedule the project under the infinite resource hypothesis, and then look at the usage ratios of each resource type.[3]

In other cases, resources are pooled across a number of different projects. Multiple-project scheduling is a very difficult task to perform using optimization algorithms. It is therefore possible to recommend simple heuristics, based on the assignment of priorities P_j to each project j. Specifically, three approaches can be considered.

The simplest approach is to make an ex-ante split of the available resources pool among the different projects, based on their priority and their resources usage, and then schedule them independently from one another. So, if the number of available resources is R and the effort required by each project is E_j, each project j will be assigned a number of resources given by

[3] This approach is valid in the case that the usage of resources is relatively constant throughout the project. There might be cases in which different resources types act as bottlenecks at different phases of the project.

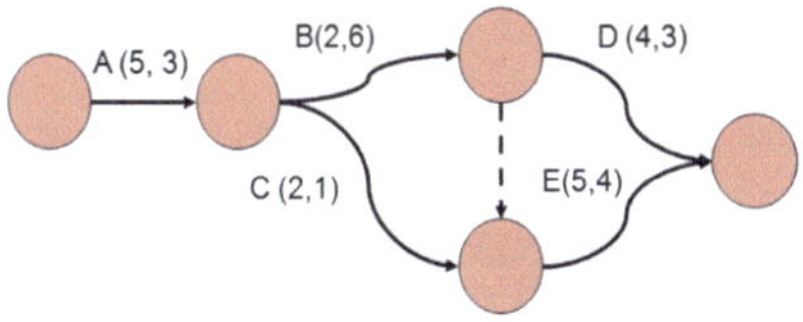

ACTIM (Activity time)

$$ACTIM\ (A) = \max\{5+2+4,\ \ 5+2+5,\ \ 5+2+5\}$$

ACTRES (Activity resources)

$$ACTRES\ (A) = \max\{3.5+6.2+3.4,\ \ 3.5+1.2+4.5,\ \ 3.5+6.2+4.5\}$$

ROT (Resources over time)

$$ROT\ (A) = \max\left\{\frac{3}{5}+\frac{6}{2}+\frac{3}{4},\ \ \frac{3}{5}+\frac{1}{2}+\frac{4}{5},\ \ \frac{3}{5}+\frac{6}{2}+\frac{4}{5}\right\} =$$

$$= \max\{4.35, 1.9, 4.4\} = 4.4$$

Fig. 12.13 Computing priorities in a simple project network

$$R_j = \frac{E_j P_j}{\sum_{j'} E_{j'} P_{j'}} \tag{12.8}$$

Another approach consists in ranking the projects according to priority, and then scheduling projects sequentially, starting from the highest-priority one. Then, the second-highest ranking project will be scheduled, based on the resource availability that has not been used by the previous project, and so on. When following this approach, the projects tend to follow a sequence, though with some overlap. The first project will be scheduled with hardly any resources constraints. The second project will be 'squeezed' in the capacity that is left over by the first, and will accelerate when the first project ends, and so on.

Finally, one can consider the entire set of projects as a single and large project to be scheduled with the usual priority rules. When using these rules, priority of an activity will be multiplied by the priority of the project the activity belongs to. This tends to allow parallel execution of projects, though the ones with higher priority will be favored and tend to start and end earlier.

References

Browning TR (2001) Applying the design structure matrix to system decomposition and integration problems: a review and new directions. IEEE Trans Eng Manag 48(3):292–306

Clarkson PJ, Hamilton JR (2000) Signposting: a parameter-driven task-based model of the design process. Res Eng Design 12(1):8–38

Drexl A (1991) Scheduling of project networks by job assignment. Manag Sci 37(12):1590–1602

Elsayed EA (1982) Algorithms for project scheduling with resource constraints. Int J Prod Res 20 (1):95–103

Elsayed EA, Nasr NZ (1986) Heuristics for resource-constrained scheduling. Int J Prod Res 24 (2):299–310
Eppinger SD, Browning TR (2012) Design structure matrix methods and applications. MIT Press, Cambridge
Gebala DA, Eppinger SD (1991) Methods for analyzing design procedures. In: ASME Design theory and methodology conference, Miami, USA, 22–25 Sept
Kroese DP, Taimre T, Botev ZI (2011) Handbook on Monte Carlo methods. Wiley, Hoboken
Kwak YH, Ingall L (2007) Exploring Monte Carlo simulation applications for project management. Risk Manag 9:44–57
Wynn D (2007) Model-based approaches to support process improvement in complex product development. Engineering Design Centre, Cambridge

From Market Research to Product Positioning

13

Abstract

This chapter discusses the main topics pertaining to the front end of a product development process, starting from the definition of the market, going through a systematic understanding of customer needs in both B2C and B2B contexts and achieving a solid positioning with respect to customer perception and pricing. Then, the chapter discusses the issue of forecasting demand, with a special focus on the case of estimating diffusion phenomena for new technologies and associated market shares. This latter discussion will build on classical models from Bass and Kotler, and add recent research findings.

Starting with this chapter, we make a deep plunge in the product development process with a very practical approach, exploring methods and tools that may support development teams in their work. The coverage of the development process will be limited to the first stages, at least up to the point where general-purpose methods can be applied, irrespectively of the industry and of the technology that underlies products or services. The subsequent stages of product development, which focus on the detailed design of products and processes and on product industrialization, will therefore be left aside.

As discussed in Chap. 11, the first phase of the product development process is concerned with product planning. During this phase, the firm must at first develop a clear understanding of the market, of customer needs, and of competitive offering. Then, it must come up with a high-level description of the product (i.e., a *product brief*) allowing its positioning in the market space. At this point, the firm will be able to make preliminary forecasts of demand and develop a first business case to study its profitability. The current chapter follows these early steps in product planning, while Chap. 14 will deal with the remainder of this phase.

M. Cantamessa and F. Montagna, *Management of Innovation and Product Development*, https://doi.org/10.1007/978-1-4471-7531-5_13

13.1 Customer-Driven Product Development

In developing these topics, and as partially discussed in Chap. 11, this book refers to a market-pull development process, in which the firm intends to deliver an incremental innovation to an existing market. In this case, firms generally follow what is called a *customer-driven product development* approach, which spells out that the main initial focus should be on understanding customer[1] needs, and the rest of development activities should gravitate around the satisfaction of these needs. This mainstream approach has been extremely popular since the '90s and is still followed today.

This customer-driven approach might not be suitable for other cases. With radically innovative products, product development must focus on matching the features that are technically attainable to generally latent needs of potential application areas and market segments. In the case of *design-driven* innovations (Verganti 2009), instead, product success will be based on a tacit interplay between the cultural and 'emotional' content proposed by firms and customers' perceptions and tastes.[2]

Customer-driven product development can be applied in both cases of products aimed at consumer markets (Business-to-Consumer, or B2C) and at corporate markets (Business-to-Business, or B2B). The two cases are clearly quite different, with the former being more complicated to manage. In the case of B2C, producers have very few contacts with their final customers, who are many and are separated by many layers of intermediaries as well as by geographical distances. This makes it difficult to go beyond a shallow understanding of customers' needs and wants, and of the purchasing process they follow.[3] In addition, consumers buy products because of a number of tacit criteria and aspects, many of which highly emotional. Therefore, consumers' purchasing process will be more difficult to understand than the generally rational analysis dominated by technical performance and features, which is often made by businesses, and will also be heavily influenced by marketing strategy, the availability of complementary products, and so on. As it has been mentioned in Chap. 4, the complexity of the purchasing process makes it impossible for consumers to carry out a rational and informed evaluation of

[1] For the time being, we will use the term 'customer' in a very broad sense, including any actor that has a role in purchasing and then using the product or service. We will later be more precise in discussing how the purchasing and usage processes include multiple actors, which will have to be considered separately and in their interactions.

[2] In the case of fine mechanical watches with multiple 'complications', for instance, customers are willing to pay significant amounts of money for an object that is at the same time a status symbol, the fruit of high-end technology and engineering skills, as well as an object of art. Similarly, most new products in the automotive industry can be associated to design-driven innovations, since aesthetic and cultural aspects have a role that is comparable—if not even superior—to vehicle performance and satisfaction of functional needs.

[3] As an example, one can think of the lack of communication between a producer of toothpaste, and the millions of people who daily purchase this product, and then compare it to the continuous interchange of information between the producers of specialized machine tools and their few and well-known customers.

competing products. In this context, brands play a major role by encapsulating and communicating a given set of tradeoffs that is coherent with the brand, so that customers may choose among competing products without actually having to learn about underlying technical details and compare them.

Based on the above, this chapter will discuss market research methods and tools that are typically used in the context of product development for B2C. This does not rule out that they may be useful for B2B products as well.

13.2 Defining the Market

The very first step in trying to understand the market lies in defining the market itself. This is not a trivial exercise, since decisions at this stage are not straightforward and can lead to important implications later on. The definition of a market implies the drawing a boundary between what is '*in*' that market (and must hence be considered when performing competitive analysis) and what is 'outside' (and will therefore be neglected).

It should also be stressed that there is a difference between the concept of 'market' (which represents demand, composed of customers) and of 'industry' (which represents offer, which is instead made up of suppliers). While this difference appears to be obvious in theory, it is always easy to confuse the two terms in practice. With this in mind, it is possible to list a number of possible criteria that can be used to define and circumscribe markets.

- **Econometric criteria**. Economists generally state that markets are defined by a set of substitute products, defined as products for which cross-price demand elasticity exceeds a given threshold.

$$E_y(x) = \frac{\Delta Q_y}{Q_y} / \frac{\Delta P_x}{P_x} \tag{13.1}$$

  In other terms, two products belong to the same market if a relative price increase for product X leads to a sufficiently large relative quantity increase for product Y. Alternative econometric criteria include measuring the correlation between price variations of products, which is an indicator of competitive dynamics, or evaluating whether cartels could be established, leading to monopolistic pricing. These criteria are widely used by economists engaged in both research and practice (for instance, when working on antitrust cases, a preliminary step usually consists in developing an objective definition of the 'relevant market'). However, it is not trivial to define thresholds for these econometric indicators in such a way that it makes sense for market research purposes. In fact, if one considers these indicators, any two products might bear *some* degree of cross-price demand elasticity and of price correlation.

- **National statistics offices** often allow an indirect definition of markets by looking at the related industries. In all countries, producers are classified by national statistics offices according to a hierarchical code[4] (e.g., NAICS in North America, NACE in Europe, and so on). Following these definitions is quite appealing because of the availability of official data that is based on these classifications (e.g., sales by industry *X*, investment made by industry *Y*, purchases made by industry *Z* from industry *K*). However, these classifications are often flawed, since it is common for individual firms to be diversified (i.e., a firm may be registered in industry *X*, but also produce goods related to industry *Y*), or to change their actual focus over the years, without bothering to change their registration code. Moreover, coding systems generally are unable to correctly represent the industries behind highly innovative products, whose very existence has not yet been captured by the statistics offices.
- **Geography**. An obvious criterion for defining markets is related to geography. The concept of geography being used in this context is generally extended to any kind of barrier that can insulate one area from another one, preventing arbitrage mechanisms that would lead to a single market. Geographical distances can matter, of course, but also can tariffs, differences in regulations, or differing consumer tastes, which can be highly product-specific. Therefore, firms may recognize the existence of a single European market for consumer electronics, but not for building materials, which are specific to each country and its architectural traditions.
- **Product similarity**. Since products pertain to the same market when they are substitutes, product similarity can be an indicator of potential substitutes. This criterion lends itself to two different interpretations. One is technical similarity (e.g., one might consider a market for 'railway transportation' that encompasses all types of trains), and the other is similarity in the needs being satisfied (e.g., the market for 'high-speed transportation' that includes short-range airways and high-speed trains).
- **Switching behavior**. This criterion can be used for fast moving consumer goods. Two products will be substitutes, and therefore belong to a same market, if customers often switch from one to the other, but not if customers remain loyal to one. For instance, biodegradable laundry soap can be considered to have a market that is independent from normal detergents if customers who purchase this kind of soap always do so, and vice-versa.
- **Demographic criteria**. Demographic criteria define segments of the population that can be relevant to a product offering. For instance, when developing a two-seat sports car, one might consider a market of 'young, urban, affluent drivers'. Demographic criteria can include age, profession, education, household structure, income, wealth, lifestyle, etc. In some cases, these segments are

[4] A hierarchical coding of economic activities is composed by a string of characters, with the first ones representing the higher-level classification and the subsequent ones representing finer subdivisions whose meaning depends on the former. So, for instance, NACE C13.9.3 defines 'manufacturers of carpets and rugs', with C representing all kinds of 'manufacturing', C13 the 'manufacture of textiles' and C13.9 the 'manufacture of other textiles'.

defined purposely for the product. In other cases, they can coincide with clusters that are uncovered independently by market research firms when performing surveys of the population.

In practice, firms will often define markets by operating on two tiers, with a broader definition of 'market' and a narrower definition of 'segment'.[5]

The definition of markets and segments depends on the type of innovation the firm is attempting to introduce. It will be narrower in the case of incremental innovations and broader in the case of radical ones, which might disrupt existing schemes. For instance, a carmaker developing a new version of an existing city car will say "we develop a new product for the passenger car market and, specifically, for the B segment". The same carmaker could at the same time develop a completely new vehicle intended for car sharing services, and should therefore redefine the concept of market (i.e., urban mobility) and the associated segments.

13.3 Understanding Customers

As previously mentioned, product development for incrementally innovative products in B2C is driven by the objective of proposing products that are able to satisfy customer needs (i.e., innovation is mainly 'demand pull'). Understanding customer needs therefore is a core activity in product development, and this effort will have two distinct (though connected) objectives.

On the one side, customer needs must be investigated to such an extent, that the company knows what he/she requires or desires so that at the end he/she will decide to buy that product or adopt the service. This perspective is necessary to ensure that product or service will effectively be purchased and represents a 'marketer's perspective' on customer needs.

On the other side, understanding customer needs is necessary to guide the design of products and services from a technical perspective, because this will ensure that, once purchased, the customer will be satisfied and happy in using the product or service he/she has adopted. This standpoint is more focused on 'user needs' and the usability of product and services. We call this as the 'designer's perspective'.

The two perspectives are obviously connected since, at their roots, customer/user needs are not different, ex-ante or ex-post the purchase. However, these two perspectives are distinct, since the designer's perspective requires a deeper investigation than the marketer's one. Moreover, the way with which a customer/user takes these needs into account may differ before and after purchasing.

In this section we will discuss customer needs mainly from the 'marketer's perspective', which is usually somewhat shallower and focuses on the perception of these needs. In the following chapter we will move to the 'designer's perspective',

[5] For instance, a firm may define its market as the 'market for urban mobility' and develop a product for the segment of 'city bicycles'. Another firm may instead define its market as 'bicycles' and operate on the segment of 'road bicycles'.

which usually is based on a significantly deeper understanding on the use process and on the actual experience in using products and services. From a practical standpoint, it is up to the firm to decide to what extent the two perspectives should be pursued (i.e., whether to privilege one over the other) and this leads to different operational choices with respect to the market research activities to be carried out in order to understand these needs.

Based on these premises, the 'marketer's perspective' on customer needs points in the direction of understanding the purchasing process followed by customers. In fact, it would be very difficult for a new product or service to sell, if the firm has not correctly understood this process.

In the case of B2C markets, the purchasing process is apparently simpler than in the case of B2B, where decisions involve a number of different actors. However, consumers' purchasing decisions are tacit and emotional and do not follow a formalized and rational process, as in B2B. This makes it difficult to achieve a deep understanding of elements that might determine product success or failure.

Anyway, given the target market or segment, the firm will have to provide a proposal that could be perceived to be of value by customers. Therefore, the firm will have to identify elements to which customers will pay attention, associate utility, and for which they are willing to pay.

Consequently, a *value proposition** will be such as to satisfy a customer 'need', either in the form of providing something customers like (i.e., 'providing a gain') or of lessening something they dislike (i.e., 'relieving a pain'). The word 'need' being used here is very significant, since it has to be stressed that—while firms tend to think in terms of product or service features and performance—customers think in terms of their own wishes/requests and of products' or service's capability of fulfilling them. Features and performance indicators are not only technically alien to consumers (e.g., who can really understand the meaning of 'in car model *X*, noise at cruising speed in the front seats is 65 dB'?), but can be misleading, if they are not linked to their specific needs and perceptions. For instance, in the previous case, the underlying need might be "I want to chat with passengers without having to raise my voice" or "don't want to be bothered by background noise, while I'm traveling and listening to music".

In order to start such an analysis, a first factor to be considered consists in looking at individuals' behavioral motivation. Abraham Maslow (1908–1970), in his 1943 seminal paper '*A theory of Human Motivation*', highlighted the existence of a hierarchy of needs, moving from 'Physiological' to 'safety', 'belonging and love', 'social' or 'esteem', and 'self-actualization'. These needs are hierarchical, in the sense that personal motivation can arise at one stage only if the needs at each previous stage have been satisfied.[6] These general needs are also shaped by specific factors that pertain to social and cultural variables (belonging, aspiration, etc.), demographic variables (age, employment, lifestyle) and psychological variables (i.e., personality). Given this general background, one can look into the *purchasing process* followed by customers.

[6] The *Hierarchy of Maslow** has been questioned by later research under a number of perspectives; nevertheless, it still remains a recognized reference for marketing applications.

Customers' purchasing decisions do not follow a formalized and rational process, and this makes it quite difficult to achieve a deep understanding of elements that might drive adoption; many models exist in the literature (for a review, Prasad and Jha 2014), of which some are explicitly oriented to being used by practitioners (Temkin 2010).

Among these models, one seminal contribution is represented by the adaptation of the '*lens model*' of perception (Brunswik 1944) to purchasing behavior (Tapp 1984; Puccinelli et al. 2009). The basic idea behind the lens model is that consumers make purchasing decisions without having a clear, objective and deterministic view of their needs, or of the way with which product features can fulfil them. To do so, they must be able to correlate technical features to their needs, and they must cope with a broad amount of data that exceeds humans' cognitive ability. Especially considering the latter, the dimensionality of the problem can become very high indeed, preventing customers from making a rational choice based on technical data.[7] In such cases, humans use perceptive processes to cope with the multidimensionality of the process and to unconsciously aggregate tacit elements and 'hard' raw data in a hierarchical way, until their decision space becomes manageable.[8] Therefore, customers will 'perceive' the mass of information involved in a product or service by abstracting it in a space of limited dimensionality, defined by high-level needs and features,[9] based on which they will make their purchasing decisions.

This perceptual process does not occur as a single event in time, but evolves along with the purchasing process, in which the consumer starts thinking about the possibility of adopting a system and ends up choosing a specific alternative and effectively buying it. By linking these two aspects it is possible to outline the overall process as in the following Fig. 13.1.[10]

---

[7] For instance, imagine a person having to choose a TV set in a store. Dozens of models are on display, each of which carries an information card that lists a dozen features or so, together with price. It would be unlikely to see that person make a choice by pulling out a sheet of paper, writing down a shortlist of TVs that exceed given thresholds, and then ranking them based on a weighted score of performance indicators. This will not happen because of two reasons: technical features alone are not enough to drive the purchasing process, and humans are not able to manage multiple dimensions in decision making.

[8] An example of perception has to do with reading. Reading this book means that your eyes are stimulated by light patterns corresponding to the characters on the page. A perceptual process allows you to isolate the characters, understand them, aggregate them in words and—hopefully—understand their meaning. The perceptual process is not innate and has to be learned, and the information thus processed can be managed only in relation to pre-existing knowledge and experience (think about the difficulty in reading a book on a familiar subject, written in an unfamiliar language, or even written with uncommon fonts).

[9] For instance, the digital cameras of the previous example may be perceived and assessed by consumers according to two dimensions, such as 'technical performance' and 'ease of use'.

[10] Before reading further, try reflecting on a recent and infrequent purchase you may have made (i.e., a relatively expensive durable product, or a consumable product that you have never bought before). Try listing the sequence of actions and thoughts that led you from the initial stage of 'I wasn't even thinking of buying *X*' all the way to actually making the purchase. What differences would there be between *X* being a car, a smartphone or new brand of toothpaste?

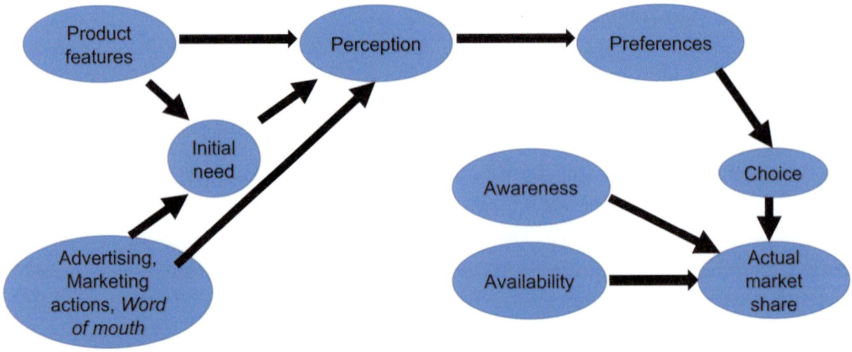

Fig. 13.1 Purchasing as a perceptual process

At the beginning, a potential customer hasn't yet developed a decision to purchase a good or adopt a service and is not even investigating this possibility (e.g., imagine someone who is not thinking about buying a new car, and casually sees a new car model on the street, or views a web banner advertising its launch). The first stage of the purchasing process brings this potential customer to the stage of feeling the need and being interested in purchasing. This can occur because of features that are inherent to the product or service (e.g., the new car models might have an outstanding fuel efficiency or might be complemented by long-term rental services) and because of influence cast on the consumer by other parties. These can convey both objective information on the product (e.g., mileage of new cars), as well as other types of information (e.g., the association of new cars to an attractive lifestyle). Such an influence can be defined as either external or internal to the market. External influence includes advertising and other media, such as articles in the press or product reviews on specialized websites. Internal influence has to do with imitative effects, which can be due to explicit word of mouth (e.g., a friend who buys a new car and tells wonders about it) as well as to tacit emulation (e.g., new cars are bought by many people and become fashionable). At this stage, the consumer has now become an active decision-maker and information seeker, and the perceptual process is at its foremost.

At the end of this process, the consumer will have defined the '*perceptual space*' in which he/she can position competing products or services, according to the degree with which these seem to allow the fulfilment of his/her needs. Eventually, this leads to the generation of preferences for each competing product or service. This will enable him/her to choose whether to purchase the system or not, in the former case deciding on which alternative system to buy and in the latter case deciding whether to abandon the idea or simply delay the decision.

If one considers a population of consumers and the choices they may make at this stage, these decisions correspond to the *potential market share* of a product or service, i.e., the fraction of consumers who—given their willingness to buy a product or service—would choose that over competing ones.

The *actual market share* will be a fraction of this potential, since it will be abated by a factor called *awareness* (i.e., due to advertising strategies, not all potential consumers might be made aware of the existence of the product or service) and *availability* (i.e., due to distribution choices, not all potential customers might actually be able to buy the product). When thinking in terms of corporate functions and processes, potential market share is clearly the responsibility of the product development process, awareness of marketing strategy and availability of distribution strategy. An integrated view of the whole process should nonetheless lead a firm to bring all of these factors under the responsibility of the product development process.

Finally, within each single purchasing process, one can observe five phases: arousal of the need, information gathering, evaluation of the alternatives, purchase decision and consumption. Within each phase, the customer performs multiple *interactions* with communication and retail channels provided by the firm (which can be physical and/or virtual), interacts with people (e.g., salespeople, friends, other users, etc.) and observes the environment (i.e., looking for similar products already being used by others, trying to imagine his life after having purchased, etc.). Each of these *states of interaction* contributes to activate his/her perception process. From this perspective, the customer is an active player involved in a process of continuous interaction with '*touch points*' within a purchasing process (Phillips and Bradshaw 1993) which determine her/his *customer experience**.[11]

At the end, when a firm would like to study customers' perceptual processes, it must understand how such broad and somewhat confused variety of information is progressively structured into a decision-making space characterized by fewer and more abstract concepts, in which the human being can more readily operate, and end up purchasing a product or a service.

This perceptual process is not only quite complex, but also tacit and—to some extent—unconscious, which implies that firms cannot model it by simply asking consumers and users about it. Therefore, in order to shed some light into the process, firms can base their analysis by hypothesizing the following three levels for customer needs:

- *Tertiary needs* are the most elementary ones, are usually quite numerous (tens or even hundreds for complex products), and can usually be elicited by interacting with customers (examples are "I want to fit a lot of baggage in the trunk of my car", "Loading and unloading heavy items in the trunk of my car must not require too much effort", etc.).
- *Secondary needs* are an aggregation of tertiary needs, achieved by clustering tertiary needs. A product can usually be defined in terms of 10–20 secondary needs, which are detailed enough that they can be effectively communicated and discussed with customers (e.g., 'Capacity and accessibility of the trunk' is a concept that is quite unambiguous and easy to grasp).

[11] This perspective has been discussed in a variety of contexts, such as B2C (Temkin 2010; Lemon and Verhoef 2016), B2B (Hague and Hague 2018; Lecoeuvre et al. 2021) and multi-channel (Barwitz and Maas 2018).

- *Primary needs* are an aggregation of secondary needs that defines the perceptual space in which products are tacitly represented and compared during the purchasing process. Primary needs are usually no more than 2–4 (e.g., a car might be perceived according to the two dimensions of 'performance' and 'convenience', where the secondary need 'capacity and accessibility of the trunk' might be one of the elements that make up the concept of 'convenience'). At this level, the perceptual process is tacit and unconscious. Therefore, this final aggregation can only be studied indirectly, by asking customers to provide clues about the affinity they perceive between secondary needs.

Marketing and Design teams can derive a very rich amount of useful information from these three levels of needs. Primary needs are the key elements for a correct positioning of the product or service. Secondary and tertiary needs are the basis for defining the design requirements, since they will define the elements of interaction and therefore impact on the purchase and use experience.

From an operational point of view, one can tackle market research aimed at developing insight on customer needs in two ways. If the main objective is the 'marketer's perspective', aimed at identifying primary needs and positioning the product, market research can be developed relatively quickly, since secondary and tertiary needs will only be instrumental to the previous objective. Conversely, if one wishes to use tertiary and secondary needs also to support product and service design later on (the 'designer's perspective'), a more thorough market research activity will be required. This will avoid having to repeat a costly activity twice, one for marketing and the other for design, at the same time risking inconsistencies between the two. In this chapter, we will focus on the former case, which implies a more cursory approach to market research, whereas in Chap. 14 we will describe methods that are required for the latter case.

A preliminary *market research* will generally be carried out according to the steps described in the following sections of the chapter.

13.3.1 Identifying Customers, Users and Stakeholders

Up to now, we have used the generic term 'customer/user' to address any actor who might be involved in the purchasing process and that generically owns needs. The first step of a market research action lies in identifying who these actors are and what is their role in the process.

A commonly used tool is that of characterizing typical profiles through the *Personas method**. This method found its beginnings with Alan Cooper in Software Design and then in the study of *Human–Computer Interaction (HCI)**; subsequently, it found wide adoption by Marketing and Industrial Design.

A *persona* is a fictional, but realistic, character created to represent a target individual whose cultural and demographic features, as well as habits and behaviors, might determine his/her needs. Each persona therefore is identified with a name, a set of demographic attributes, motivations and needs, so that analysts can better immerse themselves in the idea that they are operating for the benefit of a specific individual to infer his/her needs, and not for an anonymous group of people.[12] Marketers may use personas together with market segmentation, so that they are thought to be representative of specific and relatively distant segments[13] or they may use personas to represent different individuals who might interact with one another within more complex purchasing processes.[14]

At this phase, the definition of personas is still to be considered *preliminary* and aimed at facilitating the further market research activities to be carried out; they will lead to a clearer view of needs, segments and—therefore—a more precise definition of personas, as well.

Specifically, the marketer's perspective will consider distinct personas if they think that the two categories are so wide apart, that both their needs and the structure of their purchasing process is substantially different.

The designer's perspective is instead more nuanced and will lead to defining distinct personas even if the needs and the purchasing process are structurally the same, but there is a difference in the importance given to these needs.[15]

---

[12] For instance, one may define 'Marie, 35-year old IT professional, mom of two children (4-year old Susie and 2-year old Jean). She is the wife of Robert, an airline pilot who is often away from home. They live in an apartment in the XV Arrondissement of Paris'. To the family, one may add 'Jeanne, Robert's 70-year old mother, who lives close to Robert and Marie and sometimes borrows their car to visit her friends during the weekend'.

[13] For instance, marketing employees of a firm dealing with shaving tools may identify two personas such as 'John, a bearded man' and 'Arthur, a well-shaven man'.

[14] In the case of HVAC systems for residential use, a firm may use personas to represent the target customer, the architect managing the renovation project and the technician that will be in charge of the installation.

[15] In the shaving tool example, a designer will certainly use more than one persona per category. For instance, bearded men will include 'John the hipster, proud of his beautiful beard, interested about its softness and beauty' and 'Dave, who sports a short beard and is not willing to spend much time on it, but cares that the cut is precise and symmetrical'. Shaven men will include 'Arthur, a 40-year-old who shaves daily but, being in a hurry, often misses the razor at home when he travels', and 'Harry, a 70-year-old man with brittle skin'.

From a marketing perspective, this finer distinction would be irrelevant and even misleading, since the objective of market research is to identify market segments (and associated personas) through data analysis, and *not* to define them ex ante.

The discussion on the nature and the validity of personas is in fact quite present in the literature. Some scholars (e.g., Chapman and Milham 2006) consider them as fictional and not based on a sufficiently rigorous method, since they do not have a clear relationship to actual customer data and do not follow reproducible analysis procedures. The distinction we made in the text between preliminary personas used to support market research and more precise personas to be characterized after market research is carried out goes in the direction of a stream of literature that suggests the usage of data-driven (or quantitative) personas (e.g., McGinn and Kotamraju 2008).

13.3.2 Identifying the Network of Influences

For simple and relatively low-priced products, the purchasing process for B2C goods and services will be carried out by the individual customer, without any significant influence by other stakeholders. For goods that are technically more complex, that involve a significant financial outlay or an emotional content, the process might instead include other actors as well, such as members of the household and of other social groupings (e.g., friends, colleagues, etc.) on the one side, and members to the industry on the other (e.g., salespeople, installers, etc.). A similar situation occurs in B2B, where the purchasing decision is not a simple event, but a complex business process in which multiple stakeholders appear, act and mutually influence themselves (a simple example is personal computers in a firm, which are used by employees, but bought through a procurement procedure run by the purchasing department under the technical recommendations of the IT function).

Therefore, in both of these cases, a product or a service will be successful only if all the stakeholders will favorably align their attitude toward its purchase. So, the producer will have to understand the needs and the perceptual process that characterize this entire community of stakeholders.

In order to frame this problem in a systematic way (Cantamessa et al. 2012, 2013; Cantamessa and Montagna 2016; Cascini and Montagna 2015), it is possible to identify four main stakeholder types and three main situations in which they act (Fig. 13.2). Stakeholders can assume the role of *customers* (i.e., the actors who have formal authority over the purchasing decision), *users* (i.e., the actors who will most extensively and actively interact with the product after purchase), *beneficiaries* (i.e., those who receive benefits from the product and/or interact with it in a passive way) and *outsiders* (i.e., any other interested party).

Concerning situations, the first is *purchasing*, which is done by customers. The second is using/*delivering benefits*, which is done by users and affects beneficiaries, while the third is *creating externalities,* i.e., the effects that propagate beyond the direct beneficiaries.[16]

Each of these stakeholders operates according to a set of specific needs and ascribes importance levels to these needs. Needs can emerge from the actor itself (*native* needs) or can be the result of influences cast among actors (*reported* needs). Mutual influence therefore generates reported needs and alters the importance or the perception that an actor assigns to native needs.[17]

[16] For instance, in the case of a public transport bus, the buyer is the purchasing office of a municipal transport authority, the users will be bus drivers and maintenance crews, and the beneficiaries the passengers who ride the bus. Outsiders include citizens being affected by the emissions of the vehicle, or pedestrians who may be impacted by the bus in case of an accident.

In the case of an electric toothbrush for children, the parents are both customers and beneficiaries, the child is both user and beneficiary, and the dentist is the outsider.

[17] Referring to the public transport example, it is obvious to assume that—all the rest being equal—purchasing managers will prefer a bus that minimizes harmful emissions, even without being directly influenced by citizens. However, the importance that management will attach to this need may be altered if citizens actively cast such an influence through a citizens-rights association.

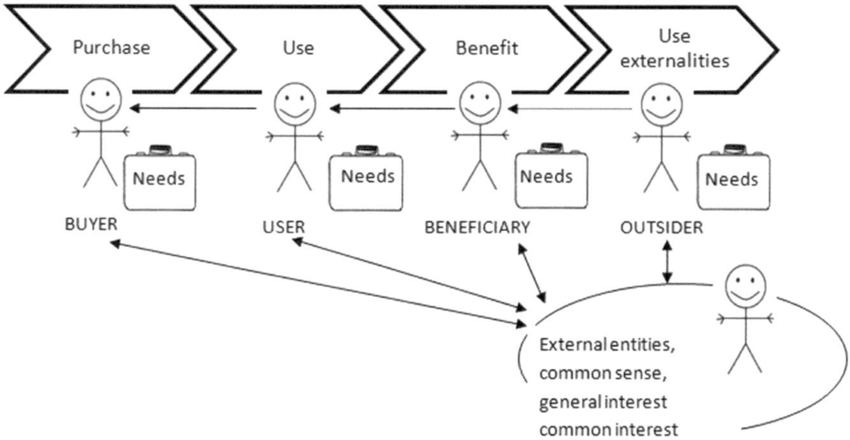

Fig. 13.2 Stakeholders involved in the purchasing and usage process of B2B products

For a firm, the ability to understand and proactively work on multiple stakeholders, on their needs and on their mutual influences can be very valuable in order to understand the purchasing process, both at the stage of product planning and when defining requirements and specifications later on, leading to a joint design of the product or service and of its go-to-market strategy.[18] In fact, the product or service will be purchased only if all the actors who are involved in the decision will see their needs being met at a degree that is that is sufficiently compelling for each of them to agree.[19]

In first approximation, a firm may analyze the purchasing process in a qualitative way and simply come out with a list of requirements coming from the various stakeholders. Rigorous approaches to represent and manage the many potential conflicts and tradeoffs can be derived from *Actor Network Analysis** (Cantamessa and Montagna 2016).

[18] It is interesting to observe salespeople interact with product development teams and notice that a common question is 'the features you are proposing for this product are cool, and they might be helpful for our customers, but how are we going to sell them?' This type of thinking may both guide minor decisions in the design of the product and strategic change in the firm's business model. For instance, should it become apparent that stakeholders are not willing to accept a product that requires a significant up-front investment (CAPEX) in exchange for reduced operational costs later on (OPEX), the firm may adopt an 'as a service' model. This often occurs for energy saving devices in buildings, medical equipment in hospitals, etc.

[19] For instance, a new bus that comes with a powerful but difficult-to-use IT-based fleet management system will probably not be adopted, because drivers will object to the procurement office and oppose the purchasing decision, or may simply avoid using it once it has been bought. Conversely, a system that is user-friendly but expensive will probably be vetoed by managers even if drivers push for its adoption, unless the latter are able to show that it will lead to financial benefits that out-weight the additional cost.

13.3.3 Eliciting Tertiary Needs from Customers/Users

Tertiary needs constitute a long and exhaustive list of *elementary necessities* that users relate to products or services. As previously mentioned, the investigation of these needs can derive from a marketer's perspective or from a designer's perspective, which is inherently broader and more detailed, since it must cover many aspects pertaining to the entire product lifecycle and multiple stakeholders, not all of which are relevant to the purchasing process.

In order to elicit the tertiary needs, it is possible to directly interact with groups of consumers, following a variety of approaches. Some of these methods can be considered to be 'marketing oriented', since they limit themselves to achieving a general understanding of the way the product is perceived by potential customers. Other methods are aimed at achieving a much deeper understanding of the interaction between user and product under different *use cases**. In this chapter we will discuss the methods that are typically associated to the former perspective, leaving to the next chapter the presentation of the other ones.

Of course, the boundary between the two types of methods is somewhat blurred, since a firm may use 'design oriented' methods earlier to avoid duplication of efforts and/or to deepen its analysis with respect to aspects that are critical for product positioning.

- **Focus groups*** are made up of a small number of consumers, who are asked to discuss the product and their needs under the guidance of a facilitator, while an analyst documents the proceedings, generally using audiovisual footage. Focus groups usually involve 8–12 participants, so to ensure that group dynamics is lively enough and that each individual has enough chances to contribute. The main objective of a focus group is to collect the widest possible collection of tertiary needs. Therefore, focus groups are not assembled looking for representative samples of the population, but striving for diversity, involving both generic users and *lead users**.[20] Given the time required to attend a focus group, there may be segments of the population who might be unwilling to participate, and therefore go underrepresented.

  In a focus group, the facilitator must be able of guiding the discussion in a way that most of the group's experience and insight is brought to the surface. Moreover, the facilitator has the role of bringing participants back to the objective of generating tertiary needs, should the discussion fall astray. To this purpose, he

[20] *Lead users* (Von Hippel 1986) are customers who face needs that will become prevalent much earlier than the general population, and which might contribute with original solutions to solve the need. Following a more extended meaning, a lead user may be a user whose particularly intense usage of the product amplifies her/him perceived needs (for instance, a taxi driver can be considered to be a lead user when it comes to driving a car in an urban environment).

must have a clear understanding of the differences between solutions, specifications, performances, outcomes and needs.[21]
Focus groups are quite expensive activities, given the costs associated to logistics, giving incentives to participants, and the effort spent by facilitators and analysts. It is therefore quite common to add new focus groups to the market research process progressively, and as long as the number of new tertiary needs uncovered at each step exceeds a given threshold. The process ends when the marginal contribution becomes too small.

- **Customer diaries**. Customers (often termed *beta testers*) are provided with products or prototypes and are left free to use them for some time. In exchange, they are required to provide feedback to the firm. The information thus gathered is generally quite shallow and is limited to finding faults, rather than uncovering needs.
- **Empathy maps**. In some cases, the product development team may need to achieve a broader and deeper understanding of customers and of their needs. This is particularly true when designing radically innovative products or business models, for which it is difficult to start from an existing offering. In these cases, the team can widen its perspective, and attempt to consider a comprehensive view of customers, their inclinations and problems to be solved. Empathy maps (Gray et al. 2010) are a simple and powerful tool that allows generating this insight in a structured way. The six areas in an Empathy Map, whose meanings can be intuitively understood by looking at Fig. 13.3, can be explored and filled in either by using direct customer involvement, or by interviewing panels of experts.
- **Participative techniques**. *Customer involvement* in the development process can also be deeper than the elicitation of tertiary needs. Participative design techniques are variations of focus groups, in which customers' creative input is also sought. Customers are therefore stimulated to provide ideas, solutions and associations between the product and concepts pertaining to everyday life.

This intense interaction with consumers can be classified in a number of ways (Fuchs and Schreier 2011). As shown in Fig. 13.4, one can look at the responsibility that the firm decides to share with its customers in the generation of ideas and designs, and in the selection of the products to be launched. Moreover, and depending on the kind of product, this involvement can cover the overall product, non-core aspects (e.g., creating and selecting external décor) or ancillary features (e.g., creating and selecting advertising campaigns).

[21] For instance, a focus group participant might say he/she needs her/him smartphone to be equipped with a low-power chipset. Since this is not a need but a technical solution, the analyst should ask her why. The customer's answer might be that she/he needs her/him phone to have 250 h of standby time between recharges but—again—this is not a need, but a performance specification. So, the analyst will have to ask a second 'why', to which the user might answer that she/he basically wants her/him phone battery to last over a weekend, should she/he forget to bring the charger with her/him. This can finally be considered as a need, since it clearly shows an *outcome* (i.e., a job to be done) that might satisfy the user, regardless of the technical means with which it is achieved.

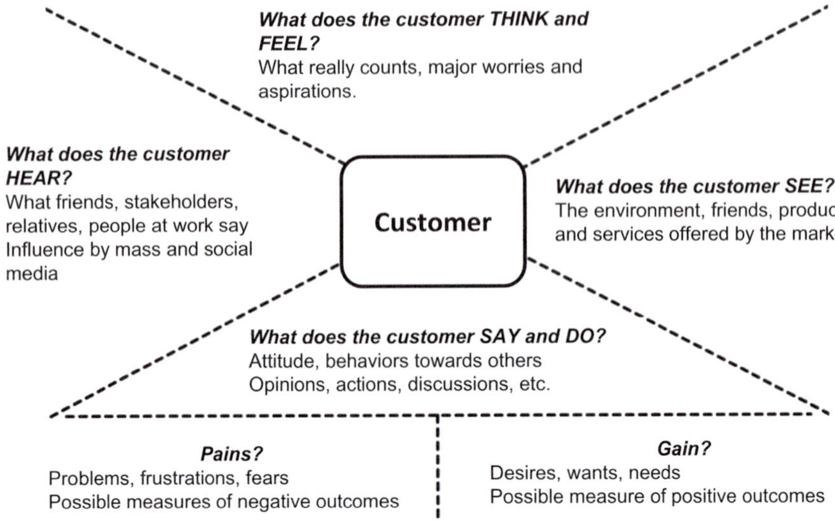

Fig. 13.3 Structure of an empathy map

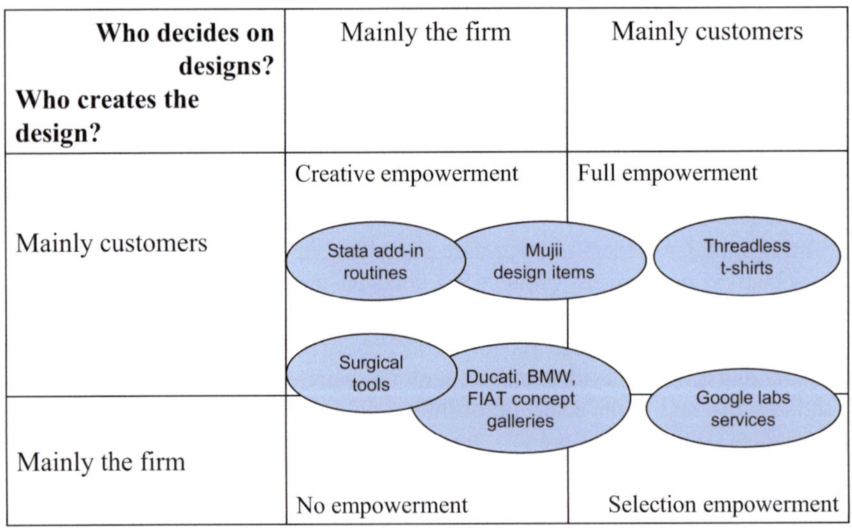

Fig. 13.4 Different forms of active customer involvement in product development

The methods discussed above allow firms to gain a deep understanding of the needs arising from the consumers and users who volunteer to take part in these activities. However, these techniques are fairly costly to run and involve only a tiny fraction of people.

Recently, digitization has allowed performing these investigations on a different scale. For instance, market research can be carried out in real time, while users actually interact with the installed base of products and with services.[22] Data analysis techniques are used to interpret user needs from this type of data by operating at different levels, be it on the aggregate population of users, on segments of users (at the limit, operating on the individual user), or on experimental samples of users to which a new function is being proposed.

Similarly, the information traditionally collected from focus groups can be sought by working on the textual analysis of the comments that users make online. This includes both spontaneous comments (e.g., user reviews on Amazon) and comments that are stimulated by the firm itself (e.g., comments made in reply to an Instagram 'teaser' for the firm's upcoming product, or to the post made by an 'influencer' that shows the product).[23]

13.3.4 Organizing Needs into a Hierarchy

13.3.4.1 From Tertiary to Secondary Needs

Secondary needs are aggregations of tertiary needs into a manageable number of higher-level items that are easy to treat and discuss explicitly. In most cases, the aggregation can be directly performed by the development team, by clustering similar tertiary needs. A common technique consists in writing down each tertiary need on a card or sticky note, and then clustering the cards on a whiteboard until a satisfactory solution is found.

This method does not require much effort, and rests on the assumption that the perspective of designers in making this aggregation will not differ much from that of customers and users. Alternatively, the development team may involve a small sample of people and ask them to individually perform the same clustering exercise. Then, by counting the % of times that need i is matched with need i', one can come up with a similarity matrix between each pair of needs, and then use hierarchical clustering[24] to obtain an objective definition of secondary needs.

Once clusters have been created, the development team will have to come up with a concise verbal description of the secondary need, such that the underlying tertiary needs are well captured and represented.

[22] A firm providing a digital service can easily monitor detailed user activity on a website or a smartphone app, looking at access in general, and at the usage of individual functions (including frequency of use, sequence with which they are run, timing with which actions are carried out, mistakes made, etc.). Sensors in a service environment (e.g., cameras in retail outlets) can capture data on the way people navigate in the physical space and act. Sensors on a product (e.g., the ECUs in a car can capture myriad information on driving habits and mission profiles).

[23] In this case, a key connection is established between the activities that have to do with web, digital and social media marketing and the product development process.

[24] As stated in the introduction, we have chosen not to provide explanations for basic multivariate analysis techniques in this textbook. We therefore suppose that readers are either already familiar with them, or they will be able to refer to specialized sources (e.g., Johnson and Wichem 2007) to gain the necessary knowledge of these techniques and of the way they can be used.

13.3.4.2 From Secondary to Primary Needs

After the list of secondary needs has been drawn, the final step consists in aggregating them into a limited number of primary needs, or *perceptual dimensions*. Attempting to do so by clustering supposedly 'similar' items—as has been suggested for the previous step—can be highly misleading since, at this stage, the perceptual process is tacit and unconscious. Moreover, it would not be wise to perform this study on a limited and possibly non-representative sample of consumers.

Therefore, tacit primary needs can be identified only by performing a survey on a broad and representative sample of consumers, using a *questionnaire** that will provide important insights on other elements as well. In general, a market research questionnaire will contain a number of sections that can collectively lead a firm to a better understanding of the market, of customer needs and of the way with which customers perceive competing products. Without going into the details of good *questionnaire design*,[25] the main sections will include the following:

- **Demographics**, which capture descriptive variables associated with the respondent, and that can be relevant for the analysis being carried out. Examples of such variables include age, gender, income, etc. These variables will also be used during the validation of the dataset, in order to check whether the sample fits with the general population or, conversely, whether the surveying process has led to any response bias.
- **Usage patterns**. Respondents can be asked a few questions on the way they use the product or service. This is not intended to obtain new ideas, but simply to classify the type of user (i.e., heavy vs. occasional, expert vs. novice) and the occasions and the environment in which usage occurs.
- **Importance of secondary needs**. At the heart of questionnaires aiming to uncover primary needs are questions in which the customer is asked to declare the importance he/she attributes to each of the $i = 1, \ldots, m$ secondary needs. 'Degree of importance' questions are usually framed so that respondents can provide their rating on a 1–5 or 1–7 scale. Later on, the variables x_i coming from these answers will be analyzed with multivariate statistics techniques, such as Factor Analysis. This is quite a critical step, since these techniques should in principle be performed on variables measured on an interval (i.e., cardinal) scale, while 1–5 ratings are inherently ordinal.[26] In order to bring this rating as close as possible to an interval scale, and at least partially overcome this problem, the response should be framed as a *semantic differential scale*. This means that respondents should be provided with a bipolar labeling of the extreme values only (e.g., '1' = not important at all and '5' = extremely important) while intermediate values should not be labeled (as it would be done in a *Likert scale*).

[25] Readers can be referred to Bradburn et al. (2004) or Brace (2008) for an in-depth presentation of this important part of market research.

[26] In other words, in an ordinal scale we know that '2' means more than '1'. However, we cannot be sure that the distance between '2' and '1' is the same as between '5' and '4' (which would be true in an interval scale).

In some cases, a 1–4 scale is used, with the objective of forcing customers away from providing a 'middle-of-the-way' rating. However, this is a questionable choice, since a 'neutral' answer may be genuine and provide a faithful reflection of the way they feel about the item being rated. Moreover, the middle level provides the respondent with an anchor to a neutral response, which can make it less critical to analyze these ratings with statistical tools meant for cardinal data.

- **Perceived quality of existing systems**. If possible, respondents can also be provided with a short list of products or services that are competing in the market and asked to provide feedback on their perceived quality. This ought to be done by using the same list of secondary needs, and asking respondent j to rate the degree with which system l fulfills secondary need i (thus leading to variables y_{il} and individual data points y_{ijl}). The usual 1–5 scale can be used to this purpose.
- **Information related to pricing**. Questionnaires can also include a few questions estimating customers' willingness to pay for the product or service being discussed. More details on this aspect will be provided in the following section.

Based on the dataset that comes out of the survey, it is possible to use multivariate statistics in order to derive primary needs from the importance ratings given to secondary needs (i.e., the variables x_i discussed above).

*Factor Analysis** is the method that is most commonly employed to this purpose. As mentioned earlier, readers are referred to textbooks on multivariate statistics for a deeper understanding of this method. The following discussion has the simple aim of introducing readers to its specific use in the context of market research for product planning.

By operating on the correlation matrix between variables x_i, Factor Analysis models the original variables as linear combinations of a small set of factors f_k (or latent variables). The coefficients of these linear combinations, which are computed with the objective of explaining as much variance as possible, are called factor loadings λ_{ik}. Factor loadings tell the analyst which original variables '*load*' strongly on each factor, thus allowing each factor to be viewed as a grouping of variables that are highly correlated with each other and with the factor. In this specific application, factors represent the latent primary needs that customers tacitly elaborate during the perceptual process, that they would be unable to express, but that can be uncovered thanks to this statistical approach.[27]

In general, Factor Analysis leads to the definition of 2–4 independent primary needs. Using a higher number of primary needs—though it would allow the explanation of a higher percentage of variance—is generally not advisable, since it would run contrary to the principle that human decision-making is carried out by using a very limited number of dimensions. Primary needs can be named, based on the constituent secondary needs. For example, the analysis carried out for a durable

[27] For instance, Factor Analysis may show high loadings on the same factor for the three secondary needs 'I can use my device with one hand', 'I can learn how to use my device quickly', 'if I make a mistake, my device warns me'. The analyst may interpret this latent variable as representing a primary need that can be named 'ease of use'.

Rotated component matrix [a]

	Component		
	1	2	3
SN_confort	-,049	,827	-,054
SN_easy_to_handle	,201	,734	-,067
SN_no_fatigue_when_pedaling	,092	,494	,188
SN_loading_capacity	,279	,311	-,029
SN_use_outside_city	,067	-,109	,448
SN_aesthetics	,011	,323	,711
SN_safe_when_in_traffic	,544	,376	-,016
SN_easy_to_lift_and_store	,696	,291	-,142
SN_does_not_dirty_clothes	,412	,458	,149
SN_no_theft	,674	,216	,027
SN_reliable_sturdy	,699	-,078	,107
SN_easy_maintenance	,665	-,025	,096
SN_accessories	,010	,104	,687
SN_brand	-,002	-,066	,807

Extraction method: Principal Component Analysis.

Rotation method: Varimax with Kaiserís normalization.

a. Rotation has reached convergence criteria in 5 iterations.

Fig. 13.5 An example factor analysis, allowing the identification of tacit primary needs

consumer product such as a city-bike might lead to the three-factor solution shown in Fig. 13.5, in which factor loadings with higher values have been highlighted.

By looking at each factor, the analyst may respectively interpret the three dimensions as 'the bike as a means of transportation' (or 'transportation'), 'the bike as something you ride' (or 'rideability') and 'the bike as an ideal and a status symbol' ('iconic'). The first component aggregates all variables that one would consider when evaluating the possibility of using a bicycle in alternative to other transportation means. The second component relates to variables that are relevant to its actual use, while the third component considers aspects that are somewhat associated to the cultural meaning of the bike.

It may happen that some variables cannot be clearly associated to a component (e.g., loading capacity in the example); from a rigorously statistical viewpoint, these variables could be pruned from the dataset in order to increase the quality of the solution, in terms of variance explained. However, the analyst may decide to keep them in and suggest that the variable pertains to more than one component. In the example, it may in fact make sense to state that 'loading capacity' is relevant to both primary needs 'transportation' and 'rideability'. Not being based on explicit judgment, this method for aggregating secondary needs into primary needs can lead to unexpected groupings. For instance, the previous example shows that the need of "a bike that can be used outside the city" is something that customers associate strongly to needs that have to do aesthetic and 'ideal values' and not (as it would be expected) to the functional need of having a means of transportation.

Factor loadings are not useful just to allow the uncovering of primary needs and to associate them to the constituent secondary needs. Factor loadings also allow computing synthetic values for latent (or tacit) variables, which represent the importance of each primary need 'had the respondent been able to articulate it'. This is important to complete the interpretation of customers' perceptual process.

Based on the ratings that each respondent j has given to secondary needs (i.e., the values x_{ij}) it is possible to compute the rating that the same respondent has implicitly given to each primary need, x'_{jk}. Finally, given the ratings assigned by each respondent j to competing alternatives l with respect to the fulfillment of secondary needs i, it is also possible to derive the rating that would be given to the same systems with respect to the primary needs, y'_{jkl}. Then, by averaging over j, the firm can obtain the implicit rating of that alternative, concerning its fulfillment of primary needs, y''_{kl}.

13.3.5 Perceptual Mapping

Thanks to the calculation of latent variables introduced in the previous subsection, the development team can come up with *perceptual maps**, which are powerful representations of the primary needs arising from the market and of the way with which the industry is fulfilling them.

A first perceptual map (Fig. 13.6) can be plotted by representing each respondent according to the values of x'_{jk}. This perceptual map shows the distribution of customers with respect to their rating of primary needs and, specifically, if there is any concentration of customers around specific areas of the perceptual map. Should this occur, it would be a clear indication of the existence of customer segments and of potential demand associated to each. Moreover, it is possible to look for correlations or associations between the position on the perceptual map to variables related to demographics or usage patterns. This may provide a much richer description of market segments and of the users that make them up.

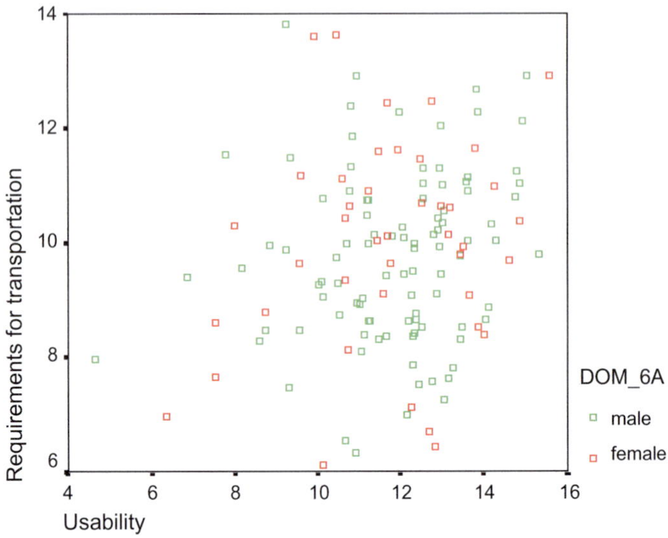

Fig. 13.6 Perceptual mapping of customers

Figure 13.6 provides the results coming from the previous example of the bicycle, and maps the first two components, checking for gender of respondents. The map shows that it is not possible to identify particularly strong groupings, but one can notice that respondents are absent in the top left area of the map, but not in the bottom right. This suggests that there might be a market for bikes with strong 'rideability' and that perform weakly as 'means of transportation', whereas a bike with the opposite features might not find customers. Moreover, observing that position of respondents does not seem to depend on gender, this could be taken as an indication that males and females do not differ with respect to the importance they attribute to primary needs and that—aside from considerations related to sizes and bicycle frame design—it might not make sense to develop separate bike models.

A second perceptual map can be overlaid to the previous one, by adding a plot of the competing products' or services' positioning, based on variable y''_{kl} (Fig. 13.7). This map adds information on how products or services are perceived with respect to the fulfillment of primary needs. This map allows spotting any 'overcrowding' that may lead to higher degrees of competition, and/or whether there is any portion of the market space that is still underserved. Products and services can be coded in order to distinguish between brands, and arrows can be added to show trends. For instance, Fig. 13.7 shows that the 'mid-market' space where most customers exist is well covered by existing alternatives, and that perceived quality is progressively increasing. The map also shows a few possibly underserved parts of the market, especially on the right-hand side, characterized by customers having very high requirements with respect to the 'driveability' need and medium–high expectations with respect to the 'transportation' one.

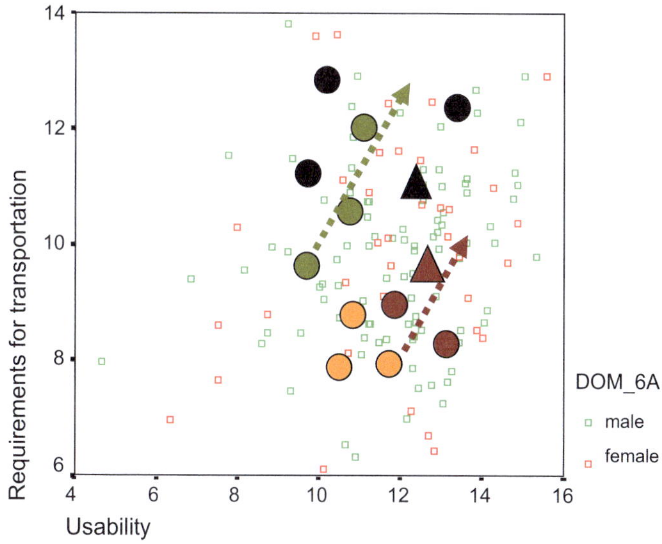

Fig. 13.7 Perceptual mapping of customers and competing products

When looking at this latter perceptual map, some products are distanced from one another by being closer to one or the other axis, but are located at roughly the same distance from the origin. These products differ because producers chose to design them in order to address different primary needs, something that is termed *horizontal differentiation** (a more precise definition will be provided in the following). Conversely, some products are distanced from one another by being located at different distances from the origin, but roughly on the same line that passes by the origin itself. These products differ because of the degree with which they satisfy primary needs, but not because of dissimilar tradeoff choices between these needs. This difference is termed *vertical differentiation** and allows to distinguish between basic (or entry-level) products and premium (or high-end) ones. Of course, basic products can survive in the market because of something that has not yet been captured in the perceptual map, i.e., price. In order to consider this key element as well, the perceptual map can be transformed in a *value map** by dividing the variables y''_{kl} by the price of product l, p_l, as in Fig. 13.8.

Assuming that the market is made up of rational individuals, one would expect the value map to locate alternatives close to a Pareto-optimal frontier, so that no product dominates any other one. Should this not be the case, the likely reason is that the benefit perceived by customers is not proportional to a simple "needs satisfaction/price ratio", but to a relationship y''_{kl}/p^{α}_{l}, where $\alpha \neq 1$. Therefore, one can try plotting the value map for y''_{kl}/p^{α}_{l} with varying values of α, until the products or services come close to the frontier and no alternative appears to be dominant. This information can be very helpful since, should $\alpha > 1$, this implies that customers are willing to pay more than proportionally for high-quality products, and vice versa.

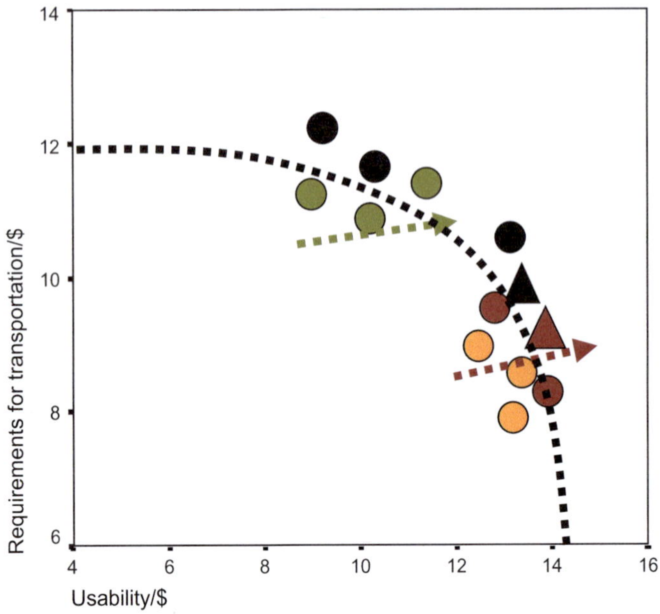

Fig. 13.8 A value map for positioning products

13.4 The Different Importance of Needs: Kano's Model

Up to now, needs have been introduced and aggregated hierarchically without suggesting any difference in their importance, from a strategic marketing perspective.

Discussing the differential importance of user needs will be done in general in this chapter, using Kano's (1984) classification, while we leave to the next chapter the discussion of methods with which needs importance can be treated in detail to support product design. According to Kano, each need and corresponding system attribute can be classified by looking at the way with which the degree of implementation of the latter leads to the satisfaction of the former. This relationship can be mapped on a two-dimensional map, as in the following Fig. 13.9.

Each need and system attribute can then be represented as a function on this graph, leading to distinct types of needs/attributes.

- **Must-be attributes/basic needs**. Customers take the fulfilment of these needs for granted, together with the presence of the associated attributes. Therefore, the former do not generate satisfaction when met, while dissatisfaction emerges if they are not fulfilled. For instance, customers expect bottles not to leak. A customer would clearly be dissatisfied by a leaky bottle, while buying a bottle that does not leak does not generate any satisfaction. Since customers expect these

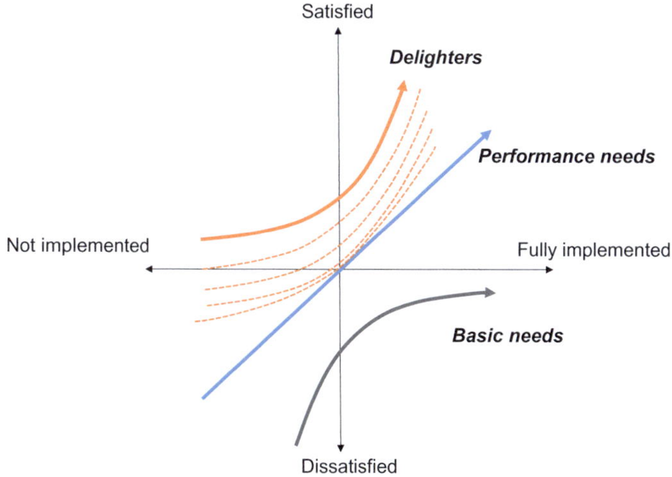

Fig. 13.9 Kano's model links the implementation of attributes to the impact on satisfaction of customer needs

attributes, it is unlikely that they will be able to express them to an interviewer during market research. As shown in the figure, these attributes and needs can be represented as non-decreasing functions in the bottom quadrants (SW and SE) of the graph.

- **Attractive attributes/'delighters'**. Customers do not usually expect the satisfaction of these needs and the presence of the related attributes. Implementing the latter generates satisfaction, but no disappointment is caused in the opposite case. For example, a guest in a hotel may not expect, but greatly appreciate, a voucher for a complimentary dinner during a stay that occurs on her birthday. Also in this case, and being unexpected, it will not be easy to elicit these needs and attributes through market research. Concerning the graph, these attributes and needs can be represented as non-decreasing functions in the top quadrants (NW and NE).
- **One-dimensional attributes/performance needs**. In this case, by not implementing these attributes and the resulting non-satisfaction of a need will generate disappointment, while their implementation will lead to the fulfilment of the need and satisfaction (for instance, the need to minimize a car's operational costs is directly linked to fuel efficiency). Customers and companies are usually quite aware of these attributes and needs and are able to articulate them. The relationship between implementation and satisfaction is a non-decreasing function that straddles the SW and NE quadrants of the graph. Figure 13.9 shows a family of such curves, in which it is made clear that the slope is not necessarily constant, since saturation effects may occur beyond a given limit both on the right-hand and left-hand sides of the graph.

- **Indifferent attributes** do not directly affect customer satisfaction. These attributes are related to technical features that customers are not able to assess, even though they might be essential to the product or service. In the graph, they are located close to the x axis.
- **Reverse attributes**. These attributes refer to features that not all customers might appreciate. For instance, in the case of high-tech products, some customers might prefer basic and easy-to-use models to high-end and complicated ones. In this case, the curve (which has not been drawn in Fig. 13.9) would be non-increasing and straddle the NW and SE quadrants.

At any given time, firms will usually compete with each other by ensuring their product or service comply with must-be attributes, will try to have an edge on their competitors on 'one-dimensional attributes' and try to differentiate themselves by providing some 'attractive attributes'. As time goes, by, customers will become more demanding and tend to get used to features and take them for granted. Therefore, what could have been considered an 'attractive attribute' in the past, may be considered to be a 'one-dimensional attribute' today, and become a 'must-be' in the future.

Kano's model leads to a classification of needs, distinguishing among ones that must necessarily be addressed and other ones that may be considered to be optional.

13.5 Pricing

Price is a key attribute for any product or service and critical to product success. For simple products, pricing can be relatively straightforward, whereas for complex products with many options, or for services, pricing can become a complicated exercise, given the variety of pricing approaches that can be used.

As shown in Fig. 13.10, there are a number of approaches to pricing that companies usually use.

Though it is well-known to be a misleading criterion, many firms tend to tackle pricing from an accounting perspective and use a *cost-plus** criterion. This is done by estimating variable costs and adding a desired margin percentage; if the margin percentage is not constant, but estimated specifically for the product, one would then speak of the *mark-up* method. The mark-up therefore becomes the same as the contribution margin (i.e., the difference between price and variable cost), whose purpose is to cover system-specific fixed costs, general overheads, and provide profit. Alternatively, the firm may work on gross margins (i.e., the difference between price and total product costs, which include both variable and system-specific fixed costs). In this case, the markup must only cover general overheads and provide profit. Though cost-plus pricing theoretically ensures a desired degree of profitability, its use is highly deceptive, since it doesn't consider neither the consumers' willingness to pay, nor the competition. If the firm is not

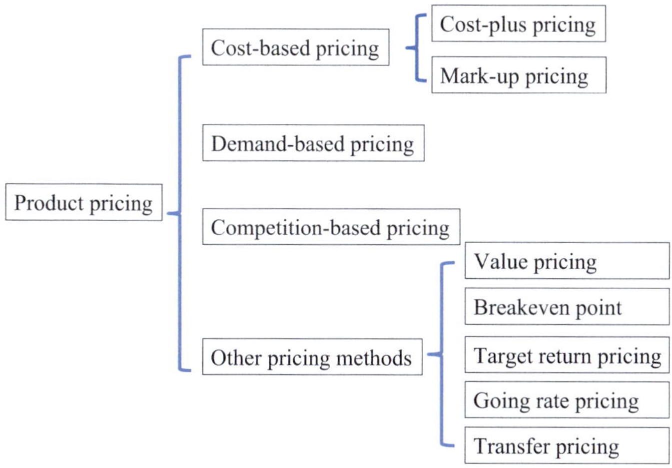

Fig. 13.10 Alternative approaches to pricing

efficient, or consumers' willingness to pay is low, the calculated price may be too high, leading to weak demand and—ultimately—to the inability to cover fixed costs as forecasted. In the opposite cases, the price may be too low, and economic value will needlessly be passed down to consumers, who will get a good bargain.

Economists would suggest setting price by equating marginal cost with marginal revenue, but this requires a precise estimation of the demand curve (i.e., how many customers would buy at each price value). Surveys can attempt to extract this kind of information from a sample of consumers by placing questions such as "what is the maximum amount would you pay for …?". Unfortunately, answers to this very direct question may not be very reliable. Consumers may either answer lightly by stating a high price (after all, they are not actually taking their credit cards out of their wallets), or 'strategically', by declaring a low price (the idea being "if we all state low values, the firm might make the product more affordable"). In some industries, firms have enough experience to amend the values provided by these answers, but this is not always possible.

A more sophisticated way to extract willingness-to-pay information is the Price-Sensitivity Meter (Van Westendorp 1976). Customers are asked to provide four different pricing values for the product or service. The first is a price at which they would not consider buying it, because it is '*too expensive*'. The second is a price at which they would consider it '*expensive*' (i.e., not so high as to rule out the purchase, but enough to require some more thoughts). The third is a price at which they would consider it a '*bargain*' (i.e., they would definitely buy), while the fourth is a price at which they would start doubting of quality ('*too cheap*'). Based on this data, one can plot the four corresponding cumulated frequencies at the different price levels (Fig. 13.11).

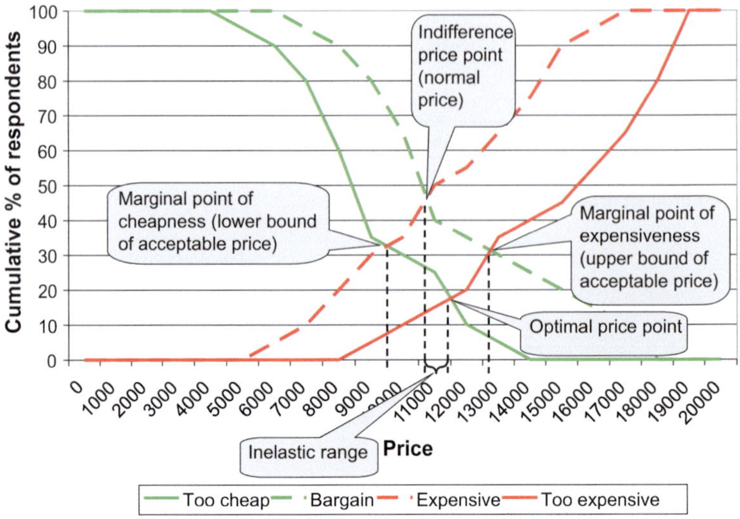

Fig. 13.11 Price Sensitivity Meter approach (Van Westendorp 1976)

The intersections between the four curves provide ranges for pricing. The intersection between the 'too cheap' and the 'expensive' curve can be considered to be the minimum acceptable price, since by decreasing price further, more customers would be lost because of the perception of 'cheapness' than the ones that are gained because of greater affordability. Similarly, the intersection between the 'too expensive' and the 'bargain' curve is an upper limit for price since, by increasing it further, more customers would be lost than gained. At the intersection between the 'bargain' and 'expensive' curves there is equilibrium between the number of respondents who consider the product or service to be cheap and the number who consider it expensive. This point is termed the 'indifference price point' (IPP). Another key point is at the intersection between the 'too expensive' and 'too cheap' curves, which is generally greater than IPP. Proponents of the method claim that the sensitivity of demand between the IPP and this latter point is very low, thus suggesting that price should be set at this very value. For this reason, it is termed optimal price point (OPP).

The above methods are applicable to products that are fairly well known and easy to describe, and for which it is possible to interview customers and ask them for reference pricing levels. When considering services or configurable products (i.e., products that can be customized with combinations of options and variants), the offering can become so highly customized that the very concept of demand curve tends to disappear. Moreover, when dealing with services such as travel, hospitality or cloud computing, customers are sold highly perishable production capacity, while variable costs tend to be negligible. In these cases, it will be very difficult for firms to make pricing decisions ex-ante. Therefore, producers will

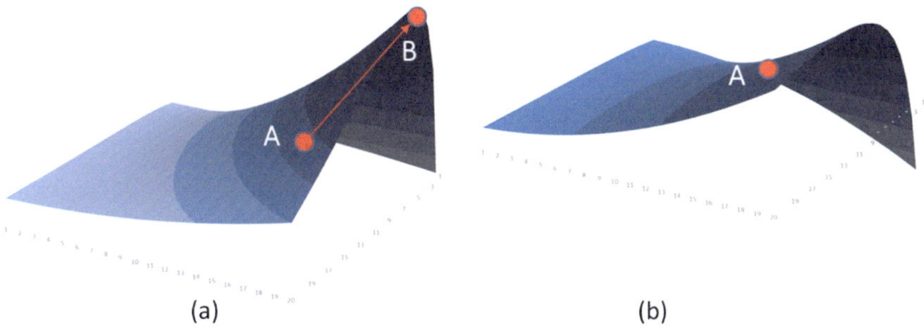

Fig. 13.12 Alternative pricing strategies according the competitive landscape theory

continuously experiment with prices, relating them to competition and to the demand they experience in near-real time.[28]

Other contributions in the literature have also attempted to model customer as 'willingness to pay', or utility, as a multivariate function of product attributes (or 'degree with which user needs are satisfied'). This approach, termed '*Competitive Landscape Theory*' (Levinthal 1997) is not easy to translate operationally, but provides interesting insights that might be of practical use. The main idea is that each customer, or each segment made up of customers that exhibit sufficiently similar utility functions, can be represented by a landscape, as in Fig. 13.12.

Firms can be represented as actors who search for a maximum point on the landscape, either by continuous experimentation, or by trying to model the landscape via market research. For instance, in the left-hand panel, a firm whose product is positioned in A will try to understand whether there is a point B, and where it is located, in order to launch its next product in proximity. In doing this, the firm will not disrupt the landscape to its competitors, who will be looking for the same point *B*.

However, as shown in the right-hand panel, an alternative strategy is possible (Vinokurova 2019) and consists in altering customers' perception of utility via marketing actions. This approach allows the firm to keep the same product and make that particular proposition the most valued. In doing this, the firm will have the additional advantage of disrupting the landscape to its competitors.

This latter strategy depends on whether, in a given landscape, it is easier to modify the product in the direction of optimality, or to modify the landscape itself. In order to implement such strategy, the firm may attempt to transform existing dimensions of

[28] The literature on optimal pricing is a key topic of both Industrial Economics and Marketing and, due to its vastness, it cannot be effectively summarized here, since it depends on the type of goods and the type of competition in the industry. Beyond the few words that have been spent here, it is convenient to remember that many goods and services require specific highly pricing strategies. Among them one can list experience goods (i.e., goods whose quality can be appreciated only after they are purchased; Shapiro 1983) and Veblen goods (i.e., 'luxury' goods, whose demand rises with price; Huang et al. 2017).

evaluation. For instance, a carmaker that aggressively passes a message like 'drive safely' will flatten consumers' reaction to driving performance and enhance their sensitivity to safety devices being adopted. Moreover, a firm may attempt to introduce new dimensions of evaluation by adopting innovative features. For instance, a carmaker might gain a competitive edge by adopting voice control features (which are incidentally quite cheap to implement) and position its products as 'the supercar that obeys to your voice without having to take your hands off the wheel').

13.6 Product Positioning

Product positioning means making a high-level decision on product or service features and on the market segment to be addressed. Thinking in terms of the perceptual maps that have been described in the previous section, it means describing the new system with respect to the way it satisfies primary needs, eventually in relation to competing alternatives, and defining its target customers.

Apart from very peculiar cases, in which the product category is served by offering a single product or service, firms and their competitors will offer a range of differentiated products. The positioning of a new product or service must therefore consider its relation to other alternatives being offered by the same firm. As previously mentioned, differentiation can be *horizontal* or *vertical*, and a basic understanding of these two concepts is quite relevant to product development choices.

13.6.1 Horizontal Differentiation

Horizontal differentiation is defined by the fact that—by varying product features—utility increases for some consumers and decreases for other ones. For instance, one may think of spiciness of food, or sportiness in the exterior design of a car body, more or less of which may be attractive for some, but less attractive for others.

Following the traditional economic discussion of horizontal differentiation (Hotelling 1929), a firm may find value in pursuing horizontal differentiation because of both demand-side and supply-side reasons. From the side of demand, a customer that finds a product or service that exactly fits her tastes will derive greater utility than from a product that is positioned at a distance to them. This greater utility implies that he/she might have a higher reservation price for that product or service, which leads to a potentially higher margin for the producer. From the side of supply, if firms spread their offering across multiple segments on a horizontal spectrum, this might lead to a thinning of competition, since not all firms will operate in each segment (Makadok and Ross 2013). In turn, less competition increases the likelihood of actually earning these higher margins.

Conversely, horizontal differentiation leads to the partitioning of aggregate demand in many segments. Lower volume in each segment might be unattractive and damaging to profitability, by forcing to give up on economies of scale that could be pursued with a single 'central' offering. This explains why firms usually try to achieve economies of scope, i.e., the possibility of reducing cost by widening the range of products being made, through shared platforms and flexible manufacturing systems.

From a practical perspective, product positioning along a horizontal dimension will require to estimate the size of each market segment, of their customers' willingness to pay and of the sensitivity of demand with respect to offerings that do not exactly match needs. The higher each of these factors, the more likely it will be that a product oriented to the segment will be economically viable, even if this entails a given amount of specific investment.

13.6.2 Vertical Differentiation

In the case of vertical differentiation, the variation of product features will lead to a corresponding variation of customer utility that will have the same sign for all customers. For instance, a car with greater mileage will always be better than a car with less mileage, though some customers might be more sensitive to this variation than other ones. System features leading to vertical differentiation will generate a segmentation that runs from 'entry level' alternatives that are perceived to be of lower attractiveness (and will have to be priced at a lower level) to 'premium' products that are highly desirable (and will be able to command higher prices).

As in the previous case, the economic reasons behind vertical differentiation have to do with both demand side and supply side aspects. Looking at demand, vertical differentiation allows price discrimination, so that customers who highly appreciate quality and are prepared to pay for it are effectively charged a high price, while customers with lower willingness to pay are charged correspondingly, and receive a lower-value product. By not practicing vertical differentiation, a firm would have to position a single product or service at an intermediate level of quality and price. The firm would therefore lose part of demand, made up of customers with a lower reservation price. Moreover, it would also lose part of the potential margins, from the customers whose reservation price is higher than the practiced one. As in the case of horizontal differentiation, by spreading offerings along vertically differentiated segments, competitive pressure in each segment might be reduced, which allows firms to retain the margins that come from price discrimination.

Vertical differentiation too leads to a smaller volume sold to each segment and, therefore, to the risk of giving up potential economies of scale. A further and particularly important risk connected to vertical differentiation is *cannibalization*. This occurs when customers who would be willing to buy a high-level product opt for a low-level alternative, which is therefore said to cannibalize the former.

Cannibalization can be explained by referring to the two cases in Fig. 13.13. The graphs show the way reservation prices change as a function of quality for two customer segments, labeled H (for 'high') and L (for 'low'). Based on these curves,

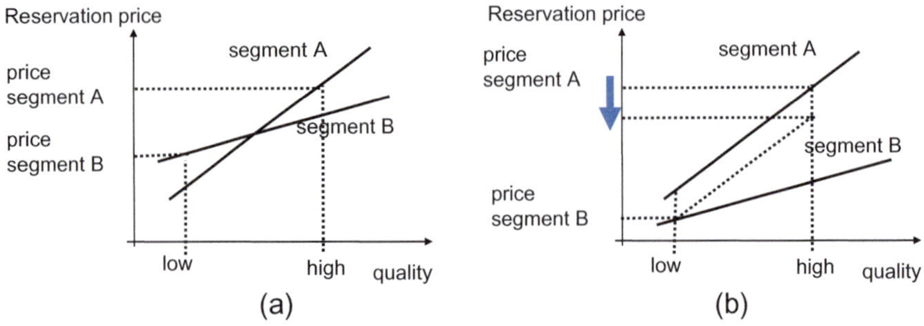

Fig. 13.13 Economics of cannibalization due to vertical differentiation

a firm can identify a 'high' price and a 'high' quality for the corresponding segment, and a 'low' price and quality for the other.

In the graph on the left-hand side, the utility curves of the two segments cross. In this case, cannibalization does not occur, and customers in each segment will buy the product or service that is aimed at them. Specifically, customers in the 'high' segment find a 'high-quality' systems at a price equal to their reservation price for that quality and will therefore buy it. These same customers will not buy the 'low-quality' alternative, since their reservation price for that product or service is less than the practiced price. If we now look at the 'low' segment, these customers will accept buying the 'low-quality' alternative at the corresponding reservation price. They will not buy the 'high-quality' product or service, since its price is greater than their reservation price for it.

In the graph on the right-hand panel, utility curves do not cross, leading to cannibalization. In this case, the 'low' segment keeps on buying the 'low-quality' product or service. However, customers in the 'high' segment will compare the two alternatives and find out that—while their reservation price for the 'high-quality' system is the same as the practiced price—their reservation price for the low-quality product or service is higher than the practiced price. They will therefore prefer the latter alternative, since it gives them come customer surplus, while the former one would leave them with none.[29] From the perspective of the producer, this is likely to be damaging, since margins coming from the high-end system are usually higher than the ones accruing from the low-end one.

Due to the risk of cannibalization, firms with a vertically differentiated alternative range must pay attention to the origin of its buyers and monitor their behavior. While clients who *expand* the customer base are of course welcome, some clients may be actually *downshifting* from higher-quality products, and therefore reduce the overall profitability of the firm. In the example shown in Table 13.1, luxury carmaker ABC tries adding a 'luxury compact' car to its product range, which cannibalizes sales from the adjoining 'business' segment and—by reducing

[29] As an example of cannibalization, think of a luxury car manufacturer that starts selling compact cars alongside to traditional sedans, and discovers that customers of the latter are moving to the cheaper model.

brand equity—leads to a small decrease in the sales to higher-level segments. Overall, volume and dollar sales increase and the new 'luxury compact' model seems to be a 'rising star' in the carmaker's product range. However, a closer analysis shows that overall margins decrease, and that the main responsible is this apparently beneficial new car model.

In order to avert cannibalization, a firm can use a number of strategies, among which (Moorthy and Png 2002; Krishnan and Zhu 2006):

- Decreasing the price of the high-level product or service, in order to leave the customers in the 'high' segment with the same surplus they would have by purchasing the low-level system. This strategy is easy to enact, but has the obvious disadvantage of reducing profits.
- Developing the low-level product or service so that the utility curves intersect. This can be done by purposely degrading the quality of the low-level alternative, until the quality and price fall to the left of the intersection between the utility curves[30] (e.g., ABC could strip the 'luxury compact' car of features and accessories that are usually desired by buyers of that brand). This strategy has the drawback of affecting the market for the 'low-level' product and bears the risk of reducing brand equity for the whole product range (e.g., "ABC is not making cars like they used to!").
- Introducing horizontally differentiating features in the low-level product or service, in order to make it more appealing to the 'low' segment and less to the 'high' segment (e.g., ABC could aim the 'luxury compact' car toward a younger audience, give it a more aggressive styling and a slightly less comfortable ride). In this case too, the firm must be careful not to spoil the attractiveness of the brand.
- Enacting marketing actions aimed at reducing the 'high' segment's perception of utility for the low-level product, thus making the utility curve for the 'high' segment steeper. This strategy is similar to the previous two ones, but it is played on customers' perceptions, instead of objective features. For instance, TV commercials for the 'luxury compact' cars could feature young people in their 20s, thus conveying the image that it is not the right model for older and more affluent customers. The risk of damaging the perception of the brand is of course always present.
- Giving up the idea of vertical differentiation and pursuing a 'niche' product strategy, producing the 'high level' or the 'low level' alternative only. A similar strategy would consist in selling the two alternatives under two different brands. However, introducing a new brand to the market is not a small endeavor and requires significant investment.

[30] In 1849, the French economist Jules Dupuit (1804–1866) noticed that railways companies were purposely making third-class carriages uncomfortable not to reduce cost, but in order to 'scare away' customers who had enough money to buy second- and first-class tickets.

Table. 13.1 Cannibalization can lead to profit-damaging vertical differentiation

Before introduction of luxury compact					After introduction of luxury compact					Δ
	Business	Premium	Top	Total	Luxury compact	Business	Premium	Top	Total	
Average unit price (Euros)	30,000	50,000	70,000		25,000	30,000	50,000	70,000		
Contribution margin	25%	30%	35%		15%	25%	30%	35%		
Unit contribution margin (Euros)	7500	15,000	24,500		3750	7500	15,000	24,500		
Volume sales (thousand units)	750	400	200	1350	600	500	350	190	1640	21.48%
	56%	30%	15%	100%	37%	30%	21%	12%	100%	
Sales (million Euros)	22,500	20,000	14,000	56,500	15,000	15,000	17,500	13,300	60,800	7.61%
	40%	35%	25%	100%	25%	25%	29%	22%	100%	
Margins (million Euros)	5625	6000	4900	16,525	2250	3750	5250	4655	15,905	− 3.75%
	34%	36%	30%	100%	14%	24%	33%	29%	100%	

- Using time as a further discriminating variable and launching the 'high-level' product or service sufficiently earlier than the 'low-level' one. Of course, this strategy has a negative impact on the sales of the latter system.

13.7 Demand Forecasting

One of the main elements to be considered when making product positioning choices is potential demand. In general, the forecasting of demand for a product j can be done by using the following simple model

$$D_j = \mathrm{DS}_j \mathrm{AW}_j \mathrm{AV}_j \tag{13.2}$$

where

D_j	is demand for the system j,
D	is a forecast of overall demand in the market segment,
S_j	is the market share that j could potentially achieve, based on its features, and is therefore directly connected to design decisions,
AW_j	(awareness) is the fraction of target customers who can be expected to know of the system's existence, which is related to marketing actions that the firm will undertake to this purpose,
AV_j	(availability) is the fraction of target customers who will be able to purchase the product or service, based on the distribution strategy followed by the firm.

When following this model, a well-focused definition of the market segment and of competing products is crucial. In fact, it would be absurd to define a broad market with hundreds of quite different competing system alternatives and to estimate a minuscule market share. On the other hand, it could be dangerous to define a very narrow segment with very few offerings (at the limit, only one), but neglecting the potential competition coming from contiguous segments.

The following subsections will outline a few elements that can be used in estimating the first two parameters, which are more directly linked to choices made in the product development process.

Recent literature from digital marketing has introduced a slightly different version of this model, which operates in reverse, since it introduces awareness and availability before (and not after) having estimated the market share achievable based on the attractiveness of the product. According to this model, one can frame the following 'markets':

- A *Potential Addressable Market* (PAM) that is a superset of D, and includes the market that is potentially reachable by the firm in perspective, after having secured a foothold in the current segment.

- A *Total Addressable Market* (TAM) that is equivalent to D, i.e., the size of the addressed market segment. So,

$$\mathrm{TAM} = D \tag{13.3}$$

- A *Serviceable Addressable Market* (SAM), which is the portion of TAM that the firm can reach, given its marketing and distribution strategy. So,

$$\mathrm{SAM}_j = \mathrm{TAM}\ \mathrm{AW}_j \mathrm{AV} \tag{13.4}$$

- A *Serviceable Obtainable Market* (SOM), which is the share S_j of SAM that the firm can conquer, given the features of its product. So,

$$\mathrm{SOM} = S_j \mathrm{SAM} \tag{13.5}$$

13.7.1 Product Demand in Stationary Markets

When dealing with products that are in the 'mature' phase of s-curves, demand can be expected to be constant. Alternatively, demand may follow some trend, or it might be correlated to other phenomena that drive sales. In all these cases, demand can be considered to be 'stationary' in the sense that is not affected by the dynamics of diffusion effects. In these cases, demand forecasting is performed by using standard time-series analysis methods coming from statistics and econometrics. We will not discuss this topic, and refer readers to specialized textbooks (e.g., Box et al. 2008; Jain and Malehorn 2005).

Here, we just mention the difference between autonomous and dependent models. In the former case, the firm observes demand over a stretch of time and tries to extrapolate it into the future by referring to the past values of the demand time series only. This may occur when demand is either constant or follows a constant trend (Fig. 13.14).

Should demand exhibit a more complex pattern, this might indicate that an underlying and independent phenomenon is driving sales. A forecast can therefore be performed by studying the correlation with this underlying phenomenon, and then using independent forecasts of the independent variables. For instance, sales of child seats for cars are dependent on birthrate and car ownership. A company having to make sales forecasts for this product can obviously derive them from demographic forecasts by national statistics offices (concerning the former factor) and from forecasts by industry bodies for the latter one (Fig. 13.15).

As a final note, it should be considered that measuring demand is not a trivial task. Sales data at market level may be difficult to access, and sources usually include specialist market research firms, or trade associations. Moreover, one can

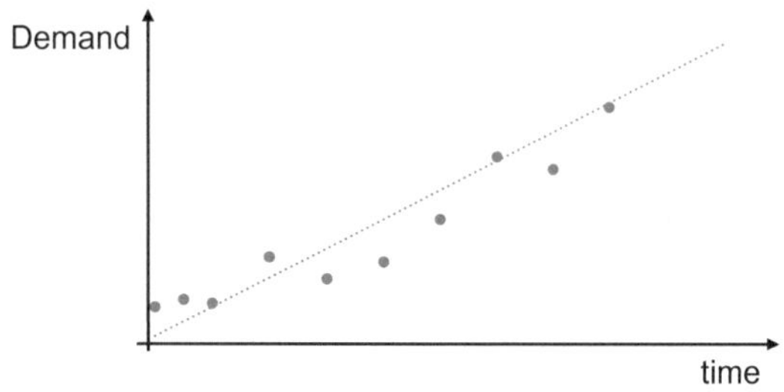

Fig. 13.14 Example of a simple time-series analysis leading to a forecast by extrapolation

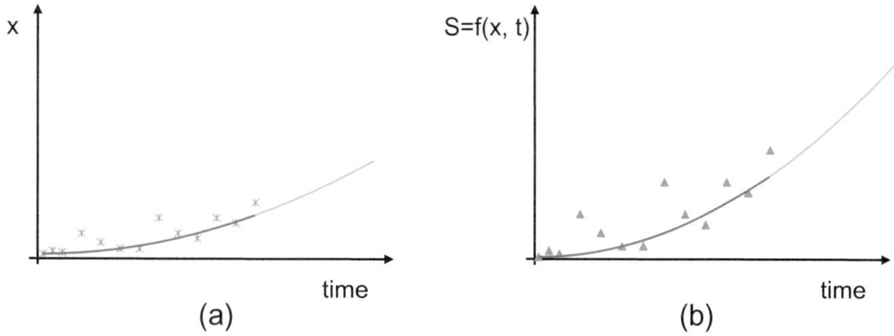

Fig. 13.15 Example of a forecast based on a trend and a correlation with an independent variable

observe sales, but not demand. In most instances the two coincide, but this is not necessarily true. If supply is restricted, or if availability and awareness are limited and do not allow to cover the whole market, potential demand could be quite larger than sales and be therefore underestimated.

13.7.2 Product Demand Subject to Diffusion Phenomena

13.7.2.1 Bass' Model of Diffusion

When dealing with innovations that are in the incubation or diffusion stage, demand must be studied by explicitly considering the diffusion phenomenon.

Early scholars who studied diffusion phenomena focused on two alternative diffusion paths, one following a *negative exponential* law (Fourt and Woodlock 1960) and one following a *logistic* (s-shaped) curve (Mansfield 1961). In 1969, Frank Bass (1926–2006) merged these two approaches into a model that bears his name and still is the most widely used representation for this kind of phenomena.

Bass' model is based on a set of apparently quite restrictive hypothesis, which are listed in the following. Variants of Bass' model, which will be discussed later on, allow the relaxation of some hypothesis, justifying a broad application of the model.

- Demand is modeled at the level of the overall market. Using the model at the level of sales for an individual firm is justifiable in the case of monopolies. In other cases, the model will require estimates for market share, awareness and availability;
- The product being studied is durable, without any substitution or additional sales. Should this not be true, the model can validly represent diffusion sales (i.e., the first sale occurring to customers who were not previously adopters of the good), leaving to further modeling the task of computing subsequent sales;
- Diffusion of the good being studied is completely independent of demand for other goods. This hypothesis is very strong, since it means that the good is so innovative that it is neither a substitute nor a complement to any other good being sold on the market;
- Marketing actions by the firm (i.e., price, promotion and advertising) are constant throughout the diffusion phenomenon. This is another very strong hypothesis, since firms normally vary marketing effort throughout the product lifecycle, but we will later show that it is practically less severe than it appears;
- Customers' adoption process is binary, in the sense that customers' purchases are limited to one item. This can be easily accepted for adoption sales in consumer markets, but rules out the possibility of using the model for corporate sales, especially if these include a limited number of very large customer companies.

Based on these hypotheses, Bass' model can be defined by the following difference equation

$$n_t = N_{t+1} - N_t = p(M - N_t) + \frac{q}{M}(M - N_t)N_t \tag{13.6}$$

where

n_t is sales at time t,
N_t are cumulated sales at time t,
M is the market size and represents the saturation level for the diffusion phenomenon,
p is the 'innovative adoption parameter',
q is the 'imitative adoption parameter'.

Bass' model posits that, at any given point in time, diffusion occurs because a subset of non-adopters, $M - N_t$, switches to the status of adopters. This switch can occur because of the sum of two phenomena.

The former is called '*innovative adoption*' and depends on non-adopters' willingness to adopt, combined with communication acts that are broadcast by producers to induce them into adoption. This phenomenon is modeled by hypothesizing that, at any time unit, a fixed proportion p of non-adopters will switch to the status of adopters.

The latter is termed '*imitative adoption*' and depends on effects that are internal to the market, including word of mouth (i.e., the influence that adopters cast on non-adopters) and network externalities. The phenomenon is modeled by stating that, at any time unit, the number of imitative adopters will be a fixed proportion q of non-adopters, weighted by the fraction of adopters (who are the ones who generate the imitative effect).

Bass' model can also be written as a differential equation:

$$n(t) = \frac{\mathrm{d}N(t)}{\mathrm{d}t} = p[M - N(t)] + q\frac{M}{N(t)}[M - N(t)] \tag{13.7}$$

which integrated, leads to a closed-form solution for the evolution of sales

$$n(t) = M\frac{(p+q)^2 \mathrm{e}^{-(p+q)t}}{p\left(1 + \frac{q}{p}\mathrm{e}^{-(p+q)t}\right)^2} \tag{13.8}$$

while the evolution of the diffusion process (i.e., cumulated sales) is given by

$$N(t) = M\frac{1 - \mathrm{e}^{-(p+q)t}}{1 + \frac{q}{p}\mathrm{e}^{-(p+q)t}} \tag{13.9}$$

The time at which the sales peak occurs is located at the extremum where $\frac{\partial n(t)}{\partial t} = 0$, and its value is given by:

$$t^* = \frac{1}{p+q}\ln\left(\frac{p}{q}\right) \tag{13.10}$$

Therefore, at this extremum, sales amount to:

$$n(t^*) = \frac{M(p+q)^2}{4q} \tag{13.11}$$

Finally, the penetration level that is reached at the sales peak is given by:

$$N(t^*) = M\left(\frac{1}{2} - \frac{p}{2q}\right) \sim \frac{M}{2} \tag{13.12}$$

In most cases, $p \ll q$, and Eq. (13.9) shows that, whatever the values of q and p, the inflection point will be reached when approaching half of the maximum penetration level. This can be seen as a useful 'rule of thumb' and allows foreseeing the position of the sales peak if M is known or, conversely, to estimate M when the inflection point has been reached.

When attempting to estimate the parameters in Bass' model, one must refer to the finite difference Eq. (13.6) and transform it in the autoregressive form (13.13)

$$N_{t+1} = pM + N_t(1 - p + q) - \frac{q}{M}N_t^2 \tag{13.13}$$

Based on (13.13), parameters can be estimated quite straightforwardly, but the strong multicollinearity between variables N_t and N_t^2, significantly weakens the statistical reliability of the results.

Therefore, one might get good estimates only when a sufficiently long time series is available, though this means that the initial part of the diffusion process is over, and there probably is limited need for a forecast of this type. Because of this contradiction, forecasts based on Bass' model are usually run by referring to analogies and by using values of p and q that occurred for similar products in similar markets (Mahajan et al. 2000; Satoh 2001).

13.7.2.2 Managerial Implications of Bass Diffusion Process

As shown in Fig. 13.16, the shapes of diffusion curves resulting from Bass' model depend on values p and q. In case of predominantly innovative diffusion (i.e., if q is negligible) as in the left-hand panel of the figure, the curve will get close to a negative exponential. This is typical of fast-moving and low-ticket consumable products, for which adoption choices are relatively frictionless (e.g., a new type of toothpaste is likely to be adopted for a first trial quite quickly by the market, essentially because of advertising).

Conversely, in case of mainly imitative diffusion (i.e., if p is very small with respect to q), the curve will resemble a logistic curve. This is typical of durable products whose adoption requires a significant outlay. In these cases, customers will hesitate and wait for confirmations of the product's validity coming from their peers, before actually 'taking the plunge' and purchasing it.

The shape of the adoption curve depends on the nature of the product, but can also be influenced by the firm. Innovative adoption can be stimulated by advertising

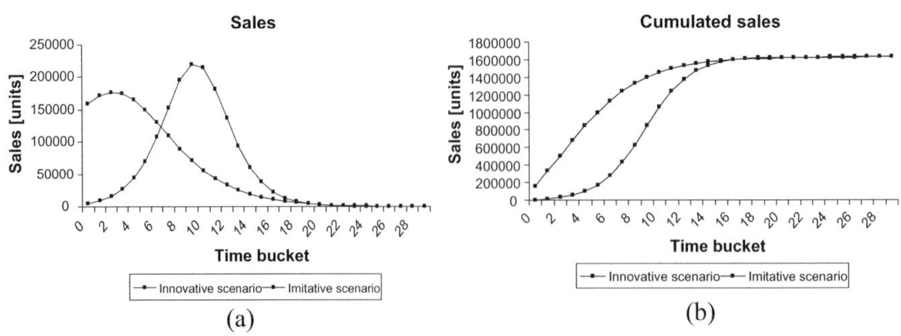

Fig. 13.16 Weight of innovative and imitative adoption determines the shape of diffusion curves in Bass' model

spending, while the imitative phenomenon can be influenced by actions that are often described as *viral marketing**. Examples range from referral coupons (e.g., 'bring a friend and gain something') to introducing features that stimulate network externalities (e.g., 'shoot a movie and share it with your friends').

Whether a firm should prefer a rapid diffusion, as in the case of mainly innovative diffusion, or a slower one, as in the case of imitative, depends on the nature of the industry, of the goods being sold and on the type of production capacity. This latter factor is quite important, since innovative diffusion requires maximum production capacity to be available at a very early stage of the diffusion process, whereas imitative diffusion allows a progressive build-up of capacity. Moreover, with imitative diffusion, the time lag between the start of the process and the sales peak allows the firm to build up inventory and/or to install a somewhat lower capacity and then increase output by relying on learning economies.

Therefore, firms will strive to achieve innovative diffusion if there is some kind of early mover advantage to be exploited, provided the investment required to acquire a customer is not excessive and provided production capacity is not an issue. This can be the case of goods that can be produced with non-specific facilities, such as books to be printed, or printed circuit boards to be assembled. It can also be the case of goods that lead to repeat business (e.g., services or consumable products), since in this latter case the required capacity depends on the repeat consumption rate and the number of adopters being reached, rather than on the maximum adoption rate. When capacity is a constraint, and if the firm aims for a quick, innovative-diffusion type of process, it is customary to stagger the launch date of the new product across geographical markets, in order to smooth out the sales peak.

13.7.2.3 Some Variants of the Bass' Model

As mentioned above, Bass' diffusion model is based on restrictive assumptions, and a number of variants have been developed over the years, each one aimed at relaxing one or the other of these assumptions. We will briefly discuss some of them in the following.

- **Variable market potential and price**. The market potential of an innovation may vary in time because price is not held constant. For instance, in a competitive market, price will tend to decline, due to the decrease in costs that is determined by learning effects. At the same time, the economic environment may change, leading to a greater or lesser willingness to pay. Horsky (1990) attempted to model this case by modifying the Bass differential Eq. (13.7) as follows:

$$n(t) = \left[\frac{M}{1+e^{-\frac{k+w(t)-\mathrm{pr}(t)}{\mathrm{var}(w(t))}}} - N(t)\right][p+qN(t)] \tag{13.14}$$

where

M is the maximum potential for diffusion,
pr(t) is the time-varying price of the good,
K is the utility of the good,
$w(t)$ is the available income within the population, expressed as a stochastic variable with expected value $E[w(t)]$ and variance $V[w(t)]$.

- **Variable price and marketing actions**. In 1994, Bass proposed what he called a 'generalized' version of his original model. As shown in (13.15), instantaneous sales depend on the relative rate of change of price pr(t) and advertising effort $A(t)$:

$$n(t) = [M - N(t)][p + qN(t)]x(t) \tag{13.15}$$

$$x(t) = 1 + \beta_p \frac{\mathrm{dpr}(t)/\mathrm{d}t}{\mathrm{pr}(t)} + \beta_A \frac{\mathrm{d}A(t)/\mathrm{d}t}{A(t)} \tag{13.16}$$

Now, if pr(t) and $A(t)$ do change, but their relative rate of change is constant over time, $x(t)$ will effectively remain constant. In this case, the generalized Bass' model falls back into the original Bass' model, only with modified values for p and q.

The generalized Bass model is seldom used as such, but provides an interesting perspective that justifies the use of the *standard* Bass' model beyond some of its restrictive assumptions, simply if the rate of change of pr(t) and $A(t)$ are constant. Concerning the former, price is generally tied to variable cost and, as it will be seen in Chap. 14, variable cost drops in time following laws that are close to the exponential (which implies a constant rate of change). Concerning advertising expenditure, we know that the evolution of sales at the beginning of the diffusion s-curve is close to an exponential. Now, if the firm defines its advertising budget as a constant percentage of sales (*bootstrapping**), advertising expenditure will follow the same evolution too.

- **Complementarity or substitution effects**. The hypothesis that the good does not exhibit any kind of economic relationship with other goods is of course very strong. A substitution effect due to competition with other goods is generally present. Similarly, the complementarity with other goods can often be a strategic element that has to be considered in order to support the diffusion process. Bass' model can be modified in order to represent the joint diffusion of related goods through Eqs. (13.17) and (13.18):

$$n_1(t) = p_1[M_1 - N_1(t)] + q_1 \frac{M_1}{N_1(t)}[M_1 - N_1(t)] + r_{21}\frac{M_2}{N_2(t)}[M_1 - N_1(t)] \tag{13.17}$$

$$n_2(t) = p_2[M_2 - N_2(t)] + q_2 \frac{M_2}{N_2(t)}[M_2 - N_2(t)] + r_{12}\frac{M_1}{N_1(t)}[M_2 - N_2(t)] \tag{13.18}$$

This model adds a cross-goods diffusion effect, in which sales of good 1 also depends, through an additive term, on the penetration reached by good 2. The phenomenon is governed by parameters r_{21} and r_{12}, which can be either positive or negative, depending on the type of relation between the goods.[31] The model is quite straightforward to understand, though its use can be hindered by the difficulty of finding a proper estimate for the two parameters.

- **Two-sided markets**. This can be viewed as a special case of the previous one. Two-sided markets are based on a platform allowing the interaction between different groups (customers, individuals or companies). Each group involved receives benefit from the use of the platform, and the demand of each group dependents on the demand of the other. In these cases, the diffusion of an innovation on one side determines diffusion on the other, and vice versa. Chun and Hahn (2008) have shown that, being—being α the coefficient of influence with which one side of the market influences the demand on the other side—the differential equation for Bass' model can be written *for both sides* as in (13.19):

$$\begin{aligned} n(t) &= \left[p + qN(t)\frac{\alpha e^{N(t)}}{\alpha e^{N(t-1)}}\right][1 - N(t)] \\ &= \left[p + qN(t)e^{[N(t)-N(t-1)]}\right][1 - N(t)] \end{aligned} \tag{13.19}$$

- **Replacement sales**. In the case of consumables or durables with finite durability, sales are not limited to adoption sales since customers will purchase the item more than once. Equation (13.20) shows that overall sales at time t will be given

[31] The two parameters are not symmetrical and depend on the specific products. For instance, in the case of BEVs and domestic recharging stations, it is likely that the former will drive the latter (i.e., people who buy a BEV will tend to buy a wallbox charger for their garage), and not vice versa. In fact, few people would buy a BEV because they have bought a wallbox charger for their garage, or because they have found one when purchasing the garage. On the opposite, in the case of BEVs and street-based recharging stations, the latter will quite heavily drive the former (i.e., people will buy BEVs when enough chargers are visible on the streets) and the former will somewhat drive the latter (i.e., recharging service operators will adapt their offer by observing the diffusion of BEVs).

by the sum of adoption, replacement and additional sales of the ith item (i.e., the sales to customers who buy their 2nd, 3rd, 4th, etc. similar product).

$$s_t = n_t + r_t + \sum_{i=1}^{l} v_{i,t} \tag{13.20}$$

Looking at (13.20), adoption sales can easily be modeled as previously discussed. Substitution sales can instead be modeled by using reliability functions R_t in (13.21), which express the probability that a good is still usable t time units after its initial purchase.

$$r_t = \sum_{t'=1}^{t} (R_{t-t'} - R_{t+1-t'}) s_{t'} \tag{13.21}$$

where the term in brackets represents the probability of a good purchased in t' to fail between t and $t + 1$, thus leading to a replacement sale. Overall replacement sales at time t will then be given by the sum of the replacement sales deriving from the sales made at all previous purchasing instants.

The modeling of additional sales is quite involved and for the sake of brevity, we will not discuss them here (more can be found in Ratchford et al. 2000). The basic approach is to estimate the probability with which customers already owning i' similar items may decide to purchase their $i' + 1$th, and then multiply it by the number of such customers.

Alternatively, a very simplified relationship for substitution and additional sales (which we can collectively call repeat sales rs_t, as in (13.22)) is given by:

$$\mathrm{rs}_t = \frac{N_t}{T} = \alpha N_t \tag{13.22}$$

where N_t is the stock of adopters at time t and T is the average time between substitutions, while its inverse, α is the rate of substitution.

This relationship is generally valid at steady state, but can be misleading during the early phases of diffusion, when not enough time has passed for products to need replacement or induce additional sales. However, if one does assume (13.22) to be valid, and by moving from discrete to continuous time, the total sales $\mathrm{ts}(t)$ will amount to

$$\mathrm{ts}(t) = n(t) + \mathrm{rs}(t) = M\left[\frac{(p+q)^2 \mathrm{e}^{-(p+q)t}}{p\left(1+\frac{q}{p}\mathrm{e}^{-(p+q)t}\right)^2} + \alpha\frac{1-\mathrm{e}^{-(p+q)t}}{1+\frac{q}{p}\mathrm{e}^{-(p+q)t}}\right] \tag{13.23}$$

It can be shown that cumulated repeat sales RS(t) are given by (13.24)

$$\begin{aligned} \mathrm{RS}(t) &= \int \mathrm{rs}(t)\mathrm{d}t = \int M\alpha \frac{1-\mathrm{e}^{-(p+q)t}}{1+\frac{q}{p}\mathrm{e}^{-(p+q)t}}\mathrm{d}t \\ &= \frac{M\alpha}{q}\left[\ln\left(p\mathrm{e}^{(p+q)t}+q\right)-pt-\ln(p+q)\right] \end{aligned} \tag{13.24}$$

and therefore, total cumulated sales TS(t) amount to

$$\begin{aligned} \mathrm{TS}(t) &= N(t)+\mathrm{RS}(t) \\ &= M\left\{\frac{1-\mathrm{e}^{-(p+q)t}}{1+\frac{q}{p}\mathrm{e}^{-(p+q)t}}+\frac{\alpha}{q}\left[\ln\left(p\mathrm{e}^{(p+q)t}+q\right)-pt-\ln(p+q)\right]\right\} \end{aligned} \tag{13.25}$$

When dealing with services, the case of *recurring revenues* (e.g., monthly subscription fees, or average monthly spending per user) allows to identify the average revenue per user, or ARPU. This case is very similar to repeat sales, since the average consumer can likened to someone that is repeatedly buying a new item with value ARPU every month. So, if one neglects the *churn rate* (i.e., the % of customers who abandon the service each month), the instant sales in value $S(t)$ will be given by

$$S(t) = \mathrm{ARPU}\,N(T) \tag{13.26}$$

And the cumulated sales in value RS(t)

$$\begin{aligned} \mathrm{RS}(t) &= \int S(t)\mathrm{d}t = \int \mathrm{ARPU}\,M \frac{1-\mathrm{e}^{-(p+q)t}}{1+\frac{q}{p}\mathrm{e}^{-(p+q)t}}\mathrm{d}t \\ &= \frac{\mathrm{ARPU}\,M}{q}\left[\ln\left(p\mathrm{e}^{(p+q)t}+q\right)-pt-\ln(p+q)\right] \end{aligned} \tag{13.27}$$

13.7.2.4 Other Diffusion Models

Recent advances in the field of innovation diffusion models have been based on an explicit representation of customer choice over time. *Choice-type models* derive from Economics; the main difference with Bass-type ones is that they focus on individual choices. The choice process is represented by a tree diagram, where the nodes are the decisional steps, and the branches are the probabilities that the users will make a specific choice.

Decker and Gnibba-Yukawa (2010) have proposed a model in which a product evolves along the s-curve, with this evolution being represented by a quality-adjusted price at time t, P_t. At each instant, potential adopters evaluate the expected utility $E[U_t]$ they would derive by adopting at time t, where U_t is a function of P_t and N_{t-1} (which includes word-of-mouth and network externality effects). Moreover, they evaluate the expected utility they would derive by postponing the adoption until $t + 1$ and discount it by an '*impatience factor*' $\delta \prec 1$. If $E[U_t] > \delta\, E[U_{t+1}]$ adoption will occur. Conversely, if $E[U_t] < \delta\, E[U_{t+1}]$, the adoption choice will be shifted to the next time unit. The authors show that the difference equation that captures this behavior is given by Eq. (13.28)

$$n_t = \frac{M - N_{t-1}}{1 + \mathrm{e}^{-[\beta_0 + \beta_1 \ln(N_{t-1}) + \beta_2(P_t - \delta P_{t+1})]}} \tag{13.28}$$

where

β_0 is a constant that represents innovative diffusion (in fact, should $\beta_1 = 0$ and $\beta_2 = 0$, Eq. (13.27) would fall back to the description of a purely innovative diffusion phenomenon);

$\beta_1 > 0$ represents imitative diffusion including network externality effects (the greater β_1 and the greater is the stock of adopters, the higher the sales);

$\beta_2 < 0$ represents the '*wait and see*' attitude followed by potential adopters. If $P_{t+1} \ll P_t$ this will reduce the number of new adopters at time t, since many will postpone adoption. Conversely, if $P_{t+1} > P_t/\delta$, potential customers will adopt at time t because their impatience is greater than the benefits coming from the expected price reduction.

The previous models show that innovation diffusion phenomena have traditionally been studied analytically, using differential equations. This approach does not allow the modeling of complex decision-making behaviors and the heterogeneity that is inherent to the population.

An alternative approach has recently emerged, with the use of agent-based simulation (Kiesling et al. 2012). In agent-based simulation, the diffusion of an innovation is modeled on the basis of a microlevel representation of a population of individual agents, with an explicit coding of their individual decision-making process leading to adoption, of their reciprocal interactions based on underlying social networks, and of their heterogeneity. Of course, as is the case for all powerful computational tools, the ability to represent very minute facets of the system's behavior leads to the challenge of identifying relevant parameters in a reliable way, which might not be an easy task.

As a final comment, demand forecasting is a completely different endeavor when diffusion depends on one-to-one marketing processes (Monat 2011) constrained by a budget. This is typically the case for B2B products, whose sales cycle generally follows a complex process such as the one represented in the following Fig. 13.17,

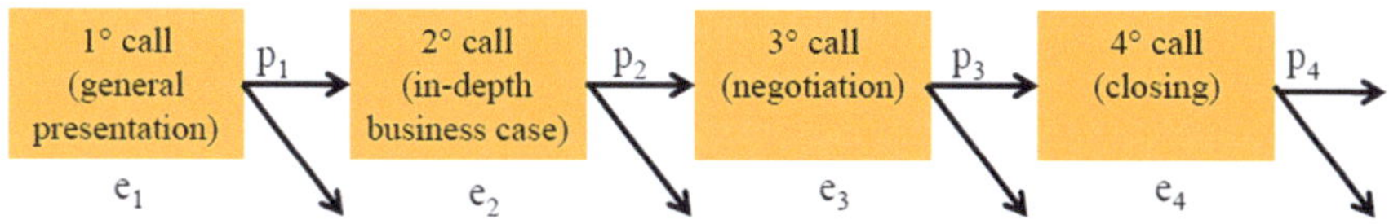

Fig. 13.17 Representation of the B2B purchasing process

in which demand is essentially driven by the effort spent on direct sales carried out by specialized personnel.

A similar process can be drawn in the case of B2C digital marketing, which can be modeled as a progressive conversion of non-users to users via targeted marketing actions, of users into paying users, etc.

In these cases, Bass' model is hardly applicable, since word of mouth is less present, while the marketing effort is not broadcast to the residual population of non-adopters at time t. The firm will therefore have to model the sales cycle as in the previous Fig. 13.17 and empirically identify the relevant parameters (i.e., conversion rates from one step to the other, and the costs and lead times required by each step). Then, given an allocation of budget to the various phases of the process, it will be possible to derive an estimate of the resulting sales.

13.7.3 Demand Estimation to Evaluate Diffusion of Services

The estimation of diffusion of innovative services is a relatively recent endeavor, partly because product innovations were dominant until recent years, and partly because diffusion in services is more challenging to model. Three main families of models exist in the literature: *Bass type*, which are built on the analogy between product and service adoption processes, *Choice type* and *Gray type*. Choice type models focus on individual choices, while gray models are used if little amount of data is available.

13.7.3.1 Bass-Type Models for Services

A first model was proposed by Giovanis and Skiadas (2007), who studied mobile plans diffusion in EU countries. They suggested that diffusion follows the imitative part of a Bass model, but introduced a delay between the time that a decision is formulated and the actual adoption. The hypothesis is that a new customer has to sign a contract, and this can require more effort and time than to adopt a new product. Service purchase is therefore viewed as less impulsive than for a product.

The authors have therefore modified the Bass equation by referring sales at time t to the situation found at time $t - \tau$, as in (13.29):

$$n(t) = \frac{\mathrm{d}N(t)}{\mathrm{d}t} = \frac{b}{M} N(t - \tau)[M - N(t - \tau)] \qquad (13.29)$$

Another variation of Bass' model was proposed by Libai et al. (2009) in order to describe the effects of customer attrition on diffusion. The idea is that, in any time period, companies can acquire customers among those who have not yet adopted the service, among those who have already abandoned it, or among those who decide to switch from a competitor (churn). At the same time, the firm can lose customers who decide to dismiss the service category altogether, or decide to switch in favor of a competitor (churn), and the sum of these two components then defines customers' attrition:

$$\begin{aligned}\text{Customer attrition} = \text{disadopting customers} \\ + \text{churning customers}\end{aligned} \tag{13.30}$$

Consequently, it is possible to compute the corresponding customer *attrition rate*, CA(t), as the ratio between attrition and current customers, as in (13.31):

$$\text{CA}(t) = \frac{\text{disadopting customers}(t) + \text{churning customers}(t)}{\text{customers}(t)} \tag{13.31}$$

Once the customer attrition rate has been defined, one can also model the customer *retention rate*, i.e., the percentage of customers that decide to continue to use the service, as in (13.32):

$$\text{CR}(t) = \frac{\text{customers}(t) - \text{new customers}(t) - \text{customers}(t-1)}{\text{customers}(t-1)} \tag{13.32}$$

or correspondingly as in (13.33):

$$\text{CR}(t) = 1 - \text{CA}(t) \tag{13.33}$$

A first version of their model focused on category-level attrition and neglected the competitive churn. It is built on to the same assumptions as of the Bass model, so that the diffusion of the new service is given by Eq. (13.34):

$$n(t) = \frac{\mathrm{d}N(t)}{\mathrm{d}t} = p[M - N(t)] + \frac{q(1-\delta)N(t)}{M}[M - N(t)] - \delta N(t) \tag{13.34}$$

where M, $N(t)$, p and q are parameters already introduced by Bass, and δ is the attrition rate.

The term δ strongly affects imitative innovation, since the positive word-of-mouth effect is spread only by those who have not dismissed the service. The closed-form solution is here defined as in (13.35):

$$N(t) = M' \frac{1 - \mathrm{e}^{-(p'+q')t}}{1 + \frac{q'}{p'}\mathrm{e}^{-(p'+q')t}} \tag{13.35}$$

The formula has the same functional form as the Bass Eq. (13.2), but with different parameters: $M' = M\frac{\Delta+\beta}{2q(1-\delta)}, p' = M\frac{\Delta-\beta}{2}$, and $q' = M\frac{\Delta+\beta}{2}$, where Δ and β are defined as in (13.36) and (13.37):

$$\Delta = \sqrt{\beta^2 + 4q(1-\delta)p} \quad (13.36)$$

$$\beta = q(1-\delta) - p - \delta \quad (13.37)$$

Due to the customer attrition rate, $p' > p$, $q' < q$ and $M' < M$.

It should be noted that this model does not consider a negative imitation effect, due to users who decide to abandon the service, and who share the reasons underlying their decisions with actual and potential customers. This may result in an overestimation of the diffusion curve obtained from Eq. (13.34).

The case of competitive attrition (or churning customers) has been considered in a second model in the same paper. However, we will not present it because it is difficult to apply, given the amount of data required, which is often not available.

13.7.3.2 Choice-Type Models for Services

Among the choice models, it is important to mention the one proposed by Landsman and Givon (2010). The model describes in Fig. 13.16 a two-stage service diffusion process. The first stage is modeled by a hazard function and is called the '*consideration stage*', where the potential customers decide whether to join the service or not. The second is the '*choice stage*' in which customers choose between the service alternatives and the no-choice option.

In particular, in the first stage, at $t = 0$, j new services are offered on the market and users do not even consider the possibility of subscribing to any of them (i.e., this state is called '*No-Service No-Consideration*'). The transition rate from no consideration to consideration state is modelled as a hazard rate which is a function of two factors: (1) the time passed since the service has been introduced and (2) a set of covariates that describes the potential market.

Authors, in order to decompose the hazard function, opted for the Proportional Hazard Model (Gupta 1991; Helsen and Schmittlein 1993; Seetharaman and Chintagunta 2003) in (13.38):

$$h_t = h_{0t}\psi(X_t) \quad (13.38)$$

where h_{0t} is the baseline hazard function that can be decomposed employing the expo-power formula (13.39) (Saha and Hilton 1997; Seetharaman and Chintagunta 2003)

$$h_{0t} = \gamma\alpha t^{\alpha-1}e^{\theta t^{\alpha}} \quad (13.39)$$

in which, gamma and alpha are respectively the scale and the shape parameter of the Weibull distribution that models initial diffusion phenomenon. Theta instead embodies a smoothing effect that reveals at the end of this stage.

The second stage of '*choice*' instead is modelled by a multinomial logit model. Once reached the choice stage, the probability that a customer chooses the alternative j at time t is given by:

$$P_{jt} = \frac{e^{V_{jt}}}{\sum_{j=0}^{J} e^{V_{jt}}} \tag{13.40}$$

where V_{jt} is the deterministic part of the utility obtained from choosing the alternative j at time t, and it is specified as follows:

$$V_{jt} = \beta' Y_t + \rho_s \mathrm{EC}_{t-1} C_{jt-1} \tag{13.41}$$

The first component of (13.41), Y_t, is a set of K covariates characterizing the alternatives of service available at t, the customers that have adopted such service at t, or the alternative-customer combination. The second component instead represents what may affect current customers in maintaining the same state during the next time period (this situation is named '*state of dependency*'). This is affected by a binary variable EC_t that represents the possibility of having purchased an alternative service in any time period up to t, while C_{jt} represents if a specific brand j was chosen before time t.

The dependency state effect then is divided into two distinct effects: a general 'stickiness' to the service relevant to all current users of the service and a 'stickiness' to an alternative option (for those customers that potentially have tried another service and that abandon the one we are considering). The binary parameter ρ_s models these two possibilities.

Another choice-type diffusion model was discussed by Shi et al. (2014). The authors considered N services competing against each other. At each time period t, the *i-th* person is a customer of the service $l = 0, \ldots, N$, where $l = 0$ means that the person has not chosen any alternative yet and is a potential adopter. Moreover, at each time period t, the same i person may choose to switch to another service $k = 0, \ldots, N$, where $k = 0$ means that the person dismisses the current service and goes back to being a potential customer. At each time period t, there are therefore five possible scenarios:

- The potential customer i chooses the service k, and utility is equal to $U_{i,t}^{k,0}$.
- The customer i switches from service l to k, and utility is equal to $U_{i,t}^{k,l}$.
- The customer i decides to continue using service l, and utility is equal to $U_{i,t}^{l,l}$.
- The potential customer i decides not to use any service, and utility is equal to $U_{i,t}^{0,0}$.

- The customer i decides to dismiss service l and not use any service; utility is equal to $U_{i,t}^{0,l}$.

Utility is here defined as the combination of two terms:

$$U_{i,t}^{k,l} = V_t^{k,l} + \varepsilon_{i,t}^{k,l} \tag{13.42}$$

where ε is the error term, and $V_t^{k,l}$ is a deterministic term changing from scenario to scenario. In the first two, it is defined as follows:

$$V_t^{k,l} = V_0^k + \sum_{t'=1}^{t} \delta_{t'}^k V_0^k \quad l \neq k,\ \ k \neq 0 \tag{13.43}$$

where V_0^k is the utility associated with the service k at time $t = 0$, and it is equal to $\beta_0^k X_0^k$. X_0^k represents the service features, such as price, service quality or network effect; β_0^k is the impact that these features have on the utility. Instead, since utility may grow or decrease over time, $\delta_{t'}^k$ represents this additional utility.

In the third scenario, the formula is equal to Eq. (13.43), plus a constant term p representing the extra utility gained by users that keep using the same service.

$$V_t^{k,l} = V_0^k + \sum_{t'=1}^{t} \delta_{t'}^k V_0^k + p \quad l \neq k,\ \ k \neq 0 \tag{13.44}$$

Finally, in the fourth and fifth scenario, the utility is assumed to be constant and given by abandoning the service.

$$V_t^{k,l} = c, \quad k = 0 \tag{13.45}$$

Given the utility computation, the authors then defined the probabilities that a potential customer starts to use service k (i.e., $P_t^{\text{First},k}$), and that a customer of service l switches to service k (i.e., $P_t^{\text{Existing},k,l}$).

$$P_t^{\text{First},k} = \frac{e^{V_t^{k,0}}}{\sum_{k=0}^{N} e^{V_t^{k,0}}} \tag{13.46}$$

$$P_t^{\text{Existing},k,l} = \frac{e^{V_t^{k,l}}}{\sum_{k=0}^{N} e^{V_t^{k,l}}} \tag{13.47}$$

As a consequence, the number of first users and switching users is defined as follows:

$$S_t^{\text{First},k} = P_t^{\text{First},k} M \quad (13.48)$$

$$S_t^{\text{Existing},k,l} = P_t^{\text{Existing},k,l} S_{l,t-1} \quad (13.49)$$

where M represents the potential market. Finally, the number of users of the service k at time t is given by the combination of the following terms:

$$S_t^k = S_{t-1}^k + S_t^{\text{First},k} + \sum_{l=1;l\neq k}^{N} S_t^{\text{Existing},l,k} + \sum_{l=0;l\neq k}^{N} S_t^{\text{Existing},l,k} \quad (13.50)$$

13.7.3.3 Gray Models

Gray models instead are used in the presence of a minimal amount of data.

GM(m, n) is the representation of a basic gray model, where m is the order of a differential equation, and n the number of variables. Lin (2013) used a first-order differential equation with one variable to propose a diffusion model based on a historical series of at least four data.

A first-order differential equation is defined as follows:

$$\frac{\mathrm{d}X^{(1)}}{\mathrm{d}t} + aX^{(1)} = b \quad (13.51)$$

where a is the sales growth coefficient, b is the gray influence coefficient, and $X^{(1)}$ is the vector of cumulative sales over time. Specifically, the gray influence coefficient here determines the contribution measure the influence degree between the cumulative sales and their variation over time (i.e., the extent with which the diffusion at a given time influences its variation).

Using the nonlinear least square method, parameters a and b can be estimated to get the curve of cumulative sales values:

$$\hat{x}^{(1)}(t+1) = \left(x^{(0)}(1) - \frac{b}{a}\right)\mathrm{e}^{-at} + \frac{b}{a} \quad (13.52)$$

The major limitation of this model is due to the absence of an upper bound, which means that the diffusion curve tends to infinity. This is conceptually wrong because the market is, by definition, a finite set of users, and the fact that there is no upper limit can lead to misleading forecasts.

13.8 Market Share

A second parameter to be determined when forecasting sales is the potential market share of the product or service. A straightforward way to estimate this parameter is based on the so-called Kotler's theorem. This theorem relates the share enjoyed by each product i, s_i, to a measure of '*attractiveness*' A_i that characterizes each product. The theorem is based on the following four axioms.

Axiom 1. $A_i \geq 0 \; \forall \, i$, $\Sigma_i A_i > 0$ (attractiveness of products is always non-negative, and at least one product has nonzero attractiveness).

Axiom 2. $A_i = 0 \rightarrow s_i = 0$ (a product with zero attractiveness will have no market share).

Axiom 3. $A_i = A_{i'} \rightarrow s_i = s_{i'}$ (two products with equal attractiveness will have the same market share).

Axiom 4. If $A_i = A_i + \Delta$, the decrease in $s_{i'}$ will be a function of Δ and not of i (market share of product i' changes as a function of the overall attractiveness of products in the market).

Axioms 1 and 2 provide a basic definition for attractiveness and its influence on market share. Axiom 3 states that market share is determined by attractiveness alone. Axiom 4 is slightly more problematic, since it states that changes to the products being offered on the market impact the market share of a product only through their combined attractiveness. Now, it might be argued that, for certain brands, the competitive impact of product change might be higher than for other ones. However, this consideration is quite inconsequential, since we have not defined what attractiveness is, and this concept may very well incorporate this 'sensitivity to the brand' aspect.

Based on these four axioms, it follows that market share of product i' will be given by the following relation:

$$S_{i'} = \frac{A_{i'}}{\sum_i A_i} \tag{13.53}$$

The consequences of Kotler's theorem are quite interesting. First, given a set of n products with similar attractiveness, the market is going to be divided into n equal shares, which is not surprising at all. However, a product whose attractiveness is much higher than that of its competitors is not going to achieve 100% market share, while a product with low attractiveness will have a small, but nonzero market share. This outcome seems to contradict the hypothesis of a rational consumer, but is quite consistent with the heterogeneity of consumer preferences and the limited rationality of consumer behavior. Finally, relation (13.53) ensures that, if new competing products are launched, or if products' attractiveness changes, market shares will self-adjust so that their sum always yields 100%.

As mentioned, Kotler's theorem does not tell us what attractiveness is and how it may be calculated. However, this vagueness can be considered to be the beauty itself of the theorem. Given current market shares from sales data, an analyst can quickly derive the implicit attractiveness of competing products, since $A_j = \alpha \; s_j$ where α is any real number. For simplicity, one can consider $\alpha = 1$ and posit that given current market share, $A_j = s_j$. It becomes straightforward to make rough estimates of market shares in a changing competitive scenario in which new

Table. 13.2 Using Kotler's theorem to estimate the impact of competitive moves

	CASE 1 (no reaction by *Z*)			CASE 2 (*Z* fights back)		
Company	Before the new competitor enters	After the entrance of *Z*	Actual market share	Before the new competitor enters	After the entrance of *Z*	Actual market share
x	0.45	0.45	0.33	0.45	0.45	0.31
y	0.3	0.3	0.22	0.3	0.3	0.21
Z	0.25	0.25	0.19	0.25	0.36	0.25
New	–	0.35	0.26	–	0.35	0.24
Total	1	1.35	1	1	1.46	1

products are defined in terms of their relation to the attractiveness of the existing ones. Table 13.2 provides a simple example of a market with three competitors, *X*, *Y* and *Z*. Based on this data, it is fairly easy to answer questions such as "what would happen if a new entrant launches a product whose attractiveness is intermediate between the one of product *X* and product *Y*?", and "in addition, what would happen if competitor *Z* fights back with a new product whose attractiveness is 20% higher than the one of product *Y*?".

In addition to these rough estimates based on implicit attractiveness indexes, it is also possible to investigate on the relationships between technical features of the products or services and attractiveness (Cooper and Nakanishi 1988). These relationships are generally nonlinear and are expressed through models that can be identified by using data on past sales. The most well-known models are the following:

- **MNL, or Multinomial Logit**. The attractiveness x_{ki} of product i is considered to be a function of a set of technical features k, of a parameter α_i that represents the influence of brand i and of parameters β_i whose estimation can be given by data on past sales.

$$A_i = \mathrm{e}^{\alpha_i + \sum_k \beta_k x_{ki} + \varepsilon_i} \tag{13.54}$$

- **MCI, or Multinomial Competitive Interaction**. This model is similar to MNL, but the function expressing attractiveness is multiplicative, similarly to a Cobb–Douglas function:

$$A_i = \mathrm{e}^{\alpha_i \prod_k x_{ki}^{\beta_k} \varepsilon_i} \tag{13.55}$$

When working on market share models, price is of course an important element to be considered. In general, it is possible to include it as one of the parameters X_{ki}. Alternatively, price could be neglected if the market is highly competitive, or if the

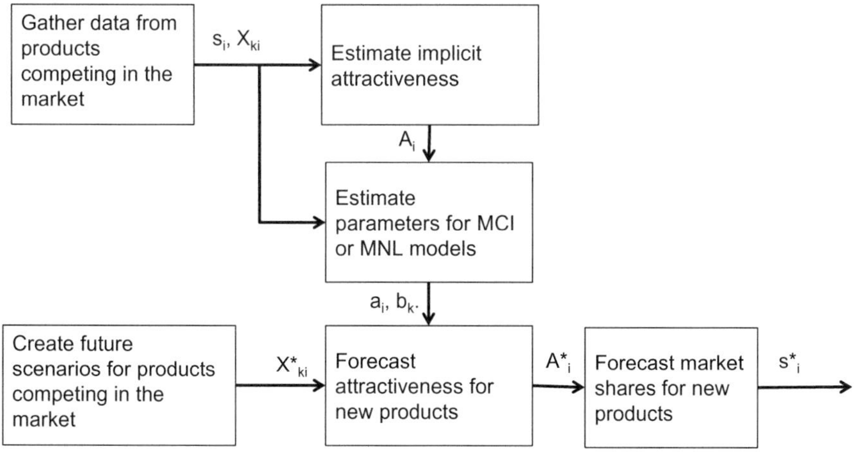

Fig. 13.18 A flowchart showing the use of market share models in practice

study is focused on a narrow segment of the market, since in these cases prices will broadly be aligned for all competing products.

Following the procedure in Fig. 13.18, the existing scenario allows the estimation of relevant parameters, which can then be used to make a forecast of future scenarios that include the launch of new products and/or modifications to the current ones.

The main difference between MNL and MCI models can be understood by computing the elasticity of the resulting market share with respect to the values of parameters X_{ki}. For simplicity, let us suppose that customers prefer high values of X_{ki} over low values.

As shown in Fig. 13.19, elasticity tends to zero for values of X_{ki} going to infinity for both MNL and MCI. This makes sense since, if one of the parameters X_{ki} becomes extremely high, customers will progressively stop appreciating this increase, and the impact on market share will tend to become negligible. Instead, MNL and MCI differ when performance X_{ki} is low. In the case of MCI, sensitivity is at first maximum and decreases with X_{ki}. This means that customers will greatly

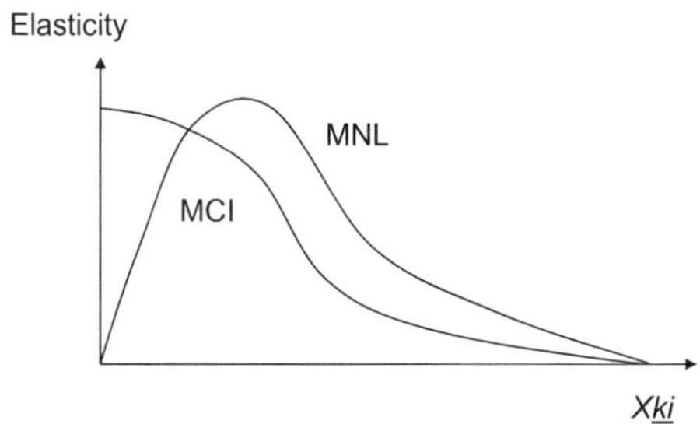

Fig. 13.19 Elasticity of market share with respect to technical parameters in MNL and MCI

appreciate changes to X_{ki} even if it starts from a very low level. On the opposite, sensitivity for MNL is at first very low, climbs to a maximum and then decreases. This means that customers will not appreciate changes to X_{ki} when the value of the parameter is low, and will start doing so only when X_{ki} becomes sufficiently high and exceeds a threshold level. Therefore, the choice between MNL and MCI depends on the nature of the parameters to be studied and on the possible existence of threshold levels.

References

Barwitz N, Maas P (2018) Understanding the omnichannel customer journey: determinants of interaction choice. J Interact Mark 43(C):116–133

Bass FM (1969) A new product growth for model consumer durables. Manag Sci 15(5):215–227

Bass FM, Krishnan TV, Jain DC (1994) Why the Bass model fits without decision variables. Mark Sci 13(3):203–223

Box GEP, Jenkins GM, Reinsel GC (2008) Time series analysis: forecasting and control. Wiley, New York

Brace I (2008) Questionnaire design: how to plan, structure and write survey material for effective market research (market research in practice). Kogan Page Ltd., London

Bradburn NM, Sudman S, Wansink B (2004) Asking questions: the definitive guide to questionnaire design—for market research, political polls, and social and health questionnaires. Jossey-Bass Publishers, San Francisco

Brunswik E (1944) Distal focusing of perception: size constancy in a representative sample of situations. Psychol Monogr 56:1–49

Cantamessa M, Montagna F (2016) Design for innovation—a methodology to engineer the innovation diffusion into the development process. Comput Ind 75:46–57

Cantamessa M, Montagna F, Cascini G (2012) Design for innovation. In: The 12th international design conference design 2012, Dubrovnik, Croatia, 21–24 May

Cantamessa M, Montagna F, Messina M (2013) Multi-stakeholder analysis of requirements to design real innovations. In: 9th International conference on engineering design (ICED09), Stanford, California, 24–27 Aug

Cascini G, Montagna F (2015) Situating needs and requirements in a multi-stakeholder context. Design Computing and Cognition (DCC 14), London, UK, 21–22 June

Chapman CN, Milham R (2006) The personas' new clothes: methodological and practical arguments against a popular method. Hum Factors Ergon Soc Annu Meet 50(5):634–636

Chun SY, Hahn M (2008) A diffusion model for products with indirect network externalities. J Forecast 27(4):357–370

Cooper LG, Nakanishi M (1988) Market share analysis. Kluwer, London

Decker R, Gnibba-Yukawa K (2010) Sales forecasting in high-technology markets: a utility-based approach. J Prod Innov Manag 27(1):115–129

Fourt LA, Woodlock JW (1960) Early prediction of market success for new grocery products. J Mark 25:31–38

Fuchs C, Schreier M (2011) Customer empowerment in new product development. J Prod Innov Manag 28(1):17–32

Giovanis AN, Skiadas CH (2007) A new modeling approach investigating the diffusion speed of mobile telecommunication services in EU-15. Comput Econ 29:97–106

Gray D, Brown S, Macanufo J (2010) Gamestorming: a playbook for innovators, rulebreakers, and changemakers. O' Reilly, Sebastopol, California

Gupta S (1991) Stochastic models of interpurchase time with time-dependent covariates. J Mark Res 28(1):1–15

Helsen K, Schmittlein DC (1993) Analyzing duration times in marketing: evidence for the effectiveness of hazard rate models. Mark Sci 12(4):339–427
Hague N, Hague P (2018) B2B customer experience: a practical guide to delivering exceptional CX. Kogan Page, London
Horsky D (1990) The effects of income, price and information on the diffusion of new consumer durables. Mark Sci 9(4):342–365
Hotelling H (1929) Stability in competition. Econ J 39(153):41–57
Huang A, Dawes J, Lockshina L, Greenacre L (2017) Consumer response to price changes in higher-priced brands. J Retail Consum Serv 39:1–10
Jain CL, Malehorn J (2005) Practical guide to business forecasting. Graceway Publishing Company, New York
Johnson R.A., Wichem D.W. (2007), Applied Multivariate Statistical Analysis, Pearson Education, Edinburgh.
Kano N (1984) Attractive quality and must be quality. Hinshitsu (Quality) 14(2):147–156
Kiesling E, Günther M, Stummer C, Wakolbinger LM (2012) Agent-based simulation of innovation diffusion: a review. CEJOR 20(2):183–230
Krishnan V, Zhu W (2006) Designing a family of development-intensive products. Manag Sci 52 (6):813–825
Landsman V, Givon M (2010) The diffusion of a new service: combining service consideration and brand choice. Quant Mark Econ 8:91–121
Lecoeuvre L, Turner R, Kuppelwieser VG (2021) Customer experience in the B2B area: the impact of age-related impressions. J Retail Consum Serv 58:102216
Lemon KN, Verhoef PC (2016) Understanding customer experience throughout the customer journey. J Mark 80(6):69–96
Levinthal DA (1997) Adaptation on rugged landscapes. Manage Sci 43(7):934–950
Libai B, Muller E, Peres R (2009) The diffusion of services. J Mark Res 46(2):163–175
Lin C-S (2013) Forecasting and analyzing the competitive diffusion of mobile cellular broadband and fixed broadband in Taiwan with limited historical data. Econ Model 35:207–213
Mahajan V, Muller E, Wind Y (2000) New product diffusion models. International series in quantitative marketing. Kluwer Academic Publishers, London
Makadok R, Ross DG (2013) Taking industry structuring seriously: a strategic perspective on product differentiation. Strateg Manag J 34(5):509–532
Mansfield E (1961) Technical change and the rate of imitation. Econometrica 29:741–766
McGinn J, Kotamraju N (2008) Data-driven persona development. In: SIGCHI Conference on human factors in computing systems (CHI'08). Association for Computing Machinery, New York, NY, USA, pp 1521–1524
Monat JP (2011) Industrial sales lead conversion modeling. Mark Intell Plan 29(2):178–194
Moorthy KS, Png IPL (2002) Market segmentation, cannibalization and the timing of product introductions. J Manag Sci 38(3):345–359
Phillips H, Bradshaw R (1993) How customers actually shop: customer interaction with the point of sale. J Mark Res Soc 35(1):51–62
Prasad RK, Jha MK (2014) Consumer buying decisions models: a descriptive study. Int J Innov Appl Stud 6(3):335–351
Puccinelli NM, Goodstein RC, Grewal D, Price R, Priya RP, Stewart D (2009) Customer experience management in retailing: understanding the buying process. J Retail 85(1):15–30
Ratchford BT, Balasubramanian SK, Kamakura WA (2000) Diffusion models with replacement and multiple purchases. In: Mahajan V, Muller E, Wind Y (eds) New product diffusion models. Kluwer Academic Publishers, Boston
Saha A, Hilton L (1997) Expo-power: a flexible hazard function for duration data models. Econ Lett 54(3):227–233
Satoh D (2001) A discrete bass model and its parameter estimation. J Oper Res Soc Jpn 44(1):1–18
Seetharaman PB, Chintagunta PK (2003) The proportional hazard model for purchase timing: a comparison of alternative specifications. J Bus Econ Stat 21(3):368–382

Shapiro C (1983) Premiums for high quality products as returns to reputations. Q J Econ 98 (4):659–679
Shi X, Chumnumpan P, Fernandes K (2014) A diffusion model for service products. J Serv Mark 28(4):331–341
Tapp SR (1984) Brunswik's lens model: a review and extension to consumer research. Adv Consum Res 11:103–108
Temkin BD (2010) Mapping the customer journey. Forrester Research Inc.
Van Westendorp P (1976) NSS-price sensitivity meter (PSM)—a new approach to study consumer perception. 29th ESOMAR congress, Venice, Italy
Verganti R (2009) Design driven innovation: changing the rules of competition by radically innovating what things mean. Harvard Business Press, Boston
Vinokurova N (2019) Reshaping demand landscapes: how firms change customer preferences to better fit their products. Strateg Manag J 40(13):2107–2137
Von Hippel E (1986) Lead users: a source of novel product concepts. Manag Sci 32:791–805

Specifying the Product

14

Abstract

In order to support the design of a product or service, it is necessary to translate the outcomes of Chap. 13 in technical terms, deriving precise requirements and specifications. This chapter discusses the approaches and methods that can be used to this purpose, with a special focus on customer-centered design methods, and on the definition of target costs, taking into account a lifecycle perspective and estimating experience economies.

Chapter 13 led us into the 'front end' of a product development process with a clear market-driven perspective. Readers have understood how firms can interpret market needs and come up with a high-level and concise description of the product or service to be developed, and of the market segment to be addressed. Additionally, they will have gained insights on demand forecasting models, thus allowing a preliminary evaluation of a product's or service's economic and financial attractiveness.

The information so far generated is however inadequate for guiding designers' efforts. It is therefore essential to move on, from the needs gathered during market research into a more detailed compilation of requirements and then, given a technical concept, to the system design specifications (or, product/service specifications).

A *requirement** represents a formal decision that a given need is to be met by the product or service, and to what extent. While needs are typically expressed in a qualitative way, requirements are defined at a definite and measurable level, without however telling how they will be technically fulfilled. A *design specification** can

M. Cantamessa and F. Montagna, *Management of Innovation and Product Development*, https://doi.org/10.1007/978-1-4471-7531-5_14

instead be defined as a list of features, functions and parameters whose values are set by designers so to comply with the requirements.[1]

In Chap. 11, a preliminary discussion has been made on the importance and the frailty of the transitions from needs to requirements. It was said that this involves a *translation* within inter-functional teams by professionals that use different languages and come from different cultural backgrounds. Such transitions do not simply involve translations, but they are the outcome of *design decisions*, though preliminary, since they could lead to different alternatives.[2] Moreover, the step from user requirements to specifications often assumes that an underlying product/service concept has been already defined.[3]

Most design activities deal with incremental innovations that will not substantially change the system concept; this being the case, firms will seamlessly move from the definition of user requirements to product specifications. In the case of radical innovations, product specification and concept generation will instead overlap and possibly lead to iterations. So, when dealing with an incremental innovation, a firm will implement the contents of Chaps. 14 and 15 sequentially, while it will apply them in parallel in the case of radical innovations.

The system specification phase is characterized by a number of challenges:

- The **inter-functional nature of the activities**, since requirements definition and product specification occur at the interface between the marketing and the technical functions of the firm.
- The **quantity of information** being generated. Moving from a list of user needs to a set of user requirements and to a complete product specification usually entails generating a significant amount of information. Product specification lists can be very long and complex, even for simple products. Companies often publish documents that list specifications for their products, either in data sheets or in user manuals. However, it must be noted that this public information constitutes only a subset of the entire product specification list, which is published because of legal obligations, to help prospective customers choose the product that best suits their needs, and users to make a proper use of the product.

[1] If one thinks about a simple product such as the remote controller for a TV set, a user need elicited during market research may be "I don't want to have to put my eyeglasses on when operating the remote control". The firm may therefore decide to cope with this need by asserting the following requirement 'button labels must be readable in 50 lx (dim light) by a user having a visual impairment of −1 diopters'. The requirement may then be translated into a specification such as 'characters for button labels must be between 6 and 8 mm tall and must be color-contrasted with their background'.

[2] In the remote controller example, it is up to the company to decide on the degree of readability of button labels with respect to illumination level and the degree of visual impairment, and to decide whether 6 mm tall and color-contrasted characters would adequately meet such requirement.

[3] When defining character size for button labels, the designer is assuming that the remote control will be operated with physical buttons, which is only one of the many possible technical solutions allowing the required functions to be carried out (for instance, a remote control might be conceived with a touch screen-based interface).

- The **uncertainty of information** generated. During this phase, the firm makes decisions without yet having cleared the technical and commercial uncertainty that characterizes product development. When specifying the product, detailed design has not been carried out yet, nor does one know how specifications will propagate along the bill of materials* into the specifications of each component. Therefore, user requirements and product specifications may either be technically unfeasible or may be too lax, leading to a product with significant opportunities for improvement. Similarly, the specifications may not correctly interpret customer needs, resulting into a product with low attractiveness and disappointing returns.
- The need to consider **system interactions with its surrounding environment (or 'supra-systems')** at different levels, the *harmful effects* arising from this interaction, the *resources the system will require* to function, and all this throughout the **phases of the product lifecycle** from production to distribution, from operation to end of life.
- Therefore, while market research focuses on customer/user needs, these needs are not enough to determine the requirements of a system. As depicted in Fig. 14.1, aspects such as manufacturability, logistics, field service and end-of-life disposal must be adequately considered by understanding the needs of the individuals, or the organizations, that represent them.
- The **role of specifications in product/service positioning**. The specifications represent the *final decision* about the technical features that will characterize the system when it is placed on the market. Customers' purchasing process, leading to the decision to buy a product and to select which of the competing alternative to prefer, will be carried out starting from specifications. As discussed in the previous chapter, the purchasing process operates at a higher level of abstraction, whose inputs nonetheless are tied to specifications (e.g., a customer shopping for a washing machine will formulate his or her preferences on the basis of information such as price, capacity and energy consumption, which are dependent on specifications. In some cases, this effort will be supported by industry-standard codifications, such as the EU energy labels for appliances, which nonetheless depend on specifications). Therefore, these have an important impact on the perception of customers and on the purchasing processes.
- The **tradeoff** between the stability of product specifications and the opportunity of adopting late commitment strategies. In theory, one might ideally wish to frame specifications clearly and firmly, with perfect adherence to technical feasibility and market needs, so that designers may operate on firm ground. In reality, as amply discussed in previous chapters, ambiguity, uncertainty and change in the technological and market environment may require a different strategy. In these cases, a subset of requirements will be kept stable, while for other ones it might be advisable to follow a '*late commitment' strategy'*, keeping options open and waiting for further information to emerge.

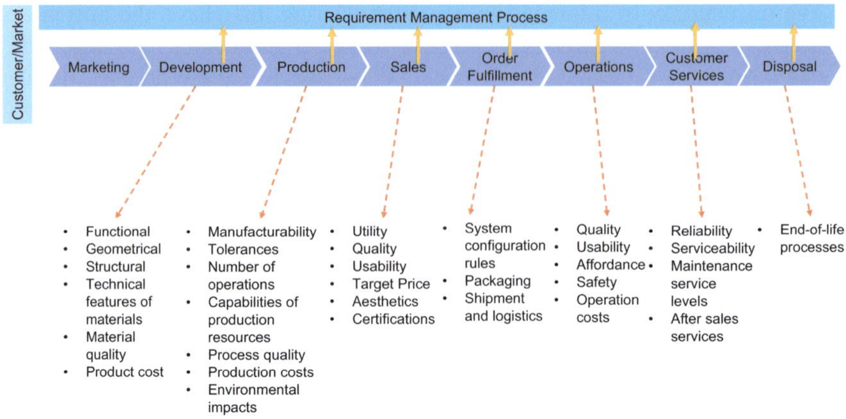

Fig. 14.1 An example list of lifecycle requirements

Given the above, a proper definition of product specifications can be considered to rely on a proper process, with appropriate information sources and a sound methodology, as will be covered in the following sections.

14.1 The Information Required for Developing System Requirements and Specifications

Product requirements and specifications are drawn from product briefs generated at the end of the market research phase but, as previously discussed, must incorporate a broader range of information.

First, having defined a primary list of requirements, the firm will often have to compile the product *mission profile*. In fact, if the firm does not understand the usage the product will be subject to by the target customer segment, it will be impossible to develop a product whose specifications will lead to the satisfaction of user needs.[4]

Moreover, and coherently with the concept of technological trajectory, this information will come from sources from the side of supply (e.g., suppliers, complementors, etc.) and from the demand side of the market (e.g., regulations).

[4] For instance, suppose users of a high-end digital camera ask for a 'reliable' product, where 'reliability' implies a defined number of years before the product fails and has to be replaced. The manufacturer will probably view a connection between this need and the expected lifetime of the mechanical shutter, expressed in thousands of actuations. However, it will be possible to connect the requirement to a specification only if the manufacturer correctly defines the 'mission profile' of the camera. In this simple example, the product brief might recite: 'this camera is aimed at photography enthusiasts and professionals who need a stand-in product along to their main camera'. The manufacturer will therefore have to define the number of yearly shutter actuations that can be expected to be made by the customers belonging to these two segments.

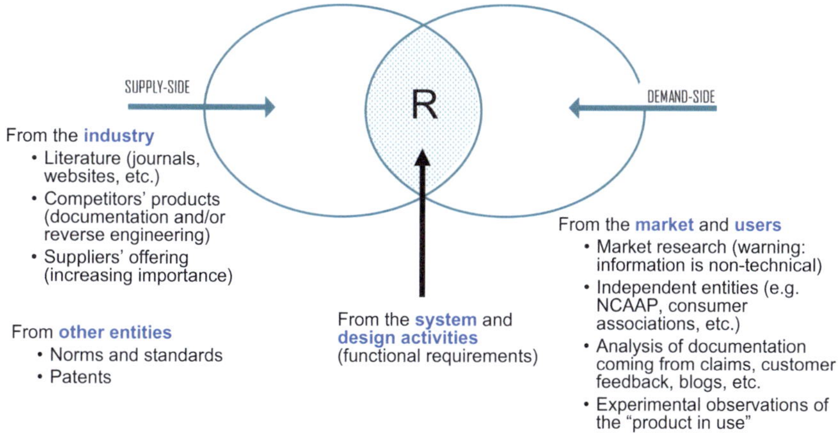

Fig. 14.2 Information sources for requirements and specifications

In order to collect this information, the firm may use the sources in Fig. 14.2:

- **Sources emerging from the industry**. Examples are trade journals and websites, together with suppliers' and competitors' catalogues. Competitors' systems can be studied either 'on paper' or actually purchased, disassembled and reverse-engineered to provide detailed information.
- **Sources associated to the market and users**. Besides market research actions as described in the previous chapter, firms can rely on a variety of other documentation: reports published by consumer associations and independent entities, internal reports on customer feedback, warranty claims, etc.
- **Sources associated to relevant regulations**. In virtually all markets, products must comply with regulations and be certified before they are legally commercialized. Regulations can be general (e.g., product liability), generic, if associated to a common technology independently of the product (e.g., low voltage devices) or product specific (e.g., toys). All these regulations, along with technical standards that the firm may choose—or be mandated—to observe, must be carefully examined, in order to derive the resulting requirements.
- **Patent databases**. Examination of patents allows the firm to understand the technical choices made by competitors and to clarify the 'freedom to operate' in view of imposed *design constraints*. These too will contribute to the requirements list, defining which technical alternatives and ranges of technical parameters cannot be tackled without the risk of infringing others' property rights.

In order to visualize the mass of information thus gathered, firms sometimes use parametric analysis on a subset of product or service features. *Parametric analysis** is based on the development of a database going back some time in the past, and recording the values of the most relevant technical features for all systems being

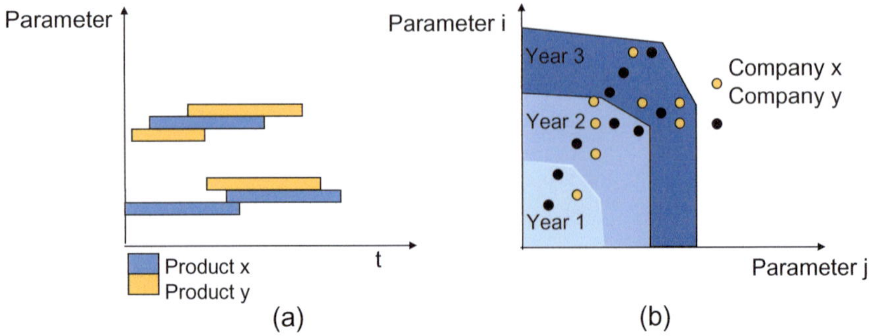

Fig. 14.3 Parametric analysis allows the visualization of the way products evolve in time

sold on the market. This database can support intuitive visualizations of competing products and services and of their evolution in time. For instance (see Fig. 14.3a), one can plot the evolution of a single parameter in time, where each system is represented as a bar that starts in the year of launch and ends in the year of withdrawal. This allows identifying entry and exit of each competing alternative in each segment. Alternatively (see Fig. 14.3b), one can plot the evolution of two parameters at the same time. In this case, each system is mapped as a point, and the technological frontier at a given date can be represented by isochrones that include all products launched at that date. This graph shows the evolution of pairs of performance indicators, eventually identifying technical complementarity (if both grow together) or tradeoffs (should growth in one be associated to a decrease in the other).

The key difference with respect to strategic maps discussed in Chap. 13 is the object of analysis: what is being monitored and visualized here are technical features, and not high-level and tacit primary needs.

14.2 From Needs to Specifications

One might think that a design problem is something certain and stable: once the needs are defined, one determines the corresponding requirements, then moves on to specifications and operates in the space of admissible solutions to conclude the design process. Actually, it's not so simple.

Deriving requirements from needs does not occur according to a biunivocal relationship, so that a specific need points to a unique requirement that is disconnected from the other requirements. In fact, designing a system entails governing a complex set of *tradeoffs* with respect to the functions to be achieved, the performances to be granted and to the resources that can be used (in terms of materials, energy, information, time and space). Defining requirements therefore implies making sure that these tradeoffs are managed in a way that they are technically

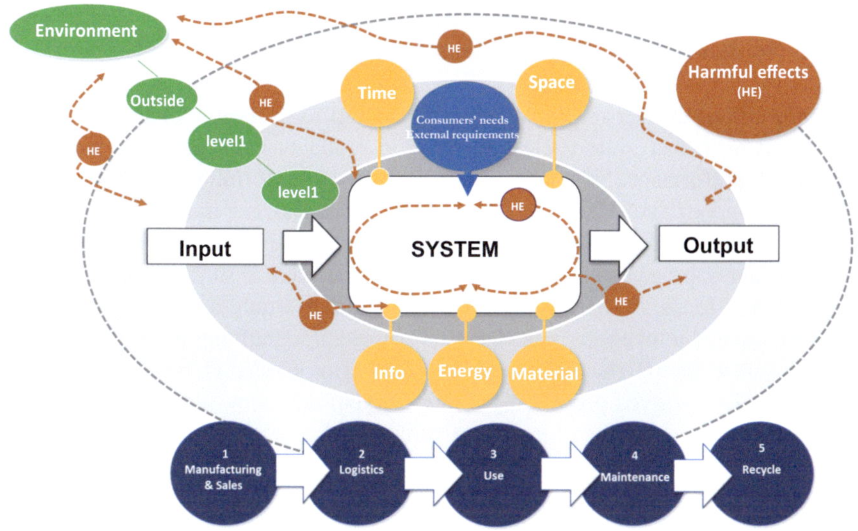

Fig. 14.4 The sources of requirements (modified from Graziosi 2012)

achievable and that they correctly represent the needs that have been spelt out. Consequently, the definition of requirements entails a *complex negotiation* concerning the system as a whole and its subsystems, and among the subsystems themselves.

Furthermore, design is a process and does take place in a single step. It requires the subdivision into subproblems, for each of which it is necessary to understand the objectives, the factors and the constraints that specify it, as well as the resources required. The solution of a subproblem is likely to influence other subproblems or even to generate new ones (also named 'harmful effects'; Altshuller 1999) which must be managed either by solving them or, in some cases, by deciding to bypassing or neglecting them.

All this can be particularly critical when dealing with complex systems and occurs within a large team drawn from different functions of an organization or, even more, when it occurs across firms operating in a value chain.

Thus, the product, service, or more generally the 'system', as in Fig. 14.4, has to be considered with respect to all the phases of its future lifecycle, to the 'mission it must perform' (transforming inputs into outputs), in a specific operational context and considering the issues emerging within the system and along the design process itself.

14.3 Identifying Requirements According to the User-Centered Design Approach

One key subset of user requirements is associated to *usability** of the product or service. It is in fact quite useless to develop products with many functions and substantial performance, if users are not in the position to fully exploit them.

User-Centered System Design[5] (or *User-Centered Design** in short, Norman and Draper 1986) can be viewed as the foundation of the customer-driven product development approach outlined in Chap. 13, which here assumes a more specific user-centered focus. In that previous context, the firm would deal with needs from what we defined a 'marketer's perspective', aimed at identifying general needs that are relevant to the purchasing process of product or service to be developed, and define a well-conceived positioning in the marketplace. Now, similar activities are implemented, but with the objective of providing fine-grained information on use process and usability, which are determinants of designing and ensuring a specific *user experience*.

User-centered design adopts the *personas* method that has been introduced in the previous chapter. However, personas are now used to provide a detailed characterization of users and of their interactions with the product or within the service,[6] often by running observational studies carried out based on methods from cognitive psychology and ethnography. In order to achieve this result, personas are imaged to operate within scenarios.

*Scenarios** are fictional narratives of the context in which a persona interacts with the system. Scenarios allow designers to frame the user-system interaction in a specific setting, thus achieving a deeper understanding of these events. One can frame multiple scenarios, thus introducing different types of situations and levels of ease or, in the case of services, multiple channels for its fruition.[7] In turn, within

---

[5] User-centered Design actually merges three different Design approaches, and namely *Cooperative Design,* from the Scandinavian tradition of Design of IT artifacts (Greenbaum and Kyng 1991), *Participatory Design**, by the North American Design school (Schuler and Namioka 1993), and *Contextual Design**, that connects customer-centered design to the actual circumstances surrounding the 'use situation' (Beyer and Holtzblatt 1997). User-centered design has become a formalized procedure through the ISO standard: the "Human-centred Design for interactive systems" (ISO 9241-210 2010).

[6] As a reminder, in Chap. 13 we made the example of a persona such as 'Marie, 35-year old IT professional, mom of two children (4-year old Susie and 2-year old Jean). She is the wife of Robert, an airline pilot who is often away from home. They live in an apartment in the XV Arrondissement of Paris'. To the family, one may add 'Jeanne, Robert's 70-year old mother, who lives close to Robert and Marie and sometimes borrows their car to visit her friends during the weekend'.

[7] For example, a scenario associated to Mary's usage of the family car might be 'Mary has collected her kids from school and now is in the grocery store's parking lot, having to move items from the shopping cart to the trunk of the car'. This scenario can be further split into a best-case scenario such as 'the sun is shining, the kids are happy and singing a song', and the worst-case scenario 'it's raining heavily, the kids are crying and a phone call is coming from Singapore, where Robert is minutes away from taking-off for a long-haul flight'.

each scenario, designers will study the interaction between users and system by making structured representations of typical usage patterns by developing use cases.

*Use cases** allow to define those specifications that are relevant to interaction modalities and, therefore, the *user interface** of the product or service.

Among the various observational methods that can be employed to carry out this work, one can mention the following:

- **Direct observation** is a variant of focus groups. While the latter are generally played out around a pure discussion format, this technique brings participants around a product, or a prototype, and the facilitator stimulates discussions around the actual interplay between users and product. This technique is more appropriate for products and also for web pages or SW applications. The observation can be helpful, since users with a non-technical background might find it difficult to articulate verbal expressions on their usage of the system, but feel at ease if they can simply show 'how things are done' or 'why this doesn't fulfill my needs'.
- **Scenarios of use** is a variant of the previous method. Instead of an open-ended discussion around the product or service, it requires participants to play roles and act according to a script (e.g., 'Mr. and Mrs. Black and the kids, loading the car at the supermarket on a Saturday morning'). Role-playing makes the participants concentrate on use cases that are considered to be relevant and—at the same time—it can lead participants to stress issues that lend themselves to comic outcomes, thus going deeper in their analysis.
- **User trials.** With this technique, groups of users are asked to perform a sequence of activities on products or prototypes and are interviewed after the session. Lack of direct observation of the user trial places the users at ease and ensures spontaneity. At the same time, it greatly reduces the richness of the information captured. Similarly, to direct observation, this technique was born for products, but it is also used for evaluating interaction within service processes.
- **Product-in-use** consists in the direct observation of the product being used in real settings by unsuspecting people, possibly with the support of audiovisual recordings. Due to privacy issues, there are obvious limitations to the types of products and use cases that can be observed. The technique is valuable, especially for uncovering wrong 'use modes' that would generally go unreported.

Designers may then use representational methods to structure and represent the outcomes of their observations. Specifically, use cases can be drawn using graphical formalisms, such as flowcharts or *swim-lane diagrams* (e.g., Fig. 14.5), which represent the sequence of actions undertaken by the user and the system. In some cases, it may be advisable to split the system in its components. At the least, this can be done by separating the *front end* (i.e., the interface that the user interacts with) from the *back end* (i.e., the portion that operates without direct user interaction). Thanks to this detailed and step-by-step description of the way with which each persona interacts with the system in each scenario, one can analyze each of these steps, raising issues associated with usability, and the consequent requirements.

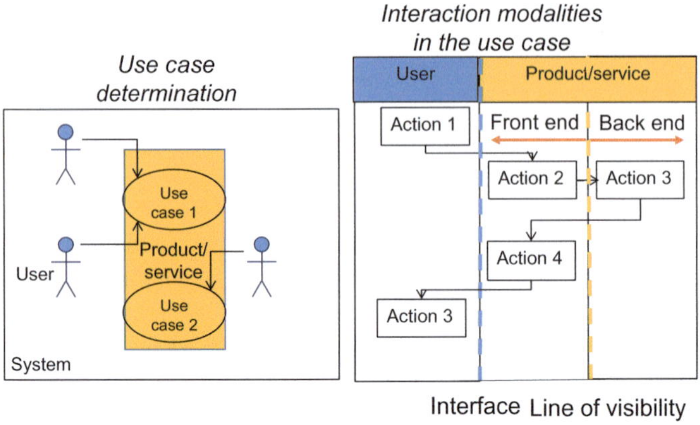

Fig. 14.5 Using use cases and swim-lane diagrams to represent interaction modalities

In the case of services, use cases are usually represented through *user or customer journey maps*. These describe the path and the activities carried out by a user within a service. Specifically, the user journey map describes the points where users interact with the front-end elements of the service (i.e., the 'touch-points'), the activities that occur in the back end of the service, and the line of visibility that separates the two parts. By analyzing these interactions, the designers can derive requirements for the process in general and for each touch point, while highlighting the points where user needs will be met, and those where users could find themselves in difficulty (pain points).

Based on this detailed investigation of user interaction, and on the well-known theory by Norman (1998, 2002), designers can systematically define requirements that impact the usability of the system. According to this theory, interaction follows a process that can be structured around the seven phases represented in Fig. 14.6. When analyzing the interaction process, one must consider that users will usually perform it in a tacit and unconscious way.

According to the Norman's model, users interact with a system with the objective of fulfilling a goal (e.g., operating an air conditioner in order to make a room comfortable and working in it). The process is made up of a first stream, which consists in evaluating the status of the system and a second one, associated to executing actions. The former can be split in the three stages of *perceiving* the state of the system (e.g., looking at the air conditioner's control panel), of *interpreting* this perception with respect to expectations (e.g., obtaining information on whether the air conditioner is working or not and on room temperature and humidity) and, finally, *evaluating* the system state with respect to intentions and goals (e.g., making up one's mind on whether the room is at the right temperature or not). The latter stream is divided into the three stages of *forming the intention to act* (e.g., deciding on whether the temperature should be raised or not), of *specifying an action* (e.g., deciding to turn the appropriate knob in order to raise the temperature level) and, finally, of *executing* the action (e.g., actually turning the knob).

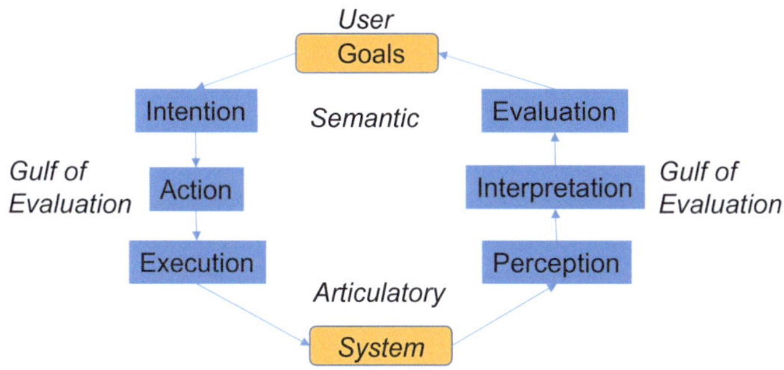

Fig. 14.6 The interaction process according to Norman

Norman highlights that each transition from one stage to another is associated to a cognitive difficulty, or *gulf*. The system and its interface are properly designed if each gulf can be easily bridged by the largest possible share of the target population. Conversely, a badly designed product is characterized by the fact that moving from one stage to another is difficult, ambiguous and/or error-prone.[8] For instance, if the air conditioner is extremely silent—which of course could be considered as a positive feature—the user may find it hard to perceive the state of the system, and some type of visual aid should therefore be provided. Similarly, the user may find it difficult to define the sequence of actions, if the knob does not clearly indicate the right direction to turn in order to increase the temperature. Finally, if the knob is too small and does not provide sufficient grip to fingers, the execution of actions may be difficult for users with some degree of physical impairment.

When studying usability, a particularly important concern is associated to the structuring of sequences of actions, especially when dealing with complex systems and many possible interaction modes. In such cases, designers must be aware that users might find it difficult to cope with the *breadth* and the *depth* of action sequences. A very broad range of possible alternatives (e.g., a webpage with dozens of links) may leave the user confused about which to choose. However, a narrow range of alternatives (e.g., a minimalistic webpage design with just two links) may leave the user with the impression that the desired action is not available at all, while in reality it might be accessible as a two-step sequence. Moreover, having to learn and perform a long sequence of actions to achieve the desired results may make interaction ineffective (e.g., think of the number of steps that are usually required to complete the purchasing process on e-commerce websites).

[8] A distinction can be made between *slips* and *mistakes*, the former being errors in the execution of appropriate actions, the latter being errors associated to the very action to be taken. For instance, a slip occurs if a user that must press a button presses both the right one and—inadvertently—another one that is located too close. It would be a mistake if the user pressed the wrong button, having misunderstood the functions associated to it.

Based on extensive research on 'good' and 'bad' designs, Norman proposed a list of eleven design principles which can improve the usability of products. These principles can be used both in the early stages of product development, when defining user requirements and product specifications, as well as later, during the design phase. These design principles are summarized in the following Table 14.1.

Among these design principles, it is convenient to provide a deeper discussion on *affordance**, since this concept has spread well beyond interaction design. The term 'affordance' was originally coined in 1977 by J. Gibson to explain *how animals were able to grasp the intrinsic meaning of an object using simply sensory perception, without resorting to any kind of reasoning*. The definition emphasizes the complementary relationship existing between the substance and the animal, a relationship that is totally independent of the effect that is generated and which is enabled through the use of the senses.

This idea was later adopted by Norman (1998) in studying the perceptions that an object transmits to an individual. Norman then reformulated the concept of affordance as "*the result of an interpretation of an object, built on the experience and knowledge that we apply to its perception*".

Further studies and definitions have been developed to describe in greater detail such perception users have of an object, the process of interacting with the artifacts, as well as the behaviors and learning mechanisms that are activated as consequences of this interaction (Vyas et al. 2006). In general, the conceptualization of affordance was used as a tool *to evaluate solutions* and *drive the design activities*, through a continuous process of testing-feedback-corrections aimed at conveying designers' meanings and interpretations to the users (Meske et al. 2023; Gaver 1991).

Nowadays, scholars are starting to reflect on how the concept of affordance changes in the case of digital artifacts and of the digital processes and services they enable (Yoo 2010; Kallinikos 2013; Colombo et al. 2022). Moreover, they are working on the definition of '*digital affordance*' as a concept that could offer a theoretical and analytical basis to support the design process for such systems (Bangle et al. 2022).

In a context of digital products, a further and important aspect of usability has to do with interaction with *autonomous products*.

When dealing with highly automated products (e.g., a robotic lawn mower, or automotive driver assistance systems, such as adaptive cruise control and lane departure warning), requirements and specifications must consider the system's capability of acting autonomously from the direct control of users. In this case, instead of having a continuous interaction characterized by the user's issuing of commands and the system's execution, one can observe two relatively independent processes carried out by the two.

Norman (2005) studied these types of product in depth, highlighting the role of different levels of reasoning and their allocation to the system. Basic automation can be equated to *visceral* reasoning, which is the unconscious behavior exhibited by humans and that does not require a particularly sophisticated or specific learning. An example of visceral reasoning is a basic cruise control, able to keep a car running at constant speed. At a higher level, one can find *behavioral* reasoning,

Table 14.1 Design principles for usability, according to Norman (2002)

Principle	Explanation	Example
Visibility	It must be easy to see the state of the system and the available controls	In an appliance, having an amber and green LED, respectively, indicating standby and operational state might be better than having a single one, indicating operational state only
Conceptual model	The system must help users understand how it works, taking into account that most people have limited technical knowledge and think in terms of 'folk physics'	In the case of an on/off thermostat, many users will tend to overshoot the desired temperature, in the mistaken assumption that this will make reaching the desired temperature quicker. This is especially true if the system does not provide quick feedback
Mappings	Relevant relationships must be easy to determine, i.e., between indicators and system state, between controls and effects	If the layout of controls and indicators follows the physical layout of the system, users will find it easier to understand their meaning
Feedback	The system must provide accurate and quick feedback on results of actions	A button that 'clicks' when depressed tells the user that the command has been received, while a silent one does not
Affordance	The mode of operation of a given control should conform to the general understanding of how it works	If a user sees a knob, he or she will assume that it should be turned, clockwise to 'close' and counterclockwise to 'open' something. It could be puzzling to the user if a knob had to be pushed or pulled
Information in the world	Relevant information should be provided directly by the system, without requiring users to make any specific effort in learning or remembering	Controls and indicators with labels and pictograms that are easy to understand can make a product easy to use without any specific instruction
Do not require precision	The system should allow users to operate with imprecise knowledge, data or actions	Buttons should be spaced from one another and shaped in a way that they accommodate the users' fingers, so that they are not pressed by mistake
Use constraints	The system should use physical and cultural constraints to induce into correct behavior	If an action moves the system state in a dangerous situation (e.g., very hot water in a shower), the device should provide some kind of resistance to make the user aware of the issue
Design for a variety of users	Designers should take into account that users vary widely with respect to technical understanding, physical traits, and abilities	The size and coloring of labels and pictograms on commonly used devices should be such that users with moderate presbyopia do not have to put their eyeglasses on

(continued)

Table 14.1 (continued)

Principle	Explanation	Example
Don't forget users	Companies generally focus on customers and other stakeholders, but users should never be neglected	Industrial equipment is bought by plant managers, but adequate attention should be paid to usability by the operators
Feature creep	Designers tend to overload products with functions that are often unnecessary and ultimately make the product difficult to use. Additional features should be inserted progressively and, if possible, through modularization	Software packages generally allow hundreds of functions, a tiny portion of which is actually used. A correct strategy might be to provide a base package and a set of add-ins catering to specific needs

which humans mostly perform unconsciously and requires significant learning. For instance, one may consider the ability to continuously manage the velocity of the car in order to chase a speed target set by the driver, taking speed limits, bends in the road and safety distances into account. Finally, *reflective reasoning* is the higher-level and conscious activity carried out by humans, associated to purpose and objectives. An example would be a self-driving car, capable of choosing the right itinerary and following it, given a destination, objectives, and knowledge of traffic conditions and road quality.

According to Norman, proper usability requires that the assignment of these three cognitive levels to the system must not be rigid, but flexible. In other words, the user must be in the position of yielding greater control to the system in normal conditions but—should conditions require it—she/he must be able to quickly regain complete command. As an analogy, it is possible to think of the interaction between a horse and a rider. When returning to the stable after a promenade, the rider can let the reins loose, and the horse will autonomously take her back home, assuming all three levels of reasoning. However, in case of danger, the rider will take the reins in and regain tight control over the mount's behavior. As with horses and riders, an effective 'symbiosis' between user and machine will emerge if the designers are able to create an appropriate pattern of communication between the two, striking the right tradeoff between the provision of a rich set of information and information overload.

Within this pattern of communication, a distinction can be made between *task automation*, in which the system assumes control, and *augmentation*, in which the task remains with the user, but with some support from the system, aimed at relieving part of physical or cognitive load.[9]

---

[9] For instance, commonly available lane departure warning systems (classified by SAE as 'level 0' devices in their classification of driving automation) *augment* drivers' perception by providing feedback if the vehicle leaves the lane, for example, by making the steering wheel vibrate. Note that these devices provide feedback that is directly tied to the human control, thus making it easy to interpret (e.g., a buzzer would not fully meet this criterion), and still allow the driver to override them by force, thus allowing to control the car in case of emergency. The situation is different for lane centering systems (SAE levels '1' and above) in which the car can automate, though under given conditions, the task of driving.

14.4 Quality Function Deployment

User requirements and system specifications can be defined following a variety of methodologies. Sometimes these methodologies are closely linked to company-specific or industry-specific standards. In this context we will present *Quality Function Deployment** (or QFD), which is a methodology that is widely used across industries either in its original form or in many adaptations.

QFD was developed in Japan in the '70s as a design support tool aimed at clarifying user needs and enabling their translation into coherent product specifications. Later on, it became one of the key techniques being proposed by the *Total Quality Management** movement for ensuring that the *voice of the customer** could be adequately 'heard' within the product development process. According to its principles, QFD should be used to support the entire development process, from the initial phase of requirements definition and product specification, all the way to the design of manufacturing processes and quality assurance systems. In practice, surveys have shown (Cristiano et al. 2000) that most companies use it in the initial phase of the development process only. Literature on QFD use (Akao and Mazur 2003) has shown that the technique is indeed beneficial, especially with respect to the decrease and the anticipation of engineering changes, which has an obvious and positive impact on product quality, as well as on development cost and time. At the same time, it has been shown that QFD essentially is an organizational tool that allows a better understanding and improved communications across corporate functions. Therefore, the method should not be treated as a tool for automated decision-making, while it should be kept in mind that successful application depends on the commitment with which its users engage in this translation process.

In practice, the main idea behind QFD is to develop incidence matrices called *Houses of Quality* (HoQ) that link the inputs, listed in the rows of the matrix, and the outputs, listed in the columns. Each translation step in the development process requires a separate HoQ, as shown in Fig. 14.7. This allows the translation of user requirements into system specifications, and then to cascade them into component specifications, process specifications and quality assurance specifications. The four steps follow the same principles and, since the process is seldom carried out in its entirety, we will describe the first House of Quality only.

A HoQ is typically structured around the sections outlined in Fig. 14.8. The following discussion describes the typical information that is recorded, but firms are of course free to make the changes they find suitable for their products.

- **Requirement analysis** is carried out on the horizontal part of the HoQ, usually under the leadership of the marketing function. The m requirements are listed on the left-hand side, while the detailed analysis is performed on the right-hand side. The first column in the analysis part is an evaluation of the average importance of each customer requirement ICR_i, indexed on the set $i = 1 \ldots m$, usually with a rating in the range [1…5]. User requirements and their importance ratings should ideally come from market research, since a development team

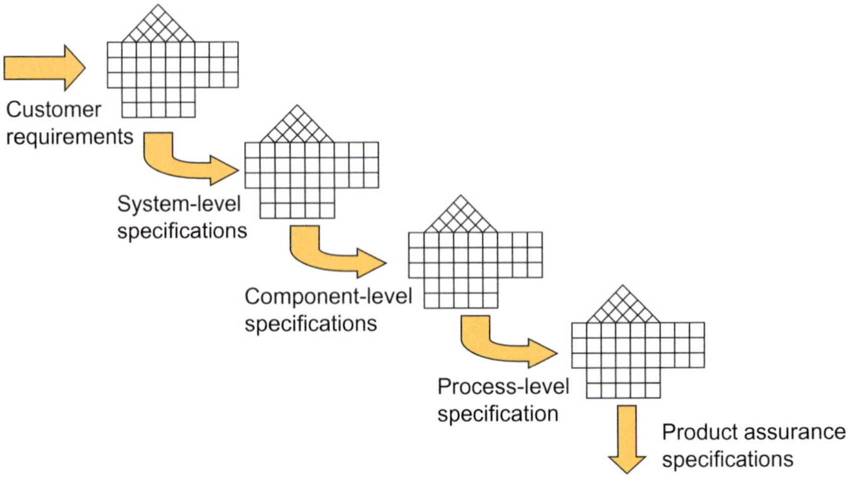

Fig. 14.7 The quality function deployment process

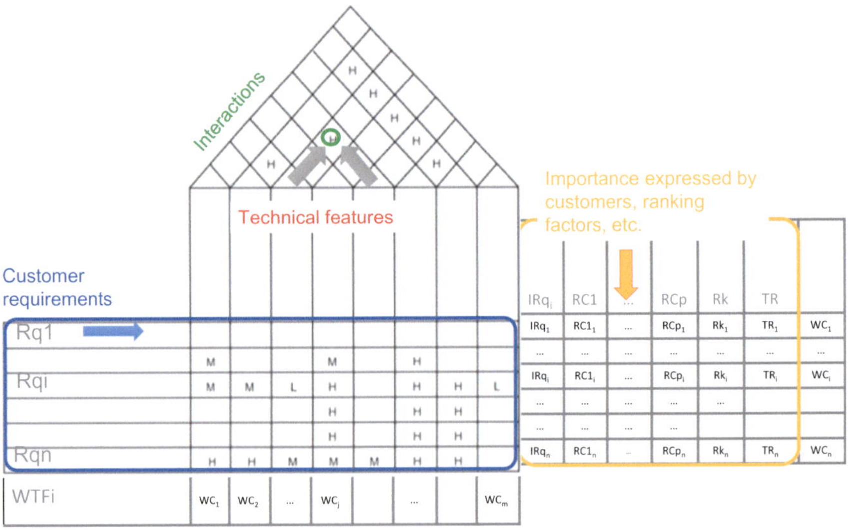

Fig. 14.8 The structure of a typical house of quality

being asked to 'estimate' weights for user requirements will unavoidably draw from their specific experience, which may be very far from what actual target customers really feel and look for.[10] By referring to market research activities

[10] The bias emerges because product developers usually have a technical competence and an inclination for technology that is far superior to users. Moreover, the target customer segment can be significantly different from the one that the product development team is used to consider (e.g., a team that has to develop a new car model for a geographical market in which the team has no prior experience).

discussed in Chap. 13, it can be considered appropriate to use secondary needs in this context, because their measures are often easier to be derived and since surveys generally provide quantitative information on the related ratings.[11] Working on primary needs would be of no use, since these are too few and are not detailed enough for supporting the definition of requirements (in fact, the purpose of identifying primary needs is to come up with a high-level market positioning for the product). Working on tertiary needs would instead lead to a very long list of qualitative descriptions, the importance of which would be difficult to assess in an objective way. However, the development team might justifiably decide to break down specific secondary needs at a finer level of detail.

The following columns on the right are used to develop competitive benchmarking in a selection of currently competing alternatives $k = 1 \ldots p$; the first of which is usually the current system being offered by the firm. Benchmarking is made by computing the average rating with which users perceive each system k to satisfy each requirement i, RC_{ik}. These ratings too ought to be obtained through objective surveys, and not by casting subjective assumptions within the development team.

Based on this competitive benchmarking, the team can define a set of target ratings TR_i for the new product being developed, one per each of the customer requirements. This decision points to the areas of improvement the development team will have to work on, in terms of customers' perception of the system. The importance that the development team will have to give to each customer requirement, WCR_i, is then computed as an absolute importance weight, which brings together the importance that customers assign to the requirement and the improvement that the new system must bring with respect to the current one:

$$\mathrm{WCR}_i = \mathrm{ICR}_i \frac{\mathrm{TR}_i}{\mathrm{RC}_{i1}} \tag{14.1}$$

In some cases, importance weights WCR_i may be further altered by multiplying them by additional factors. For instance, the firm may wish to increase the weighting of a few customer requirements for which it is easier to communicate improvements to the market, thus translating them into increased sales. In this case, a 'bonus' of 10–20% may be added to the weights of these particular requirements. For simplicity, importance weights WCR_i are then normalized WCR'_i, so that their sum adds up to 100.

- **Technical features definition** is carried out on the vertical part of the HoQ, usually under the leadership of technical functions. A list of $j = 1 \ldots m$ technical

[11] These average ratings should not be computed on the entire sample used for market research, but only on the subset of respondents that can be associated to the customer segment the product is being targeted to.

features are placed at the top of the matrix, at an appropriate level of detail. These technical features must in fact be easy to relate to system-level customer requirements, but must not be so generic that they cannot be measured.[12] The technical team will then collect values of these technical features from the same panel of alternatives for which a comparison was made during customer requirements analysis.

- **Technical features weighting.** At this point, the development team can relate customer requirements to technical features, in order to understand how the weights given to the former lead to priorities concerning the latter. At first, one must define the strength of the relationships r_{ij} between each customer requirement and each technical feature. These relationships, which are placed in the central part of the HoQ, express the degree with which customer requirement i is affected by technical feature j. These relationships are usually expressed on the basis of expert judgment, using a {0, 1, 3, 9} scale, where 0 means 'no relationship', 1 a 'very weak' relationship, 3 a 'weak' relationship and 9 a 'strong' relationship. This scale may seem to be quite arbitrary, but it is quite effective in discriminating technical features that are essential to the fulfillment of customer needs from the ones that are not so important. At this point, it is possible to compute a weight for each technical feature, WTF_j

$$\mathrm{WTF}_j = \sum_i r_{ij} \mathrm{WCR}'_i \tag{14.2}$$

  Weights WTF_j can also be normalized to WTF'_j, so that their sum adds up to 100.

  Thanks to this exercise, the design team now has a clear indication of the importance that each technical feature has 'in the eyes of the customer' and according to the improvements that are needed to make the product more attractive. These weights can therefore provide guidance for the definition of target technical features for the new product. Specifically, more ambitious targets will be set for those technical features which carry a higher weight.

- **Relationships between technical features**. Quite often, the HoQ is enriched by a further section that is located on the triangular 'roof', above the list of technical features. In this section, the development team can place information on relationships between technical features, using a parameter in the range [$-M$, …, $+M$]. A negative relationship will indicate a technical tradeoff, while a positive relationship implies that an improved value on one technical feature makes it easier to improve the other. For simplicity, relationships can also be indicated by using the discrete set {−1, 0, 1}. All this is usually quite well known to

[12] For instance, in a passenger car it makes sense to specify maximum speed and acceleration, but not engine horsepower and torque, which would be the corresponding specifications at engine level.

designers, since they represent technical tradeoffs and harmful effects that are at the heart of their work. It is essential to make these tradeoffs explicit, thus making sure that they are properly documented, both to make them visible to non-technical members of the product development team and since significant opportunities for innovation arise by tackling such tradeoffs in a radical way (see the TRIZ method in Chap. 15).

From the description made above, the pros and cons of QFD should be quite evident. Looking at drawbacks, QFD clearly rests on quite weak methodological basis, since the algebraic tools with which priorities on customer needs are computed and then translated into priorities for technical features are quite disputable (Franceschini and Rossetto 2002). Specifically, the method relies on a number of judgmental indicators that are inherently based on ordinal scales and—as such—are ill suited to be treated as real numbers. Furthermore, typically the importance associated with requirements is entrusted to the responsibility of designers who attribute them on the basis of their experience and sensitivity.

On the positive side, QFD is a simple and powerful tool for rationally collecting, communicating and connecting the elements that define a product from the two perspectives of customer requirements and technology. By enforcing discipline into the specification of technical features and making sure they are somehow connected to customer needs, QFD is indeed a valuable tool in the hands of inter-functional product development teams.

14.5 Product and Lifecycle Costing

14.5.1 Traditional Costing and Target Costing

Cost estimation is a key element of the development process and, as discussed in Chap. 13, it obviously affects strategic choices and impacts the firm's performance. It should be intuitive that inaccuracies in this activity can lead to mistakes that have strategic relevance, since an underestimation of costs leads to financial losses, while an overestimation will lead to a high price and the loss of market opportunities.

Depending on the approach being followed, it is possible to view cost as either a *dependent* or an *independent variable* of the development process. If one follows a somewhat traditional approach, cost will be treated as a function of the technical features that define the product. In this case, cost is studied and managed according to the flowchart on the left-hand side of Fig. 14.9.

Once product requirements have been defined, product cost is estimated in a relatively rough way and then, as system design evolves, it is computed more precisely. If, at any given point, the firm realizes that cost estimates are becoming too high with respect to the future competitiveness and profitability of the product, this will lead to an iterative modification of product requirements, until cost becomes acceptable again. After product launch, the firm will run cost reduction

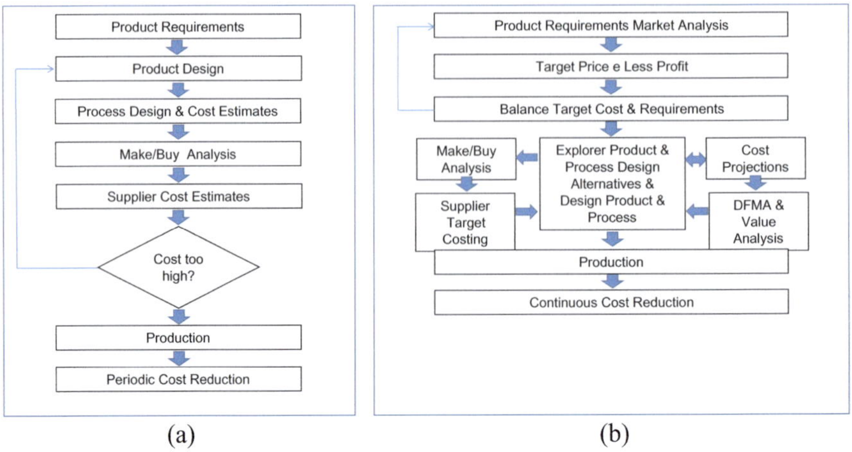

Fig. 14.9 The traditional (**a**) and target costing (**b**) approaches

initiatives periodically, or when market conditions require it. This approach has the drawback of viewing cost as secondary feature and managing it only if it exceeds limits that are however defined in a somewhat arbitrary fashion.

The alternative approach, termed *Target Costing**, is depicted on the right-hand side of Fig. 14.9. Coherently to pricing methods discussed in Chap. 13, this procedure immediately matches system features with a *target price*, which is estimated as the price that customers would be willing to pay for a system able to provide the same value. Target pricing is usually defined by comparing the new and improved system with an existing one and then estimating customers' willingness to pay for these improvements. In some cases, and especially in B2B industries, the target price can be computed by considering the benefits the system provides to customer, assuming that customers are able to objectively quantify them.[13] Given the target price, the firm applies a desired contribution margin, thus deriving the *target cost*, which then becomes one of the product specifications. The target cost is then divided into subsystem-level target costs, which will drive the development of each subsystem and component. In the case of systems that are incremental improvements of existing ones, the target cost must be allocated by keeping the sources of improvements into account. The team will then allow a higher target cost than the current one to those subsystems whose improved performances are leading to an increased target price. Conversely, the same (or even a reduced) target cost will be imposed for the subsystems that are essentially unchanged.

For instance, Fig. 14.10 shows a stylized example of a passenger car, with the three main subsystems 'body', 'powertrain' and 'infotainment functions'. We can suppose the new model is improved over the previous one because of greater fuel

[13] For instance, a firm may set the price of a product by computing the annual savings or profits P it determines on a customer firm and hypothesizing that customers will purchase it if its price $PR \leq P\ PBT$, where *PBT* is a payback time.

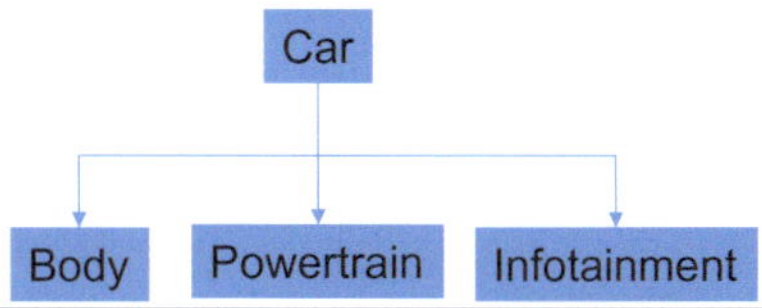

	COST				MARGIN	PRICE
	body	Powertrain	infotainment	TOTAL		
Old model	2000	2000	1000	5000	20%	6000
Improvement value perceived		500	1000	1500		
Target for new model	1900	2500	1850	6250	20%	7500

Fig. 14.10 An example of target costing

efficiency and richer infotainment functions and that market research assigns a price premium of 500 € for the former improvement and 1.000 € for the latter. The development team will allocate target costs to the three subsystems by following the logic of 'added value', as shown in the picture. In doing so, it might happen that a higher target cost may amply accommodate the required technical improvements in one subsystem (e.g., for the infotainment system) but not in another one (e.g., for the powertrain). The development team should as far as possible resist a further renegotiation of cost allocation between the two subsystems. In other words, it should accept that some of its suppliers (or internal functions) might earn higher margins, and not be afraid of engaging in tough negotiations with the ones that provide low added value. This is an important element of target costing, whose underlying logic is that cost should follow value creation. If one takes a long-term view, granting some added profitability to suppliers who create significant value cannot be but beneficial, since this will improve future relationships and their capability of creating further value by reinvesting profits.

14.5.2 Cost Estimating

The literature presents many contributions on methods for *Product Cost Estimation* (PCE). Niazi et al. (2006), as shown in Fig. 14.11, propose a hierarchical classification that explores possible methodological alternatives for cost estimation.

Qualitative methods are divided into *intuitive* and *analogical*, the former ones based on rules, while the latter ones are based on exploiting similarities with existing systems. In fact, products or services being developed are generally quite similar to existing ones, save for given areas of improvement. By analyzing historical data on previous alternatives, and by focusing on the 'cost of improvements', the design team may come up with a relatively accurate estimation of the costs of the new system.

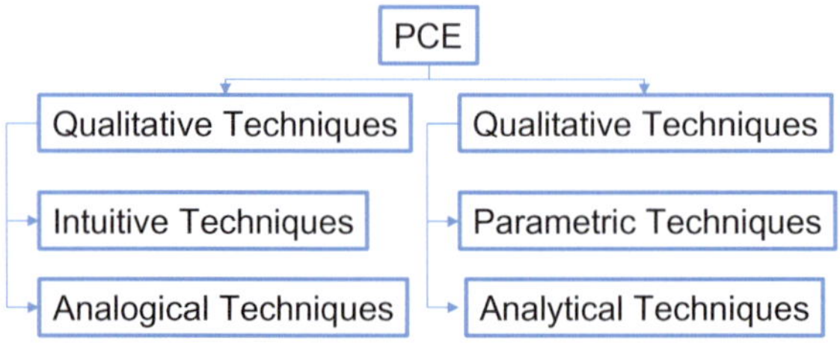

Fig. 14.11 A classification of product cost estimation methods (from Niazi et al. 2006)

Quantitative methods are categorized into *parametric* and *analytical*. Parametric models consider cost as a function of specific product variables (e.g., geometry, volume, etc.) that drive the economics of a product and its process.[14] These models can be developed bottom up throughout the system architecture, based on a detailed analysis of the process plan of each component and its cost structure. For instance, Fig. 14.12 shows three components of a single architecture, whose physical features (e.g., gross weight, diameter, volume), manufacturing processes (e.g., material waste) and operations (e.g., typology of supplier, demand) affect the overall costs of the product.

These models are also very useful for evaluating the effect of the design variables associated to the individual components of the architecture, as well as for analyzing design choices, such as the effect of modularity or the level of standardization/customization in the architecture (as in Fig. 14.13).

Alternatively, they can be developed by following a 'black box' approach, by creating a database of components and then searching for statistical relationships between relevant features and cost.

In this case, *analytical models* are based on a relatively detailed description of the product, which leads to a greater precision in the cost estimate. Specifically, the Activity Based Costing (ABC) approach identifies cost drivers not by looking at system features, but by focusing on the activities and the resources involved in the process (Tornberg et al. 2002; Park and Simpson 2005). In order to have a more precise estimate, or whenever there is not enough experience in similar components or products, cost estimation must be performed by actively involving the actors that are likely to supply the part. This means issuing a *Request for Quotation* (RFQ) based on a preliminary technical description. Internal departments of the same firm or suppliers will then respond to the RFQ with preliminary bids that—depending on the case—might or not be binding.

Choosing one cost estimation method or the other depends on the goals, on the decisions to be made and on the surrounding uncertainty. Exact system costs can be computed only at the end of the development process, when production is effectively started. However, firms must make estimates on product cost as early as

[14] For instance, parametric models can be based on rules, such as 'surface treatment *X* costs 22 €/m^2', or 'with technology *Y*, the cost of producing parts like this is about 12 €/cm^3, adding 2 € per each geometric feature to be machined'.

Internal quenched component

Final average price	Average Gross weight	Average order volumes	complexity	Observ number	Outsourcing production
0,653 Euros	0,2920 Kg	296 458	Low	886	Shaping and Turning

Internal rough shaped component

Final average price	Average Gross weight	Average order volumes	complexity	Observ number	Outsourcing production
3,569 Euros	2,81 Kg	156 014	High	135	Shaping

External component

Final average price	Average Gross weight	Average order volumes	complexity	Observ number	Outsourcing production
2,542 Euros	1,93 Kg	241 646	Medium	110	Shaping

		Total Cost
year	2011	-0,179 ns
	2012	-0,1025 ns
	2013	0,118 ns
Gross Weight		1,154***
Raw material costs (€/Kg)		5,315***
Cumulated quantity (log)		-0,157**
Supplier Bifrangi		0,284**
Supplier Dong-Eun		0,601***
Diameter difference (external-internal)		-0,192 †
Diameter difference ^2		0,009 †
Raw material waste		-2,707*
Raw material waste^2		4,215*
Volume of the component		0,011**
R^2		85,43%
N		90

**** p-value<0.1%; ** p-value<1%: * p-value<5%; † p-value<10%*

Fig. 14.12 Physical features, manufacturing and operations aspects of three components affecting the overall cost of a product

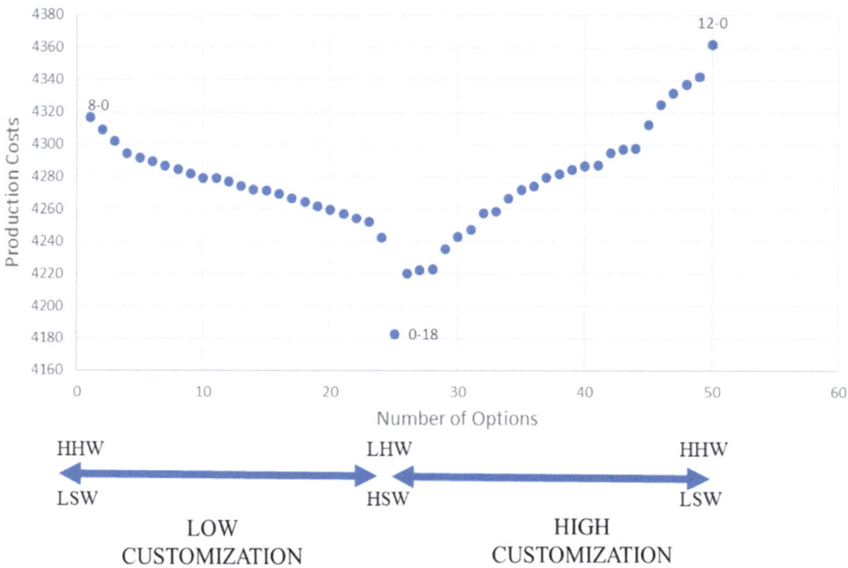

Fig. 14.13 The effect on costs of more or less HW and SW modules into a system architecture, reaching more or less customized systems

possible, when preliminary choices are being made, and even before any design activity has actually taken place. Therefore, the choice of the method will be influenced by the stage of the design process and by the amount and quality of data that has been collected. As shown in Fig. 14.14, qualitative cost estimation techniques are usually appropriate during the early stages of design, as they are based

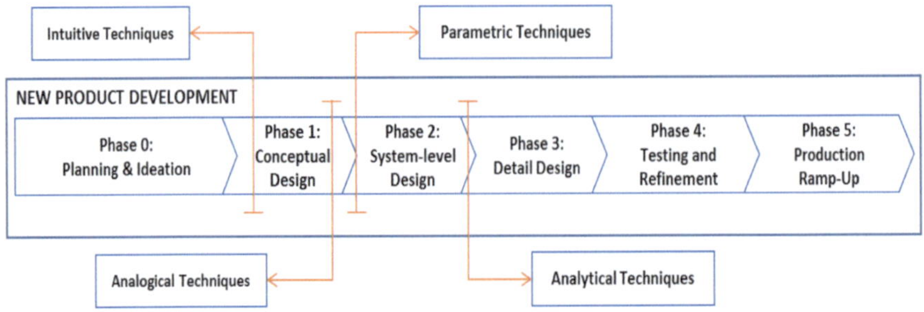

Fig. 14.14 Use of cost estimation techniques in the product development process

on the comparison of the new product with previous ones, using historical data, past design experience, or manufacturing knowledge. Instead, quantitative techniques provide more accurate estimations and their usage is often restricted to the final stages of the development process, when detailed information on system features, production and other ancillary processes (e.g., maintenance, post-sales, etc.) are available.

When estimating costs, care should be given in providing a precise definition of the elements that make up overall cost. Specifically, one should be able to distinguish between *variable* and *fixed*, and between *direct* and *indirect costs*. Readers ought to refer to cost accounting textbooks for a detailed description of this topic (Horngren et al. 2008). However, just to provide a few essential definitions, fixed costs include costs whose overall value does not change when production volume varies. Conversely, variable costs are the ones whose total amount changes proportionally with production volume. It should be noted that fixed costs are never such in absolute terms, but only in reference to a given production range.[15]

On the other side, direct costs can be directly attributed to the good or service, while indirect costs are associated to activities or items that are generally enablers of the production process, but do not directly intervene in it. The classification of a cost as direct or indirect is not only related to its nature, but also to the ease with which the firm would be able to trace the actual usage of the corresponding activity or asset to a specific system. Table 14.2 provides a few examples of costs and their classification according to the direct/indirect and variable/fixed costs categories.

In general, firms use a simple job costing approach based on the following three elements:

- **Direct material cost** includes all the costs associated to the materials used to make the component. This requires accounting inefficiencies such as scrap material and disposal of residual material. Direct material cost is by definition direct and is obviously variable.

[15] For instance, a die used in an injection molding process will be a fixed cost up to a given production level. After that threshold is passed, the firm will have to invest in a second die.

Table 14.2 A few examples of costs and their allocation to cost categories

Type of costs	Direct costs	Indirect costs
Variable costs	Materials used in making a product	Cost of ordering raw materials by the purchasing department
Fixed costs	The cost of machinery used to produce a part	Cost of the building where manufacturing takes place

- **Direct labor cost** includes the use of personnel who is directly engaged in the production process. It is given by the hourly salary rate of such personnel, multiplied by the time they are engaged in this activity. In general, direct labor costs are treated as variable, in the assumption that staffing can be varied according to demand and/or that personnel not being used for a given product can be diverted to other types of production. There might be cases in which they should be treated as fixed (e.g., a manufacturing cell that requires the continuous presence of a supervisor, regardless of production volume).
- **Use of assets.** A distinction can be made between indirect, generic direct and specific direct assets. Indirect assets, such as the building in which production occurs, are generally treated as a fixed and indirect cost, being part of what is called an overhead cost pool. Costs in an overhead cost pool will generally be allocated to the system according to some *cost driver*.[16] Generic direct assets include general-purpose machinery that is directly involved in production (e.g., a metal stamping press that can be used for any part requiring this technology). This is a direct cost that can be treated as either variable or fixed, in the former case assuming that this asset will be used at saturation level by the firm. Such a cost will generally be considered by defining an hourly rate, given by the annual cost of the machine (including depreciation, maintenance, etc.) divided by the budgeted production hours. Finally, specific and direct assets are the ones that require an investment and that can only be used for manufacturing the specific product (e.g., the die made for processing a given part on the metal stamping press). This cost is direct and fixed, within the range of production volume that can be made with the asset.

The calculation of these costs is extremely sensitive to the allocation rule being used and can have non-negligible effects on pricing, positioning and other strategic choices in general. So, as already discussed for project portfolio decisions in Chap. 9, it is necessary to ensure coherency between the objectives of the analysis, the level of detail obtainable, and the modeling choices.

[16] For instance, if the cost driver used is direct labor hours, the firm will first compute the total annual cost of the overhead cost pool, *TAC*, and the total labor hours worked, *TLH*. It will then determine an 'overhead rate' of *TAC/TLH* €/hour, which will be added to the direct costs of the product.

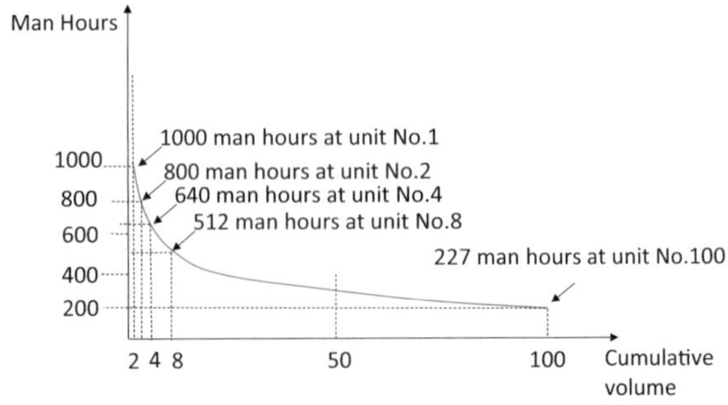

Fig. 14.15 The evolution of product cost observed by T. P. Wright

14.5.3 Cost Estimating with Learning Effects

14.5.3.1 The Learning Effect

As mentioned in previous chapters, production cost is usually not constant over the product lifecycle. As experience builds up, personnel engaged in production will progressively discover inefficiencies and propose improvements to the system and to the process, thus reducing costs. These improvements may in principle be connected to all of the elements that make up product cost, from the use of direct materials, to the man-hours and/or machine-hours that are required to complete production.[17]

This phenomenon has been studied by the proponents of scientific management since the beginning of the twentieth century, but a major contribution came in 1936 from T.P. Wright (1895–1970), who was working in the aircraft manufacturing industry. Wright observed that the time and cost required for producing an airplane decreased in time. However, he also noticed that this learning effect could be modeled more precisely by relating it not to time but, rather, to cumulated production volume. By doing so, he realized that the data would neatly fit around a regular curve *similar* to a hyperbole or a negative exponential (Fig. 14.15). By displaying data on logarithmic scales, the phenomenon closely followed a straight line, which suggests a slightly different mathematical model, that will be explored in detail in the following subsection.

When estimating production costs, neglecting the learning effect can lead to serious strategic errors (Henderson 1984). For instance, one can imagine an engineering company making a bid for the design and production of a large batch of components. A precise estimation of future production costs is essential both to win the bid and to ensure profitability. In fact, should the firm neglect the 'experience curve effect', it will quote a price that will be significantly higher than the one proposed by competitor who does consider it, and the contract will be lost. However, should the firm be overconfident of its capability of progressively reducing

[17] A difference is sometimes made between *learning effects* and *curves* (Wright 1936) and *experience effects* and *curves* (Henderson 1974), the former related to labor productivity and the latter to overall efficiency. The two phenomena follow mathematical models that are equivalent, and we will therefore consider the two interchangeably in the following discussion.

costs, the price quoted will be very low and will very likely lead to a winning bid. However, should the expected cost reduction not materialize, the firm might be forced to supply the parts at a loss.

The experience curve effect has strategic implications also for firms that are engaged in an open market, where market share depends on product quality and price. Firm *X* may be very conservative in considering the experience curve effect and therefore decide to price its products according to a cost-plus approach, progressively lowering the price when it realizes that cost is indeed decreasing. On the opposite side, another firm called *Y* may make an aggressive move, based on its forecast of cost reductions tied to experience. *Y* will therefore start with a low price, sometimes accepting that margins may initially be negative, but in the expectation that this low price will determine higher market share and demand. In turn, this large production volume will drive costs down as expected. Such a move can lead to major strategic benefits, especially for markets characterized by first-mover advantage. Firm *Y* will also enjoy competitive advantage with respect to competing firm *X* which, not having followed the same strategy and having been left behind in the experience curve, will have no way to recover. Of course, the impact of an aggressive pricing behavior based on the expected experience curve will also depend on competitors' behavior (Gilbert and Lieberman 1987). If *X* and *Y* simultaneously follow the same strategy, the outcome will be a low price, with no gains in market share for any of the two firms, and a consequent lower-than-expected decrease in costs. As a matter of fact, the profitability of many capital-intensive industries is routinely undermined by this type of behavior (Armstrong and Green 2007).

14.5.3.2 The Boeing-Crawford Experience Curve Model

As discussed above, empirical analysis has shown that cost reductions as a function of volume usually show up as a straight and downward sloping line, if the two variables are plotted on logarithmic scales. The corresponding model, which is generally known as the Boeing-Crawford[18] experience curve, therefore is:

$$C(n) = C(1)n^{-b} \tag{14.3}$$

where:

- $C(n)$ is the cost of the product being manufactured or, depending on the case, a portion of the total cost that is subject to learning effects. Alternatively to cost, one might consider the time required to manufacture the part. This cost is viewed as a function of cumulated volume, so that $C(n)$ represents the cost of the n-th part that has been produced (one might therefore consider it as a marginal cost when moving from the $(n - 1)$-th part to the n-th);
- b is called the learning rate and represents the rate with which costs decrease;
- $C(1)$ is the cost of the first part

[18] For a review of learning curve models, readers may refer to Anzanello and Fogliatto (2011).

If one takes logarithms, (14.3) becomes (14.4), which is clearly a straight line whose downward slope is given by b.

$$\ln C(n) = \ln C(1) - b \ln(n) \tag{14.4}$$

Because of the linear relationship on a log–log scale, it is straightforward to use linear regression in order to identify parameters from empirical data. However, it would be misleading to use the actually recorded cost of the first part for $C(1)$ and therefore use the regression to identify b only. The reason is that the cost of the first part is subject to many possible incidental factors, which might lead to unexpectedly high or low values. Secondly, it has been noticed (Li and Rajagopalan 1998) that experience often suffers from a '*decay factor*' due to obsolescence of knowledge and employee turnover. This decay can be modeled with a parameter $\delta < 1$, which decreases the impact of cumulated volume, so that (14.3) becomes

$$C(n) = C'(1)(\delta n)^{-b} \tag{14.5}$$

In Eq. (14.5), $C'(1)$ does not incorporate the knowledge decay effect. However, (14.5) can fall back into the usual form (14.3) if one only considers the value $C(1)$ to be estimated from empirical data to include knowledge decay too, as in (14.6)

$$C(1) = C'(1)\delta^{-b} \tag{14.6}$$

Once parameters have been estimated—or drawn by analogy from similar products the firm has worked on in the past—learning curves can be used to forecast future cost. By using the Mean Value Theorem, and approximating the numbering of part 1 as part 0, one can derive the average cost of parts 1 through n

$$\begin{aligned} \overline{C}(n) &= \frac{1}{n}\int_0^n C(n')\mathrm{d}n' = \frac{1}{n}\int_0^n C(1)n'^{-b}\mathrm{d}n' \\ &= \frac{C(1)}{n}\frac{n^{1-b}}{1-b} = C(1)\frac{n^{-b}}{1-b} = \frac{C(n)}{1-b} \end{aligned} \tag{14.7}$$

Similarly, if one has to foresee the average cost of a batch of products ranging from part n_1 to part n_2,

$$\begin{aligned} \overline{C}(n_1, n_2) &= \frac{1}{n_2 - n_1}\int_{n_1}^{n_2} C(n)\mathrm{d}n \\ &= \frac{1}{n_2 - n_1}\int_{n_1}^{n_2} C(1)n^{-b}\mathrm{d}n = \frac{C(1)}{n_2 - n_1}\frac{n_2^{1-b} - n_1^{1-b}}{1-b} \end{aligned} \tag{14.8}$$

Equation (14.8), if coupled to a forecast of production volume over time, can be very useful to foresee the average production cost in the time period that will see the cumulated production of parts from n_1 to n_2.

An interesting property of the Crawford-Boeing model is that each time cumulated production increases by a factor k (e.g., if $k = 2$, when going from the 2nd to the 4th part, or from the 120,000th to the 240,000th), production cost will be reduced by the same relative amount $1 - \alpha$. In fact,

$$\frac{C(kn)}{C(n)} = \frac{C(1)(kn)^{-b}}{C(1)(n)^{-b}} = k^{-b} = \alpha, \tag{14.9}$$

$$b = -\frac{\ln \alpha}{\ln k} \tag{14.10}$$

Firm usually consider $k = 2$ and evaluate relative cost reduction at each doubling of production. In such case, the learning rate b can be readily computed via (14.10). The fact that the same relative cost reduction occurs at constant multiples of cumulated production suggests that savings are easy to achieve at the beginning, but become quite harder as experience builds up.

In order to understand this point in greater depth, one may take Eq. (14.3) and at first take its derivative with respect to n

$$\frac{\mathrm{d}C(n)}{\mathrm{d}n} = \frac{\mathrm{d}C(1)n^{-b}}{\mathrm{d}n} = -bC(1)n^{-b-1} = -\frac{b}{n}C(n) \tag{14.11}$$

Equation (14.11) is the instantaneous *absolute* cost decrease at each increase of n. Then, the instantaneous *relative* cost decrease at each increase of n can be found by dividing (14.11) by $C(n)$, yielding

$$\frac{\mathrm{d}C(n)/\mathrm{d}n}{C(n)} = -\frac{b}{n} \tag{14.12}$$

This can be compared to the behavior of the exponential learning curve, which is quite often used by psychologists and in the social sciences (Bills 1934; Teplitz 1991; Ritter and Schooler 2001):

$$C(n) = C(1)e^{-bn} \tag{14.13}$$

and

$$\frac{\mathrm{d}C(n)}{\mathrm{d}n} = \frac{\mathrm{d}C(1)e^{-bn}}{\mathrm{d}n} = -bC(1)e^{-bn} = -\mathrm{b}C(n) \tag{14.14}$$

and hence:

$$\frac{\mathrm{d}C(n)/\mathrm{d}n}{C(n)} = -\mathrm{b} \tag{14.15}$$

If a learning phenomenon follows the exponential learning curve, the relative cost decrease remains constant with n. This is equivalent to stating that—if learning has to do with identifying and solving areas of improvement (or inefficiencies)—the slope of $C(n)$ will decrease simply because the number of residual inefficiencies becomes progressively lower. Instead, in the Crawford-Boeing model, the relative cost decrease shrinks hyperbolically with n. This is like saying that residual inefficiencies to be tackled are progressively not only fewer, but also either harder to spot or to solve.[19]

Students of learning phenomena (Dar-El 2000; Liao 1988) tell us that, while the exponential model holds well when learning occurs at individual or small-group level, the Crawford-Boeing model is better at representing learning phenomena at organizational level. It is not surprising to see that learning at organizational level is more complex and difficult to pursue than for individuals.

These considerations have interesting practical implications. For instance, firms who launch cost reduction programs in order to improve their competitiveness often set targets such as 'we will reduce cost by $x\%$ per quarter over the next two years', and will tend to do so during a time of crisis. This approach is inherently flawed because of two reasons. First, it is not time, but volume that can drive a decline in costs (so, ambitious objectives are easier to meet when demand is going well, than when it is shrinking). Second, even if demand and time were proportional, organizational learning would usually not lead to a constant relative change in cost. Therefore, achieving a constant relative cost reduction per time unit when demand is decreasing is impossible, if not accepting to degrade product quality.

14.5.4 Lifecycle Costing

Traditionally, product costing had to do with computing the costs that were associated to the manufacture of a product, but not beyond. However, products usually determine numerous other costs during their operation *until the end* of their lives. If one thinks of a passenger car, the money spent in fuel, repairs, insurance, and toll roads is likely to be of the same order of magnitude than the money spent to buy the car. Energy-intensive appliances (e.g., light bulbs, electric heaters and air conditioners), or durable goods used by businesses, such as commercial planes and trucks, easily have an acquisition cost that is only a fraction of the overall costs associated to the product throughout its lifetime. In the case of a service, production

[19] When dealing with continuous improvement projects, it is customary to say that it is easy to 'pick low-hanging fruit' at first, but that things become quite harder as time goes by. Following on this analogy, exponential learning would be as if fruit had already fallen on the ground, and one simply had to find it hidden in the grass.

and consumption occur simultaneously, so that the majority of costs will occur during the delivery of the service itself. For instance, in a food delivery service, on must consider IT and organizational infrastructures needed to run the service, the cost of a call center for customer care activities, the cost of delivery personnel, etc.

As it has been discussed in previous chapters, the majority of these costs are determined by design decisions made during the early stages of the development process (Ulrich and Eppinger 1995). In the past, designers would have aimed at complying with target manufacturing costs, somewhat leaving the other lifecycle costs as a relatively uncontrolled consequence of their decisions. This approach is nowadays seen as too myopic, since the lifecycle consequences of design decisions can have significant impact on product or service success. Conversely, taking them into account in order to reduce the overall lifecycle cost ('*designing for cost*' in the words of Dean and Unal 1992) can have significant strategic value for the firm.

For instance, a decision that leads to reduced manufacturing costs may make field service extremely difficult and expensive, thus reducing the attractiveness of the product, while a badly designed service could lead to high costs due to claims, refunds, etc. Conversely, a design with higher manufacturing cost that must be sold at a higher price might be attractive, but provided it requires significantly lower operation, maintenance and disposal costs (Dunk 2004), as well as a well-designed service would see less errors in the service process, thus reducing the probability of costly remedial actions (Gawdat et al. 2020).

In order to work properly, this tradeoff mechanism must be objectively advantageous and easy to communicate and requires customers willing to accept making the investment. In some cases, benefits and tradeoffs must be played out by involving multiple parties or stakeholders. For instance, if a carmaker wants to incorporate features leading to a reduced pedestrian impact in case of accident, the added cost will be easier to justify in case insurance companies were ready to decrease the insurance premium.

From a strategic perspective, taking a broad perspective on lifecycle cost allows firms to innovate their business models. For instance, a manufacturer may decide to shift from the simple production of goods to the provision of services, in order to internalize different lifecycle phases and the profits accruing from a proper management of tradeoffs.[20] Similarly, producers can purposely shift their revenue streams and their profits from one lifecycle phase to the other and decide to gain from the future use of their products, rather than from their initial adoption. In the so-called *razor and blades** business model, the firm sells the product at or below cost and then profits from sales of consumables (e.g., razor blades in the case of shaving razors, cartridges for inkjet printers, and coffee pods for espresso machines).

*Lifecycle Costing** (LCC), defined as the "*systematic analytical process of evaluating various designs or alternative courses of action in relation to their*

[20] For example, if customers are unwilling to pay a higher price for a system that has very low maintenance costs, a producer might find it highly profitable to start leasing it. This is especially true if the producer is able to secure finance at low interest rates, while customers base their purchasing decisions on short-term criteria such as Payback Time.

lifecycle impact" (Fabrycky and Blanchard 1991), is a well-known approach for achieving the goal of 'designing for cost'. LCC is defined as the overall cost associated to the product, from its inception to its withdrawal at the end of its life. The overall costs that are experienced by the customer are instead termed *Total Cost of Ownership** (TCO). Making a systematic assessment of LCC and TCO is based on the following steps:

- The lifecycle costs associated to the product can be analyzed and arranged hierarchically in a Cost Breakdown Structure (or CBS). In general, the first level of the CBS distinguishes between R&D cost, manufacturing (or 'investment') cost, and operations and maintenance cost (left-hand side of Fig. 14.16).
- Stakeholders that are involved in the product lifecycle are identified, and costs in the CBS can be allocated to stakeholders according to the current (or *as-is*) state (right-hand side of Fig. 14.16). When doing so, each element may be analyzed by looking at both the simple cost to the stakeholder who directly bears it, and by studying the economic exchanges that occur between stakeholders, including their margins.
- After performing a critical analysis of the *as-is* cost-stakeholder matrix, the firm can work on the new product at different levels, both tactical and strategic, and define a *to-be* matrix that will serve as a requirement for product design.

Figure 14.17 reports a toy example related to a washing machine. Seven lifecycle phases have been considered, and five stakeholders have been identified. The matrix on the top reports the pure cost of each phase and its allocation to the stakeholders who directly bear it. The matrix on the bottom reports the overall economic flows between stakeholders, including their revenues (which will appear as a cost to some other stakeholder). For instance, the pure cost of distribution appears to be 80 € and is borne by the distributor. However, since the distributor

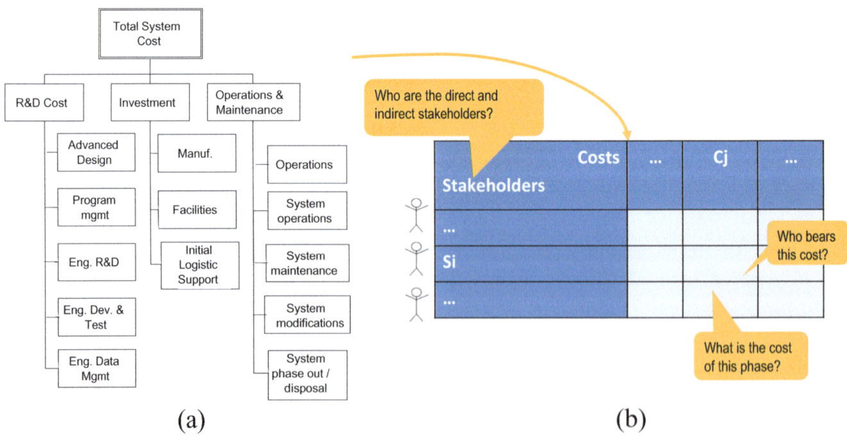

Fig. 14.16 Cost Breakdown Structure and cost allocation to stakeholders in lifecycle costing

(a)

Internal cost	Production	Distribution	Installation	Operation	Consumables	Servicing	EOL	Total	
Producer	150							**150**	**14%**
Distributor		80	20				100	**200**	**19%**
Maintenance						200		**200**	**19%**
Consumables vendor					400			**400**	**38%**
Customer				100				**100**	**10%**
Total	**150**	**80**	**20**	**100**	**400**	**200**	**100**	**1050**	
	14%	**8%**	**2%**	**10%**	**38%**	**19%**	**10%**	**100%**	

(b)

Overall flows		Production	Distribution	Installation	Operation	Consumables	Servicing	EOL	**Total**		
Producer	*Revenue*	200	0	0	0	0	0	0	**200**		
	Cost	150	0	0	0	0	0	0	**150**		
	Margin	50	0	0	0	0	0	0	**50**		14%
Distributor	*Revenue*	0	380	0	0	0	0	40	**420**		
	Cost	0	280	20	0	0	0	100	**400**		
	Margin	0	100	-20	0	0	0	-60	**20**		5%
Maintenance	*Revenue*	0	0	0	0	0	300	0	**300**		
	Cost	0	0	0	0	0	200	0	**200**		
	Margin	0	0	0	0	0	100	0	**100**		27%
Consumables vendor	*Revenue*	0	0	0	0	600	0	0	**600**		
	Cost	0	0	0	0	400	0	0	**400**		
	Margin	0	0	0	0	200	0	0	**200**	**370**	**54%**
Customer	*Revenue*	0	0	0	0	0	0	0	**0**		
	Cost	0	380	0	100	600	300	40	**1420**		
	Margin	0	-380	0	-100	-600	-300	-40	**-1420**		
Total	***Revenue***	**200**	**380**	**0**	**0**	**600**	**300**	**40**			
	Cost	**150**	**80**	**20**	**100**	**400**	**200**	**100**			
	Margin	**50**	**300**	**-20**	**-100**	**200**	**100**	**-60**			

Fig. 14.17 A simple example for lifecycle costing analysis

also buys the washing machine from the producer at a price of 200 €, his total cost will be 280 €. Similarly, the distributor completely bears the full cost of installation (20 €) and 60% of the 100 € cost of withdrawal of the product at the end of its life (the rest being paid for by the customer).

The matrix can be read by cost element (by columns) and by stakeholder (by rows). From the former perspective, it appears that the majority of costs and margins are associated to dealing with consumables. So, a producer may see the cost of consumables as an interesting area of improvement (e.g., developing a washing machine that requires less use of detergent) or even as a strategic direction in business model innovation (e.g., proposing a 'full service' contract including the supply of detergent). Instead, by reading the matrix from a stakeholder perspective, one realizes that the distribution activity is only apparently profitable. The profits earned by distributors are in reality quite meager given that—in the current business model—this actor subsidizes both installation and decommissioning. The producer may therefore become much more attractive to distributors by introducing innovations aimed at reducing these costs.

14.6 Lifecycle Analysis

There is a growing attention toward environmental issues, and product development is ever more involved to ensure that the impact made by new products and services on the environment is limited and coherent with regulations and with the firm's strategy.

One key tool available to perform this type of analysis is *Lifecycle Assessment** (LCA, Crawford 2011). We do not have the possibility of providing a full discussion on LCA in this textbook; therefore, we will limit the discussion to its foundations, in the context of new product development. The main idea behind LCA is to carry out a *systematic analysis of the impact of a product or service on the environment throughout its lifecycle*, from the sourcing of materials to production, assembly, distribution, use, and end of life.[21]

Carrying out one such exercise would seem next to impossible, given the complexity of products and value chains, but the possibility is real, if one considers each value-adding (and also resource-consuming and emission producing) phase on top of the previous ones. Referring to Fig. 14.18, one can carry out LCA within each phase (System Boundary A, in LCA jargon) and perform a '*Gate to Gate*' analysis. Now, the firm may have the availability of a similar analysis on the upstream and the downstream phases of the value chain (either because the suppliers have made one, or because a public database provides reliable information). If so, it will be possible to carry out LCA on the aggregation of a number of phases (System Boundary B) and perform both '*Cradle to Gate*' analysis and '*Cradle to Grave*' analysis. If the firm wishes to push the analysis beyond, it may also include the LCA of enabling infrastructure (System Boundary C).

An LCA analysis is usually carried out in phases. After having decided objectives and scope (and supporting a specific new product or service development project can be one example) and having identified the main flows to be considered with respect to resources being used and emissions being produced, the firm must perform the so-called *inventory analysis*.

Inventory analysis consists in a computation of the resources and emissions associated to each product unit, by merging knowledge from the firm's own processes and knowledge deriving from processes that are situated upstream and downstream.[22] Figure 14.19 shows a simple example of the main flows that might be studied for a washing machine.

After Inventory Analysis, the firm will operate an *Impact Assessment*, in which the previously computed inputs and outputs will be analyzed in terms of the potential impact on the environment. This requires the choice of a set of relevant impact categories (e.g., climate change, land use, etc.) and to characterize the

[21] More precisely, ISO 14000 standards define LCA as the "*compilation and evaluation of the inputs, outputs and potential environmental impacts of a product system throughout its life cycle*".

[22] Public databases are often used to this purpose. For instance, a washing machine manufacturer may wish to compute the energy consumed to produce the casing of each machine. The team will multiply the weight of steel used in each machine times a standard parameter that expresses the specific energy that goes in each kg of steel (kWh/kg). To this, they will add the kWh used internally to stamp and weld the steel casing of each machine.

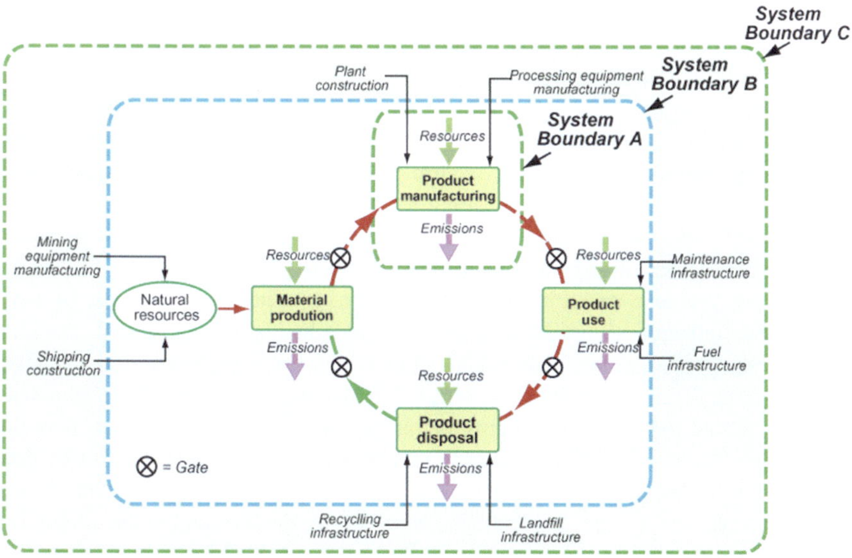

Fig. 14.18 The rationale of lifecycle assessment

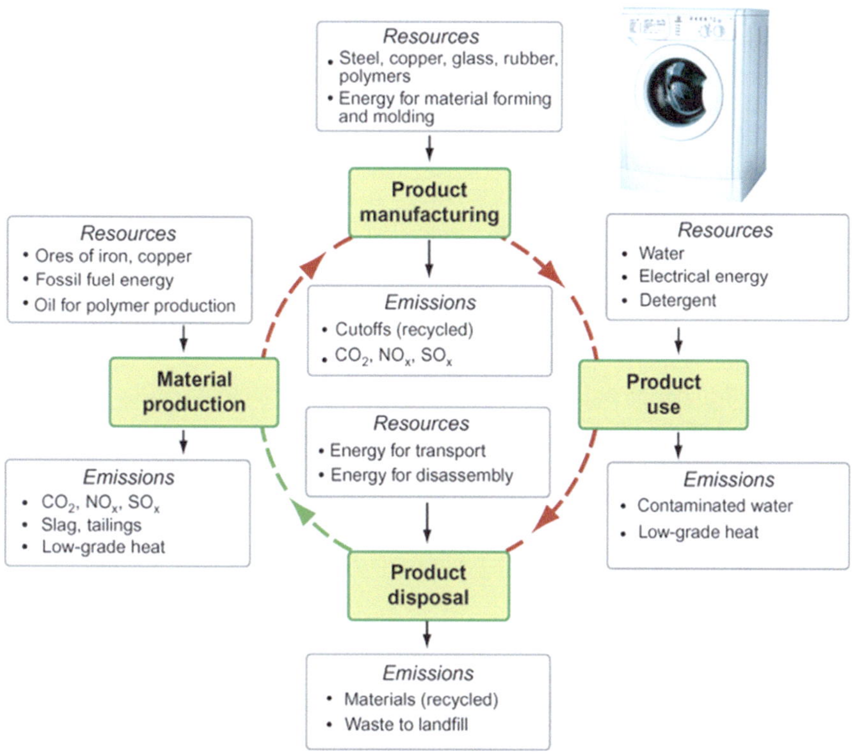

Fig. 14.19 The inventory analysis of a lifecycle assessment

impact of each input and output on each impact category.[23] This will eventually lead to computing an aggregate score on a standardized indicator, such as Eco-indicator 99 (Goedkoop 2007).

In the context of a growing attention to environmental impact, data coming from LCA can become an important subset in the requirements list that characterizes products and services. For instance, a firm may explicitly posit a high-level target such as 'the new version of our washing machine must reduce overall CO_2 emissions by *x*%' and then deploy this into specific targets such as 'it must reduce energy consumption during each washing cycle by *y*%'.

LCA analysis also can drive radical change to products and services, by supporting the redesign of traditional linear value chains into so-called *circular economy* strategies and business models (Lewandowski 2016; Bocken et al. 2016; Salvador et al. 2021). In such cases, the LCA becomes a 'cradle to cradle' analysis. The idea behind the circular economy is that products must be conceived from the beginning as parts of feedback loops in which one minimizes the amount of new resources used and emissions generated. This can entail changes to the firm's business model, the creation of linkages with other industries and value chains (so that each input and output in one industry becomes output and an input for another industry) and new ways to manage the tradeoffs in product requirements. Among the typical strategies suggested by the circular economy approach, one can list:

- minimizing the use of materials in general, with a special focus on new materials (using recycled materials or remanufactured components is obviously fine), on materials that are particularly harmful on the environment at any stage of the lifecycle, and on materials that are hard to recycle at the end of life;
- minimizing the effort required to reuse, remanufacture, disassemble and recycle products at the end of life;
- minimizing the use of energy in production and use, which includes the minimization of transport activities occurring along the value chain;
- minimizing the number of units to be manufactured, thus moving to 'as a service' business models based on rental and sharing, instead of selling and ownership;
- maximizing the durability of products and their repairability, at the same time avoiding planned obsolescence.

[23] In the previous example, the firm may translate the kWh/unit in terms of CO_2 emissions, based on the country's energy mix (again, using publicly available data).

References

Akao GH, Mazur Y (2003) The leading edge in QFD: past, present and future. Int J Qual Reliab Manage 20(1):20–35
Altshuller G (1999) The innovation algorithm: TRIZ, systematic innovation, and technical creativity. Technical Innovation Center, Worcester, MA
Anzanello MJ, Fogliatto FS (2011) Learning curve models and applications: literature review and research directions. Int J Ind Ergon 41(5):573–583
Armstrong JS, Green KC (2007) Competitor-oriented objectives: the myth of market share. Int J Bus 12:117–136
Bangle CE, Rosso M, Montagna F, Cantamessa M (2022) Design methods review for smart product: objectomy, a new approach. In: Design 2022, Dubrovnik, Croatia (online), May 23–26th, pp 2045–2054
Beyer H, Holtzblatt K (1997) Contextual design: defining customer-centered systems. Morgan Kaufmann Publisher, San Francisco
Bills AG (1934) General experimental psychology. Longmans psychology series, Longmans, Green and Co, New York
Bocken NMP, de Pauw I, Bakker C, van der Grinten B (2016) Product design and business model strategies for a circular economy. J Ind Prod Eng 33(5):308–320
Colombo S, Montagna F, Cascini G, Palazzolo VF (2022) Digital artefacts and the role of digital affordance. In: Design 2022, Dubrovnik, Croatia (Online), May 23–26th, pp 11–20
Crawford RH (2011) Life cycle assessment in the built environment. Taylor and Francis, London
Cristiano JJ, Liker JK, White CC III (2000) Customer-driven product development through quality function deployment in the U.S. and Japan. J Prod Innov Manag 17:286–308
Dar-El E (2000) Human learning: from learning curves to learning organizations. Kluwer Academic Publishers, Boston
Dean EB, Unal R (1992) Elements of designing for cost, aerospace design conference, AIAA 1992, Irvine, USA
Dunk AS (2004) Product life cycle cost analysis: the impact of customer profiling, competitive advantage, and quality of IS information. Manag Account Res 15(4):401–414
Fabrycky WJ, Blanchard BS (1991) Life-cycle cost and economic analysis. Prentice Hall, Englewood Cliffs, New Jersey
Franceschini F, Rossetto S (2002) QFD: an interactive algorithm for the prioritization of product's technical characteristics. Integr Manuf Syst 13(1):69–75
Gaver WW (1991) Technology affordances. In: Proceedings of the SIGCHI conference on human factors in computing systems, ACM, New Orleans, LA pp. 79–84
Gawdat AN, Bassily MA, Elamerab AA (2020) Do life-cycle costing and assessment integration support decision-making towards sustainable development? J Clean Prod 267:122056
Gilbert RJ, Lieberman MB (1987) Investment and coordination in oligopolistic industries. Rand J Econ 18:17–33
Goedkoop M (2007) The eco-indicator 99 methodology. J Life Cycle Assess Jpn 3(1)
Graziosi S (2012) From requirements classification to design problem analysis. Alta Scuola Politecnica (ASP) - Design Methods, 15th May
Greenbaum J, Kyng M (1991) Design at work—cooperative design of computer systems. Lawrence Erlbaum, Hillsdale
Henderson BD (1974) The experience curve reviewed: the experience curve reviewed: why does it work? Boston Consulting Group, Perspectives 128
Henderson BD (1984) The logic of business strategy. Harper & Row, Cambridge
Horngren CT, Foster G, Bhimani A, Datar SM (2008) Management and cost accounting. Prentice Hall, London
ISO 9241-210, (2010) International Organization for Standardization. http://www.iso.org/iso/catalogue_detail.htm?csnumber=52075. Retrieved 23 Oct 2022
Kallinikos JA (2013) The ambivalent ontology of digital artefacts. MIS Q 37(2):357–370

Lewandowski M (2016) Designing the business models for circular economy—towards the conceptual framework. Sustainability 8(1):43
Li G, Rajagopalan S (1998) Process improvement, quality, and learning effects. Manage Sci 44 (11):1517–1532
Liao SS (1988) The learning curve: Wright's model vs Crawford's model. Issues Account Educ 3:302–315
Meske C, Amojo I, Thapa D (2023) A conceptual model of feedback mechanisms in adjusted affordances—Insights from usage of a mental mobile health application. Int J Inf Manage 69:102597
Niazi A, Dia JS, Balabani S, Seneviratne L (2006) Product cost estimation: technique classification and methodology review. J Manuf Sci Eng 28:563–575
Norman DA (1998) The invisible computer. MIT Press, Cambridge
Norman DA (2002) The design of everyday things. Basic Books Inc, New York
Norman DA (2005) Emotional design: why we love (or hate) everyday things. Basic Books Inc, New York
Norman DA, Draper SW (1986) User centered system design; New perspectives on human-computer interaction. Erlbaum Associates Inc, Hillsdale
Park J, Simpson TW (2005) Development of a production cost estimation framework to support product family design. Int J Prod Res 43(4):731–772
Ritter FE, Schooler LJ (2001) The learning curve. International encyclopedia of the social and behavioral sciences. Pergamon, Amsterdam, pp 8602–8605
Salvador R, Barros M, Freire F, Halog A, Piekarski CM, De Francisco AC (2021) Circular economy strategies on business modelling: identifying the greatest influences. J Clean Prod 299:126918
Schuler D, Namioka A (1993) Participatory design: principles and practices. Lawrence Erlbaum, Hillsdale, New York
Teplitz CJ (1991) The learning curve deskbook: a reference guide to theory, calculations, and applications. Quorum Books, New York
Tornberg K, Jamsen M, Paranko J (2002) Activity-based costing and process modeling for cost-conscious product design: a case study in a manufacturing company. Int J Prod Econ 79(1):75–82
Ulrich KT, Eppinger ED (1995) Product design and development. McGraw-Hill, New York
Vyas DC (2006) Affordance in interaction. In: Proceedings of the 13th eurpoean conference on cognitive ergonomics trust and control in complex socio-technical, pp 20–22
Wright TP (1936) Factors affecting the cost of airplanes. J Aeronaut Sci 3(4):122–128
Yoo Y (2010) Computing in everyday life: a call for research on experiential computing. MIS Q 34(2):213–231

Designing Products and Services 15

Abstract

This Chapter focuses on the design process that is at the heart of product development. Following Herbert Simon's view of design as the 'science of the artificial', it discusses the main ways with which design has been studied from a descriptive and normative perspective. The chapter then proceeds to discussing the design of technical systems and focuses especially on the conceptual design phase, viewed as the phase in which requirements are translated into a functional design, followed by a highly creative activity, in which solution principles are identified, product concepts are proposed and then selected. Methods supporting this activity, such as TRIZ, are discussed at some length.

With this chapter, our journey in product development enters the design phase, a stage that requires us to adopt a deeper technical perspective. In Chap. 14, system specification was mainly concerned about what the product or service should be like, do and perform. In this chapter we will move on to discussing how technical solutions to these desiderata may actually be found. The discussion will start from the theoretical underpinnings of the fundamental activities and cognitive processes of human design thinking, as well as on the ways with which they can be supported by methods and tools that are specific for this type of problem-solving.

The discussion will focus on the initial (or conceptual) phase of design, but will not enter the stages of embodiment and detailed design, which would require discussing discipline-specific (or industry-specific) methods and perspectives and which would lead us beyond the scope of this textbook.

Designers are commonly considered to be professionals involved in developing new products and services. However, the cognitive approach they apply to their profession is applicable to all decision-making contexts and types of problems. A "*Designerly way of knowing and thinking*" (Cross 1982, 2001) can in fact be applied to define a strategy, an organization, or to solve any complex and/or poorly

M. Cantamessa and F. Montagna, *Management of Innovation and Product Development*, https://doi.org/10.1007/978-1-4471-7531-5_15

structured problem, characterized by uncertainty and scarcity of elements or information, in both private and public, profit or non-profit sector.

In view of that, '*design thinking*' (Rowe 1987; Buchanan 1992; Kumar 2012) has become, in recent years, an increasingly important *problem-solving** approach in all sectors and at all levels. Design thinking is no longer the prerogative only of those who design and produce products, machinery, software, services or buildings, but of anyone who has to face complex decision-making problems and generate innovative solutions in organizational and business, or social contexts (Simon 1969; Cross 2011; Brown 2015). Design Thinking has now become a recognized corpus of transversal skills, competences and soft skills with respect to the different branches of technology and human action (Brown and Katz 2011).

For this reason, this chapter also takes on a broader and deeper value, defining those basic concepts that constitute a professional and cultural corpus for 'designers' of any type, including—but not exclusive to—managers involved in innovating products and services.

15.1 Some Theoretical Foundations of Design

15.1.1 Design as Problem-Solving—The Cognitive Process of Technology

From an engineer's perspective, *Design** is a quintessentially technical activity, in which the designer puts technology to practical use, in order to achieve a desired outcome. However, if one focuses on the *subject*—rather than on the *object*—of the design activity, one quickly realizes that design is a very 'human' activity that occurs in the brains of individuals and through social interaction. Therefore, it is quite apparent that design cannot be completely understood if one adopts a merely technological perspective. Instead, the comprehension of the design process requires adopting multiple 'lenses', which may come from fields such as Philosophy, Cognitive and Social Sciences. According to this perspective, Pahl and Beitz stated that "*Design places at the crossroads between Science and Technology, Art and Social Sciences*" (as depicted in Fig. 15.1).

To this purpose, one can even go back to Aristotle, when the ancient philosopher identified the dualism between Science and Technology, at the same time being one of the first to recognize design as a distinctively human and philosophically relevant activity. Elaborating on Aristotle's view, Science and Technology can be roughly classified as in Table 15.1.

The essence of Design is *to devise a solution that will work in practice*. This implies creating an artifact that is able to perform as intended in a given situation, thus answering to specific goals and intentions. Conversely, a design that is generically applicable to a host of situations, but does not really perform adequately in any of them, would be of no real use. A good design would be able to tackle at least one given situation well, while a 'perfect' design would perform successfully in a wide range of circumstances.

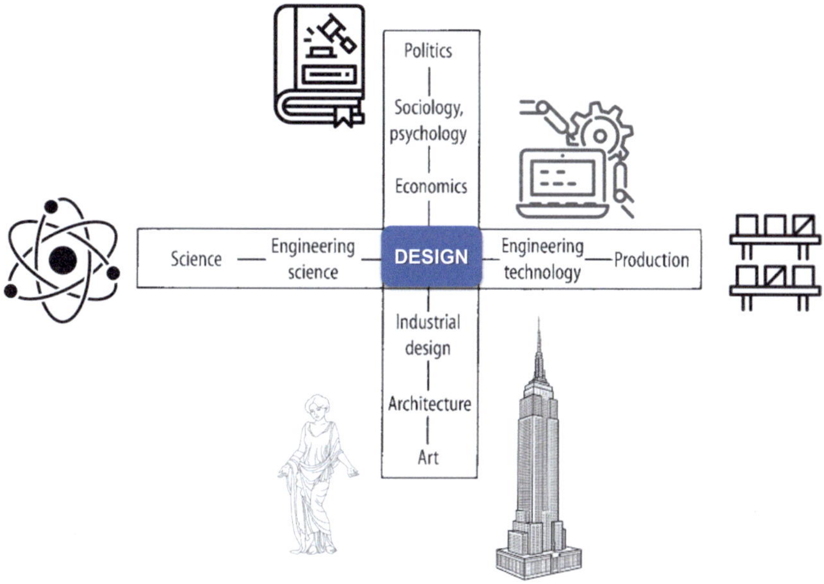

Fig. 15.1 Design, at the crossroads between Science and Technology, the Arts and Social Sciences (Pahl and Beitz 2007)

Table 15.1 The distinction between Science and Technology

	Science	Technology
Purpose	Science (episteme) produces knowledge and theory (theoria)	Technology (techne) has the purpose of creating and creating (poiesis)
Cognitive approach	Science moves from specific facts to create generalized knowledge (inductive approach)	Technology moves from generalized knowledge to determine specific actions (deductive approach)
Reference frame	The scientist defines a 'space of phenomena' and seeks rational explanation for them	The technologist generates a 'solutions space' and seeks a suitable and workable one

Another typical aspect of Design is its *creative* nature, which further reinforces the previous statement that it must be studied in relation to its human foundation. *Creativity** does not occur in a single moment but—rather—through complex cognitive processes. Designers must first of all frame and understand the problem (*problem formulation** or *problem setting**). Then, designers will generate a space of potential solutions, which is the core creative phase by using '*divergent thinking*' processes. Finally, they will select and refine the solutions until a suitable one emerges. This requires making a synthesis through '*convergent thinking*'.[1]

[1] The Design Council expressed in 2005 this sequence of divergent and convergent processes into the *Double Diamond** design process model.

Especially in the early divergent and creative parts of the design process, one can use a wide range of approaches, methodologies and algorithms, which provide guidance and support but—at least at the current state of the art—cannot deliver solutions autonomously. In the subsequent steps of design activity, the final and detailed solution may emerge through a more structured and somewhat easier-to-automate problem-solving process.

The debate on whether the creative phases of design can really be automated is quite old and, since the 1990s, a variety of efforts have been made to automate these portions of the design process (e.g., Wang et al. 1995; Seidel et al. 2018). Digitalization and the availability of large amounts of data, along with cheap computational power and sophisticated data analysis algorithms, are allowing the automation of small design decisions (Anderson et al. 2018). However, it is yet to be understood whether the early, more creative and less routinely design phases may really be automated and to what extent (Kwon et al. 2018).

15.1.2 The 'Science of the Artificial' and Decisional Processes

Moving fast forward 23 centuries, we can jump from Aristotle to Herbert Simon* (1916–2001), an author that has made key contributions to the way with which we nowadays look at Design. Simon was originally an organizational scientist and economist, who won the Nobel Prize for economics in 1978 thanks to his studies on the nature of human reasoning and decision-making and—specifically—as a key contributor to the '*theory of bounded rationality*'*.[2] In the second part of his career, Simon decided to continue working on decision-making and human cognition with a broader perspective. In his landmark book "The Sciences of the Artificial" (1969), he advocated ending the dualism between Science and Technology by giving scientific relevance to the latter, and moving beyond the traditional representation of Technology as a field based on empiricism and tacit knowledge. Simon justified this position because of both intellectual and practical reasons.

Concerning the former, he observed that *the world we live in is shaped by humankind's design activity*. Therefore, while scientists have traditionally dedicated most of their efforts to studying the natural world, it should make as much sense to research the artificial world and the way it evolves. Such a research should be carried out by using a scientific approach and should focus on Design, which is the 'phenomenon' that generates and progressively modifies this pervasively artificial world.

[2] The theory of bounded rationality challenges traditional models in Economics, in which human beings were represented as utility-maximizing and perfectly rational decision-makers who operate on perfect information. According to Simon, humans operate on imperfect information (since information may not be available or may be costly to obtain) and do not have the ability or the time to act in a way that is perfectly rational and utility-maximizing. So, instead of making 'optimal' decisions, they rather look for 'satisficing' solutions, i.e., solutions that—given the information and the means available—will provide them with a utility that passes a given threshold and is therefore acceptable.

Under this perspective, Design is the human activity *par excellence*, since human beings spend most of their time '*designing courses of action*'. Such a view of Design is very broad indeed, since it encompasses a very wide spectrum of artifacts and not only ones having to do with technology, as is commonly understood. In fact, Simon has the merit of having been one of the first to identify the strong cognitive similarities between seemingly quite different idea generation activities and having viewed 'design' as their common trait. According to this view, the activity performed by a mechanical engineer involved in the design of a gearbox is not at all dissimilar from the one carried out by a manager who is designing a professional development path for the employees in her company, or by a group of friends who are planning their next holiday journey.

The connection between *Decision-Making** and Design therefore becomes quite evident, since Simon recognizes problem-solving and design as fundamental activities in any decision process (see Fig. 15.2a). He identifies three stages in his rational decision-making model:

- **intelligence**, represented by data and information gathering that is required to clarify the problem;
- **design**, intended as the conception and development of alternatives. It comes after the problem has been formulated and before the alternatives for the decision are selected;
- **choice,** that considers the evaluation of the previously generated alternatives, and the selection of the consequent course of the decision.

These three phases are clearly similar (see Fig. 15.2b) to the three phases studied by researchers engaged in analyzing design processes (i.e., problem formulation, idea generation and selection).

A corollary to this viewpoint is that Design cannot be understood from the perspective of a specific technology, and its comprehension requires investigating it as a *cognitive phenomenon*. As a further implication, the design activity must be considered to be *largely independent of technical domains* and *economic sectors*.[3]

Aside from the purely academic interest in studying this phenomenon, Simon suggested that the scientific study of Design would lead to desirable outcomes, such as the elevation of its social and intellectual status, and this would allow the attraction of talented individuals to the field. Moreover, a better understanding of the cognitive processes that occur within design process could lead to the development of better and scientifically grounded methods and tools, which would improve the performance of design activity. In turn, this would have had a substantial and positive impact on the competitiveness of firms and on the economy in general. The focus on Design as a cognitive phenomenon led Simon to viewing the then emerging

[3] This uprooting of design from its technical and industry-specific origins has not been easily accepted by those who have a 'bottom-up' and strongly mono-disciplinary perspective. Curiously enough, this is more common in academic circles than in industry, where the interdisciplinary nature of design activities is ever more seen as a fact and as a challenge to be embraced.

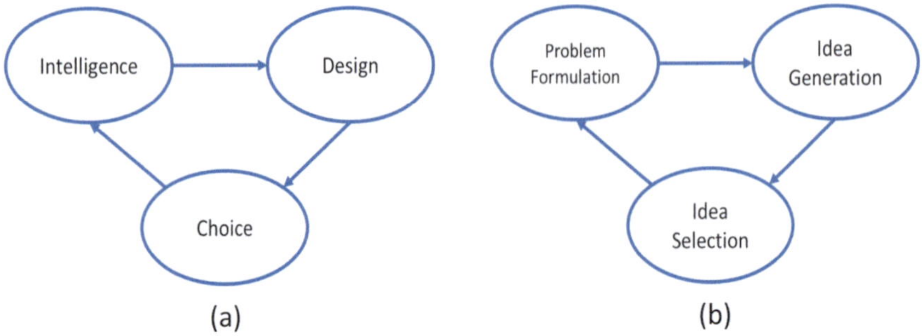

Fig. 15.2 **a** the decision-making process; **b** the design process model

field of *Artificial Intelligence**[4] as his main area of interest. In hindsight, advances in the field did lead to the development of many powerful new methods and tools supporting design, but not (or at least not yet) to the fulfillment of Simon's original vision of 'intelligent algorithms' enabling and automating design activity.

In Simon's wake, academic research on Design has been carried out by many authors from different fields and pursuing quite a wide range of objectives. Following Cross' classification (2001), design research is a broad field of investigation that attempts to move beyond the traditional effort of '*Scientific Design*', which is the use of scientific concepts rooted in a given discipline in order to support design activities (i.e., using Ohm's law to design an electric circuit).

As represented in Fig. 15.3, design research can be classified as a '*Science of Design*' whenever design is viewed as a phenomenon to be studied *scientifically*. In this case, researchers view designers and the design process as objects of investigation and try to understand how they work, assuming academic perspectives that can range from psychology to ethnography and from cognitive to organizational science. This type of research is generally carried out through experimental or empirical studies.

On the other side, design research can be viewed as '*Design Science*[5]', when it has the aim of developing scientifically based methods and tools to improve design action with a *normative* approach. What sets this apart from 'Scientific Design' is that design science methods are generally domain-independent and derived from the outcomes of the 'Science of Design', and not of natural sciences. As practical outcomes of Design Science, one can think of design methodologies that can be taught to novice designers in order to accelerate their professional development, the

[4] Artificial Intelligence in the '60s and '70s followed a completely different approach to the one followed today, and was not able to achieve substantial results. Instead of looking for patterns in data and extrapolating this experience to new cases, it focused on creating 'expert systems' that would combine a complex set of declarative rules and use them to infer conclusions from specific cases.

[5] This objective of 'Design Science' can be viewed in analogy to the one of 'Management Science', which is usually defined as the application of scientific insight to improve managerial action.

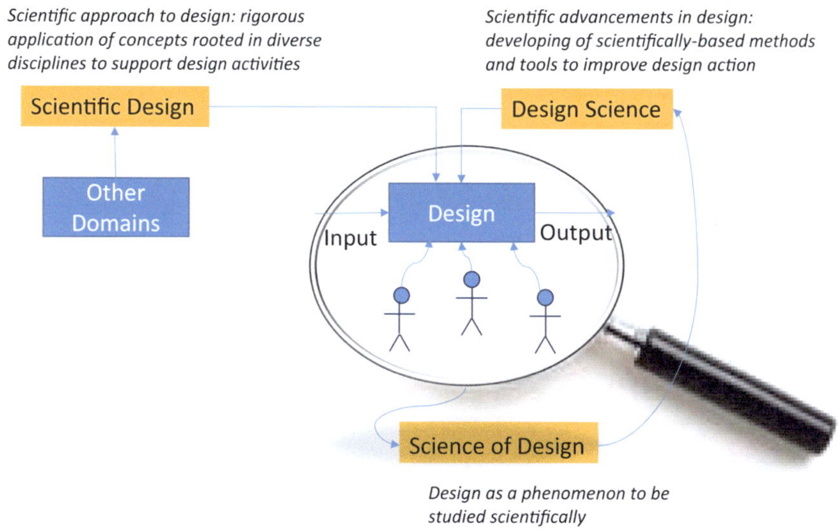

Fig. 15.3 The relationships between Design and Science according to Cross (2001)

definition of methodological standards and norms defining the design process and its constitutive activities and, finally, the development of software tools supporting the design process at individual, group or organizational level.

A complete survey of *Design Research** goes well beyond the objectives of this textbook, and for this purpose, readers may be referred to excellent surveys such as Cross (2021), Bayazit (2004), Clarkson and Eckert (2005), Finger and Dixon (1989), Mehalik and Schunn (2006), Birkhofer (2011), and Finkie et al. (1992). Moreover, interested readers may find it fruitful to connect literature on Design Research to literature on Problem-Solving in general (Eide et al. 2002; Sanabria and Orozco Pulido 2009; Liedtka et al. 2013). The following subsections will therefore provide just a few key findings that can be considered to be of greater significance for teaching purposes and more relevant to the management of innovation. Perspectives coming from the 'Science of Design' will be proposed at first and will be followed by a discussion on normative approaches to design.

15.1.3 Design Thinking as Investigated by the 'Science of Design'

15.1.3.1 Descriptive Models of Design at the Individual Level

At individual level, a first and notable contribution comes from Donald Schön* (1930–1997), who was already introduced in Chap. 11. By studying the unfolding of activities in a number of professions, among which designers, Schön examined the way with which experts devise 'courses of action' when engaged in problem-solving. He realized that—exception made for very simple and highly

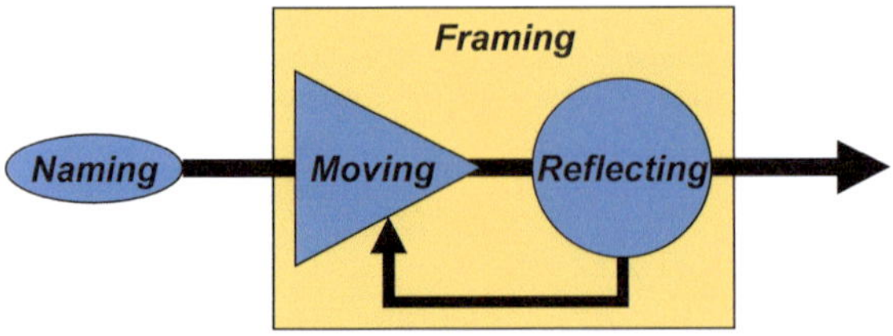

Fig. 15.4 A simplified representation of design as 'reflective practice'

routinary problems—this process cannot be viewed as a rational and predefined sequence of steps in which a subject (i.e., the professional) operates on an object (i.e., the problem) by drawing on the knowledge she/he possesses. Instead, the problem and the professional can be considered as a '*universe*' within which both co-evolve. In this joint evolution, both the solution process and the solution progressively and iteratively emerge out of a continuous '*reflective conversation with the problem*'. By running protocol studies,[6] Schön was able to frame his theory as described in Fig. 15.4.

The process starts with the '*naming*' of elements that are relevant to the problem. In this context, naming is not simply a matter of providing verbal labels to these elements, but also has to do with their identification. This also implies setting a boundary to the problem, thus excluding what can be considered to be irrelevant elements and reaching a precise understanding of the ones to be considered, including the terminology with which they must be addressed.

Among the elements to be identified, one can list *design parameters* (i.e., the elements that have to be decided on, such as 'what material should be chosen for

[6] A *protocol study** is an experiment aimed at measuring and validating a theoretical framework on cognitive processes at individual or group level. The subjects involved are assigned a problem to work on and are asked to 'think aloud' and provide as much visual support to their actions as possible (i.e., by sketching). The session is recorded with audiovisual means and then transcribed, assigning each element in the process (e.g., fragment of conversation, a non-verbal act of communication, a sketch, etc.) to the categories that were defined by the framework ('coding'). The study allows researchers to evaluate the degree with which the observed process 'fits' in these categories and to perform quantitative analysis on them (e.g., the frequency with which one category occurs, the time spent on each category, transition probabilities from each pair of categories, etc.).

Protocol studies have a long-standing tradition in the context of experimental research on design processes. This method has led to important insights on the way designers think and act, despite the limitation of having to rely on explicit acts of expression and communication.

Nowadays, traditional protocol studies are being complemented by investigation methods based on capturing neurophysiological signals relevant to the cognitive processes being carried out (Jauk et al. 2012; Fink and Benedek 2014; Fink et al. 2009; Fink et al. 2011; Gero 2014; Colombo et al. 2020). For instance, cognitive processes can be studied by using brain signals (via electroencephalography, functional magnetic resonance, or functional near-infrared spectroscopy) or, somewhat more indirectly, via eye tracking. Other interesting physiological signals, such as heartbeat and galvanic skin response, may allow to trace other aspects, such as arousal and stress, which can be related to a highly engaging cognitive process, such as design.

part X'), *specifications* (i.e., the functions or performance levels that have to be reached, such as 'the product must switch on in less than 2 s') and *constraints* (i.e., the physical, organizational or legal elements that may influence decisions, such as 'we cannot require tighter tolerances than our manufacturing equipment allows').

When a design activity is carried out within a team that belongs to an organization and, even more, in the case of a team that spans multiple organizations, reaching an agreement on the elements to be dealt with and on the related terminology[7] is not trivial. The issue also has significant implications for ensuring collaboration and for providing effective support by the means of information systems.

After naming, problems are *'framed'*, i.e., broken down into elements and tackled frame by frame. Similarly to before, the criterion with which a problem should be decomposed is not trivial and requires both experience and organizational-level agreement. Within each frame, the professional will act on the problem by alternating *'moves'* (i.e., making or suggesting decisions) and *'reflections'* (i.e., observing the impact of moves) until a satisfactory solution is identified, so that the process can shift to another frame.

When designing, professionals usually engage in a variety of cognitive mechanisms. As noted by some scholars (Nakajima 1995; Zeng and Cheng 1991; Gero and Kannengiesser 2004), designers make use of multiple cognitive and logical approaches, as shown in Table 15.2. Specifically, in their creative activity (as already anticipated in Chap. 1), designers do not only use traditional *inductive* and *deductive* reasoning, but also operate on an *abductive* approach, which allows to make an effective—though potentially imprecise—use of empirical observations.

Following Schön, the design process is strongly *iterative* and based on feedback loops that occur between the decisions taken and the analysis of their consequences. However, a designer does not simply 'move around in circles' until a given condition is met. If one observes closely, the conclusion of each loop allows him/her to make steps forward, adding details at each step, which leads to an always crisper definition of the solution. From a high-level perspective, one can therefore see design as a process that is initially concerned with 'task clarification' (i.e., understanding the problem and defining its boundaries and its objectives). Then, the process moves into a creative phase during which ideas are generated and broad decisions are taken on the system and its architecture. Finally, the designer delves into 'detailed design', in which he will define all the facets that are needed to provide a complete description of the previously identified alternatives. By combining the two perspectives, iterative and progressive, one can visualize the design

[7] Anyone who has approached a sport such as mountaineering or sailing has experienced that technical terminology is quite complex, and learning it is necessary to advance in the sport. Similarly, anyone entering a professional domain is usually overwhelmed by the amount of technical jargon that is used. This vocabulary is often specific to individual firms and—if design has to be carried out in cooperation between different companies—it may also require some sort of translation, or the drafting of an official and shared glossary.

Table 15.2 The three types of reasoning used in design activity

Type of reasoning	Structure	Example
Inductive	Facts → general statements and rules	'Given lab results, component X fails at high temperature. Therefore, it should not be used above 70 °C'
Deductive	General statements and rules → solutions	'Given our design guidelines on component X at high temperature, and given that product Y is going to Africa where we foresee operating temperatures approaching 80 °C, we should therefore not use component X in product Y'
Abductive	Experience → a plausible and economic justification (i.e., sufficient but not necessary)→ solutions	'Field failures of component X are occurring in hot climates. It is likely that X has issues at higher temperatures. Since product Y is being designed for Africa, we should therefore avoid using component X'

process as a *spiral-like progress,* in which greater or lesser emphasis can be placed on one dimension or the other.[8]

A similar view of the design process has been formulated by Smith and Browne (1993), who concentrated on the cognitive effort made by designers in managing elements such as goals, constraints and solutions. According to these authors, and as described in Fig. 15.5, designers continuously alternate between the level of reality they are ultimately interested in (i.e., what is actually required, what is actually working, etc.) and the level of knowledge and abstraction in which they cognitively operate (i.e., what they think is required, their mental models of relevant constraints, working solutions, reasons why solutions failed, etc.). Following this model, it is apparent that design activity does not only rely on the capability of generating solutions to perceived problems, but also on the capability of properly operating a continual translation between the two levels of 'reality' and 'knowledge'.

This continuous shift between knowledge and reality can be explicitly captured into the concept of '*situated cognitio*n' or 'situatedness in designing' (Gero and Kannengiesser 2004). The basic idea, as shown in Fig. 15.6, is that designing involves interactions between three worlds (the *external*, the *interpreted* and the *expected world*). The external world is represented by the physical environment in

[8] Design researchers are often engaged in significant controversy over the nature of the 'progressive' component of the design process. Some simply view it simply as the emerging result of the iterative component.

Others instead ascribe a 'normative' meaning to it, implying that design 'must' progress along the path from unstructured and broad-based decisions to the structured and detailed ones. In this latter case, the progressive path can be required and planned, while the iterative component will be viewed as 'incidental' and as something to be avoided as far as possible.

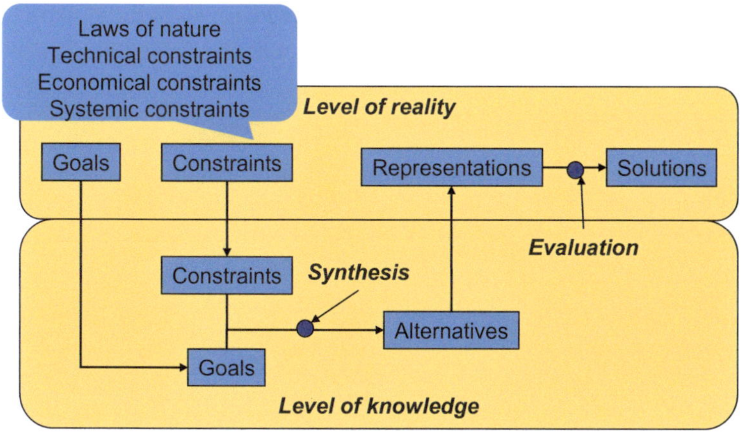

Fig. 15.5 The design activity iterates between the levels of reality and abstraction

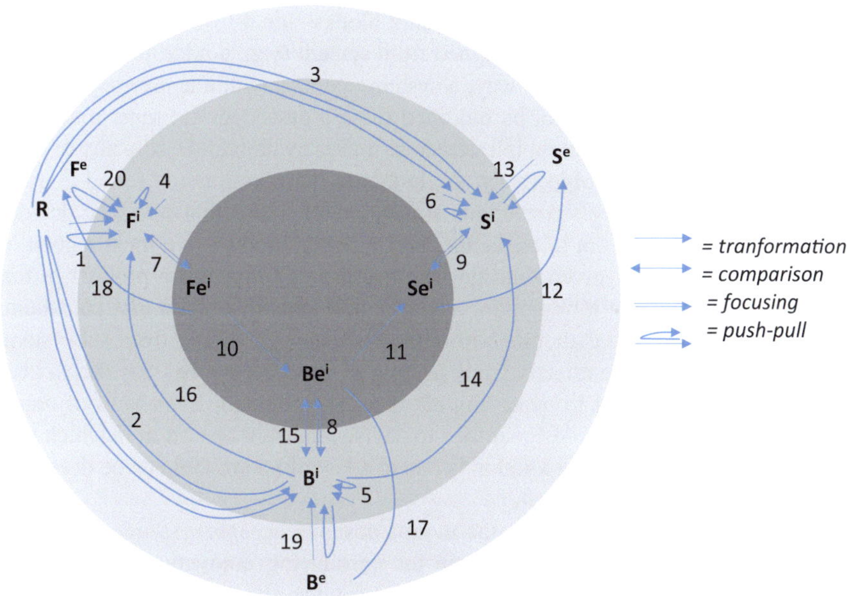

Fig. 15.6 The situated FBS framework (Gero and Kannengiesser 2004)

which the designer is involved, the interpreted world refers to experiences, percepts and ideas the designer has built on his/her interactions with the external world. Finally, the expected world is made of the expectations about results from the designer's actions, driven by goals and hypotheses about the current state of the world. Designers move within and interact across these three domains by interpretation, selecting subsets of variables in the interpreted world and using them as goals in the expected world (focusing), by operating changes to the systems (action) and by producing memories as a result of reflecting and reinterpreting past experiences (constructive memory).

15.1.3.2 The Role of Constraints

Throughout this discussion, it is apparent that constraints are pervasive in the design activity. Constraints have an ambivalent role in designing. When the design task is aimed at incremental innovation, constraints may be viewed as helpful elements that provide guidance and restrict the variety of admissible solutions to a manageable subset, thus speeding up the design process. Conversely, if one aims at introducing radical innovation, constraints will become obstacles that must be removed, either by changing them, or by finding solutions that 'circumvent' them.

Constraints may be due to laws of nature, to the technical environment and state of the art that characterizes the firm or the industry, to the economic conditions cast by target costing analysis and to systemic influences. This latter type of constraints comes from the hierarchic nature that is a distinguishing feature of all technical systems (Simon 1962).

A system is usually made out of building blocks—or *subsystems*—that may be chosen, but may not always be designed from scratch (e.g., a microprocessor in an electronic product). In the same way, a system operates within a *supra-system* that is taken as a given and cannot be modified (e.g., a passenger car must run on the fuel that the local transportation infrastructure makes available). Hardly any of these systemic constraints are 'objectively' so. In fact—and as commonly found in many cases of decision-making—constraints simply are elements that one has 'decided not to decide upon', or not to challenge. For instance, abiding to a local regulation is mandatory, but only given that the firm has decided to sell the product in that specific country without lobbying for a change in regulations. Similarly, constraints due to the technical infrastructure within the company, or arising from subsystems and/or supra-systems are effective only as long as the company has decided to keep them as 'givens' and not to make any effort to change them. Even laws of nature may be 'suspended', if the designer tries to devise modes of operation in which they reduce their impact (e.g., evaporation is due to a law of nature, but can be decreased by increasing ambient pressure).

Following this line of thought, the more a designer accepts systemic constraints as givens, the lesser will be the scope for introducing innovations. Conversely, broader and more radical innovations will occur when the designer attempts—and is given the freedom—to 'change the givens' and challenge commonly accepted constraints.

15.1.3.3 Descriptive Models of Design in Groups and Organizations

If one moves from the individual to the organizational level, an interesting contribution comes from the field of ethnography. When studying Design as a social process, ethnographers (Bucciarelli 1988) have mainly focused on the symmetry that arises between the structure of the artifact (i.e., product architecture) and that of the organization that carries out the design. The reader should remember that this observation has already been made in Chap. 3, when dealing with architectural innovation. Ethnographers also observe that—especially when working on complex systems—the designing organization becomes larger and involves numerous specializations, each of which will have its own specific way of modeling the product and its behavior (or 'object world') and will use its own specific language to represent it.[9] When managing the design process and when trying to support the related communication flows, a major challenge will therefore consist in reconciling these different object worlds and languages, so that different 'engineering subcultures' may effectively work together.

Organizational psychologists have also carried out numerous studies of the design process at both individual and organizational levels. Among them it is worthwhile to mention Stempfle and Badke-Schaub (2002), who studied design activity in group settings, with the aim of uncovering the main cognitive activities involved. At first they noticed—coherently with Schön's perspective—that these cognitive activities can be both content-related and process-related. The former have to do with the artifact, while the latter cope with the design process itself. Then, they further detailed such activities in subcategories (top panel of Fig. 15.7) and performed group-level protocol studies with the objective of measuring the time spent on each (bottom panel of Fig. 15.7).

The first element that emerges is the effort spent on process-related activities, which amounts to a quite substantial 1/3 of overall time. This finding supports the hypothesis we have now and again repeated that the design process is not a predefined sequence of steps that can simply be followed as a given recipe. Another key finding is that design teams spend most of their time analyzing the ideas and proposals they have come up with. This is a further evidence of the explorative nature of the design process, in which iterations are central in defining and exploring the solution space. Finally, the percentage of time spent on solution generation (i.e., what is usually considered to be the core and 'creative' part of design) is less than 5% of the total. These results may lend themselves to two opposite interpretations. One interpretation accepts the complexity of the design process and the central role carried out by iterations and considers them to be inherent properties of the design activity. On the opposite, one may conclude that the design process is—as of today—carried out in an extremely inefficient way, with close to 95% of the time being spent in activities that do not directly 'add value'. Following this latter perspective, activities other than 'solution generation' are necessary only because we have not yet been able to develop any other better and more efficient way to carry out the design process. Even though this latter perspective is quite extreme, it is impossible to challenge the idea that methods or

[9] For instance, when working on a car body, an expert in aerodynamics will mostly think about the way with which surfaces impact on the flow of air, while a manufacturing engineer will concentrate on the geometrical features and their radiuses of curvature, trying to figure out whether the metal can be deformed without excessive strain. The technical 'languages' used to model product behavior and represent these thoughts will be quite different from one another and not be easily understood across disciplines.

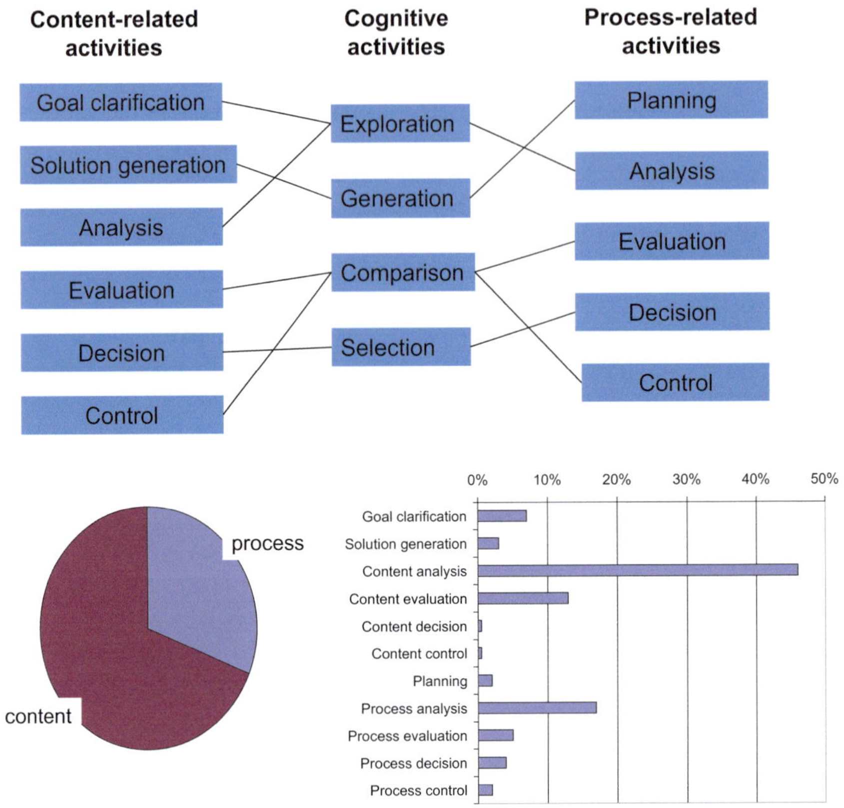

Fig. 15.7 The share of time spent by design teams in different activities, following Stempfle and Badke-Schaub (2002)

tools able to streamline the analysis phases, and to provide a focused direction to designers' 'reflective conversations', would make the design process much more effective and efficient.

15.1.4 The Objects of Design as Determinants of Design Process

The previous discussion has been based on reviewing a few notable investigations that have focused on the question 'what happens, when people design?' Now, this subsection will attempt to provide some key tools related to another question, 'what do people who design deal with?' which is closely associated to a further question 'how should design be carried out?'.

Most descriptive and normative (or prescriptive) models of design are based on the key concepts of '*function*' and '*form*'. The role of these two concepts can quite

Fig. 15.8 The Guggenheim Museum in New York as an example of functionalist design

easily be understood if we look at the evolution that occurred in architecture and in industrial design during the early twentieth century.

Up to the nineteenth century, buildings were usually designed by focusing on formal stylistic elements that were typical of their time, geographic location and local culture, while their intended function was somewhat neglected. When visiting European cities one can notice that—probably with the exception of military architecture and churches—buildings of a given epoch tended to look very similar to one another, regardless of their usage. With the twentieth century, functionalist architects, starting from Louis Sullivan* (1856–1924), started positing that "*"form [ever] follows function*" (Sullivan 1896), i.e., that the geometry of the building should be designed with the aim of supporting the intended purpose of the building itself. In Le Corbusier's words, "*a house is a machine for living*". As an example of this way of thinking, one can consider Frank Lloyd Wright's Guggenheim Museum in New York (Fig. 15.8). In this well-known building, the architect had to reconcile the function of allowing effective visits to museum exhibits (which would call for a circular path without diversions into halls) with the constraints due to the limited space available and came out with the helical structure that has become one of Manhattan's key landmarks. The theorization of the function-form dualism gradually spread from architecture to industrial design thanks to the work of functionalist architects who were also active in this field, such as Le Corbusier in France and Walter Gropius in Germany, who founded the Bauhaus movement.

From here, students of Engineering Design (Kesselring, Tschochner Niemann, Matousek, Leyer, as cited in Pahl et al. 1988) started applying the same concepts to the design of technical systems and machines[10] in general. Following Pahl et al. (1988), an artifact can be viewed (Fig. 15.9a) as a set of *functions* to be performed,

[10] The term 'machine' is here to be considered in its most general sense, i.e., as any artifact capable of transforming inputs into outputs, based on controls. Machines are therefore not only mechanical devices, but also electrical, electronic, or computational.

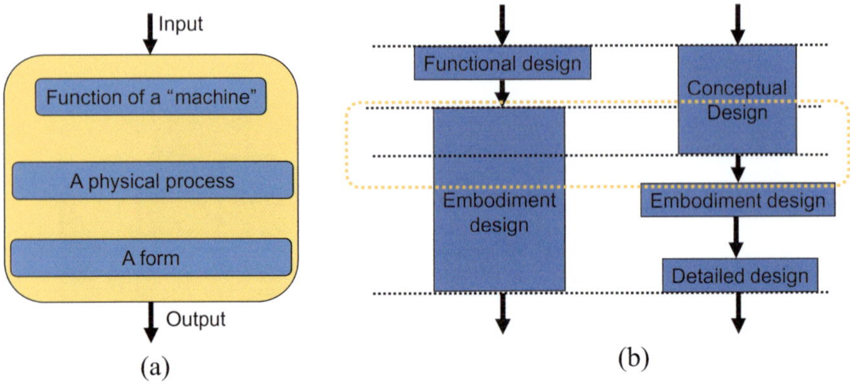

Fig. 15.9 A technical system (**a**) and its design process (**b**) following Pahl et al. (1988)

as a physical *working principle* (or *process*) capable of fulfilling them, and as a detailed *form* implementing the working principle.[11]

Nearly all contributions to Design Science follow the idea that a technical system can be viewed under the triad defined by *function*, *working principle* and *form*, though the exact definitions may at times vary between authors. If one assumes a *normative* perspective, these three elements lead to designating the phases of the design process, since the designer must first of all define the artifact's function, then choose a suitable working principle able to fulfill it, and finally proceed to describe the artifact's form in detail.

Consequently, one possible subdivision of the design process (Fig. 15.9b) splits the design process in the two phases of functional design (in which the designer concentrates on the artifact's functions) and embodiment design (in which the designer picks a working principle and then determines form).

This subdivision is seldom used, though it has the merit of making a clear separation between activities dealing with the space of functions and those dealing with the space of physical embodiments. Instead, it is quite common to segment the design process in the following three phases as suggested by Pahl and Beitz:

- At first, the identification of functions and the choice of a process are performed together in the **conceptual design** phase. Conceptual design involves drafting functional descriptions, drawings and block diagrams, defining high-level lists of components and modules, and comparing and selecting alternative solutions;

[11] For instance, a hammer has the function of 'driving a nail in a solid body'. The physical process used to fulfill this function is to 'supply a few intense impacts by using the momentum created by a mass that is fixed at the end of a handle to which a small rotation is imparted'. Finally, the hammer's form is the description of its components, geometry, masses, materials, etc. One should be aware that this is by no means the only process capable of driving a nail into wood, since this could also be accomplished by using multiple small impacts, by supplying a continuous pressure, and so on.

- Conceptual design is followed by **embodiment design**, in which the system's Bill of Materials is defined in greater detail. Components are defined at the level of preliminary technical drawings, while a preliminary selection is made for their materials and manufacturing processes;
- Finally, **detailed design** copes with the dimensioning of the previous results, technical validation through extensive engineering calculations and the detailing of all of the information that will be required to develop the ancillary systems that will allow the firm to produce the system and support it throughout its lifecycle.

These three design phases are substantially different from one another. For instance, creativity has a central role in conceptual design but hardly any role in detailed design. Moreover, information being produced and exchanged during the conceptual design process is relatively little and mostly unstructured, being limited to high-level descriptions of the product. During the embodiment design phase, the amount of information increases substantially and becomes somewhat more structured, with a focus on the architectural composition of the system. Finally, the detailed design phase witnesses a true explosion in the amount of information being generated and exchanged, involving the description of the most minute and detailed elements of the product and of its behavior.

These differences have a clear impact when one thinks of IT support to the design process, which must progressively shift from flexible tools aimed at supporting cooperative work to structured tools that focus on representing product Bill of Materials and, finally, to the provision of a complete product modeling system.

Slightly different frameworks can arise if one assumes a *descriptive* perspective, whose focus will not be to define what a designer should do. Rather, the aim will be to observe what happens when a designer acts, therefore taking into account not only the linear path leading from the requirements to the technical solution, but also the iterative dynamics with which the design process evolves.

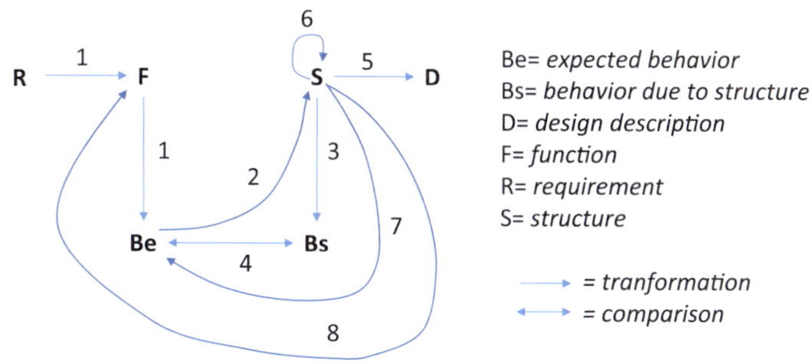

Fig. 15.10 The FBS ontology (Gero 1990)

Table 15.3 Navigation paths among the three domains FBS

Path	Description
1—Formulation	Defines the problem space by defining the functions, based on requirements, and prescribing the behavior to be expected
2—Synthesis	Proposes a structure able to operate according to the expected behavior
3—Analysis	Derives the actual behavior that the structure will have
4—Evaluation	Compares the expected and actual behavior
5—Documentation	Produces a description of the structure that has been proposed
6—Reformulation 1	Re-interprets and revises the structure
7—Reformulation 2	Revises the expected behavior, following a reinterpretation of structure
8—Reformulation 3	Revises the functions, following a reinterpretation of structure and behavior

A well-known model that assumes this perspective is the *Function Behavior Structure (FBS) framework** (Gero 1990). The main idea behind FBS is that systems can be analyzed by referring to the three perspectives of *Function* (i.e., the purpose of the system), *Behavior* (i.e., what the system does) and *Structure* (i.e., what is the system made up of). This 'FBS ontology' rests on the idea that linkages exist between F-B and B-S, but not between F-S (if not mediated by B), which implies the centrality of 'Behavior' in any design activity.

Under this premise, the FBS framework depicts design activity as a 'navigation' among the three domains identified by the ontology, following a number of possible paths (see Fig. 15.10 and Table 15.3).

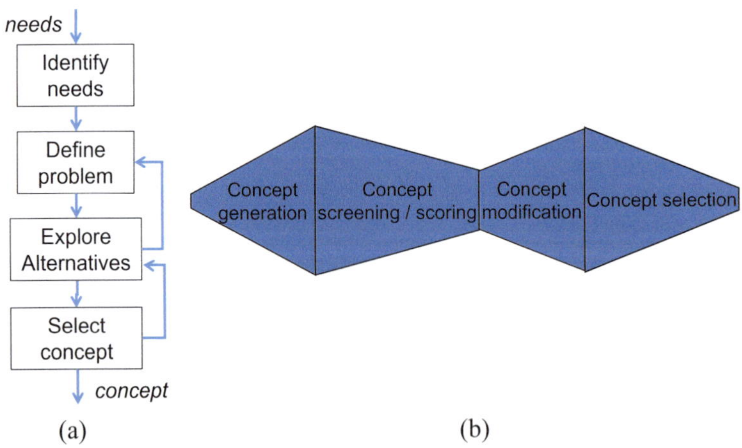

Fig. 15.11 The activities (**a**) and the divergent and convergent phases (**b**) in conceptual design

15.2 Conceptual Design

The *conceptual design** phase, which sees the translation of functions into an embodiment, is at the heart of design processes, and a number of methodologies have been developed in order to support it. Following Ulrich and Eppinger (1995), conceptual design can be described to occur as in the following Fig. 15.11a.

At first, the design team formulates the design problem, identifying the functional elements that must be designed. The design process then enters its most creative phase, in which alternative technical solutions are explored and generated. This usually consists in the identification of a number of technical solutions, each of which can—at least in principle—fulfill the required functions. At this point, following Pugh (1981), each feasible combination of technical solutions makes up what is called a *concept*. Alternative concepts are then compared, so that the less promising ones are discarded, while the most promising ones are subject to further analysis and improvements. Finally, the candidate concepts are subject to a rigorous comparison and selection exercise, so that a single one may be chosen and be further developed during the subsequent embodiment design phase.

Figure 15.11b shows what has been already stated more generally, for the design process as a whole: the conceptual design process alternates between *divergent* and *convergent* phases that are, respectively, characterized by relatively unbridled creativity and rigorous analysis. In general, a key rule to be observed in the conceptual design phase is that designers must come out with and work on multiple concepts. This is because only rarely will the 'best' concept emerge at once, especially if the design team is trying to introduce a significant degree of innovation. On the opposite, it is quite likely that the first concept to emerge will just be a slight modification to currently prevailing technical solutions.

The degree with which the divergent and convergent phases will be allowed to challenge the current state of the art will depend on the brief that has been assigned to the design team. In the case of radical innovations, the process might be allowed to proceed quite unbounded and possibly be iterated more than once. Conversely, in the case of incremental change, the process might be governed tightly, and allowed to introduce modifications only in the subsystems that have been identified as the most critical and in need of innovation.

Though conceptual design is often seen as a task to be carried out as a team, it is imperative that individuals come to the group session after having performed some preliminary research on their own, so that they can provide a richer and broader contribution. The individual search for solutions can be carried out by using a number of possible sources, such as:

- Interviewing **lead users** in order to gain insight on solutions they may have generated. Lead users (von Hippel 1986) are users of a product or service who—because of their intense exposure to the product and capability to modify it—are likely to have come out with creative solutions to problems. Following Berthon et al. (2007) lead users are "*customers who adapt, modify, or transform a proprietary offering*".

- Interviewing **domain experts** such as suppliers and academics, who may have anticipated exposure to the emerging state of the art.
- Reading **technical and scientific literature,** in order to keep abreast of new solutions to technical problems that are arising in the industry.
- Domain-specific **sourcebooks**. Sourcebooks are repositories that classify problems that are typical of a given domain and provide a collection of known solution principles that a designer may choose from and adapt. For instance, a mechanical engineering sourcebook such as Artobolevsky (1975) may list dozens of ways to "*transform linear into circular motion*". Historical sources can also be helpful. Manuscripts by Renaissance inventors such as Leonardo da Vinci and Francesco di Giorgio Martini provide an impressive coverage of technical solutions that designers can still use for inspiration.
- **Product benchmarking** and **reverse engineering***. By looking at how competitors are dealing with technical problems, and studying the solutions they have come up with, usually provides valuable insight for conceptual design. In order to be proficient in this activity, the firm must set up a regular process for collecting competitors' products from the market, assessing their performance in a lab and disassembling them in order to study the solutions being used. In some industries, this service may be provided by independent firms.
- **Patent search**. The widespread availability of full-text patent databases has become a very important resource for conceptual design. A patent database may be used in two ways. The first consists in observing the state of the art in the same industry, in a way that is quite close to reverse engineering competitors' products. However, compared to reverse engineering, patent search also allows looking into technical solutions that have not yet been launched on the market. Given the monopoly-granting nature of patents, patent analysis also allows the firm to define its '*Freedom to Operate*', i.e., to understand which solutions are legally pursuable and which are not.

  Patent analysis may also be used to look for solutions to analogous problems arising in other industries. For instance, a company that is working on pipeline inspection devices may be inspired by solutions developed for endoscopes used to inspect the human circulatory system. It is quite likely that the claims of these 'analogous' patents will not be so extensive as to prevent the use of the same solution principle in a completely different application and industry. This being the case, the idea can be used without infringing the patent holder's rights. In those cases in which the patent office has granted a very wide patent, it is unlikely that the holder of this asset is a competitor to the firm, and this increases the chances that it will accept to license it in an industry it is not operating it.

The creative core of conceptual design lies at the interface between the exploration of the function space and the definition of embodying components. This is a critical area of design, since it determines most of the future performance of the product and represents the area in which innovative content may be introduced.

Moreover, the definition of functional embodiments defines product architecture and—as we well know from previous chapters—product architecture will henceforth tend to 'lock-in' and lead to significant inertia. When defining product concepts, a tradeoff will have to be struck between creativity and conservation of previous designs, in a way that is coherent with the brief given to designers at the beginning of the process.

The following section will lead to the discussion of a number of methods that support the management of such complexity; they are the object of study of the Design Science and are generally named '*Design Methods*'.

15.3 Methods Supporting Conceptual Design

Methods supporting conceptual design are general purpose and domain independent because, in accordance with Simon's observations, design activities, elements and metaknowledge are common to any design problem, regardless of the domain. Figure 15.12 depicts how these methods act in a dual supporting role: on the one hand, they represent a resource that aids design decisions, and on the other hand, they constitute the tool through which designers derive the constraints to the design problem or the means by which they validate their activity execution.

This textbook will not deal with all possible methods for Concept Design, but only the ones that have the broadest diffusion. These are highlighted in red in Fig. 15.12 and located close to the phase to which they are typically applied.[12] The methods supporting design task clarification are in common with the ones already introduced for needs identification and requirement analysis. Finally, we will not discuss methods for Embodiment Design and Detailed Design, given the domain-independent nature of this textbook.

The methods shown above are usable in any design process; however, it should be reminded that they were originally developed in specific domains of Design. In fact, engineers, architects and industrial designers have over time defined methods to cope with the typology of problems they typically deal with, each according to the methodological approaches that distinguish them. Therefore, design methods from engineering disciplines tend to be more systematic and quantitative, while disciplines that are closer to the creative arts will focus more on qualitative methods supporting creativity (Fig. 15.13).

15.3.1 Functional Analysis

The objective of *functional analysis** is to define the set of functions that the system will have to carry out, in order to meet its requirements.

[12] Value Analysis will not be covered in this chapter, but in the next.

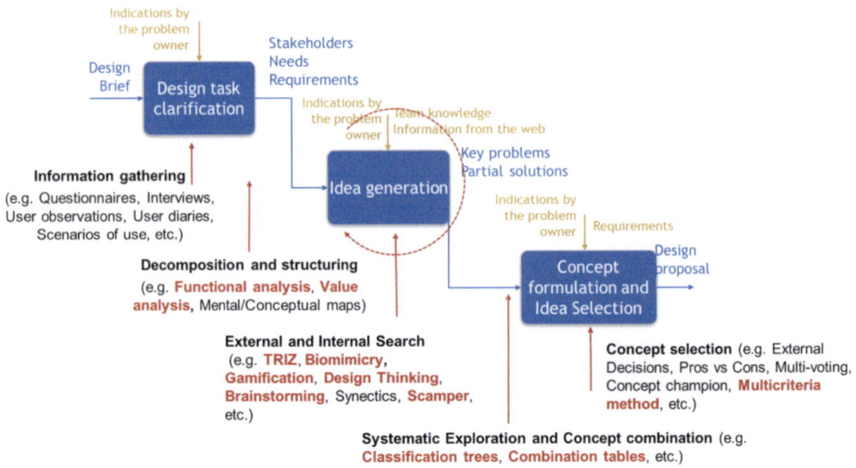

Fig. 15.12 The most relevant methods for conceptual design

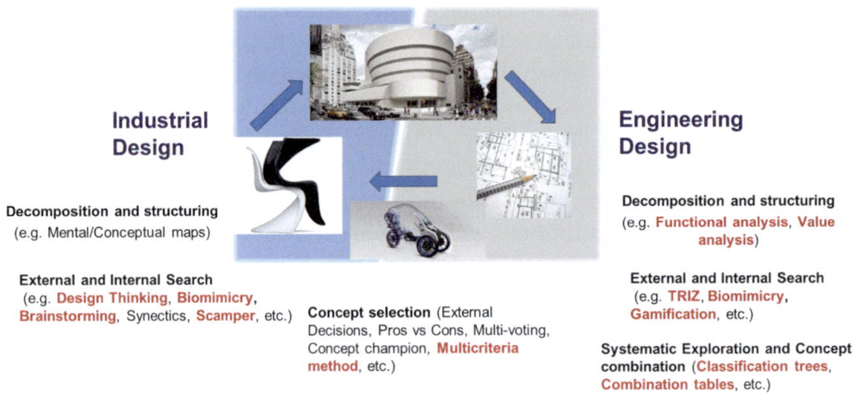

Fig. 15.13 Design methods are grounded into design cultures and traditions

In defining the boundaries of the technical problem (i.e., in problem framing), designers typically consider the interactions that the system must have with the user and with the surrounding environment. However, they also must consider that part of the intended functions can be performed by the user. Therefore, the *Theory of Technical Systems* (TTS)* considers that any *technical system* (TS) capable to deliver a function is made of four elements (Fig. 15.14):

- a **Tool**, which is the working element that delivers the function of the TS;
- an **Engine**, which is the element providing the energy required by the tool to produce the expected effect of the function;
- a **Transmission**, which brings energy from the Engine to the Tool;

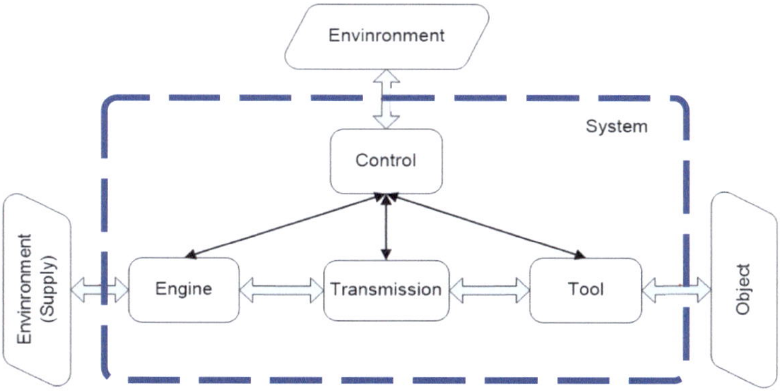

Fig. 15.14 The components of a technical system

- a **Control**, which governs the way with which the other three elements operate.

A preliminary design decision therefore lies in defining which of the three elements is to be performed by the system and which by the user. For instance, the function 'slicing vegetables' can be carried out by a technical system consisting of a double-handle crescent chopper as tool, the human hand as transmission, the arm muscle as engine and the users' eyes and brain as control. A food processor would have the cutting tool, transmission and engine incorporated in the mechanical system, but the control would still be entrusted to the user.[13]

Defining the useful functions of a technical system can generally be done by working with *function trees** or with *block diagrams**.

Function trees (Fig. 15.5) are built on the necessity of dividing the design problem in subproblems and are based on the idea that a system can be described as a hierarchy of functions. The analysis starts from the root function that describes the core purpose of the system and progressively leads to detailing each function by exploding it into subfunctions. When exploding the function tree, subfunctions can be viewed as the answer to the question 'how can you?' Conversely, when back-tracking toward the root function, the aggregation of subfunctions is the answer to the question 'why do you?'.

Defining functions is not a trivial task and requires a significant effort in providing precise verbal definitions, which ought to be expressed as a verb accompanied by complements. For instance, the core function of a corkscrew driver can be defined as 'extract + cork + from bottle', which can be exploded according to the function tree in Fig. 15.14. As shown in the figure, at some point in the function tree, the designer may suggest alternative ways for fulfilling a higher-level function. This can be expressed in the function tree hierarchy by placing the functions on top

[13] The fact that functions can be split in these four elements and distributed among system and user is the theoretical element that justifies the importance of analyzing user-system interaction and is the ontological foundation of User-centered Design. In fact, both system and user are part of the technical system and contribute to the fulfillment of the required function.

of each other, instead of across. The combination of alternative working principles will ultimately lead to alternative concepts, as shown by the two different corkscrew designs depicted in the figure.

When one progressively explodes the function hierarchy, there may be a point at which the 'how can you?' question is no longer answered by other lower-level functions, but by the description of a physical process (and associated components) able to fulfill or 'embody' the functions (Fig. 15.15). The question therefore becomes 'what can you use to?' Functional analysis is now completed, and the envisaged solution is defined by matching the function tree with the components list.[14]

Figure 15.16 shows that, when analyzing a solution alternative, the link between lower-level functions and components can exhibit different cardinalities.

- In the case of a 1:1 association there is a clear identification of which component fulfills which function, and vice versa. As we will see in Chap. 16, this functional separation between components leads to *modular* architectures. This significantly facilitates detailed design activities, and the use of standard and off-the-shelf components. For instance, a modification to the performance required of function F1.1 will require changing component C1, but it is unlikely that this change will propagate and require any redesign of other components.
- In the case of an m:1 association, a single component is able to fulfill m functions at the same time, and independently from other components. This is called *function sharing* (Ulrich 1988). In general, this choice leads to greater complexity in the detailed design phase and to the use of *ad-hoc* components. However, a single component able to fulfill more than one function can often lead to substantial savings concerning cost, weight, etc.
- With 1:n associations, a single function is fulfilled jointly by n components who do not serve other functions. This can be viewed as a shorthand representation of multiple 1:1 associations, since it is likely that the function could have been exploded into n subfunctions, each leading to a 1:1 association with the corresponding components.

[14] Having identified the components list, it is customary to aggregate the components into a hierarchy, thus completing the definition of the product's Bill of Materials* (BoM). This aggregation is not strictly part of the design process since, as mentioned in previous chapters, it can follow different criteria. For instance, one may generate an engineering BoM that replicates the structure of the function tree, so that 'functional modules' may be mapped to higher-level functions.

Another way to generate the engineering BoM would be to aggregate components by technical discipline (e.g., by aggregating mechanical, electrical and computational devices). A production BoM will instead combine components by considering the way with which they will be assembled. Finally, a purchasing BoM may be generated by clustering components by supplier, or by vendor type.

Each of these BoMs is justifiable from an operational and logical perspective. The firm will therefore have to make sure that its information systems are able to manage multiple views of the same BoM and to guarantee its consistency, so that changes made (or required) by one view are correctly translated into the other ones.

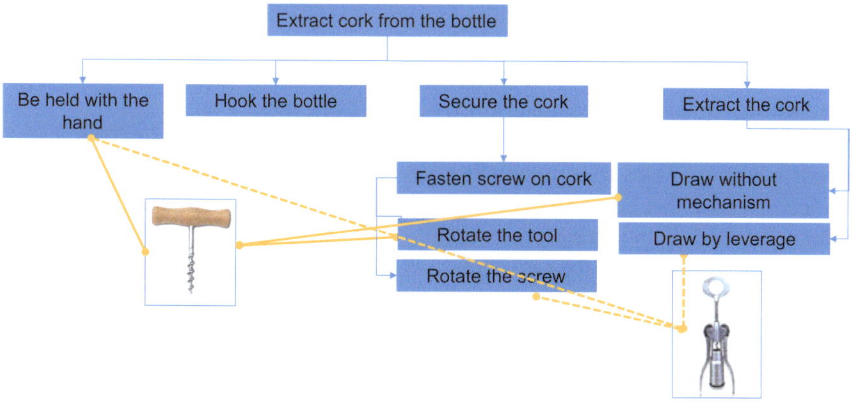

Fig. 15.15 An example of function tree analysis

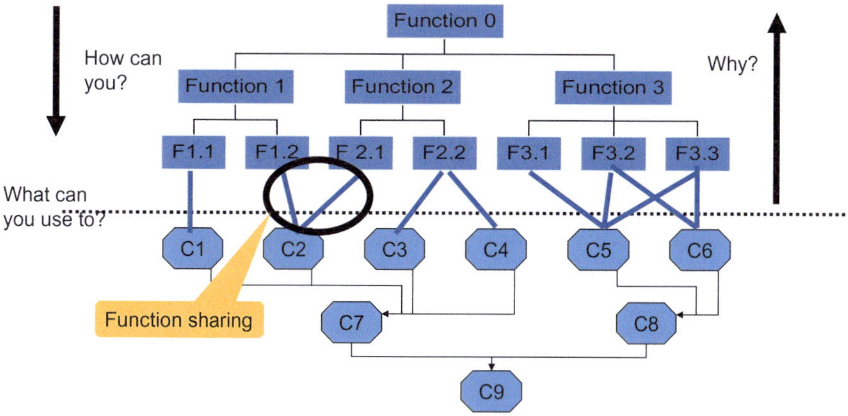

Fig. 15.16 Defining a concept by matching a function tree with the embodying components list

- Finally, the case of an m:n association exhibits an arrangement in which components are functionally interdependent. As it will be seen in Chap. 16, this leads to an *integral architecture,* in which detailed design becomes quite complex. In fact, any change carried out on one function or component will propagate to the other ones, thus triggering a chain of design iterations that will end only when a solution that satisfies the requirements cast by each function is found. Managing the detailed design process for products characterized by this type of architecture requires significant experience and—to this purpose—specific techniques have been developed (a review is given in Wynn and Clarkson 2005, while an in-depth analysis of the '*signposting method*' is provided by Clarkson and Hamilton 2000).

When going through the previous discussion, the attentive reader may have noticed a potential ambiguity. Functional analysis should operate in a way that is neutral with respect to potential embodiments, at least until one reaches the level of components. However, in Fig. 15.15, the 'securing the cork' function is achieved by 'inserting screw in cork', which implies that there will be a screw in the components list. This is not a mistake, since functional analysis is an integral part of the design process and can involve design choices of a functional nature, which may of course hint toward specific embodiments. However, the designer must be aware that the term 'screw' being used in the functional context is not to be seen as a component yet, but as a working principle.[15]

Taken to the extreme, even the definition of the root function is the outcome of a design choice and depends on the innovative content that one wishes to introduce in the process. In fact, 'extracting a cork from the bottle' is not the only way to allow the spillage of its contents, that is the final outcome being sought when using a corkscrew. Alternative ways to achieve this same result would be its dissolution with an acid (which is unlikely to be very practical), the puncturing of the cork (as in Coravin™ systems), or the clear cutting of the bottle at its neck (as, for instance, is done with drugs in ampoules, or when opening a champagne bottle with sabers).

Functional analysis can also be carried out through diagrams that highlight the functional relationships between functional elements according to the typical representation of technical systems, showing the transformation of inputs into outputs, from left to right. For instance, *Function Analysis System Technique** (FAST) diagrams follow the blueprint shown in Fig. 15.17, in which the system's functions are described by showing how a set of inputs is progressively transformed in a set of outputs by a network of functional elements. With respect to function trees, a FAST diagram does not make the functional hierarchy explicit, but it provides much greater insight on the way with which the system works and on how its structural elements interoperate.

A FAST diagram requires the definition of system boundaries as well, and this analysis leads to distinguish between primary (or core) and secondary (or auxiliary) functional elements: the former are directly involved in the conversion flow of inputs into outputs; the latter are required to enable the operation of primary functional elements, or to provide ancillary functions. For instance, a projector may have a primary 'provide light' functional element, which will usually be embodied by a halogen lamp. The projector will also have a secondary 'generate flow of air' functional element (probably embodied by a fan) that enables the operation of the 'provide light' element.

A FAST diagram is not simply a block diagram. In fact, its building blocks are not components, but *functional elements*, though of course they will be embodied by physical components and subsystems in the subsequent phases of the design

[15] In this example, 'screw' is to be intended as shorthand for a 'helical inclined plane that enables the easy insertion of an element of the corkscrew driver inside the cork, in such a way that it will lead to a solid connection between the two, enabling future extraction by traction'. An alternative —though quite fanciful—way to achieve the same result of 'securing the cork' would be to 'connect with adhesives', where the focus would be on the principle of adhesion, rather than on the adhesive that will be used.

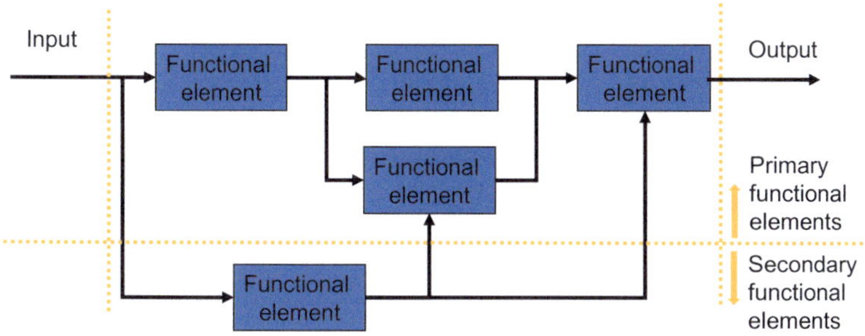

Fig. 15.17 A FAST diagram

process. Just as in the case of function trees, there will always be a latent ambiguity between the 'purity' of the functions that are being represented and their implied embodiments.

Looking back at the previous example, the 'generate flow of air' functional element has a role only if the embodiment of the 'provide light' functional element is known to generate a significant amount of unwanted heat. In the case of fan-less projectors using low-power LED light sources, the 'generate flow of air' functional element might either disappear or be substituted by an 'allow flow of air' element (that will be embodied by simple vents in the casing).

When designers are not only interested in analyzing useful functions, but also in investigating the *possible usage of resources* within the system, *Rodenacker diagrams* (Fig. 15.18; Rodenacker 1971) can be helpful, since relationships and flows between functional elements are graphically coded depending on their nature. Through a Rodenacker diagram, the designer can distinguish among flows of material, energy and information. Additional relationships can be coded in the form of spatial or temporal constraints, dictating either physical adjacency or separation between functional elements, as well as a required temporal span between their activation. For instance, the battery of a car and high energy-consuming apparatus, such as the starter motor, might be constrained to be located close to one another in order to reduce energy losses and/or minimize the amount of high-diameter wiring connecting the two. Conversely, the car engine and an accelerometer may be constrained to be set physically apart, in order to decrease the disturbance generated by the former's vibrations on the latter.

Functional analysis is quite helpful in defining a systematic path from functions to their embodiments. However, it neither allows to identify tradeoffs among functions, nor it provides suggestions on potential embodiments of the desired functions. These two aspects are central to the TRIZ method, that will be described in the following.

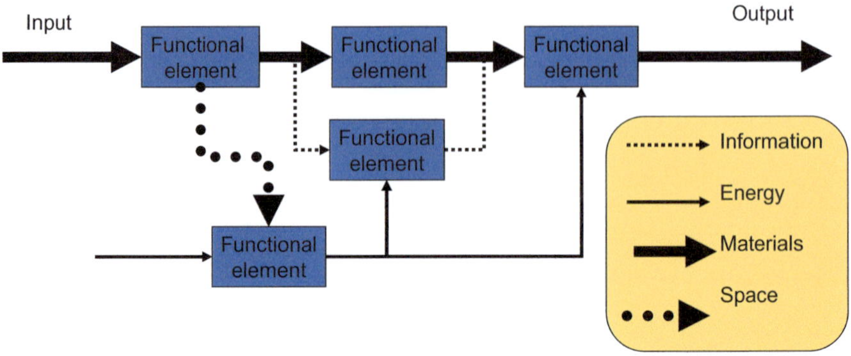

Fig. 15.18 A Rodenacker diagram

15.3.2 The Theory of Inventive Problem-Solving (TRIZ)

TRIZ, or the *Theory of Inventive Problem-Solving**, comes from the intuition of a Russian engineer, Genrich Altshuller* (1926–1998). While working for the technical corps of the Soviet Navy, Altshuller (1999) tried to develop a general theory on inventions and their underlying principles, thanks to a systematic study of a huge number of patents. After being interned in a labor camp, he was freed and able to publish and teach the results of his work. Since the end of the Cold War, the TRIZ method has been the subject of many training courses and embedded in software tools that have made the method easier to approach, thus assisting its diffusion throughout the world.

TRIZ is based on the analysis of a vast number of inventions and on the attempt to induce general rules and principles that explain how a designer can creatively tackle a problem in order to generate inventive solutions. Altshuller originally worked on 40,000 patents, but the number analyzed by the TRIZ community has now passed one million.

TRIZ is grounded into the already cited Theory of Technical Systems and it is built on three postulates, the first of which regards the *laws of evolution of technical systems*. Technical systems evolve in time, according to specific evolution paths that involve system static elements, cinematics and dynamics. These 'laws' attempt to find general patterns in the way technical systems evolve, considering both incremental improvements and radical change, the tendency to re-arrange in a hierarchy of systems and subsystems, and the progressive shift from mechanical to electrical energy.

These 'laws' exhibit interesting analogies with the discussions that have been previously made on the dynamics of innovation (and, specifically, with the Dominant Design theory). However, the purpose was markedly different, since Altshuller's aim was essentially normative, carrying the advice that designers should direct their efforts toward the 'natural' evolution of the systems they are asked to design.

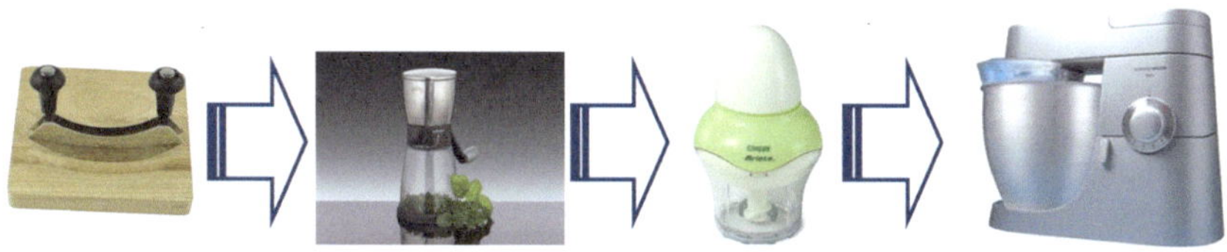

Fig. 15.19 The evolution of a double-handle crescent chopper

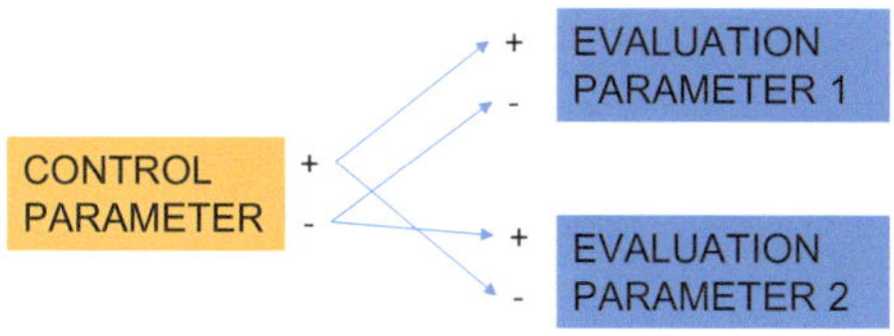

Fig. 15.20 The concept of contradiction in TRIZ

One of the most interesting of Altshuller's 'laws', for instance, states that technical systems usually evolve in the direction of reducing human involvement. Therefore, going back to the previous example on a technical system for slicing vegetables, the traditional technical system is made up of a tool (double-handle crescent chopper), a transmission (the hand), an engine (arm muscles) and a control (eyes and brain), while the modern food processor is the result of design decisions aimed to increase technical performance and progressively substitute human activity with artificial means (consider the way with which the four elements were progressively embodied in a modern food processor, as in Fig. 15.19).

Each design problem can be divided into subproblems (as similarly expressed by the function tree), but the attempt to solve subproblems will open up other decision trees creating what Altshuller names a *Network of Problems* (NoP)*.

The second postulate of TRIZ states that the solution of design problems has to do with the management of contradictions. *Contradictions** are conflicts that designers meet, due to conflicts between the system and its environment, or between the components of the system itself. So, when a designer solves one aspect of the contradiction, she/he generates a problem on another one.

Contradictions can be described as a design tradeoff and expressed through three elementary elements, as shown in Fig. 15.20: one '*control variable*' on which the designer can act and two '*evaluation parameters*' whose values depend on the control variable. A contradiction arises if, by changing the control variable, the effect on one parameter is beneficial and the other is harmful with respect to the specifications.

For instance, a table should be sturdy so that one can safely work on it, but also light enough so that it can be easily moved around. A control variable that affects both of these parameters is the thickness of the tabletop, and this leads to a contradiction, since a greater thickness ensures sturdiness, but makes the table heavier.

Table 15.4 Contradiction instances according to TRIZ

ID	Instance	ID	Instance
1	Weight of moving object	21	Power
2	Weight of non-moving object	22	Waste of energy
3	Length of moving object	23	Waste of substance
4	Length of non-moving object	24	Loss of information
5	Area of moving object	25	Waste of time
6	Area of non-moving object	26	Amount of substance
7	Volume of moving object	27	Reliability
8	Volume of non-moving object	28	Accuracy of measurement
9	Speed	29	Accuracy of manufacturing
10	Force	30	Harmful factors acting on object
11	Tension/pressure	31	Harmful side effects
12	Shape	32	Ease of manufacture
13	Stability of object (resistance to change)	33	Ease of use
14	Strength	34	Ease of repair
15	Durability of moving object	35	Adaptability (to external conditions)
16	Durability of non-moving object	36	Complexity of device
17	Temperature	37	Complexity of control
18	Brightness	38	Level of automation
19	Energy spent by moving object	39	Productivity
20	Energy spent by non-moving object		

After having analyzed a very large number of inventions, Altshuller realized that contradictions occurred among a relatively limited number of performance instances, or *contradiction instances,* and came out with a list of 39 (Table 15.4). If contradictions are asymmetrical, the number of possible pairwise contradictions will therefore amount to $39 \times (39 - 1) = 1482$, which is a high number indeed, but finite. Therefore, TRIZ considers that *the space of problems to be solved in design is finite*.

The technical systems analyzed by Altsuller predate the 'digital revolution' that characterizes contemporary technology. So, while it is possible to expect that the number of contradiction instances (and, therefore, of contradictions) may increase with the development of new technology, at the same time it remains sensible to think that this could remain a 'large, but finite' set.

If the space of problems is finite, creative problems are universal, and every specific technical problem can be traced back, by abstraction, to a finite set of standard problems. In a way, this approach is not too dissimilar from Plato's theory that a finite 'world of ideas', could constitute the reference for reality. Because of the finite number of problems *similar solution principles can be applied to analogous technical problems,* and this implies that *the number of solution principles will be finite.*

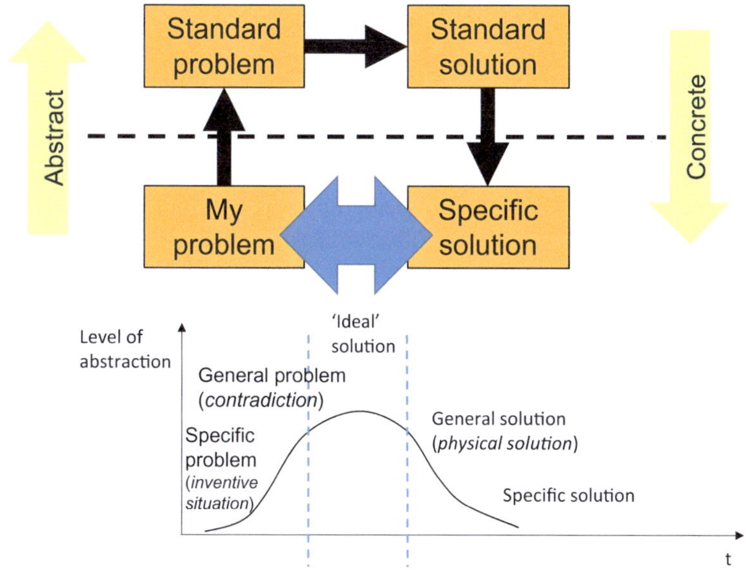

Fig. 15.21 The 'hill model' of Problem-Solving

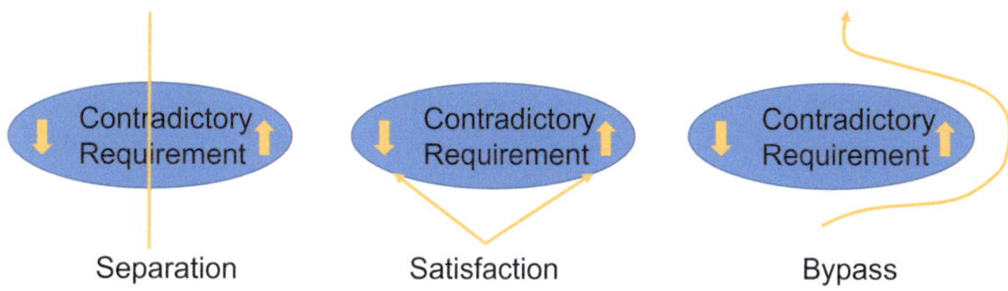

Fig. 15.22 The three approaches for dealing with contradictions

Therefore, TRIZ suggests that designers should derive the solution of specific cases by deducing them from a finite set of standard solutions. To do so, they must first move from lower to higher levels of abstraction, in order to identify the standard problem of which the specific problem being tackled is an instance. Then, they must choose one of the standard solutions relevant to the standard problem and move down, throughout the constraints posed by the physical properties of the system, to apply that solution to their specific application. Going up and down among lower and higher levels of abstraction is like going up and down the ridges of a hill (Fig. 15.21); in view of that, this model, which is among the most interesting in the Problem-Solving* literature, is called the 'hill model'.

According to Altshuller's analysis, the space of solution principles is based on three main approaches, which are shown in Fig. 15.22.

- **Satisficing** the contradictory requirement consists in understanding the contradiction and looking for a compromise solution. This is the approach most frequently used by designers, since it does not challenge the current structure of the system, and since technical disciplines are grounded on it, to the point that it is taught by engineering educators. Most of the time, this approach will simply lead to finding an acceptable tradeoff, without neither making any substantial performance improvement, nor introducing any change in the technical solution. This leads to the least inventive of the results and to incremental innovations that do not depart from the currently prevailing solutions.
- **Bypassing** the contradiction. This is an explicit decision not to deal with the contradiction at all, and to focus design effort on other contradictions that characterize the design problem.
- Overcoming the contradiction by finding technical means that allow the **separation** of the contradictory requirements.[16] This approach can lead to highly innovative solutions, typically consisting in change to product architecture and, potentially, to radical innovation.

The concept of '*separation*' is at the heart of Altshuller's contribution. In his review of patents, he concluded that the most significant 'inventive steps' were the ones in which the inventor had been able to reconceive the product in a way that contradictions were removed, instead of simply managed. Given the definitions shown in Fig. 15.20, the only way to remove a contradiction consists in *separating the elements that are associated to the conflicting evaluation parameters*.

Such a separation can be made in a finite number of ways, the main four being 'in space', 'in time', 'between parts and the system' and 'between states'. For instance, '*separation in space*' means physically splitting the two components that lead to a contradiction. The previous tabletop problem can therefore be solved by deciding for a sandwich structure, with two thin layers of a sturdy and heavy material and an interposed lighter material (or even air). With 'separation in time', instead, the designer looks at the process of using the artifact and restructures it so that the system adapts to the different operating conditions.[17]

Thanks to his analysis of patents, Altshuller was able to identify 40 technical 'separation principles', which are presented in Table 15.5. Albeit dated, this table represents the finite set of solution principles that may solve the universe of technical problems.

Altshuller studied the applicability of each of these 40 separation principles to *each* of the 1482 contradictions and this allowed the compilation of a 'contradiction

[16] The idea is that intricate interdependencies can be solved either by a strenuous effort in untangling the components, or by cutting them at the root, reminds of the legend of Alexander the Great cutting the Gordian Knot with his sword and, following the associated prophecy becoming ruler of Asia.

[17] An example from Aeronautical Engineering can be used to illustrate this principle. At low speed and when maneuvering, an ideal airplane wing should be extended, while it should be swept back when cruising at high speed. A variable-sweep wing used in military aircraft can be adjusted according to the need, and is therefore an example of 'separation in time'.

Table 15.5 The 40 separation principles in TRIZ

Principle	Subprinciple	Examples
1. Segmentation	Divide an object into independent parts	Replace mainframe computer by personal computers Replace a large truck by a truck and trailer Use a work breakdown structure for a large project
	Make an object easy to disassemble	Modular furniture Quick disconnect joints in plumbing
	Increase the degree of fragmentation or segmentation	Replace solid shades with Venetian blinds
2. Taking out	Separate an interfering part or property from an object, or single out the only necessary part (or property) of an object	Locate a noisy compressor outside the building where compressed air is used Use fiber optics or a light pipe to separate the hot light source from the location where light is needed Use the sound of a barking dog, without the dog, as a burglar alarm
3. Local quality	Change an object's structure from uniform to non-uniform, change an external environment (or external influence) from uniform to non-uniform	Use a temperature, density or pressure gradient instead of constant temperature, density or pressure
	Make each part of an object function in conditions most suitable for its operation	Lunch box with special compartments for hot and cold solid foods and for liquids
	Make each part of an object fulfill a different and useful function	Pencil with eraser Hammer with nail puller Swiss army knife
4. Asymmetry	Change the shape of an object from symmetrical to asymmetrical	Asymmetrical mixing vessels or asymmetrical vanes in symmetrical vessels improve mixing (cement trucks, cake mixers, blenders) Put a flat spot on a cylindrical shaft to attach a knob securely
	If an object is asymmetrical, increase its degree of asymmetry	Change from circular O-rings to oval cross-section to specialized shapes to improve sealing Use astigmatic optics to merge colors
5. Merging	Bring closer together (or merge) identical or similar objects and assemble identical or similar parts to perform parallel operations	Thousands of microprocessors in a parallel processor computer Veins in a ventilation system Electronic chips mounted on both sides of a circuit board or subassembly

(continued)

Table 15.5 (continued)

Principle	Subprinciple	Examples
	Make operations contiguous or parallel;	Link slats together in Venetian or vertical blinds
	Bring them together in time	Medical diagnostic instruments that analyze multiple blood parameters simultaneously Mulching lawnmower (leaves grass clippings on the ground)
6. Universality	Make a part or object perform multiple functions	Handle of a toothbrush contains toothpaste
	Eliminate the need for other parts	Child's car safety seat converts to a stroller Mulching lawnmower (demonstrates principles 5 and 6) CCD (Charge-coupled device) with microlenses formed on the surface
7. Nested doll	Place one object inside another	Portable audio system (microphone fits inside transmitter, which fits inside amplifier case)
	Make one part pass through a cavity in the other	Extending radio antenna Zoom lens Seat belt retraction mechanism Retractable aircraft landing gear stowed inside the fuselage
8. Anti-weight	To compensate for the weight of an object, merge it with other objects that provide lift	Inject foaming agent into a bundle of logs, to make it float better. Use helium balloon to support advertising signs
	To compensate for the weight of an object, make it interact with the environment	Aircraft wing shape reduces air density above the wing and increases density below wing, to create lift (also demonstrates principle 4) Hydrofoils lift ship out of the water
9. Preliminary anti-action	If it will be necessary to do an action with both harmful and useful effects, this action should be replaced with anti-actions to control harmful effects	Buffer a solution to prevent harm from extremes of pH
	Create beforehand stresses in an object that will oppose	Prestress rebar before pouring concrete
	Known undesirable working stresses later on	Masking anything before harmful exposure (use a lead apron on parts of the body not being exposed to X-rays, use masking tape to protect the part of an object not being painted)

(continued)

Table 15.5 (continued)

Principle	Subprinciple	Examples
10. Preliminary action	Perform, before it is needed, the required change of an object	Prepasted wallpaper Sterilize all instruments needed for a surgical procedure on a sealed tray
	Pre-arrange objects such that they can come into action from the most convenient place and without losing time for their delivery	Component kits in a Just-In-Time factory
11. Beforehand cushioning	Prepare emergency means beforehand to compensate for the low reliability of an object	Back-up parachute Spare tire
12. Equi-potentiality	In a potential field, limit position changes (e.g., change operating conditions to eliminate the need to raise or lower objects in a gravity field)	Spring-loaded delivery system for parts in a factory or for dishes in a buffet restaurant
13. The other way round	Invert the action(s) used to solve the problem	To loosen stuck parts, cool the inner part instead of heating the outer part
	Make movable parts (or the external environment)	Rotate the part instead of the tool
	fixed, and fixed parts movable)	Treadmill (for walking or running in place)
	Turn the object (or process) 'upside down'	Turn an assembly upside down to insert fasteners (especially screws) Empty grain from containers by inverting them
14. Spheroidality—curvature	Instead of using rectilinear parts, surfaces, or forms, use curvilinear ones; move from flat surfaces to spherical ones; from parts shaped as a cube to ball-shaped structures	Use arches and domes used for strength in architecture
	Use rollers, balls, spirals, domes	Ball point and roller point pens for smooth ink distribution
	Go from linear to rotary motion, use centrifugal forces	Produce linear motion of the cursor on the computer screen using a mouse or a trackball Replace wringing clothes to remove water with spinning clothes in a washing machine Use spherical casters instead of cylindrical wheels to move furniture
15. Dynamics	Allow (or design) the characteristics of an object, external environment or process to change to be optimal or to find an optimal operating condition	Adjustable steering wheels, seats, rearview mirrors

(continued)

Table 15.5 (continued)

Principle	Subprinciple	Examples
	Divide an object into parts capable of movement relative to each other	The 'butterfly' computer keyboard
	If an object (or process) is rigid or inflexible, make it movable or adaptive	The flexible boroscope for examining engines The flexible sigmoidoscope, for medical examination
16. Partial or Excessive Actions	If 100% of an object is hard to achieve using a given solution method then, by using 'slightly less' or 'slightly more' of the same method, the problem may be considerably easier to solve	Over spray when painting, then remove excess (or, use a stencil) Fill, then 'top off' when filling the gas tank of your car
17. Another dimension	To move an object in two- or three-dimensional space	Five-axis cutting tool can be positioned where needed
	Use a multistory arrangement of objects instead of a single-story arrangement	Cassette with 6 CD's to increase music time and variety Electronic chips on both sides of a printed circuit board
	Tilt or re-orient the object, lay it on its side	Dump truck
	Use 'another side' of a given area	Stack microelectronic hybrid circuits to improve density
18. Mechanical vibration	Cause an object to oscillate or vibrate	Electric carving knife with vibrating blades
	Increase frequency	Distribute powder with vibration
	Use an object's resonant frequency	Destroy gall stones or kidney stones using ultrasonic resonance
	Use piezoelectric vibrators instead of mechanical ones	Quartz crystal oscillations drive high accuracy clocks
	Use combined ultrasonic and electromagnetic field oscillations	Mixing alloys in an induction furnace
19. Periodic action	Instead of continuous action, use periodic or pulsating actions	Hitting something repeatedly with a hammer. Replace a continuous siren with a pulsed sound
	If an action is already periodic, change the periodic magnitude or frequency	Use Frequency Modulation to convey information, instead of Morse code
		Replace a continuous siren with sound that changes amplitude and frequency
	Use pauses between impulses to perform a different action	In cardio-pulmonary respiration (CPR) breathe after every 5 chest compressions
20. Continuity of useful action	Carry on work continuously; make all parts of an object work at full load, all the time	Flywheel (or hydraulic system) stores energy when a vehicle stops,
	Eliminate all idle or intermittent actions or work	Print during the return of a inkjet printer carriage

(continued)

Table 15.5 (continued)

Principle	Subprinciple	Examples
21. Skipping	Conduct a process or certain stages (e.g., destructible, harmful or hazardous operations) at high speed	Use a high-speed dentist's drill to avoid heating tissue Cut plastic faster than heat can propagate in the material, to avoid deforming the shape
22. Blessings in disguise	Use harmful factors (particularly, harmful effects of the environment or surroundings) to achieve a positive effect	Use waste heat to generate electric power Recycle waste (scrap) material from one process as raw materials for another
	Eliminate the primary harmful action by adding it to another	Use a helium–oxygen mix for diving, to eliminate both nitrogen narcosis and oxygen poisoning from air and other nitrox mixes
	Harmful action to resolve the problem	Use a backfire to eliminate the fuel from a forest fire
23. Feedback	Introduce feedback (referring back, cross-checking) to improve a process or action	Automatic volume control in audio circuits
	If feedback is already used, change its magnitude or influence	Change sensitivity of a thermostat when cooling vs. heating, since it uses energy less efficiently when cooling
24. Intermediary	Use an intermediary carrier article or intermediary process	Carpenter's nailset, used between the hammer and the nail
	Merge one object temporarily with another (which can be easily removed)	Pot holder to carry hot dishes to the table
25. Self-service	Make an object serve itself by performing auxiliary helpful functions	Halogen lamps regenerate the filament during use—evaporated material is redeposited
	Use waste resources, energy or substances	Use heat from a process to generate electricity: 'Co-generation' Use animal waste as fertilizer Use food and lawn waste to create compost
26. Copying	Instead of an unavailable, expensive, fragile object, use simpler and inexpensive copies	Computer simulation Instructional videos instead of attending seminars
	Replace an object, or process with optical copies	Do surveying from space photographs instead of on the ground Measure an object by measuring the photograph
	If visible optical copies are already used, move to infrared or ultraviolet copies	Make images in infrared to detect heat sources, such as diseases in crops, or intruders in a security system

(continued)

Table 15.5 (continued)

Principle	Subprinciple	Examples
27. Cheap short-living objects	Replace an expensive object with a multiple of inexpensive objects, comprising certain qualities	Plastic cups in motels, disposable diapers, many kinds of medical supplies
28. Mcchanics substitution	Replace a mechanical means with a sensory (optical, acoustic, taste or smell) means	Replace a physical fence to confine a dog or cat with an acoustic 'fence' (signal audible to the animal) Use a bad-smelling compound in natural gas to alert users to leakage, instead of a mechanical or electrical sensor
	Use electric, magnetic and electromagnetic fields to interact with the object	To mix 2 powders, electrostatically charge one positive and the other negative
29. Pneumatics and hydraulics	Use gas and liquid parts of an object instead of solid parts	Comfortable shoe sole inserts filled with gel Store energy from decelerating a vehicle in a hydraulic system, then use the stored energy to accelerate later
30. Flexible shells and thin films	Use flexible shells and thin films instead of three-dimensional structures	Use inflatable (thin film) structures as winter covers on tennis courts
	Isolate the object from the external environment using flexible shells and thin films	Float a film of bipolar material (one end hydrophilic, one end hydrophobic) on a reservoir to limit evaporation
31. Porous materials	Make an object porous or add porous elements	Drill holes in a structure to reduce the weight
	Fan object is already porous, use the pores to introduce a useful substance or function	Store hydrogen in the pores of a palladium sponge
32. Color change	Change the color of an object or its external environment	Use safe lights in a photographic darkroom
	Change the transparency of an object or its external environment	Use photolithography to change transparent material to a solid mask for semiconductor processing
33. Homogeneity	Make objects interacting with a given object of the same material (or with identical properties)	Make the container out of the same material as the contents, to reduce chemical reactions
34. Discarding and recovering	Make portions of an object that have fulfilled their functions go away (dissolving, evaporating, etc.)	Use a dissolving capsule for medicine. Sprinkle water on cornstarch-based packaging and have it reduce its volume Lost-wax casting
	Conversely, restore consumable parts of an object directly in operation	Self-sharpening lawn mower blades

(continued)

Table 15.5 (continued)

Principle	Subprinciple	Examples
35. Parameter changes	Change an object's physical state (e.g., to a gas, liquid or solid)	Transport oxygen or nitrogen or petroleum gas as a liquid, instead of a gas, to reduce volume
	Change the concentration or consistency	Liquid hand soap is concentrated and more viscous than bar soap at the point of use
	Change the degree of flexibility	Vulcanize rubber to change its flexibility and durability
	Change the temperature	Raise the temperature above the Curie point to change a ferromagnetic substance to paramagnetic Lower the temperature of medical specimens to preserve them for later analysis
36. Phase transitions	Use phenomena occurring during phase transitions	Heat pumps use the heat of vaporization and heat of condensation of a closed thermodynamic cycle to do useful work
37. Thermal expansion	Use thermal expansion (or contraction) of materials	Fit a tight joint together by cooling the inner part to contract, heating the outer part to expand, putting the joint together and returning to equilibrium
	Use multiple materials with different coefficients of thermal expansion	The leaf spring thermostat: (2 metals with different coefficients of expansion are linked so that it bends one way when warmer than nominal and the opposite way when cooler.)
38. Strong oxidants	Replace air with oxygen-enriched air	Scuba diving with nitrox for extended endurance
	Replace air with pure oxygen	Cut at a higher temperature using an oxy-acetylene torch Treat wounds in a high pressure oxygen environment to kill anaerobic bacteria and aid healing
	Use ionized oxygen	Ionize air to trap pollutants in an air cleaner. Speed up chemical reactions
39. Inert atmosphere	Speed up chemical reactions by ionizing the gas before use	Prevent degradation of a hot metal filament by using an argon atmosphere
	Add neutral parts or inert additives to an object	Increase the volume of powdered detergent by adding inert ingredients

(continued)

Table 15.5 (continued)

Principle	Subprinciple	Examples
40. Composite materials	Change from uniform to composite (multiple) materials	Composite epoxy resin/carbon fiber golf club shafts are lighter, stronger and more flexible than metal Same for airplane parts Fiberglass boats instead of wooden boats

matrix'.[18] Altshuller originally suggested to use the contradiction matrix as a prescriptive reference showing which separation principles should be used to solve a specific type of contradiction. Nowadays this normative aspect has lost its value and practitioners tend to use the contradiction matrix more as a general inspiration than as a recipe.

However, the theoretical underpinnings of TRIZ remain valid, suggesting that technical problems should not be solved simply by looking for a compromise in a known tradeoff, but also for radical solutions based on separation principles. This general approach has also been proposed in other fields, including the business domain (Ruchti and Livotov 2001; Souchkov 2007).

15.3.3 Other Methods for Creative Generation

Other creativity-fostering techniques have been proposed in the literature, with the aim of stimulating people to think '*out of the box*' and concentrate on the generation of new ideas, deferring critical analysis and selection. The main techniques used in Conceptual Design are the following.

- **Brainstorming*** (Osborn 1953) is the most well-known creativity-fostering technique and—in its many variations—is amply used outside the design environment as well. Brainstorming requires participants to freely engage in the proposal of new ideas, with an explicit prohibition of making critical remarks on any. The aim is to collect a diverse group and generate as many ideas as possible with the involvement of all participants, trying to avoid social or hierarchical inhibition between people of different age or status. A brainstorming session usually requires a *facilitator* in charge of maintaining discipline, keeping up enthusiasm and stimulating the association and combination of ideas into new ones.

  In general, brainstorming is supported by visual tools such as sticky notes to be placed on a board, ideas galleries to be shown to other people and magazines for clipping pictures that may be associated to the ideas. The brainstorming session

[18] The contradiction matrix is obviously too large to be presented on paper, but can be found online at http://www.innovation-triz.com/TRIZ40/TRIZ_Matrix.xls.

may be organized according to different techniques, with ideas being openly proposed (*Classic*), or written privately on cards before making them public (*Crawford slip writing*). Moreover, the facilitator may or not propose some kind of exchange mechanism between individuals, so that new ideas can be generated by building on ideas set forth by others (*Brain Writing or Crawford x-3-5*). Finally, the brainstorming session can be carried out in a neutral environment that does not bear any reference to the product, or in a workshop setting that is strongly associated to it (*Imaginary Brainstorming, Negative Brainstorming*). In this latter case, designers can see and touch the product, are shown videos of the product in use, etc.

The choice of the brainstorming technique depends on the problem complexity and clarity, as well as on the environment the facilitator aims to create.

- **Synectics*** (Gordon 1961; Prince 1970), though originally proposed as a more general and ambitious methodology, can be viewed as an evolution of brainstorming that is aimed at generating ideas through metaphors and analogies. Synectics also advocates the principle of setting known elements aside ('*make the familiar strange*') while embracing strange and seemingly irrelevant solutions ('*make the strange familiar*').
- **Wish and wonder** and **law-breaking**. These variants to brainstorming may be used to make designers break loose from constraints and allow a freer flow of creative ideas. The team is invited to pretend they can operate in an ideal world in which phenomena occur quite differently than usual. The solutions thus generated cannot usually be applied straight away, but might work with some adaptation. In order to bring participants out of their 'comfort zone', the facilitator might pick on well-known references in literature or films. For instance, a design team having to develop medical devices may be invited to pretend they are living in Alice's Wonderland and have to fulfill the desired functions after having been shrunk to the size of a pinhead.
- **SCAMPER** (Eberle 1996) is a complementary technique to Brainstorming, and provides some additional guidance to the idea generation process. Participants are required to suggest ideas belonging to the categories for which the term SCAMPER is an acronym. Therefore, ideas are formulated by using the verbs 'substitute', 'combine', 'adapt', 'modify' (or 'magnify', 'mignify', 'multiply'), 'put to other use', 'eliminate' and 'reverse' (or 're-arrange'), while the facilitator tries to ensure that each category is sufficiently represented.
- **Bio-inspired design*** (or **Biomimicry, Biomimetics,** etc. Goel et al. 2014; Macnab 2012) is another approach that uses analogies between problems, but looks for inspiration from nature. The power of biomimetics lies in the remarkable way with which evolutionary dynamics have progressively shaped living beings and allowed them to successfully adapt to the environment. Well-known examples of biomimetically inspired solutions are Velcro straps, which were conceived by observing at burdock burrs, and surface finishing that minimizes water drag, designed by emulating the structure of shark skin.

- **Gamification**. Playing games is an essential part of human behavior. Placing game-like elements in products has been used since the early twentieth century to stimulate consumer purchasing activity and to foster their loyalty; more recently, game-like elements have been introduced in products not only as an additional gadget that has little to do with the product, but as a part of the actual experience. By doing this, firms expect to achieve a greater degree of engagement in using the product, and even changing customer/user behavior (Takahashi and Kang 2020; Michie et al. 2011). Well-known examples of this approach (for a more general reference, Deterding et al. 2011; Robson et al. 2015) are the Odenplan metro station in Stockholm, where the design of the exit staircase simulated a piano keyboard and led people to give up the use of the escalator. Gamification can be used for products or services and has found interesting applications, especially when these involve a complex customer journey, with many undesired situations or tedious actions (e.g., airport, queues at the supermarket, etc.).[19]

As noted by Girotra et al. (2010), the idea generation process is not easy and straightforward to carry out. In fact, the purpose of the process should not be to generate a large number of ideas of medium or even high average quality. The aim should be to come up with a significant number of exceptional ideas, among which at least one might be viable and lead to the desired degree of innovation. Moving from simply 'good' to 'great' (though possibly a bit crazy) ideas depends on the facilitator's experience and on participants' disposition and capability.

In many organizations, working with creativity techniques is strongly resisted because of cognitive inertia and people's dislike of challenging the status quo, not counting the hesitation many would have before 'making themselves a fool' by proposing potentially silly ideas. Because of this, creativity techniques require an appropriate organizational culture, some training and a correct balancing between the work done in groups and individual research, during which each designer has greater freedom to think and generate new ideas.

15.3.4 Concept Combination and Classification

Both Functional Analysis and the Network of Problems from TRIZ suggest that designers mainly divide the problem into subproblems. Therefore, the solutions that are generated will usually cover part of the problem and not its entirety. So, at the end, designers will have to recombine all partial solutions into the overall solution and therefore generate a number of variants.

[19] Frequently, gamification was adopted to find a number of interesting applications for pushing consumers toward the adoption of eco-friendly behaviors. For instance, many carmakers, from Toyota to Ford, have developed user interfaces designed in such a way that the driver is enticed to improve his driving style and reduce fuel consumption. In some cases, the data is collected so that users can also compare themselves against their peers, then engaging in a virtuous game that make them improve their behavior.

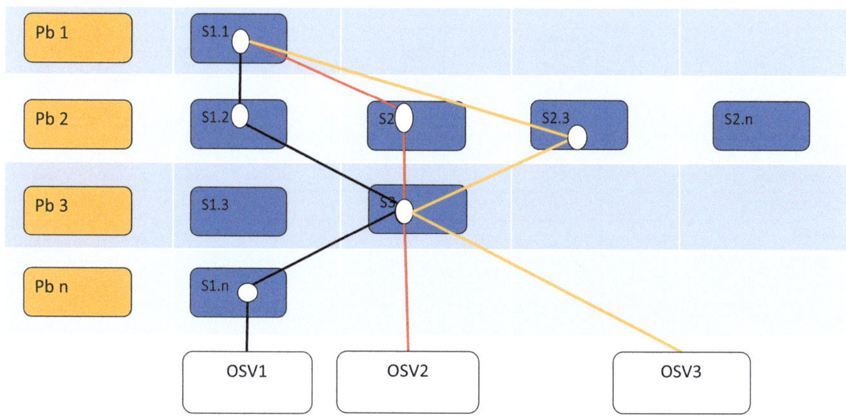

Fig. 15.23 Combination table to derive overall solution variants

This activity typically makes use of a morphological chart like the one in Fig. 14.23 and is composed of three steps:

- **selecting** one solution among the partial ones generated for each subproblem (by row in Fig. 15.23);
- **combining** the selected partial solutions to generate overall solutions (by column in Fig. 15.23);
- **repeating** the process for all the possible combinations to generate all the possible overall solution variants.

This combination is not automatic and represents a critical task for designers, who must be aware that some combinations may not be feasible at all. The final outcome of this activity is a set of feasible alternative overall solutions that will be then screened and selected in view of different criteria, which should depend on the requirements and/or the primary needs.

15.3.5 Screening and Selecting Concepts

As it has been mentioned, conceptual design requires a continuous alternation between divergent phases, in which concepts are created or improved, and convergent phases, which reduce the size of the concept pool. In general, one can distinguish between *screening* and final *selection*.

Screening has the purpose of reducing the number of candidate concepts generated by a previous divergent phase to a manageable number, so that they can later be developed and improved, often by recombining them. Screening therefore does not have the aim of selecting the best candidate concept(s), but of *weeding out the ones that are very unlikely to be fit for purpose*. Conversely, concept selection has

Safety risk	---	--	--	--	+	--	+	+++	+++	-
False operation	---	-	-	+	++	+	+	+	++	++
Required force	---	--	--	-	+	+	++	++	+++	+++
Compactness	+	+++	++	+++	+++	++	++	++	-	-
Usability	-	+	-	++	+++	++	+	++	+++	++
Volume	+++	+	+	+	+	+	+	+	+	+
Weight	+++	+	+	+	+	+	+	+	+	+
Production cost	++	++	+	+	+	+	-	+	--	--
$\sum +$	9	8	5	9	13	9	9	13	13	9
$\sum -$	10	5	6	3	0	2	1	9	3	4

Fig. 15.24 A Pugh comparison table for concept selection

the objective of *singling out the 'best' concept,* which will be developed during the subsequent embodiment design phase.

The screening and selection of concepts will depend on a number of different criteria. These will have some relation to the key technical features and to the customer/user needs that emerged during the requirements analysis phase as, for instance, through a QFD analysis. If this analysis has been properly carried out, the design team should have a quite clear and 'objective' picture of the criteria and of their weights. So, a method for concept screening and selection will necessarily be based on some sort of multicriteria analysis.

Given the limited time available, one might suggest a relatively straightforward approach such as AHP (which has already been mentioned in Chap. 9, when dealing with project selection problems, Saaty 1980) and leave aside more formal and structured multicriteria methods (Roy 1996; Belton and Steward 2002). However, in practice, firms tend to use even simpler tabular multicriteria methods, as will be described in the following.

Screening and selection can be performed by using *Pugh comparison tables* (Pugh 1981, Fig. 15.24) in which each concept is evaluated on each relevant criterion on a relatively simple five-level evaluation scale ('−2' if the concept is likely to perform significantly worse than the average product on that criterion, '0' if it will probably perform close to average, '2' if it will perform significantly better, and '−1' and ' +1' in case of intermediate evaluations). The overall score of each concept is simply given by the sum of these criterion-level scores. When screening concepts, this method allows creating three groups of candidates, i.e., 'very promising' concepts that have good to high scores on all criteria, 'interesting' concepts that have high scores on some criteria, but low scores on other ones and, finally, 'unpromising' concepts that have low scores on all criteria and are therefore dominated by concepts in the 'promising' set. When used properly, the design team will use the Pugh comparison table to select the 'promising' group of concepts and to carry out a more thorough analysis on the 'interesting' group. In the subsequent

	weight				
Safety risk	15	1	3	3	5
False operation	12	3	4	3	4
Required force	20	2	3	5	5
Portability	6	5	4	2	2
Usability	15	4	5	3	5
Volume	8	5	3	2	1
Weight	12	5	3	3	1
Production cost	12	5	5	2	1
$\sum$		341	372	314	342

Fig. 15.25 A weighted scoring table for concept selection

steps, it would be advisable to 'prune' the ideas that determine the strengths of this latter group and try to 'graft' them onto the concepts belonging to the former, in order to further reduce their weaknesses.

Conversely, when making the final selection, the design team will simply pick the highest-ranking concept and bring it to the embodiment design phase.

As a variant of Pugh comparison tables, one may also use weighted scoring matrices, with criterion weights normalized so that they sum to 100, and using a 1–5 scoring scale for evaluating each concept on each criterion (Fig. 15.25). The difference between these matrices and Pugh tables is due to the fact that criteria are weighted, thus allowing a finer comparison of alternatives. Therefore, weighted scoring matrices are generally used for the final selection of the product concept, and not in the earlier screening phase.

As already discussed in Chap. 9, these tables suffer from methodological limitations associated with the use of scales and the difficulty in choosing the weights, which can heavily influence the final outcome of the evaluation.

15.4 Linking the Design Process to the Innovation Process

Given the multidisciplinary approach that underlies this textbook, and given the fact that design processes are at the heart of innovation activities, it is convenient at this point to try to integrate the two perspectives (Montagna and Cantamessa 2019). Specifically, one can wonder how design processes fit within the broader innovation process or, conversely, whether the type of innovation that is being pursued can shape the underlying design processes.

In principle, the three phases of design outlined above (i.e., conceptual, embodiment and detailed) are always present in any design process. However, the weight and the effort cast on each will greatly depend on the innovation content that the design team has been tasked to introduce. In the case of significant innovations (i.e., radical and/or architectural), conceptual design will have a major role, since the designers will be asked to conceive completely new technical solutions. In the case of modular or incremental innovations, conceptual design will have a passing role, or be restricted to a few subsystems. In fact, a complete conceptual redesign of the system might lead to architectural change that would fall outside the scope of the design task that has been set up, or would challenge a dominant design that the firm does not wish to modify.

In order to provide a fuller discussion, we propose to expand Table 4.4 of Chap. 4 to the following Table 15.6. Table 4.4 focused on the business-level challenges that are typical of the fluid, transition and specific phases of the technological lifecycle, and is now enriched by adding a discussion on the ensuing type of design activities at product and process level, and of the role of designers.

During the fluid phase, technology-push forces are predominant and lead to radical and paradigm-changing innovations. Previous technology has reached its limit and is unable to address emerging or increasing customer needs. At the same time, advances in research create new opportunities and prospectively enable previously unimaginable functions or performances.

During this phase, designers will engage with researchers in order to carry out technology foresight activities, by understanding the practical (i.e., technical and business) potential emerging from research results and match it with current or prospective needs arising in the market and society. When moving from basic to applied research, the effort of matching technological potential with market needs becomes more focused and application-directed, and allows the definition of more definite strategies (namely *Technology Road mapping**; Firat et al. 2008) in which designers can provide a significant contribution.

In this context, the development of technology demonstrators often sees an interesting interplay between the efforts of applied researchers, who provide a scientifically grounded perspective, and that of designers, who provide an application-oriented one. At this phase, creativity has a key role. From designers' perspective, this implies a high-level instance of conceptual design, which allows radically new solutions to emerge. In this particular context, conceptual design can operate with high levels of creativity, since one is still quite far away from the

Table 15.6 Business-level challenges in the technological lifecycle, the ensuing design activities at product and process level, and the role of designers

Lifecycle phases		Supply-side issues			Demand-side issues	Implications on the phases of the technology and product development process				
		Supra-system	System	Subsystem	Customers	Technology process		Product development process		
						Basic research	Applied research	Conceptual design	Embodiment and detailed design	Process design
Fluid phase	Still immature technology, with low performance Many firms enter the industry, because of its future promise Each entering firm brings prior competencies and assets, which leads to a diversified range of technical solutions High rate of product innovation, but progress is limited, since efforts are not concentrated in the same direction	In many cases, the system cannot operate unless inserted in a coherent supra-system. It is therefore likely that the supra-system too must undergo ***radical change***	The focal system undergoes *radical innovation*, which implies change in the underlying technical principles and in the architectures	In some cases, components and subsystems might be carried over from existing stock, but it is likely that they too must undergo ***radical change***	Innovators and technology enthusiasts with very specific needs, and tolerant to shortcomings of the technology Can act as lead users and/or suggest technical improvements The technology often becomes overhyped	Radical innovation typically emerges because advances in research create new opportunities and prospectively enable previously unimaginable functions or performance in a variety of products and industries (*technology push*) The role of the designer is to contribute to the understanding of the potential emerging from research results and identify the most promising outcomes for a given field (***Technology Foresight***)	Applied research activities aim at directing outcomes of basic research to the solution of practical problems with market potential (*technology push*) The role of the designer is to contribute to the assessment of the technical and commercial feasibility of applications for emerging technology (***Technology Road mapping***)	Conceptual design is at the heart of radical innovation, since all technical and architectural options are open Depending on the degree with which one must deal with supra- and subsystems, the scope of conceptual design will be narrower or broader, eventually reaching into 'system-wide' solutions. The understanding of functional and performance requirements is weak and fuzzy, because the market is still in its initial phases. ***Creativity thinking*** is predominant	The role of embodiment and detailed design is to bring to completion the hypothesized conceptual designs. Design choices are known to be transitory and not yet optimized since the dominant design is still not established The complexity of embodiment and detailed design will depend on the degree of vertical integration	Process design does not have a significant role in this phase, since volumes are low, willingness to pay can be high, and most of all, the system architecture is not stable yet

(continued)

Table 15.6 (continued)

Lifecycle phases		Supply-side issues			Demand-side issues	Implications on the phases of the technology and product development process				
		Supra-system	System	Subsystem	Customers	Technology process		Product development process		
						Basic research	Applied research	Conceptual design	Embodiment and detailed design	Process design
Transition phase	One product architecture emerges as the Dominant Design This event triggers the take-off of product performance, market diffusion and shakeout of firms	Unless it has already been established, the freezing of a dominant design in the system enables the design and diffusion of a complementary and specific supra-system	Once the dominant design is established, innovation in the system will typically shift to ***incremental*** In some cases, there might be ***architectural innovations*** that 'optimize' the dominant design	Unless they have already been established, the freezing of a dominant design in the system enables the design and diffusion of complementary and specific subsystems	Early adopters who want to gain early experience with emerging (and possibly still immature) technology Word of mouth supports diffusion The hype cycle starts to deflate	Basic research does not have a specific role	Functional and performance requirements arising in the newly defined market start to dictate an applied research agenda (*demand pull*) The role of the designer is to translate maturing user needs into performance requirements and then into well-defined demands on technology development (***Technology Road mapping***)	Creativity processes are constrained, and the latitude of conceptual design become limited to architectural choices; the scope of conceptual design becomes narrower, due to the freezing of supra-system and subsystems (System Design)	Embodiment and detailed design decisions are aimed at improving performance according to emerging user needs ***(Design Optimization)***	The freezing of the dominant design at all levels allows the design and development of production processes specific to the dominant design itself It is necessary to innovate production processes to ensure increasing production capacity ***(Process Design)***
Specific phase	Firms compete on cost and quality while demand increases	Due to path dependency, *incremental innovations* prevail	Due to path dependency, *incremental innovations* prevail At times, *modular*	Subsystems might be subject to *radical innovation* when new technology	Early majority and later customers require mature technology	Basic research can be relevant for optimizing required performance (*demand pull*)	Applied research can be relevant for optimizing required performance (*demand pull*) and to allow	The latitude and scope of conceptual design becomes limited at system level. Conceptual design either focuses on	Embodiment and detailed design decisions are oriented to improve performance and reduce cost	Design decisions are devoted to incremental innovations to the production process in order

(continued)

Table 15.6 (continued)

Lifecycle phases		Supply-side issues			Demand-side issues	Implications on the phases of the technology and product development process				
		Supra-system	System	Subsystem	Customers	Technology process		Product development process		
						Basic research	Applied research	Conceptual design	Embodiment and detailed design	Process design
	Economies of scale kick in, leading to process innovation (specific to the dominant design) and to oligopolistic competition		*innovations* that do not alter system architecture might be introduced	emerges, as long as they can be easily integrated within the dominant design	and adopt based on costs and benefits (eventually postponing adoption until they feel safe) Additional and replacement sales become prevalent with respect to adoption sales	Basic research also may allow radical change at component level (*technology push*)	radical change at component level (*technology push*) The role of the designer is to translate matured user needs into well-defined demands of technology development The role of the designer is also to understand the potential of promising technology at component level (***Technology Road mapping***)	minute and specific details of the system or with the integration of new subsystems in the architecture The role of conceptual design can be significant for subsystems when new technology emerges	Incremental innovations are used to stimulate product replacement ***(Design Optimization)***	to reduce production costs ***(Process Design Optimization)***

constraints cast by a product aimed at a specific and narrow application. Moreover, being in the fluid phase, and due to the possibility that the innovation may lead to a completely new technological paradigm, this allows ample freedom from any technical and architectural constraint.

Moving to actual product development, conceptual design becomes the most important design phase. Designers will now take charge of the constraints arising from the initial application domain and market segment (usually that of innovators and then early adopters), but still with ample degrees of freedom, since technical and architectural design decisions are still open. Given the uncertainty on customer needs and the impossibility of deriving them from users who have not yet adopted and experienced the new and radically innovative product, designers will mainly have to operate based on their vision of the emerging technology and product. Depending on the degree with which one must (or decides to) deal with supra- and subsystems (something that runs in parallel to decisions on vertical integration of both the system and the firm), the scope of conceptual design can become narrower or broader.

A dominant design is still not established yet, and design choices are known and accepted to be transitory and non-optimized. This implies that relatively less effort will be placed on embodiment and detailed design, since the aim will be that of bringing a working product to the initial market, and not striving for perfection. Finally, process design will not have a significant role in this phase; low volumes, early customers' higher willingness to pay and the lack of a stable product architecture do not justify significant investments in this area.

In this context, conceptual design allows the definition of functions the artifact must fulfill, and the identification of an innovative combination of principles and processes that will allow the artifact to behave accordingly. It therefore follows that the traditional Engineering Design literature (Rodenacker 1971; Pahl et al. 1988; Ulrich and Seering 1988) can provide a clear frame of reference and a rigorous support to concept creation, screening and selection, using a wide variety of tools and methods.

During the transition phase, basic research will have a lesser role given that—following the emergence of a dominant design—the main technological choices have been already made. Functional and performance requirements arising in the newly defined and rapidly increasing market start to dictate an applied research agenda, according to a typical demand-pull perspective. In this context, the role of the designer is to translate maturing user needs into performance requirements and discuss them with researchers.

During this transition phase, the latitude of conceptual design becomes much more constrained, and the scope of inventiveness narrower. Creativity processes are bounded by the now-stabilized architecture and System Engineering Design approaches become prevalent (Ramo and St. Clair 1988). At the most, conceptual design consists in making the dominant design evolve before it freezes definitively, generally by introducing architectural innovations and adjustments to the system, in order to enable improved fit with user needs. The scope of conceptual design will also become narrower, because supra-systems and subsystems will tend to freeze quite rapidly and—quite often—their development will be allocated to other firms in the value chain, following a process of vertical disintegration.

With respect to user needs, it becomes easier to elicit them from current users, who now are belonging to the 'early adopters' segment. These users have already gained some experience with the products, and their needs are closer—though not perfectly aligned—to those of the mainstream population. Designers can therefore look for support from *user-centered approaches** (Norman and Draper 1986), though the outcomes must still not be taken at face value, but interpreted with the consciousness that—when moving toward the Early Majority—these needs might be different.

Finally, the freezing of technological choices at all levels allows the design and development of production processes that are specific to the dominant design and its underlying technologies. It is therefore time for (re)designing production and support processes, in order to support growing demand and reduce production costs by achieving economies of scale.

During the specific phase of the technology lifecycle, which can last for decades, basic and applied research follows an interesting interplay between demand-pull and technology-push forces. Concerning demand pull, an established industry dominated by large oligopolistic players, and serving a well-known and stable market can determine a clear (and well-funded) R&D agenda, with the aim of improving technology in order to fulfill progressively growing user needs and therefore support incremental innovation. On the technology-push side, this may lead to radical innovation at component level, provided that this can be easily incorporated in the dominant design without disrupting it (therefore, at the level of the entire system, we would talk about 'modular' innovation). This complexity, together with the often significant amount of money involved, calls for an important role of *Technology Road mapping**.

Given that the dominant design is now frozen, conceptual design at system level ceases to be significant. However, conceptual design becomes particularly important at the level of subsystems and components, which from time to time witness radical innovations and—therefore—the introduction of completely new concepts. Given that the specific phase often leads to vertical disintegration of the value chain, the locus of component-level inventive activity involving conceptual design will tend to shift from the producer of the system to its suppliers. The specific phase will instead lead to a growing importance of embodiment and detailed design, who will progressively engage in *Design Optimization** within the dominant design, in order to improve performance and reduce cost.

Moreover, during the specific phase, the diffusion process has been well exhausted and sales will nearly entirely be associated with the replacement of the installed base. It follows that embodiment and detailed design will be entrusted with the objective of stimulating these replacement sales, by providing an endless stream of incremental innovation, new features (often involving modular innovations as described above) and minor adjustments and facelifts. Process innovation will also be quite important, in order to further improve product quality and reduce cost.

References

Altshuller G (1999) The Innovation algorithm: TRIZ, systematic innovation, and technical creativity. Technical Innovation Center, Worcester, MA

Anderson D, Perez KB, Xuereb Z, Otto K, Wood K (2018) Design processes of design automation practitioners. In: 30th International conference on design theory and methodology, American society of mechanical engineers

Artobolevsky I (1975) Mechanics in modern engineering design, a handbook for engineers, designers and inventors. Mir Publishers, Moscow

Bayazit N (2004) Investigating design: a review of forty years of design. Des Issues 20(1):16–29

Belton V, Steward TJ (2002) Multiple criteria decision analysis, an integration approach. Kluwer Academic Publishers, Norwell, MA

Berthon PR, Pitt LF, McCarthy I, Kates SM (2007) When customers get clever: managerial approaches to dealing with creative consumers. Bus Horiz 50(1):39–47

Birkhofer H (2011) The future of design methodology. Springer-Verlag, London

Brown T, Katz B (2011) change by design. J Prod Innov Manag 28(3):381–383

Brown T (2015) When everyone is doing design thinking, is it still a competitive advantage? Yes, but only if you master it. HBR Publisher, Boston

Bucciarelli LL (1988) An ethnographic perspective on engineering design. Des Stud 9(3):159–168

Buchanan R (1992) Wicked problems in design thinking. Des Issues 8(2):5–21

Clarkson PJ, Hamilton JR (2000) Signposting: a parameter-driven task-based model of the design process. Res Eng Des 12(1):8–38

Clarkson PJ, Eckert C (2005) Design process improvement: a review of current practice. Springer-Verlag, London

Cross N (1982) Designerly ways of knowing. Des Stud 3(4):221–227

Cross N (2001) Designerly ways of knowing: design discipline versus design science. Des Issues 17(3):49–55

Cross N (2011) Design thinking: understanding how designers think and work. Berg Publishers, Oxford

Cross N (2021) Engineering design methods: strategies for product design. Wiley, New York

Colombo S, Mazza A, Montagna F, Ricci R, Dal Monte O, Cantamessa M (2020) Neurophysiological evidence in idea generation: differences between designers and engineers. In: 16th international design conference (DESIGN 2020), Online, October 26–29

Deterding S, Dixon D, Khaled R, Nacke L (2011) From game design elements to gamefulness: defining "gamification,". In: 15th International academic Mindtrek conference, Tampere, pp 28–30

Eberle B (1996) Scamper: creative games and activities for imaginative development. Prufrock Press, Waco

Eide AR, Jenison RD, Mashaw LH, Northrup LL (2002) Introduction to engineering design and problem solving. McGraw-Hill, Boston

Finger S, Dixon JR (1989) A review of research in mechanical engineering. Part I: descriptive, prescriptive, and computer-based models of design process. Res Eng Des 1(1):51–67

Fink A, Benedek M (2014) EEG alpha power and creative ideation. Neurosci Biobehav Rev 44:111–123

Fink A, Grabner RH, Benedek M, Reishofer G, Hauswirth V, Fally M, Neuper C, Ebner F, Neubauer AC (2009) The creative brain: Investigation of brain activity during creative problem solving by means of EEG and FMRI. Human Brain Mapp 30(3):734–748

Fink A, Schwab D, Papousek I (2011) Sensitivity of EEG upper alpha activity to cognitive and affective creativity interventions. Int J Psychophysiol 82(3):233–239

Finke RA, Ward TB, Smith SM (1992) Creative cognition: theory. MIT Press Cambridge, MA, Research and Application

Firat AK, Woon WL, Madnick S (2008) Technological forecasting—a review, composite information systems laboratory (CISL). MIT Press Cambridge, MA

Gero JS (1990) Design prototypes: a knowledge representation schema for design. AI Mag 11 (4):26–36
Gero JS, Kannengiesser U (2004) The situated function-behaviour-structure framework. Des Stud 25(4):373–391
Gero JS, Kannengiesser U, Williams CB (2014) Designing have a common cognitive behavior independent of domain and task: a meta-analysis of design protocols. Human Behavior in Design Conference, Ascona, October 4–17
Girotra K, Terwiesch C, Ulrich KT (2010) Idea generation and the quality of the best idea. Manage Sci 56(4):591–605
Goel A, McAdams D, Stone R (2014) Biologically inspired design: computational methods and tools. Springer, London
Gordon WJJ (1961) Synectics: The development of creative capacity. Harper and row Publishers, New York
Jauk E, Benedek M, Neubauer AC (2012) Tackling creativity at its roots: Evidence for different patterns of EEG alpha activity related to convergent and divergent modes of task processing. Int J Psychophysiol 84(2):219–225
Kwon H, Park Y, Geum Y (2018) Toward data-driven idea generation: application of Wikipedia to morphological analysis. Technol Forecast Soc Chang 132:56–80
Kumar V (2012) 101 design methods: a structured approach for driving innovation in your organization. Wiley, New York
Liedtka JM, King A, Bennett K (2013) Solving problems with design thinking: ten stories of what works. Columbia Business Press, New York
Macnab M (2012) Design by nature: using universal form and principles in design. New Riders, Berkeley
Mehalik MM, Schunn CD (2006) What constitutes good design? A review of empirical studies of the design process. Int J Eng Edu 22(3):519–532
Michie S, van Stralen MM, West R (2011) The behaviour change wheel: a new method for characterising and designing behaviour change interventions. Implementation Sci 6(42)
Montagna F, Cantamessa M (2019) Unpacking the innovation toolbox for design research and practice. Des Sci 5:E8
Nakajima M (1995) Chishiki Shi San no Saikochiku (Reconstruction of Knowledge Assets). Nikkan Kogyo Shinbun, Tokyo
Norman DA, Draper SW (1986) User centered system design; new perspectives on human-computer interaction. Erlbaum Associates Inc., Hillsdale
Osborn AF (1953) Applied Imagination: principles and procedures of creative thinking. Scribner, New York
Pahl G, Beitz W, Feldhusen J, Grote K-H (1988) Engineering design: a systematic approach. Springer-Verlag, London
Pahl G, Beitz W (2007) Engineering design: a systematic approach. Springer-Verlag, London
Prince GM (1970) The Practice of creativity: a manual for dynamic group problem-solving. Collier, New York
Pugh S (1981) Concept selection: a method that works, first international conference on engineering design, Rome, August 9–13
Ramo S, St. Clair RK (1988) The systems approach: fresh solutions to complex civil problems through combining science and practical common sense. KNI Incorporated, Anaheim
Robson K, Plagger K, Kietzmann JH, McCarthy I, Pitt L (2015) Is it all a game? Understanding the principles of gamification. Bus Horiz 58:411–420
Rodenacker W (1971) Methodisches konstruieren. Springer-Verlag, Berlin
Rowe PG (1987) Design thinking. MIT Press Cambridge, MA
Roy B (1996) Multicriteria methodology for decision aiding. Kluwer, Dordrecht
Ruchti B, Livotov P (2001) TRIZ-based innovation principles and a process for problem solving in business and management, proceeding of European TRIZ Association
Saaty TL (1980) The analytic hierarchy process. McGraw-Hill, New York

Sanabria LB, Orozco Pulido LH (2009) Critical review of problem-solving processes traditional theoretical models. Int J Psychological Res 2(1):67–72
Schön DA (1983) Reflective practitioner—how professionals think in action. Basic Books Inc., New York
Seidel S, Berente N, Lindberg A, Lyytinen K, Nickerson JV (2018) Autonomous tools and design: a triple-loop approach to human-machine learning. Commun ACM 62(1):50–57
Simon HA (1962) The architecture of complexity. Proc Am Philos Soc 106(6):467–482
Simon HA (1969) The science of artificial. MIT Press, Cambridge
Smith GF, Browne GJ (1993) Conceptual foundations of design problem solving. IEEE Trans Syst Man Cybern 23:1209–1219
Souchkov V (2007) Breakthrough thinking with TRIZ for business and management. ICG Training and Consulting Press
Stempfle J, Badke-Schaub P (2002) Thinking in design teams-an analysis of team communication. Des Stud 23(5):473–496
Sullivan LH (1896) The tall office building artistically considered. Lippincott's Magazine (March 1896), pp 403–409
Takahashi M, Kang N (2020) Proposal and Kansei evaluation of a garbage can to change human behaviour based on fun theory evaluation of a garbage can to change human behaviour based on fun theory. In: International Conference on Applied Human Factors and Ergonomics, Online, July 25–29
Ulrich KT, Eppinger ED (1995) Product design and development. McGraw-Hill, New York
Ulrich KT, Seering WP (1988) Function sharing in mechanical design. In: Seventh national conference on artificial intelligence (AAAI-88), St. Paul, USA, August 21–26
Von Hippel E (1986) Lead users: a source of novel product concepts. Manage Sci 32(7):791–805
Wang Q, Rao M, Zhou J (1995) Intelligent systems approach to conceptual design. Int J Intell Syst 10(3):259–293
Wynn DC, Clarkson PJ (2005) Models of designing. Design process improvement: a review of current practice. Springer, Berlin Heidelberg, pp 34–59
Zeng Y, Cheng GD (1991) On the logic of design, Des Stud 12(3):137–141

Design and Redesign of System Architecture

16

Abstract

This chapter deals with the topic of product architecture, highlighting its central role and its manifold implications at both strategic and operational level. It then discusses a few design support methods based on product architecture, such as platform design, Value Analysis and engineering, design to value and variety reduction.

This chapter deepens the discussion on the design of products and systems that started in Chap. 15. At first, we will consider the design of product architecture and then propose techniques that are used to this purpose, especially when redesigning existing products.

16.1 Product Architecture Defined

The concept of product architecture has now and again appeared in the pages of this text. It has been initially defined as '*the components of the product and their mutual relationships*' when discussing Henderson and Clark's taxonomy of innovation (Chap. 3), highlighting the role that product architecture has in defining the type innovation, in shaping organizational routines and eventually leading to organizational inertia. Also dominant designs (Chap. 4) have been described as architectures which have a significant role on innovation dynamics. When looking at a product from a deeper technical perspective, as it is being done now, the definition of

M. Cantamessa and F. Montagna, *Management of Innovation and Product Development*, https://doi.org/10.1007/978-1-4471-7531-5_16

product architecture can be rephrased, adding some details and discussing the manifold implications that architecture has on the product, on the firm and on its supply chain.[1]

According to a widely accepted definition, product architecture is defined by '*the relationships between its functional elements, the mapping among functional elements and physical components and the interfaces among physical components*' (Ulrich 1995). With this new definition, the focus now extends from components to functional elements, and centers on the interface between the function tree and the lower echelon of the product Bill of Materials (BOM).

With this definition in mind, architectures can be classified according to the mapping between functional elements and physical components:

- **Integral architectures** are characterized by functional interdependence between components, with each function being fulfilled by multiple components and by components embodying multiple functions.
- **Modular architectures** are, instead, characterized by functionally independent components. Each component therefore is in charge of implementing a single function, and each function is fulfilled by a single component. In turn, modular architectures can be *slot-based*, if the interconnections between pairs of components are not based on a standard interface. A modular architecture will be *sectional* if the interface between each pair of interconnected components is standardized. Finally, in a *bus-based* modular architecture, components will not be directly connected with each other, but through a common component (called *bus*) and using a standardized interface.

Figure 16.1 provides examples of two products that might be implemented by means of alternative architectures. Of course, some of these alternatives are practically relevant, while others may be less so.

This classification has been originally proposed to products, but can easily be applied to services as well (e.g., Voss and Hsuan 2009; Brax et al. 2017). In principle, the classification of product architecture should apply to the entire

[1] It is important to distinguish system architecture from a Bill of Materials, or BOM. Both can be represented with a hierarchical tree diagram, but the architecture does not represent the physical membership of a component to a subassembly of the system, while it represents the physical and functional interactions between the components. For this reason, the system architecture can be represented by block diagrams, such as the ones introduced in Chap. 15. Differently from a tree-like representation, a block diagram loses the decomposition of the system 'by levels' and instead shows the links between the components of the same level. Moreover, the hierarchical aggregation defined for a BOM usually follows the aim of supporting operational decisions, such as purchasing and production planning, which might not be consistent with an understanding of how the system works. For instance, a purchasing BOM may organize items by type of commodity or supplier, while a production BOM may organize items by looking at assembly sequences or at the facilities in which items will be produced.

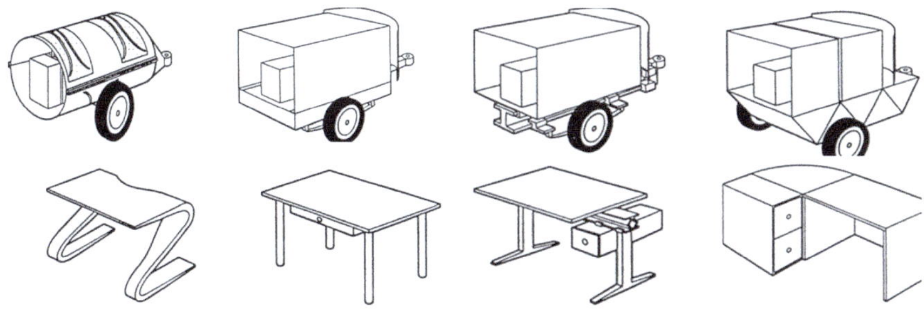

(a) *integral* (b) *slot* (c) *bus* (d) *sectional*

Fig. 16.1 Alternative architectures for two products (as proposed by Ulrich 1995)

product. However, it is possible for complex artifacts to exhibit portions that appear to be integral, while other ones may instead be modular. When dealing with portions of a product that exhibit an integral architecture, the term '*chunks*' is often used.

Concerning interfaces in a given product architecture, these may be *coupled* or *uncoupled*. In the former case, a parametric change in one of the connected components will propagate to the other component and require at least a partial redesign (e.g., thickness of component (a) in Fig. 16.2, left-hand side). Conversely, in an uncoupled interface, a change in one component will not affect the other (Fig. 16.2, right-hand side). In general, coupled interfaces provide a tighter connection among components but—at the same time—make detailed design of the artifact significantly more complex because of change propagation.

The type of architecture that is used by a firm casts a strong influence on aspects that are relevant at both tactical and strategic level. The following discussion explores these influences and may be used either to foresee likely consequences that will emerge because of a given product architecture, and as a guideline for choosing a product architecture that may lead to desired outcomes.

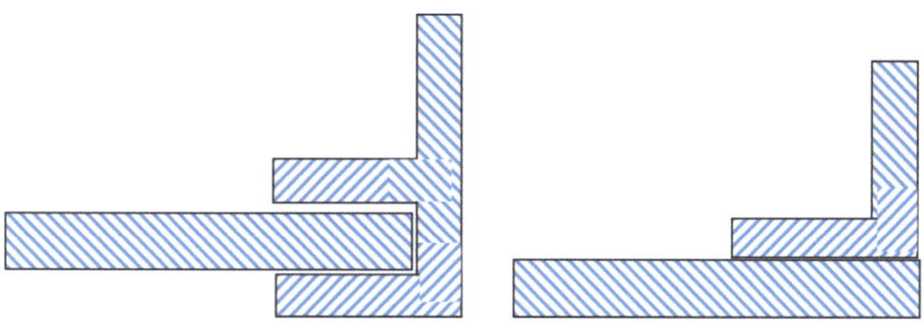

(a) (b)

Fig. 16.2 Coupled (**a**) and uncoupled (**b**) interfaces

16.1.1 Product Performance

Product performance depends on either *localized* or *general* performance variables. The former variables are relevant at system level, but their value can be directly traced to a corresponding performance in a specific subsystem (e.g., the mass memory capacity of a computer is given by the capacity of its hard disk drive). The latter are performance variables whose values depend on a corresponding performance that is distributed across multiple subsystems. For instance, mass and cost of a personal computer are given by the sum of the corresponding values for all subsystems. Similarly, operating time when working on batteries will jointly be given by the capacity of its batteries (in Ampere-hours) and by the combined power consumption (in Watts) of its components.

With a modular architecture, it will be easy to design a system with the objective of enhancing localized performance variables, since this simply requires choosing and/or designing components that match these requirements. However, the 1:1 relationship between functions and components that characterizes a modular architecture will prevent function sharing, which is generally a good way to enhance general performances. Conversely, an integral architecture will be likely to lead to an improvement in general performances, but at the expense of localized ones.

16.1.2 Product Change

Product change can be associated to variations that occur within the lifetime of an individual product in the hands of its user, as well as changes introduced by the firm over the product lifecycle. The first perspective involves introducing actions such as:

- **upgrades**, which consist in the substitution of individual components in order to enhance localized performances (e.g., inserting a hard disk drive with greater capacity in a computer system);
- **add-ons**, or additions of a component, so that the product can deliver new functions (e.g., addition of an external lighting source on a video camera);
- **functional variety**, consisting in the substitution of components, in order to increase the range of activities that the product can fulfill (e.g., exchangeable lenses in cameras);
- **adaptations**, or substitution of components that allow the product to operate in different conditions of use (e.g., an adapter that allows powering a device from an AC socket or in a car);
- **substitution** of worn-out or exhausted parts (e.g., toner cartridges).

Change is generally easier for systems with modular architectures, since components can be added or changed without redesigning the whole product. Moreover, modular architectures can open the road for these components to be developed, manufactured and sold by third-party producers. The firm developing the main product or service must be fully aware of the strategic consequences of this opportunity, given the obvious tradeoff between losing part of the margins, at the same time increasing customer value without having to make investments. In some cases, the tradeoff leans heavily toward 'closing the system off' to third parties (e.g., ink cartridges for a printer maker that relies on a razor–blade business model). In other cases, the tradeoff may hint at greater opening (e.g., the plethora of accessories that can make a smartphone appealing to consumers).

The other perspective on system change looks at modifications introduced by the firm over its lifecycle. This has to do with the concept of '*component carryover*' between successive product releases. In general, component carryover can be expressed as the percentage of components (or component value) that a new product inherits from its predecessor. It is intuitive that a modular system will make it relatively easy to redesign a product or service by substituting part of its components and without having to redesign the other ones. Conversely, the tight inter-component relationships that characterize an integral architecture could require the redesign of all—or most of—its components.

On a strategic level, the adoption of modular architectures allows firms to develop 'platform-based systems', in which a common platform enables the quick and effortless development of product derivatives catering to specific and contingent market needs. Thanks to modularity, derivative-specific components can be designed without having to redesign the core and 'in-platform' components.

If a firm becomes very quick in exploiting modular architectures and modifying its product or service offering, it may reduce its reliance on market research and use actual market acceptance as a guidance tool. Taken to the extreme, firms can choose to adopt the concept of 'real-time market research' and gradually modify their products and services, based on the feedback they progressively receive from the market.

16.1.3 Product Variety

Modular architectures and platform products or services do not only allow a quick change between successive system releases and generations, but also the simultaneous offering of differentiated variants based on the common design platform. This strategy can lead to three main advantages:

- The firm is able to address multiple *horizontally* and *vertically differentiated* market segments, with obvious benefits for its pricing power.
- In-platform components are produced in significant volumes, which can lead to significant economies of scale.
- The overall effort spent in developing the platform and product or service derivatives can also lead to economies of scale, because of the high volume that comes from the aggregate of the product variants being offered.

Moreover, modular architectures can enable the offering of products characterized by *combinatorial variety*, in which a core system can easily be customized by adding or changing of number of components. For instance, if a customer can configure her product by choosing among two variants for n of its modules, the number of potential product variants becomes 2^n. Thanks to combinatorial variety, the firm can make a strategic upstream shift of its *Customer Order Decoupling Point** (i.e., the instant at which customer orders are accepted) and move from 'Make-To-Stock' to 'Assemble-To-Order' operations.

Going even further, and given that major economies of scale usually occur at component manufacturing level rather than at final assembly, the firm may opt to fragment assembly operations and locate them closer to the customer. For instance, a firm may decide to centrally retain the manufacturing of the core part of the product, but let to local dealerships the task of completing customization and assembly of the final product. In some cases, these local '*miniplants*' may be given the task of creating their own supplier base and of locally sourcing non-core components with high transportation costs, for which centralized manufacturing would not be justified (e.g., shovels and scoops used in earth moving machines).

Finally, a highly modular architecture can lead a firm to commercialize a broad variety of elementary and interoperable building blocks, leaving the task of structuring the final 'system' to customers or vendor/installers. This is the typical case of computer networking and electrical equipment, which are assembled into an actual system by installers.

In general, integral architectures do not lend themselves well to supporting product variety. However, there might be exceptions. For instance, in the case of very simple products that are produced in very high volumes, an integral architecture may allow an extreme simplification of the BOM (e.g., ballpoint pens, disposable razors, etc.) and thus facilitate product variety. Moreover, if one thinks of the flexibility granted by modern manufacturing technology (e.g., NC machining centers, additive manufacturing techniques, etc.), one-off products with complex shapes that embody multiple functions can be produced with relative ease, though not in high volumes.

16.1.4 Standardization

Because of component decoupling, modular architectures also make it easier to use standard—instead of purpose-designed—components. Standardization can here be conceived both in its extensive and 'industry-wide' meaning, as discussed in previous Chap. 4, as well as in a narrower and company-wide sense.

Standardization of components leads to obvious advantages because of economies of scale and reduction of complexity. However, these advantages will be balanced by additional costs, since having to choose among a limited set of standardized components will lead to overdimensioning with respect to the needs of the specific application. In some cases, the overall product requirements may be so stringent (e.g., the weight of airborne systems) that this becomes unacceptable.

16.1.5 Influence on the Organization and on the Supply Chain

In general, modular architectures lead to a significant decrease in the intensity of information flow between the designers that are in charge of developing each component. If functional relationships between components are relatively loose, the same component can readily be used in different systems, thus leading to economies of scale. Moreover, a designer who is developing a component can focus on its performance, rather than spending time on issues pertaining to its integration in the product or service, and this also favors the rise of experience economies.

Therefore, modular architectures will often be associated with a low degree of vertical integration between producers of the end product and producers of modules and components. This leads to a clear separation in the roles and competencies developed by each party. Producers of the final system will have to develop systems engineering skills and focus on coordination and integration issues, while component producers will concentrate on the specific requirements of the individual components.

Depending on the degree of separation between producers of the final product or service and suppliers, it is common to distinguish among '*black box*', '*white box*' and '*gray box*' product development. In the former, the producer of the end system does not look into the inner workings of the components at all, but limits himself/herself to specifying and testing the features and performances he/she thinks are relevant. With 'white box' development, the producer has a direct influence on the development of components and—of course—this requires her to be proficient in the related technology. 'Gray box' development can be viewed as an intermediate approach, in which the producer has a sufficient understanding of the issues that are relevant to the development of the components, but restrains herself from having a direct influence on suppliers' choices.

Functional interdependencies strongly influence also the strategic choice made by firms in a supply chain, concerning which product or service components they will develop, produce and deliver. Empirical research in the automotive supply chain (Cantamessa et al. 2006) has shown that firms can decide to specialize in developing and producing components according to the three criteria of similarity of underlying technology, adjacency of components in the final assembly and functional interdependencies. However, the strongest of these three factors in explaining firms' specialization choices appears to be the pattern of functional interdependencies.

The previous discussion has focused on the impact of product architecture on the supply chain. If one looks at the level of project management, architectural choices lead to quite different management approaches.

In the case of modular architectures, the project manager behaves as a '*system architect*'. His major focus is on the specification of components, while detailed design is delegated to designers and design teams. These designers engage in a customer–supplier relationship with the project manager and may work in parallel without significant interaction among each other. Verification and testing can be performed toward the end of the development process, with the main objective of validating system performance and of identifying unexpected interactions between components.

Conversely, in the case of an integral architecture, the project manager will have to operate as a '*system integrator*'. Most of her effort will be dedicated to the continuous coordination and negotiation with the designers in charge of developing components. Designers will have very close reciprocal interactions, both at a formal and an informal level, in order to 'iron out' all inconsistencies among their design outcomes. Verification and testing are performed continuously as a 'fine tuning' of the system, in order to bring performance in line with the expected specifications.

16.2 The Design of the System Architecture

The previous discussion has given an indication on the type of architecture—i.e., integral vs. modular—that could be preferable, given the system, strategy and supply chain the firm is operating in. Especially when dealing with products with moderate to high complexity, designing the product architecture means something more. Given the complex web of relationships between functional elements, the designer will have to look for an aggregation of the components into modules that are relatively independent of one another and that might exhibit significant inter-dependencies within.

This activity can be performed either intuitively or analytically. In the former case, the designers can use functional block diagrams and visually identify '*chunks*' of components (or modules) that exhibit a high degree of interrelationships, at the same time with limited interactions with components belonging to other chunks. When analyzing functional block diagrams, one may use rules of thumb such as the following (Stone et al. 1998):

- **Dominant flow**. Functional elements crossed by a single 'flow' and that do not 'branch out' to other functional elements are good candidates for being assigned to a same chunk.
- **Branching**. When one observes 'branching' in a flow, this is an indicator that each branch should probably be assigned to separate modules.
- **Conversion and transmission**. Elements whose function is the conversion and transmission of material, energy or information, will generally go in the same, dedicated, module.

Figure 16.3 shows an example of a visual analysis carried out on a product block diagram, leading to a definition of product architecture.

When products are quite complex and/or when the firm wants to carry out a rigorous analysis, inter-component relationships can be represented with adjacency matrices (Fig. 16.4a). In the adjacency matrix $A = [a_{ii'}]$, $a_{ii'}$ represents the strength of the relationship between component i and component i'. This can be expressed in

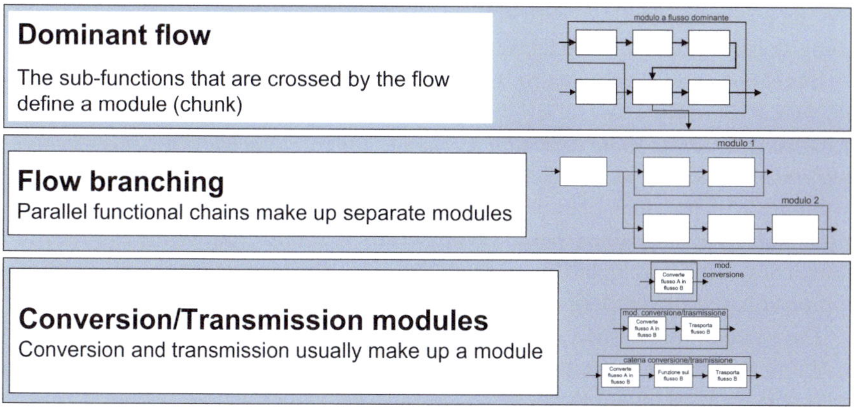

Fig. 16.3 Visual identification of component chunks, based on a functional block diagram

	1	2	3	4	5
Element 1		2			
Element 2	2		1		
Element 3		1		2	2
Element 4			2		2
Element 5			2	2	

(a)

	1	2	3	4	5
Element 1	3 7 3	2 1 -1 0			
Element 2	2 1 -1 0	8 8 7	1 1 1 0		
Element 3		1 1 1 0	5 5 4	1 0 -1	2 1 0 1
Element 4			2 0 -1	1 2 3 1 4	2 3
Element 5			2 1 0 1	2 3	3 2 1

s=space e=energy
i=information m=materials

s	e
i	m

On the diagonal: properties A, B, C, D, E, F

A	B	C
D	E	F

(b)

Fig. 16.4 Identification of inter-component relationships with adjacency matrices

as a continuous and nonnegative parameter in the range [0, …, M], with 0 meaning there is no connection between the two components and M that there is a very strong relationship. For simplicity, this can be reduced to a Boolean parameter indicating presence or absence of a relationship between components.

Parameter $a_{ii'}$ can also be computed as the aggregation of inter-component relationships associated to different perspectives, such as functional interactions, reciprocal disturbances, ease of assembly. In some instances, relationships between two components may lead to opposing conclusions, suggesting on the one side to place the two together in the same chunk and, from another perspective, to keep them apart. This typically occurs when one of the relationships is harmful (e.g., if component 1 generates heat and component 2 must operate at a low temperature). In order to represent this possibility, parameters in the adjacency matrix may be assigned both positive and negative values in the range [$-M$, …, M], with negative values indicating a harmful relationship (Fig. 16.4b). It may be useful to remind that representing multiple perspectives in inter-component relationships can be easily done by using the Rodenacker notation discussed in Chap. 15.

Based on the adjacency matrix A, components can be assigned to chunks, so that the *sum of adjacency measures between components belonging to the same chunks is maximized, at the same time minimizing the sum of adjacency measures between components that are assigned to different chunks*.

Figure 16.5 provides a simple example of this kind of calculations, showing how a functional decomposition based on adjacency measures (top right) can lead to a completely different architecture than the one obtained through an 'intuitive' approach based on similarity of underlying technology (bottom right).

The value of this approach does not only lie in providing a mathematical support to defining system architecture. Rather, its value lies in the dialogue that the design team will be forced into having when defining the values that go in matrix A. This dialogue can lead to a strong and valuable mutual understanding of design issues and can easily lead to challenging established ways for conceiving the product or service. In the previous example, the 'optimal' and counterintuitive decomposition at the top is due to the '− 1' adjacency value between 'switch' and 'cord'. In turn, this value may have been decided by the team in order to affirm the opportunity of allowing last-minute fitting of different cords to the appliance.

Inter-component relationships, which—as we have seen—can be represented by means of block diagrams or through matrices, are generally associated to a given technological state of the art. These relationships can be studied when working at a strategic level on technology roadmaps, in order to foresee the architectural impact of an upcoming technology, which may create or break architectural linkages between components. Conversely, the matrices can be used to identify which

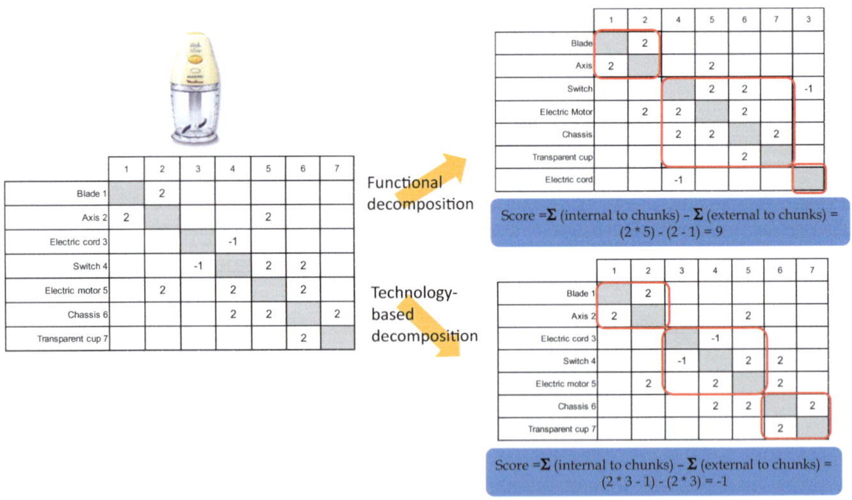

Fig. 16.5 Decomposition in chunks based on adjacency measures

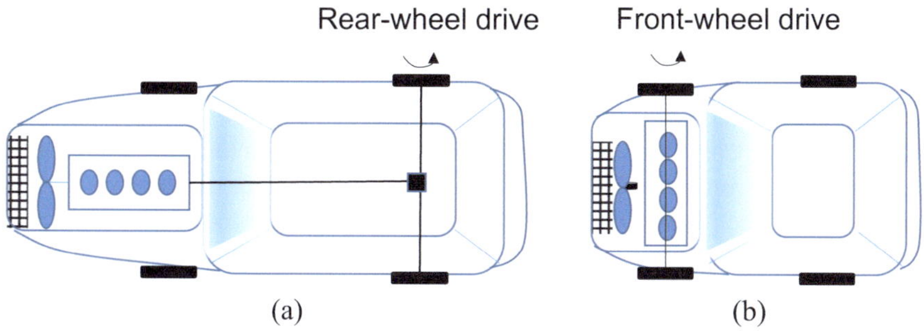

Fig. 16.6 Example of removal of architectural constraints

technology might be desirable for modifying inter-component linkages, and therefore lead to a desired architecture.

As a historical example, one can think of automobiles between the '70s and the '80s. At that time, radiator fans were engine-driven, i.e., directly powered by the crankshaft. Since the fan and the radiator have to be mounted at the front of the car, this casts a constraint on vehicle architecture: powering the fan is easier if the engine is placed longitudinally and—in turn—a longitudinal engine naturally lends itself to adopting rear wheel drive (Fig. 16.6, left-hand side). These constraints made it quite difficult to design compact cars with ample floorplan in the cockpit, which would have required a transverse engine layout and front-wheel drive. The architectural coupling was broken thanks to the introduction of electric thermostatic fans, which made it easier to place engines transversally (Fig. 16.6, right-hand side). Readers can work out, by using a similar line of thought, the implications on vehicle architecture that could be enabled by electric powertrains (especially in the case of 'electric wheels') and steer by wire technology (instead of using traditional steering wheels and steering columns).

16.3 Designing Platform-Based System Architectures

Platform-based product design has already been outlined as a strategic approach that firms may use in order to efficiently develop a wide variety of products or services, based on a common set of components and technologies. It goes without saying that a system platform will be closely connected to the architecture that the underlying products or services will have in common. Therefore, the design of platform-based products will be closely associated to the design of such architecture. This section aims at providing a preliminary presentation of the approaches that can be used when designing the architecture underlying platform products or services.

At first, it is possible to define a *platform* as a '*set of common assets*' on which a product family can be based (Meyer and Lehnerd 1997). When speaking about common assets shared by the products or services belonging to the family, the most obvious ones are components and subsystems. However, one can widen the scope of 'assets' and include common customers and customer relations, common values pertaining to a brand and common resources involved in product or service development. The identification of which type of common assets should be part of the platform allows the firm to define the strategic objectives of a platform-based product. In the case of common components, the firm will generally be able to leverage them in order to serve different market segments. In such a case, designers may find it useful to come up with a visual representation of the roadmap showing how the platform will support the launch of a number of horizontally and/or vertically differentiated products (as discussed in Chap. 9). Besides the choice of customer segments, the roadmap must show the timing at which each segment will be addressed. In fact, the firm may decide for a near simultaneously entry in all customer segments, while in other cases this will occur along a longer time span, starting from an initial 'beach-head' segment.

In its most basic form, designing a system platform around a set of common components implies deciding which components must be 'in platform' (i.e., common to all products based on the platform) and which should be 'off platform'. A simple way to operate this choice consists in assigning each component (or subsystem) two scores, one measuring the 'value of commonality', the other the 'value of specificity', the former expressing the advantages that may come to the firm by sharing the component across the product family, and the latter the importance attributed by customers of having a specific component for each product. These scores can be assigned by expert evaluation or using objective data on cost-volume curves and customer preferences, respectively.

A two-dimensional map, as in Fig. 16.7, makes it possible to make 'visual' decisions on platform-based component sharing. Components with high value of commonality and low value of specificity will be obvious candidates for being shared.

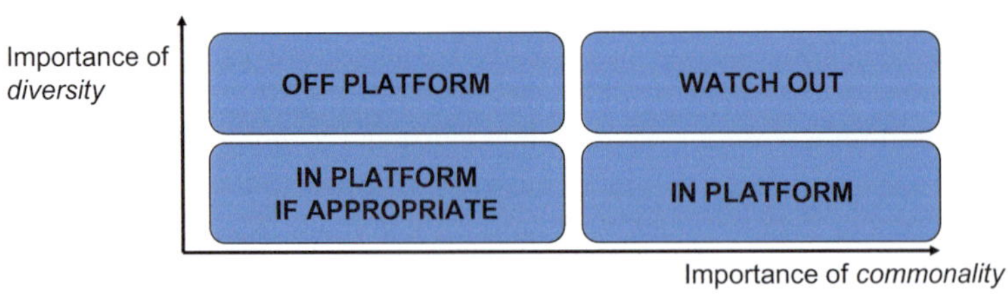

Fig. 16.7 Mapping components in order to decide platform composition

For instance, having a single heating and ventilation system in an automotive platform can lead to significant economies of scale, while customers will probably not even be aware that this component is the same for all models in the range. Components with low value of commonality and high value of specificity ought to be taken ‘off platform’. For example, customers will make a strong association between a specific car model and components having a mainly esthetical value (think of the steering wheel), but the firm will have limited cost disadvantage in producing a different one for each model. Components with low value of commonality and low value of specificity can usually be shared. Finally, critical decisions can arise with components having high values of commonality and specificity. The firm will generally have to decide whether to seek a cost advantage or lean on customer preferences. In some cases, the components may lend themselves to hybrid solutions. For instance, the component can be shared across products, but some of its features and functions could be modified and/or disabled according to the product it will be installed in (e.g., a low-end digital camera might have the same microprocessor as a high-end one, but use a different firmware that provides image processing algorithms of lesser quality).

In order to properly manage component sharing, a number of metrics have been developed in order to measure the degree with which components are being used in different products (Thevenot and Simpson 2006). The basic measure is the number Φ_j of parents to each component *j*, i.e., the number of instances in which component *j* appears in the firm’s Bills of Material (end items are obviously not considered).

For instance, given the BOMs in Fig. 16.8a and b, one can compute values of Φ_j (Fig. 16.8c) per each component and then calculate aggregate indicators. Named *d* the total number of parents, one **quite simple indicator** is the *Degree of Commonality Indicator*, expressed by (16.1):

$$\mathrm{DCI} = \frac{\sum_j \phi_j}{d} \geq 1 \tag{16.1}$$

The greater the value of DCI, the greater is the degree of component sharing, while DCI = 1 means there is no component sharing at all. In the previous example, DCI = 11/5 = 2.2.

Another similar indicator is the *Total Constant Commonality Index*, as formulated in (16.2):

$$\mathrm{TCCI} = 1 - \frac{d-1}{\sum_j \phi_j - 1} \in [0, 1] \tag{16.2}$$

The TCCI indicator is bounded in the interval [0, 1], with TCCI = 0 implying no component sharing and TCCI = 1 the maximum commonality possible. In the previous example, TCCI = 1 − (5 − 1)/(11 − 1) = 0.6.

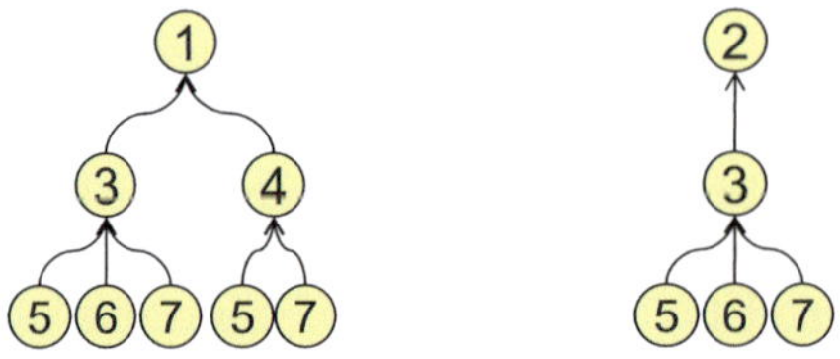

(a) (b)

Component	Φ_j
1	n/a
2	n/a
3	2
4	1
5	3
6	2
7	3

(c)

$$TCCI = 1 - \frac{d-1}{\sum_j \phi_j - 1} \in [0,1]$$

TCCI = (5-1)/(11-1) = 0.6.

Fig. 16.8 Computing commonality measures between products

16.4 Designing Modular System Architectures

Modularity and platforms have become key enablers of customization and personalization (Porter and Heppelmann 2014; Rossit et al. 2019), and especially in the case of digital product and services. However, when engaging in platform and modular design, firms must face the issue of understanding the tradeoff between cost, production constraints, differentiation issues and openness and flexibility of the product architecture (Krause et al. 2017).

Due to the competitive and strategic impact that can derive from modular and platform-based architectures, researchers have proposed a number of techniques aimed at supporting designers in their choices. Among them, *Modular Function Deployment** (or MFD, Ericsson and Erixon 1999) has gained some diffusion in industry, thanks to its intuitive use and to the value provided by relating design decisions to customer and user needs.

MFD starts with an analysis of customer needs that is very similar to the one performed in QFD. Each of these customer needs is then associated to a degree of relevance of modularity, i.e., a 1–9 score expressing the degree with which modularity might contribute to the satisfaction of each customer need. In turn, this potential contribution of modularity can be systematically analyzed with respect to a number of '*modularity drivers*'. The drivers originally cited by the authors are (i) economies of scale arising by using the same component across a product family; (ii) economies of scale and economies of learning due to the use of the same component across successive product generations; (iii) absence of specifications forcing the use of a specific component in each system; (iv) styling and aesthetic issues; (v) risk of obsolescence due to planned changes or to technological evolution; (vi) better use of bottleneck resources in R&D or production; (vii) possibility of outsourcing; (viii) possibility of testing the component separately; (ix) easier servicing and maintenance; (x) easier upgrading and (xi) easier end-of-life treatment.[2]

After this first step, the product is analyzed from a functional perspective, and this leads to a list of 'technical solutions' (or, functional elements that are independent from one another).

At this point, the core step of MFD is performed by developing the 'Module Indication Matrix' (Fig. 16.9). This is a table that represents the relevance of each modularity driver (in rows) to each technical solution (in columns), with the same {1, 3, 9} scale that is generally used in QFD (where 1 indicates a weak relationship, 3 a significant one and 9 a very strong one). By summing up these scores across columns, one can understand which of the modularity drivers appear to be more relevant to the system. Conversely, by summing the scores over rows, one can identify the technical solutions for which modularity appears to be of greater importance. Moreover, by analyzing the matrix, the designer can observe clusters of technical solutions that are relevant to a same modularity driver. This can suggest ways for aggregating technical solutions into modules that obey to a same driver.

The MFD method also provides some guidance on the number of modules m that should be planned for a system with n components. In general, each component within a module requires a given effort EC for the various activities associated to it (e.g., design, assembly, testing, etc.), while the development of each module requires an effort E. As a rule of thumb, $E = 1 \div 5$ EC. This, being the case, the overall total effort TE associated to the product will be given by

$$\mathrm{TE} = m\mathrm{EC} + \frac{n}{m}E \tag{16.3}$$

the value of m that minimizes E will therefore be given by

$$m^* \left| \frac{\mathrm{dTE}}{\mathrm{d}m} = 0 \right. \tag{16.4}$$

$$m^* = \sqrt{n\frac{E}{\mathrm{EC}}} = 1 \div 2\sqrt{n} \tag{16.5}$$

16.5 Value Analysis and Value Engineering

When designing product or service architecture, firms must constantly base their decisions on customer and user needs. One widely-known and time-tested method for doing this is *Value Analysis* (VA)*, a technique developed by Lawrence Miles* (1904–1985) at General Electric during World War II. Miles had the insight that

[2] Of course, each firm is free to adapt the list of drivers to make it more relevant to its own case.

Technical solutions / **Driver**	**TS1**	**...**	**TSn**	**Sum**
Common unit	1 | 3 | 9			
Carryover				
Different specification				
Styling				
Planned product changes				
Technology evolution				
Process / Organization				
Available from supplier				
Separate testing				
Service / Maintenance				
Upgrading				
Recycling				
Sum				

Which module drivers are most relevant in defining modularity?

Which TSs have similar or compatible drivers? → find the possible m modules by aggregating TSs

To what extent should TS be an independent module?

Fig. 16.9 Module indication matrix used in modular function deployment

customers choose and purchase products because of their ability to produce a useful outcome, and this can be considered to be the 'value' of the product.

The value of a product or a service is quite a complex concept, and an economist could in first instance associate it with the 'reservation price', i.e., the price at which a given customer would be indifferent between purchasing and not purchasing the product (or, in layman's terms, the maximum amount the customer would be willing to pay). In order to have a better grasp on value as a concept, it is possible to tackle it from three different perspectives. *Use value* comes by estimating the value of a system from its quantifiable benefits (e.g., time saved, profit accrued, etc.). Use value is generally appropriate for instrumental goods bought by businesses (e.g., a machine tool), but can be applicable to consumer goods as well (e.g., time saved for cleaning procedures when you shift from a traditional espresso machine to a capsule coffee machine) and services (e.g., time, activation and transportation costs saved in having food delivered at home). Moreover, in the case of consumer goods or services, customers base their choices on a wider concept of value, called *esteem value*, in which other intangible elements come into play (taste, aesthetics, the role as a status symbol, etc.). Finally, one may consider the concept of *market value*, which is the price a customer would expect to pay, given the intrinsic value of the system (which is associated to use and/or esteem value) and the existence of competing alternatives. The difference between use and/or esteem value and market value (or price) is the surplus that the consumer is left with, after purchasing the product. The difference between price and cost is—of course—the margin that goes to the firm.

If one understands the concept of value, it should be quite clear that a firm should not design and redesign its products with the simple objective of reducing cost, since this action might actually lead to reducing value as well. Instead, firms should maximize the difference between value and cost. This can also be said to be equivalent to reducing 'waste', intended as any element that has a cost but does not add value.

By referring to Fig. 16.10, and starting from the baseline case on the left, one can see three ways with which a firm may redesign its products. Alternative (a) exemplifies a cost reduction that destroys value: the cost reduction makes the product less attractive to such an extent that value is reduced by an even greater amount. This ultimately leaves the customer with less surplus, and the firm with less profit. Alternative (b) shows a cost reduction that does not modify the perception of the product. Its use/esteem value is unaltered, so that the firm can keep the same price, thus internalizing the value created as profit. Finally, alternative (c) shows a cost increase that actually creates value, since it leads to an improved product that exhibits an even greater increase in value.

The VA method aims at guiding firms through the process of making the right design and redesign decisions with respect to these issues. In general, the method is implemented by development teams with representatives coming from different functions, who meet regularly (i.e., once a week) for a sequence of sessions dedicated to analyzing the product, brainstorming for solutions and then deciding on the changes to be made (Fig. 16.11). In preparation to the meetings, participants are usually assigned tasks for gathering data on the existing system (prior to analysis) and on the viability of proposed solutions (prior to decision-making). As in all inter-functional activities, it is critical that the organization at large and the people involved are sufficiently committed, and share a common objective (e.g., "we will redesign product *X* in four weeks' time, with the objective of reducing variable cost

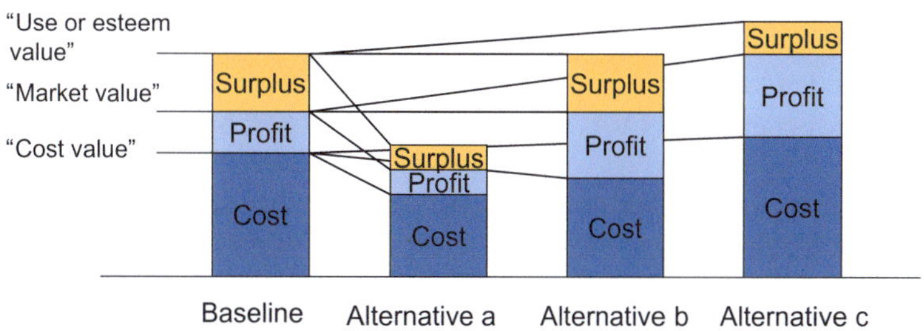

Fig. 16.10 Redesign alternatives and their impact on value

Fig. 16.11 Product redesign process in value analysis

by 7.5%"). Absent such commitment both during the meetings and in the in-between activities, it would be quite likely to end up with a failed project.

VA originated in the field of manufacturing, and we will present it in this context. However, the method is readily adaptable to services as well. It is usually carried out in four phases of growing complexity. The first two phases usually go under the actual name of '*Value Analysis*' and cope with the redesign of a current system, which implies that modifications will be somewhat constrained. The third phase is better known as '*Value Engineering*' and is usually aimed at new product development, thus allowing a more substantial degree of system redesign. The fourth phase aims at rationalizing and extending the outcomes of the first three. These phases are described in the following.

16.5.1 Phase 1—Obtaining the Same Product at a Lower Cost

During this phase, the VA team is tasked with analyzing the items in the system BOM one by one, and proposing changes that will alter neither the functions, nor the system architecture, but only the individual components. Examples of such proposals include changing materials, suppliers, process technology, reversing previous 'make or buy' choices and altering the detailed design of components (e.g., shapes, dimensions, tolerances). The rationale behind this phase lies in the uncertainty that might have led to specific choices when the product had been initially designed. For instance, in choosing between a labor-intensive process (with low fixed cost and high variable cost) and a capital-intensive process (with high fixed cost and low variable cost) a firm might have gone for the former, expecting a low level of demand. After some time, the firm may realize that the product is quite successful, and it would therefore be advisable to reverse the original decision. Similarly, designers unsure of the reliability of a part might have originally chosen a costly and high-resistance material from a very reputable supplier. Experience gained in the field might later suggest going for a cheaper solution.

When tackling phase 1 of VA, the time spent on analysis, brainstorming, checking the viability of the solutions and deciding on them can be quite substantial (consider that, considering current labor rates, a 1-hour discussion carried out by a group of 10 people on a single component can cost around 1000 €). Therefore, it will be worthwhile to engage in such an activity only if the expected benefits—given by unit percentage savings, multiplied by unit cost and volume—are worthwhile. It comes as a consequence that VA should be carried out in the case of products whose volume—or whose unit value—is high enough to justify the effort. Since Bills of Materials are usually characterized by a Pareto distribution, with a minority of components being responsible for the majority of total cost, the analysis can be carried out for the higher-value components of the BOM only and neglecting the rest.

The application of this phase of VA to the case of services is quite straightforward, since it consists in making a systematic analysis of process steps, instead of the BOM.

16.5.2 Phase 2—Obtaining the Same Functions at a Lower Cost

Phase 2 of VA involves a slightly deeper redesign of the system. The overall functions of the product or service will not be changed, and neither will its architecture. Components are to be analyzed one by one, with the aim of substituting them with '*functional equivalents*', i.e., different and cheaper components that may deliver the same low-level functions. Referring to the discussions in Chap. 15, this implies changing embodiment choices made at the lower echelons of the function tree. As an example, a mechanical fastener connecting two components could be substituted with adhesive bonding, while an electromechanical relay could be substituted by a low-cost electronic solid-state relay.

The discussions that can be carried out in phase 2 of VA are typically quite narrower than the ones in phase 1, since not all components are actually amenable to substitution, and the range of options available is typically quite small. In order to stimulate the discussion, the VA team can try comparing the actual cost of a component to its theoretical minimum value, i.e., the lowest-cost alternative embodiment, without considering its actual applicability. For instance, a 0.50 $ spring latch having the function of joining *X* and *Y* may be compared to the 0.01 $ value of a drop of glue. Even if using adhesives is not viable, since *X* and *Y* might have to be separated at some point in the product lifecycle, this huge value difference could lead the team to debate on whether separation of *X* and *Y* is really an issue and how often it occurs and—in the end—the team might decide for a 0.05 $ bolt.

Because of the relatively limited range of redesign opportunities, this second phase of VA usually doesn't last for a long time, since teams will focus on the few higher-value components that can actually be substituted with functional equivalents. However, it would be a mistake to skip this phase altogether, since it usually provides a good baseline understanding of the system and its functional structure, which can be useful when engaging in the subsequent phase 3.

16.5.3 Phase 3—Obtaining the Required Functions at a Low Cost

This third phase of VA (which is usually called Value Engineering) involves a substantial redesign of the system and is therefore suitably carried out during the initial development process, rather than for redesigning an already launched product.

The main idea behind Value Engineering was that customers are not interested in the embodiment of the product and of its components (which determines product cost) but in the functions that these components deliver (which determines the product's utility and value).

So, if one focuses on functions, each function will 'natively' be characterized by a degree of utility (which matters to the customer), while its embodiment will be characterized by cost (which instead matters to the producer). Conversely, if one focuses on components, each will be characterized by cost, but can be associated to the function it delivers, which is instead characterized by utility. So, by translating between the two spaces of functions (and associated utility) and components (and associated costs) the development team may be able to study and compare cost and utility of both functions and components.

The starting point of this analysis consists in drafting a function tree and comparing it to the bottom tier of the BOM for products, or functional elements related to process stages for services (as shown in Chap. 15). The utility of functions is usually computed in two steps (Fig. 16.12). At first, relative utilities are computed, assigning a [0, 1] value to each function, which expresses the degree with which that function is necessary in order to fulfill the corresponding higher-level function (in Fig. 16.12, these values are placed in the boxes). Since this utility is relative to the higher-level function and not to its peers at the same level, it is not required that the values at a same tier sum up to 1. Conventionally, the top-level function is given a value of 1. Then, absolute utilities are computed, by progressing top down and multiplying the relative utility of each function by the absolute utility of the function that is placed immediately at a higher level[3] (in Fig. 16.12, these values are placed next to the boxes).

After having defined utility levels for the lower-level functions, U_i, it is possible to compute their cost CF_i by associating them to the cost cc_j of the components that fulfill them. In the case of modular (i.e., 1:1) relationships the task is trivial, while it is less so in case of more complex architectures. In this case, one will have

$$CF_i = \sum_j cc_j cf_{ij} \tag{16.6}$$

where cf_{ij} is the share of the cost of component j that fulfills function i. In first approximation, cf_{ij} can be computed as $1/k$, k being the number of functions to which i contributes. This computation is often misleading, since components usually are designed around a main function and are then modified to fulfill additional ones. For instance, the casing of a consumer electronics product may have the main function of 'containing other components', with the added function of 'providing shock protection' to the product.

In these cases, a better estimation would be made by determining the cost of the main function as the cost that the component would have if it only had to fulfill that function. Then, the cost of the other function would be computed as the additional cost that has to be incurred because of function sharing (for instance, the use of a different material in order to improve shock absorption).

Conversely, it is possible to compute the utility of each component, UCj as in (16.7):

[3] The evaluation of utility can be performed either by considering purely technical criteria (i.e., 'the degree with which function F2.1 is necessary to fulfill F2') or a mixture of technical issues and user preferences (i.e., 'the degree with which the customer wants function F2.1 in order to achieve functn F2').

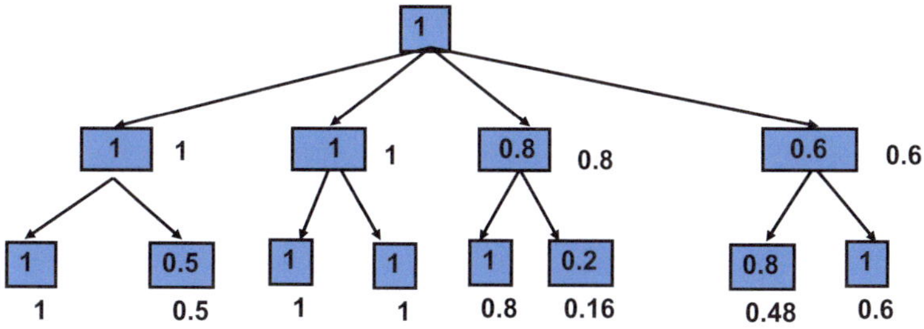

Fig. 16.12 Computing utility of functions in value engineering

$$UC_j = \sum_i uf_i fc_{ij} \tag{16.7}$$

where uf_i is the utility of function i and fc_{ij} is the contribution given by component j to function i on a [0, 1] interval.

Having performed these calculations, the VA team will be able to generate the maps of utility and cost for both functions and components (Fig. 16.13). Elements with low cost and high utility will of course be kept and considered as benchmarks. On the opposite side, elements characterized by high cost and low utility will be subject to intense debate and considered as interesting candidates for being pruned away. Expensive and high-utility elements will be kept, but the team will dedicate some time in order to find solutions for further cost reductions. Finally, low-cost and low-utility elements will be left alone, dropped or subject to integration and function sharing in order to squeeze some more costs out of the system.

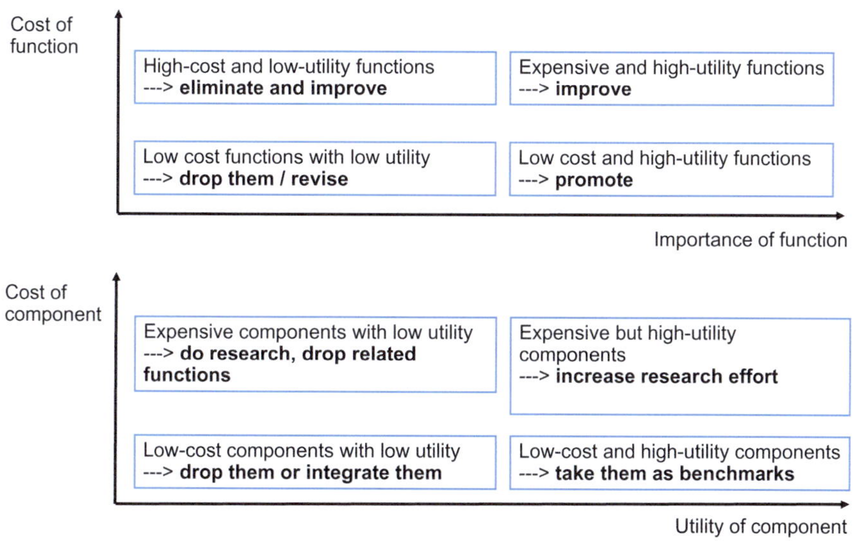

Fig. 16.13 Mapping functions and components in value engineering

16.5.4 Phase 4—Extending the Results Horizontally

As previously mentioned, applying VA to a single product often requires spending significant time and effort, which can easily amount to more than the potential benefits, given the unit savings and the production volume of that system. This would limit the applicability of VA.

However, many firms offer a very broad catalogue of relatively similar and simple systems. In such cases, the ideas coming from a 'vertical' VA activity performed on a single product can be easily—and without any additional effort—'horizontally' extended to all the products in the same family. Moreover, ideas that might have not been applicable if applied to a single product may turn out to be interesting if they were applied to all the products in the family.

Based on these considerations, phase 4 of VA consists in revising the proposals that have arisen in the first three phases of VA for a single system and studying their applicability to all similar ones in the company's assortment. A firm following this approach may therefore continuously alternate vertical and horizontal actions of VA. This approach to VA can be extremely beneficial because it can become an established and efficient routine, at the same time creating a diffuse 'culture' of value creation throughout the organization.

16.6 Design to Value

In recent years, consultancy companies have tried to revise the traditional, and relatively old fashioned, Value Analysis methodology in order to make it more up to date with the needs arising from contemporary business challenges and technology. *Design to Value** or DTV (Henrich et al. 2012) is one of these revised approaches and deserves a short presentation, though it must be mentioned that diffusion is still relatively limited, and the method has not yet reached a significant validation by empirical research.

The main elements that have been added to Value Analysis are the following:

- The progress of the DTV project is shaped around an **Agile Approach** to project management, with a sequence of 'sprints'.
- Due to the lower degree of vertical integration that characterizes contemporary organizations and to the strategic value of defining the right degree of vertical integration, DTV implies a **stronger involvement of nontechnical functions and top management**.
- It tends to include a **broader amount of aspects** relevant to product development, including a detailed analysis of user needs (e.g., Kano analysis, etc.), a strategic assessment of suppliers' potential contribution (so-called purchasing chessboards, that can be considered as an evolution of the Kraljic matrix used in supply chain management).

- The **concept of 'cost'** being used in DTV is **not limited to the traditional principle of 'manufacturing cost'**, but is managed according to the principles laid out in Chap. 14 (e.g., target costing, lifecycle costing, etc.).
- There is greater emphasis on **benchmarking with competing products**, with the aim of comparing their BOM, looking both at the structure (to find technical ideas) and at their cost. This benchmarking exercise allows to compare products 'BOM-to-BOM' (i.e., comparing the cost breakdown of the focal product with that of each of its competitors) and 'BOM to best' (i.e., comparing the cost breakdown of the focal product with an ideal product that would incorporate the lowest-cost subsystems, handpicked from each of the competitors).
- DTV usually includes statistical analysis of the cost of similar components and subsystems being used in the product portfolio, carried out using the parametric approaches seen in Chap. 14. Specifically, these analyses are often carried out by developing **price-performance** scatterplots in order to identify the most efficient solutions and/or suppliers, to spot anomalies and to observe trends.

16.7 Variety Reduction

In many cases, the redesign of products and their architecture has the primary aim of reducing excessive variety in the product range. Many manufacturers witness a proliferation of product variants, components and processes that is relatively unjustified and certainly quite costly. *Variety Reduction Programs** (or VRP in short) are approaches aimed at managing and reducing this phenomenon.[4]

Excessive variety is usually due to lack of communication and the tendency to adopt solutions that are 'locally optimal', rather than searching for solutions that are better for the organization as a whole. In general, variety is due to both commercial and technical reasons.

Concerning the former, Marketing and Sales usually try to segment the market as finely as possible and push the product development process to closely match the needs of each segment, if not of each individual consumer. This tendency can certainly be advantageous by making the product more attractive and allowing the firm to increase sales and/or margins. However, excessive customization can lead to a disproportionate and quite costly degree of variety. This is often compounded by

[4] Some authors (Kōdate and Suzue 1990) consider VRP to be an antagonist method to VA, since they assume that VA's efforts to cost-optimize individual products can easily lead to technical solutions that are highly product-specific. As we have seen above, this is not really true, since a proper use of VA should generally take both individual products and the entire product range into account, by alternating vertical and horizontal analysis. So, we can consider VRP to be a useful complement to VA and—at the most—a way for correcting an improper use of this latter technique.

strategic mistakes made by the company in setting its Customer Order Decoupling Point, which should never be placed after this variety is generated.[5]

Concerning technical reasons, designers often tend to pursue a high level of perfectionism, therefore looking for the 'perfect' component to fit a given function or performance level, thus disregarding the possibility of using a common component that would still perform satisfactorily. This tendency to 'overengineer' is partly psychological and partly due to uncertainty over the future behavior of the product. Moreover, designers are seldom given tools, such as component databases, that could support the 'carryover' of currently used components to new products. Sometimes, they may be provided with tools that create an even greater variety (e.g., a spreadsheet algorithm that automates the design of a simple component, but does not check whether the outcome is a close match to an existing part number).

Variety can lead to both advantages and disadvantages, thus generating a U-shaped curve, as is qualitatively sketched in Fig. 16.14. Advantages are associated to a tighter fit with user needs (and therefore value) of the product. Disadvantages are instead due to lower returns on investment made on component-specific assets (e.g., a same die will support smaller production runs) and because it forfeits the possibility of looking for production techniques that may exploit economies of scale. In addition, variety will add to organizational complexity and will therefore have multiple impacts on the cost of logistics, information processing, supplier management and so on.[6]

The problem with variety management is that the curve that is shown in Fig. 16.14 is only qualitative and it is impossible to compute for a given firm. Therefore, managers will have to work on rough indicators of variety.

One first and easy step will be to realize whether the firm is located on the left-hand side of the optimum and should therefore increase variety or (far more likely) whether it is on the right-hand side and reduce it. Then, rough indicators for variety can be introduced, with the objective driving them down to a given threshold.

For instance, a 'bought component variety indicator' could be introduced by multiplying the number of externally sourced components, times the number of suppliers, times the number of countries from which the suppliers come. A 'process variety indicator for internally manufactured components' could be defined as the number of different technological processes used, times the number of different machines per process, times the number of different toolings being used.

[5] For instance, if customers want products with many different types of decor, the firm should stock unpainted goods and add decorations only after individual orders have come in (or, as in the case of smartphone covers, make it easy for customers to add their own decoration). Similarly, variety in high-end car manufacturing is due to the addition of a high number of combinations of options. So, car manufacturers tend to adopt Assemble-To-Order strategies in which they stock the main subsystems and then assemble each car to order.

[6] As a rough guideline coming from inventory management, cost of variety increases with the square root of variety itself. In other words, managing a set of n similar items with demand D costs $\sqrt{n}$ times more than managing a single item with demand nD.

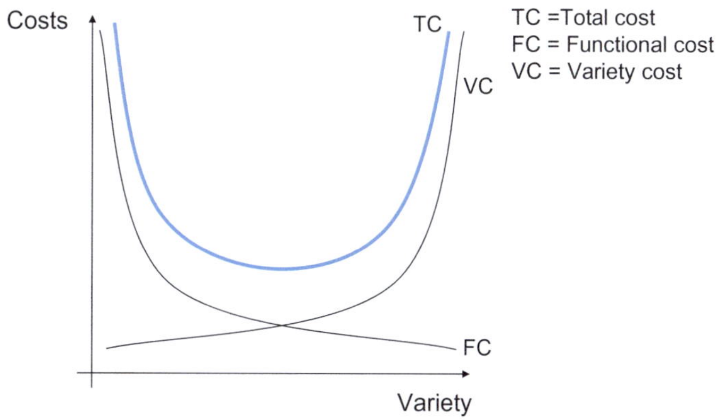

Fig. 16.14 Costs of variety

Having set objectives, management can then propose and study solutions for variety reduction such as the following:

- Reducing the number of components by **working on** their **similarity**. This task is best performed by using dedicated information system modules, called *Component Supplier Management** (CSM).

  These operate on verbal descriptions and physical features of components and—by highlighting similarities—enable the firm to compact their number by defining a limited number of 'approved' components. Verbal analysis is carried out by working on spelling mistakes, abbreviations and synonyms. For instance, a component category called '*helical spring*' should incorporate all components described as '*coil spring*', '*coil sprng*' or '*coil sprung*'. Then, statistical tools can be used to look for physical similarities on key parameters.

  In the case of a single parameter, it is possible to use simple histograms. As shown on the left-hand side of Fig. 16.15, the histogram reports the number of product instances that use a component of a given size (alternatively, the number of occurrences could be weighted by the demand of each product using the component). This histogram provides an intuitive indication of which component sizes should be considered as standard (the red bars in the histogram). A firm may then introduce a policy that considers the standard sizes to be 'pre-approved', so that a designer working on a given application and who has come up with an 'ideal' size is nudged into using the closest standard size. In order to provide some more flexibility, non-standard sizes might be used, but only if explicitly authorized by a manager.

  In the case of components defined by two or more key parameters, a firm may use cluster analysis to highlight groupings of similar components and identify standard ones (right-hand side of Fig. 16.15).

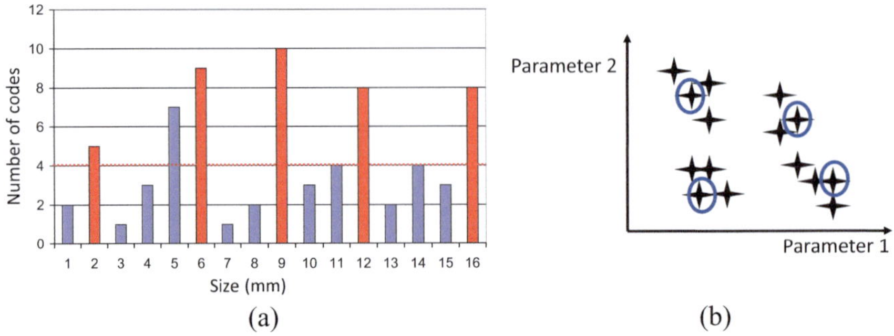

Fig. 16.15 Using clustering to reduce component variety

- Defining which parts of a product family should be considered to be fixed and common to all members of the family, and which may be product-specific. In a way, this solution is equivalent to defining a **product platform strategy**.
- Reducing part count by integrating multiple functions in a same component (i.e., **function sharing**).

Making an **appropriate use of combinatorial variety** and revisiting the position of the Customer Order Decoupling Point.

References

Brax SA, Bask A, Hsuan J, Voss C (2017) Service modularity and architecture—an overview and research agenda. Int J Oper Prod Manag 37(6):686–702

Cantamessa M, Milanesio M, Operti E (2006) Value chain structure and correlation between design structure matrices. In: ElMaraghy HA, ElMaraghy WH (eds) Advances in design. Springer, New York

Ericsson A, Erixon G (1999) Controlling design variants: modular product platforms. Society of Manufacturing Engineers, Dearborn, USA

Henrich J, Ashish Kothari A, Makarova E (2012) Design to value: a smart asset for smart products. Operations extranet. McKinsey & Company

Kōdate A, Suzue T (1990) Variety reduction program. A production strategy for product diversification. Productivity Press, Tokyo

Krause D, Gebhardt N, Greve E, Oltmann J, Schwenke J, Spallek J (2017) New trends in the design methodology of modularization. In: 11th International workshop on integrated design engineering, Magdeburg

Meyer MH, Lehnerd AP (1997) The power of product platforms: building value and cost leadership. Free Press, New York

Porter ME, Heppelmann JE (2014) How smart, connected products are transforming com-petition. Harv Bus Rev 92(11):64–88

Rossit DA, Tohmé F, Frutos M (2019) An Industry 4.0 approach to assembly line resequencing. Int J Adv Manuf Technol 105(9):3619–3630

Stone BR, Wood LK, Crawford HR (1998) A heuristic method to identify modules from a functional description of a product. In: ASME Design engineering technical conferences, Atlanta, USA, 13–16 Sept

Thevenot HJ, Simpson TW (2006) Commonality indices for product family design: a detailed comparison. J Eng Des 17(2):99–119

Ulrich K (1995) The role of product architecture in the manufacturing firm. Res Policy 24:419–441

Voss CA, Hsuan J (2009) Service architecture and modularity. Decis Sci 40(3):541–569

MIX
Papier aus verantwortungsvollen Quellen
Paper from responsible sources
FSC® C105338
FSC
www.fsc.org

If you have any concerns about our products,
you can contact us on
ProductSafety@springernature.com

In case Publisher is established outside the EU,
the EU authorized representative is:
Springer Nature Customer Service Center GmbH
Europaplatz 3, 69115 Heidelberg, Germany

Printed by Libri Plureos GmbH
in Hamburg, Germany